99-WWF-3

METHODS FOR WASTEWATER CHARACTERIZATION IN ACTIVATED SLUDGE MODELING

by:

Henryk Melcer
Brown and Caldwell

Peter L. Dold
Richard M. Jones
Christopher M. Bye
Imre Takacs
EnviroSim Associates Limited

H. David Stensel
University of Washington

A. Warren Wilson
WPC Solutions Inc.

Paul Sun
Shell Global Solutions, US Inc.

Scott Bury
Dow Chemical Company

2003

Water Environment Research Foundation

The Water Environment Research Foundation, a not-for profit organization, funds and manages water quality research for its subscribers through a diverse public-private partnership between municipal utilities, corporations, academia, industry, and the federal government. WERF subscribers include municipal and regional water and wastewater utilities, industrial corporations, environmental engineering firms, and others that share a commitment to cost-effective water quality solutions. WERF is dedicated to advancing science and technology addressing water quality issues as they impact water resources, the atmosphere, the lands, and quality of life.

For more information, contact:
Water Environment Research Foundation
635 Slaters Lane, Suite 300
Alexandria, VA 22314-1994
Tel: (703) 684-2470
Fax: (703) 299-0742
www.werf.org
werf@werf.org

This report was co-published by the following organizations. For non-subscriber sales information, contact:

Water Environment Federation
601 Wythe Street
Alexandria, VA 22314-1994
Tel: (800) 666-0206
Tel: (703) 684-2452
Fax: (703) 684-2492
www.wef.org
pubs@wef.org

IWA Publishing
Alliance House, 12 Caxton Street
London SW1H 0QS, United Kingdom
Tel: +44 (0) 20 7654 5500
Fax: +44 (0) 20 7654 5555
www.iwapublishing.com
publications@iwap.co.uk

Library of Congress Catalog Card Number: 2003100316

WERF ISBN: 978-1-893664-71-5   IWAP ISBN: 978-1-84339-662-8   WEF ISBN: 978-1-57278-214-3

Brown and Caldwell

The research on which this report is based was funded, in part, by the United States Environmental Protection agency through Cooperative Agreement No. CR-824468-01 with the Water Environment Research Foundation (WERF). Unless an EPA logo appears on the cover, this report is a publication of WERF, not EPA. Funds awarded under the Cooperative Agreement cited above were not used for editorial services, reproduction, printing, or distribution.

ACKNOWLEDGMENTS

The project team would like to acknowledge and thank the staff of the King County South Plant, Seattle, Washington; the City of Phoenix, Arizona, 91st Avenue Multi-Cities Wastewater Treatment Plant; the Metropolitan Council Environmental Services, St. Paul, Minnesota; the City of Calgary, Alberta; the Cities of Greater Sudbury and Hamilton, Ontario; the Ontario Ministry of Environment and Energy, Canada; Shell Global Solutions, US Inc.; and Dow Chemical Company for their support of this project, and Richard Ke, graduate student at the University of Washington, for his extensive experimental work.

Report Preparation

Principal Investigator:
Henryk Melcer, Ph.D., P.E.
Brown and Caldwell

Project Team:
Patricia Tam, M.S., P.E.
Brown and Caldwell

Peter L. Dold, Ph.D.
Richard M. Jones, Ph.D., P.Eng.
Christopher M. Bye, Ph.D., P.Eng.
Imre Takacs, Ph.D.
EnviroSim Associates Limited

H. David Stensel, Ph.D.
Richard Ke
University of Washington

A W. Wilson, Ph.D., P.Eng.
WPC Solutions, Inc.
(Dr. Wilson was with EarthTech Canada, Inc. during part of this project)

Paul T. Sun, Ph.D., P.E.
Shell Global Solutions, US Inc.

Scott Bury, Ph.D.
Dow Chemical Company

Project Subcommittee & Reviewers

M. Aitken, Ph.D., Chair
University of North Carolina

J. Barnard, Ph.D.
Black and Veatch

C. P. L. Grady, Ph.D.
Clemson University

R. L. Irvine, Ph.D.
University of Notre Dame

J. C. Kabouris, Ph.D., P.E.
Carollo Engineers
(Dr. Kabouris was with Miami-Dade Water and Sewer Dept. during part of this project)

D. L. Sheridan, Ph.D., P.E.
Gannet Fleming Inc.

Additional Peer Reviewers

L. A. Carrio
New York City of Environmental Protection

G. Daigger, Ph.D., P.E.
CH2M-Hill

M. Henze, Ph.D.
Technical University of Denmark

J. B. Copp, Ph.D.
Hydromantis, Inc.

Reviewer & Advisor

D. S. Parker, Ph.D., P.E.
Brown and Caldwell

Water Environment Research Foundation Staff

Project Manager: Mary Strawn

ABSTRACT AND BENEFITS

Abstract

This report presents the state-of-the-art on peer-reviewed methods for measuring key influent wastewater characteristics and kinetic/stoichiometric parameters used in IWA-type activated sludge models. Practical guidance is provided to aid practitioners in the application of simulators incorporating these models to wastewater treatment plant design, operation and optimization.

The status of IWA model development and of commercially available simulators that incorporate IWA models is reviewed. The reliability of wastewater treatment plant design and analysis based on simulations depends upon well-calibrated models and, in turn, upon good estimates of model parameters. For example, estimates are that using conservative values for the maximum specific nitrifier growth rate parameter alone will cost agencies approximately $2 billion in excess expenditures over the next 20 years for over-design of nitrification facilities.

The IWA-type models are based on fractionating influent COD and TKN into various components – that is, characterizing the influent. The report presents methods for measuring these (e.g., RBCOD, SBCOD and inert soluble and particulate portions).

The maximum specific nitrifier growth rate (μ_{AUT}) was identified as one of the critical kinetic parameters because of its impact on sizing and control of nitrifying activated sludge systems. Different methods for measuring μ_{AUT} were evaluated, and compared to existing methods. Two new methods are recommended to agencies that need to implement nitrification and/or BNR so that they might optimize their designs and minimize capital expenditures.

A new method for estimating nitrifier decay rate (b_{AUT}) was developed; a value of $0.17 d^{-1}$ (at 20°C with a temperature dependency factor of 1.029) was estimated. Using these values, estimates of μ_{AUT} from all three bioassay methods tested at several facilities across North America corresponded closely, in the range of 0.90 to 0.95 d^{-1} (at 20°C). This finding questions the widely held perception that μ_{AUT} varies significantly from plant to plant, suggesting that the variability in the rate reported previously perhaps was an artifact of the incorrect assumption of a low nitrifier decay rate.

It was concluded that higher, more conservative temperature dependency factors for μ_{AUT} reported in the literature may be a mathematical artifact of calculating the factor from changes in (μ_{AUT} - b_{AUT}) with temperature rather than changes in μ_{AUT}, and/or incorporating the assumption that b_{AUT} is "negligible." A temperature dependency factor for μ_{AUT} of 1.072 is suggested.

Examples of activated sludge model calibration are presented for municipal and industrial wastewater treatment applications. Guidance is offered on model calibration procedures and the use of process simulators incorporating IWA-type models.

Benefits

- Provides a compendium of peer-reviewed methods for measuring important wastewater characteristics and International Water Association (IWA) activated sludge model parameters.
- Reviews status of methods for measuring maximum specific nitrifier growth rate (μ_{AUT}), the most critical parameter in the design of nitrifying activated sludge systems.
- Recommends the most appropriate methods of measuring μ_{AUT}. More accurate measurement of μ_{AUT} will reduce the capital cost of nitrifying activated sludge systems.
- Identifies limitations in current methods of measuring nitrifier decay rate and recommended modified procedures. Use of more appropriate values of nitrifier decay rate provides improved accuracy in measuring μ_{AUT}.
- Assesses methods of measuring readily biodegradable COD (RBCOD). Improved measurement of RBCOD will improve the design of anaerobic and anoxic selectors.
- Assesses the status of IWA activated sludge models and model simulators. Increased use of activated sludge model simulators will improve system design and operation, reducing capital and operating costs.
- Provides guidance and case histories on the use of activated sludge model simulators.

Keywords: Activated sludge models, model simulators, model parameter measurement, nitrifier growth rate (μ_{AUT}), nitrifier decay rate, readily biodegradable COD (RBCOD).

TABLE OF CONTENTS

ACKNOWLEDGMENTS iii
ABSTRACT & BENEFITS v
LIST OF TABLES xx
LIST OF FIGURES xxiv
LIST OF ACRONYMS xxviii
EXECUTIVE SUMMARY ES-1

1.0 INTRODUCTION 1-1
1.1 Background 1-1
1.2 Activated Sludge Models 1-2
1.3 Models versus Simulators 1-2
1.4 Simulator Calibration 1-3
1.4.1 Influent Wastewater Characteristics 1-4
1.4.2 Activated Sludge Model Kinetic and Stoichiometric Parameters 1-4
1.5 Industrial Wastewaters 1-5
1.6 Objectives 1-6
1.7 Summary of Peer Review Workshop 1-6
1.8 Report Outline 1-7
1.9 References 1-8

2.0 REVIEW OF IWA ACTIVATED SLUDGE MODELS 2-1
2.1 Historical Background 2-1
2.2 The Basis of IWA Activated Sludge Models 2-4
2.3 Developments in IWA Activated Sludge Models 2-5
2.4 References 2-6

3.0 ACTIVATED SLUDGE PLANT SIMULATORS 3-1
3.1 Introduction 3-1
3.2 Survey of Activated Sludge Model Simulators 3-2
3.3 References 3-4

4.0 WASTEWATER CHARACTERISTICS 4-1
4.1 Background 4-1
4.2 Nature of Wastewaters 4-2
4.3 Important Note on Filter Type and the Term "Soluble" 4-3
4.4 Organic Material 4-6
4.4.1 Biodegradable and Unbiodegradable Portions 4-6
4.4.2 Quantifying the Division 4-8
4.4.3 Biomass COD in the Influent 4-10
4.4.4 Alternative Approach for Quantifying the COD Components 4-10
4.5 Nitrogenous Material 4-11
4.5.1 Free and Saline Ammonia and Organically Bound Portions 4-11
4.5.2 Quantifying the Division 4-12
4.6 Phosphorus Content 4-13
4.6.1 Orthophosphate and Organically Bound Portions 4-15

4.6.2 Quantifying the Division 4-15
4.7 Typical Magnitudes for Wastewater Fractions 4-16
4.8 Measurement Techniques 4-17
4.8.1 Direct Methods 4-18
4.8.2 Bioassay Methods 4-18
4.9 Closing Comments on Colloidal COD 4-19
4.10 References 4-23

5.0 SOLUBLE UNBIODEGRADABLE (INERT) COD 5-1
5.1 Background 5-1
5.2 Method 5-1
5.3 Example 5-1
5.4 Considerations 5-2
5.5 References 5-3

6.0 PARTICULATE UNBIODEGRADABLE (INERT) COD 6-1
6.1 Background 6-1
6.2 Method 6-1
6.3 Example 6-2
6.4 Considerations 6-4
6.5 Influent COD/cBOD$_5$ Ratio as an Indicator for f_{XI} 6-5
6.6 Mixed Liquor VSS/TSS Ratio as an Indicator for f_{XI} 6-6
6.7 Influent VSS Concentration as an Indicator of f_{XI} 6-6
6.8 Estimation of f_{XI} from Influent COD, cBOD$_5$ and VSS Concentrations 6-7
6.9 References 6-8

7.0 READILY BIODEGRADABLE COD 7-1
7.1 Background 7-1
7.2 OUR-Based Respirometric Methods 7-1
7.2.1 Aerobic Batch Method 7-1
7.2.2 SBR Method 7-3
7.2.3 Automated Respirometric Methods 7-5
7.2.4 Batch Test Without (External) Seed 7-5
7.3 OUR Response in Batch Test 7-5
7.3.1 OUR for Readily Biodegradable COD Utilization 7-6
7.3.2 OUR Associated with Slowly Biodegradable COD Removal 7-6
7.3.3 OUR for Endogenous Respiration 7-7
7.3.4 OUR for Nitrification 7-7
7.4 Physicochemical Methods 7-8
7.4.1 Flocculated and Filtered COD (ffCOD) Method 7-9
7.4.2 Ultrafiltration Method 7-10
7.5 Considerations 7-11
7.6 References 7-12

8.0 SLOWLY BIODEGRADABLE COD 8-1
8.1 Background 8-1
8.2 Method 8-1
8.3 Example 8-2

8.4 Particulate and Colloidal SBCOD 8-3
8.5 Alternate Method for Determining SBCOD 8-6
8.6 Considerations 8-7
8.7 References 8-8

9.0 ACTIVE BIOMASS COD 9-1
9.1 Background 9-1
9.2 Method 9-1
9.3 Example 9-2
9.4 Considerations 9-4
9.5 References 9-4

10.0 VFA COD 10-1
10.1 Background 10-1
10.2 Method (1) – Gas Chromatograph 10-2
10.3 Method (2) – Ion Chromatograph 10-3
10.4 Considerations 10-3
10.5 References 10-3

11.0 INFLUENT TKN FRACTIONS 11-1
11.1 Background 11-1
11.2 Ammonia Fraction of TKN: f_{SNH} 11-1
11.3 Soluble Unbiodegradable Organic Nitrogen Fraction: f_{NSI} 11-2
11.4 Nitrogen Content of the Particulate Unbiodegradable Influent COD: f_{NXI} 11-3
11.5 Division of Biodegradable Organic Nitrogen into Soluble and Particulate: f_{XNB}.. 11-3
11.6 Considerations 11-4

12.0 INFLUENT SOLIDS CHARACTERIZATION 12-1
12.1 Background 12-1
12.2 Example of Influent Solids Data Collection 12-2
12.3 Inorganic Suspended Solids (X_{ISS}) 12-4
12.3.1 Significance of Inorganic Suspended Solids (X_{ISS}) 12-4
12.4 Influent Solids COD/VSS 12-5
12.5 Influent Solids Nitrogen Content 12-7
12.6 Influent Solids Phosphorus Content 12-9
12.7 Considerations 12-9
12.8 References 12-9

13.0 MIXED LIQUOR COD/VSS 13-1
13.1 Background 13-1
13.2 Biomass COD/VSS 13-2
13.3 Method for Measuring Mixed Liquor COD/VSS 13-2
13.4 Examples 13-3
13.5 Considerations 13-4

14.0 MIXED LIQUOR NITROGEN AND PHOSPHORUS CONTENT 14-1
14.1 Background 14-1
14.2 Method for Measuring Mixed Liquor N/VSS and P/VSS 14-2
14.3 Example 14-2

14.4 P/VSS in Biological Phosphorus Removal Systems 14-3
14.5 Considerations 14-4
14.6 References 14-4

15.0 DENITRIFICATION STOICHIOMETRY 15-1
15.1 Background 15-1
15.2 Heterotroph Yield Coefficient in Anoxic Growth 15-1
15.3 Reduced Anoxic Yield: Modeling Implications 15-4
15.4 Biodegradable COD Available for Denitrification 15-5
15.5 References 15-5

16.0 NITRIFICATION RATE 16-1
16.1 Background 16-1
16.2 Modeling Nitrification 16-1
16.3 Approaches to Measuring or Estimating μ_{AUT} 16-4
16.4 Low F/M Bioassay 16-5
16.4.1 Estimating μ_{AUT} from Nitrite/Nitrate Response 16-6
16.4.2 Calculation of Nitrifier Concentration 16-8
16.4.3 Implementation of Low F/M Bioassay in SBR 16-9
16.5 Rapid Bioassay Methods 16-10
16.6 High F/M Batch Test 16-11
16.6.1 Brief Equation Formulation 16-12
16.6.2 High F/M Batch Test – pH Control Issue 16-13
16.6.3 Inhibition/Toxicity Evaluation 16-15
16.7 Washout Bioassay 16-16
16.7.1 Brief Equation Formulation 16-17
16.7.2 Washout Test Experimental Issues 16-19
16.8 Data Analysis for High F/M and Washout Methods 16-19
16.9 Development of High F/M and Washout Equations 16-20
16.10 Estimating μ_{AUT} by Calibration with a Simulator 16-23
16.11 Impact of Nitrifier Decay Rate 16-24
16.12 Growth Rate Temperature Dependency 16-24
16.13 Comparison of the Methods 16-26
16.14 Concluding Remarks 16-29
16.15 References 16-30

17.0 NITRIFIER DECAY RATE 17-1
17.1 Introduction 17-1
17.2 Initial Empirical Finding in This Study 17-2
17.3 Decay Rate in the Measurement of Maximum Specific Growth Rate 17-3
17.3.1 Low F/M Tests 17-3
17.3.2 High F/M Tests 17-6
17.4 Experimental Methods for Determining Decay Rate 17-7
17.5 Simulation of Batch Test Response 17-12
17.6 Estimation of Nitrifier Decay Rate from Batch Test NPR Data 17-14
17.7 Decay Rate and Growth Rate Temperature Adjustment Interaction 17-16
17.8 Conclusions 17-18
17.9 References 17-19
17.10 Appendix – Derivation of Equations 17-20

18.0 DENITRIFICATION RATE **18-1**
18.1 Background 18-1
18.2 Anoxic Growth - η_G 18-2
18.2.1 Approaches to Measuring η_G 18-4
18.2.2 Aerobic and Anoxic Batch Tests 18-4
18.3 Anoxic Hydrolysis – η_{HYD} 18-5
18.4 Considerations 18-8
18.5 References 18-8

19.0 PROCESS MODELING USING A SIMULATOR **19-1**
19.1 Background 19-1
19.2 Approach to Using a Simulator 19-2
19.3 Configuration of the Plant in the Simulator 19-4
19.4 Data Gathering 19-5
19.5 Calibration of the Simulator 19-7
19.6 Verification of the Calibrated Simulator 19-8
19.7 Application of the Simulator 19-8
19.8 References 19-8

20.0 SENSITIVITY ANALYSIS IN PRACTICE **20-1**
20.1 Background 20-1
20.2 Parameters and States in the Sensitivity Analysis 20-3
20.2.1 Model Parameters 20-3
20.2.2 Calculated Variables 20-4
20.3 Sensitivity Analysis Example 20-4
20.3.1 The Wastewater Treatment Plant 20-4
20.3.2 Selected Parameters 20-5
20.3.3 Sensitivity Analysis–the Procedure 20-7
20.3.4 Steady-State Sensitivity Analysis–Results 20-8
20.3.5 Dynamic Sensitivity Analysis 20-12
20.4 Conclusions 20-15
20.5 References 20-16

21.0 GUIDANCE FOR MODEL CALIBRATION **21-1**
21.1 Introduction 21-1
21.2 Distinguishing Between Categories of Data 21-2
21.3 Tiered Approach to Data Collection and Calibration 21-2
21.3.1 Objective of Calibration 21-2
21.3.2 Plant Type 21-3
21.3.3 A Generalized Procedure of Calibration 21-3
21.3.4 Calibration Level 1: Defaults and Assumptions Only 21-4
21.3.5 Calibration Level 2: Historical Data Only 21-5
21.3.6 Calibration Level 3: On-Site, Full-Scale Tests 21-9
21.3.7 Calibration Level 4: Direct Parameter Measurements 21-15
21.3.8 Steps in Calibration – A Typical Procedure 21-16
21.4 Common Pitfalls in Calibration 21-18
21.4.1 Sampling and Analysis Problems 21-19
21.4.2 Misleading Results from Deficiencies in Plant Operation 21-21
21.4.3 Waste Activated Sludge (WAS) Measurements and Sludge Age 21-22
21.5 References 21-23

22.0 ILLUSTRATION OF MODEL CALIBRATION TO BENCH AND PILOT SCALE SYSTEMS 22-1
22.1 Background 22-1
22.2 Example 1: Bench Scale Aerobic System 22-1
22.2.1 System Description and Operation 22-1
22.2.2 Operation Results 22-2
22.2.3 Model Calibration 22-3
22.2.4 Calibration Results 22-5
22.3 Example 2: Anoxic-Aerobic Pilot Scale System 22-6
22.3.1 System Description and Operation 22-6
22.3.2 Model Calibration 22-7
22.3.3 Calibration Results 22-8
22.4 Summary 22-11
22.5 References 22-11

23.0 EXAMPLES OF MODEL CALIBRATION IN MUNICIPAL PLANTS: AEROBIC SYSTEM 23-1
23.1 Background 23-1
23.1.1 Plant Description 23-1
23.2 Calibration Approach 23-2
23.2.1 Simulator Process Flow Configuration 23-2
23.2.2 Data Collection 23-3
23.3 Calibration Results 23-6
23.4 Calibration Discussion 23-10
23.4.1 Effluent $cBOD_5$ 23-10
23.4.2 Plant Loading and Sludge Production 23-10
23.4.3 RAS and WAS Flow Rates 23-11
23.5 Conclusions 23-12

24.0 EXAMPLES OF MODEL CALIBRATION IN MUNICIPAL PLANTS: ANOXIC SYSTEM FOR NITROGEN REMOVAL 24-1
24.1 Background 24-1
24.1.1 Plant Description 24-1
24.2 Calibration 24-3
24.2.1 Process Flow Schematic 24-3
24.2.2 Wastewater Characterization 24-3
24.2.3 Model Calibration 24-8
24.3 Application 24-12
24.4 Acknowledgments 24-12
24.5 References 24-12

25.0 CASE STUDY: APPLICATION OF THE MODEL TO AN N & P REMOVAL MUNICIPAL WASTEWATER TREATMENT PLANT 25-1
25.1 Introduction 25-1
25.2 The Bonnybrook Wastewater Treatment Plant 25-1
25.3 Configuring Bonnybrook Secondary C in the Simulator 25-4
25.4 Plant Process Performance Data 25-5
25.5 Initial Model Calibration 25-11

25.6 Initial Modeling Results 25-13
25.7 Model Refinements 25-14
25.7.1 Wasting Rate 25-14
25.7.2 Raw Sewage TKN 25-16
25.7.3 Raw Sewage COD Data 25-16
25.7.4 COD Fractions 25-16
25.8 Concluding Remarks 25-19
25.9 Acknowledgments 25-19
25.10 References 25-19

26.0 INDUSTRIAL WASTEWATERS—GENERAL DISCUSSION 26-1
26.1 Background 26-1
26.2 Selection of Model Structure and Complexity 26-2
26.3 Gathering of Experimental Data 26-3
26.4 Model Formulation 26-3
26.5 Model Calibration 26-4
26.6 Summary 26-4

27.0 CASE STUDY: MODEL DEVELOPMENT FOR PETROLEUM REFINERY WASTEWATER TREATMENT 27-1
27.1 Background 27-1
27.2 Selection of Model Structure and Complexity 27-1
27.2.1 Gathering of Experimental Data 27-1
27.3 Model Formulation 27-2
27.3.1 Model Components 27-3
27.3.2 Model Processes 27-4
27.3.3 Matrix Representation 27-8
27.4 Model Calibration 27-8
27.4.1 Stoichiometric Parameters 27-8
27.4.2 Kinetic Parameters 27-9
27.5 Model Verification 27-11
27.6 References 27-12

28.0 CASE STUDY: MODEL DEVELOPMENT AND CALIBRATION FOR PETROCHEMICAL WASTEWATER TREATMENT PLANT (1) 28-1
28.1 Background 28-1
28.1.1 Plant Description 28-2
28.2 Model Formulation 28-3
28.2.1 Model Components 28-3
28.2.2 Model Processes 28-5
28.2.3 Matrix Representation 28-7
28.3 Calibration 28-7
28.4 Application 28-9
28.4.1 Case 1: Influence of Operating Configuration on Volatile Organic Compound (VOC) Emissions 28-9
28.4.2 Case 2: Spike Loading 28-10
28.5 References 28-11

29.0 CASE STUDY: MODEL DEVELOPMENT AND CALIBRATION FOR A PETROCHEMICAL WASTEWATER TREATMENT PLANT (2) 29-1
29.1 Background 29-1
29.2 Plant Description 29-2
29.3 Model Formulation 29-3
29.3.1 Model Components 29-3
29.3.2 Model Processes 29-4
29.3.3 Matrix Representation 29-5
29.4 Calibration 29-5
29.4.1 Calibration Results 29-8
29.5 Model Verification 29-11
29.6 Application 29-14
29.7 References 29-15

30.0 CASE STUDY: MODEL DEVELOPMENT AND CALIBRATION FOR A CHEMICAL WASTEWATER TREATMENT PLANT 30-1
30.1 Background 30-1
30.2 Plant Description 30-1
30.3 Model Formulation 30-3
30.4 Calibration 30-5
30.4.1 Steady State Calibration 30-5
30.4.2 Dynamic Calibration 30-7
30.5 Application 30-9
30.6 References 30-9

31.0 SBR: MCMASTER UNIVERSITY INVESTIGATIONS 31-1

32.0 SBR: KING COUNTY INVESTIGATIONS 32-1
32.1 Background 32-1
32.2 SBR Operation 32-1
32.2.1 Influent Wastewater 32-2
32.2.2 Selection of SBR Sludge Age 32-2
32.2.3 SBR Start-up Period 32-2
32.2.4 Monitoring during Intensive Testing 32-3
32.3 SBR Data Validation: Mass Balances 32-8
32.4 Characteristics from Direct Measurements 32-10
32.5 Characteristics from Modeling of SBR Operation 32-12
32.5.1 Particulate Unbiodegradable COD Fraction f_{XI} 32-12
32.5.2 Nitrification Rate (μ_{AUT}) 32-14
32.6 References 32-15

33.0 SBR: CITY OF PHOENIX INVESTIGATIONS 33-1
33.1 Background 33-1
33.2 SBR Operation 33-1
33.2.1 Influent Wastewater 33-2
33.2.2 Selection of SBR Sludge Age 33-2
33.2.3 SBR Start-up Period 33-2
33.2.4 Monitoring during Intensive Testing 33-3

33.3 SBR Data Validation: Mass Balances 33-8
33.4 Characteristics from Direct Measurements 33-9
33.5 Characteristics from Modeling of SBR Operation 33-11
33.5.1 Particulate Unbiodegradable COD Fraction (f_{XI}) 33-11
33.5.2 Nitrification Rate (μ_{AUT}) 33-12

34.0 LOW F/M SBR METHOD 34-1
34.1 Background 34-1
34.2 Impact of Influent Feeding Regime 34-1
34.2.1 Operating Conditions 34-2
34.2.2 SBR Performance 34-4
34.2.3 Nitrification Rate Tests 34-5
34.2.4 Estimation of μ_{AUT} 34-8
34.3 Impact of Temperature 34-9
34.3.1 Operating Conditions 34-9
34.3.2 SBR Performance 34-10
34.3.3 Nitrification Rate Tests 34-11
34.3.4 Estimation of μ_{AUT} 34-13

35.0 SUMMARY OF HIGH F/M INVESTIGATIONS 35-1
35.1 Background 35-1
35.2 University of Washington Investigations 35-2
35.2.1 Operating Conditions and Data Collection 35-2
35.2.2 Evaluation of pH Control Method 35-3
35.2.3 Evaluation of Test Reproducibility 35-4
35.2.4 Evaluation of Initial Ammonia Concentration 35-5
35.2.5 Evaluation of Wastewater Source 35-7
35.3 McMaster University Investigations 35-8
35.3.1 Evaluation of Test Reproducibility 35-9
35.3.2 Evaluation of Influent Inhibition 35-11
35.3.3 Evaluation of Seed Sludge Source 35-12
35.3.4 Evaluation of Alkalinity Addition Method 35-14
35.3.5 Evaluation of Aeration Method 35-15
35.3.6 Summary of McMaster University Results 35-18
35.4 City of Phoenix Investigations 35-20
35.4.1 Test #1 – April 18, 2001 35-20
35.4.2 Test #2 – May 14, 2001 35-21
35.4.3 Test #3 – May 26, 2001 35-21
35.5 References 35-22

36.0 SUMMARY OF WASHOUT INVESTIGATIONS 36-1
36.1 Background 36-1
36.2 University of Washington Investigations 36-1
36.2.1 Apparatus and Operating Conditions 36-1
36.2.2 Washout Test Results 36-4
36.3 McMaster University Investigations 36-5
36.3.1 Washout of the McMaster Low F/M SBR 36-6
36.3.2 Washout of Dundas WWTP Sludge 36-7

36.4 City of Phoenix Investigations 36-8
36.4.1 May 26, 2001 36-8
36.4.2 September 17, 2001 36-9
36.4.3 September 24, 2001 36-10
36.4.4 October 1, 2001 36-11
36.4 References 36-11

37.0 NITRIFIER DECAY RATE: UNIVERSITY OF WASHINGTON INVESTIGATIONS 37-1
37.1 Background 37-1
37.2 Apparatus and Operating Conditions 37-1
37.3 Data Analysis 37-2
37.4 Washout Test Results 37-3
37.4.1 Estimate of b_{AUT} Accounting for Supplementary Nitrifier Growth 37-3
37.4.2 Estimate of Temperature Dependency for b_{AUT} 37-4
37.4.3 Estimate of b_{AUT} Using Method by Katehis et al 37-4
37.4.4 Comparison with Established Literature Values 37-6
37.5 Impact of b_{AUT} on μ_{AUT} 37-6
37.6 References 37-8

38.0 COMPARISON OF μAUT RESULTS FROM DIFFERENT METHODS 38-1
38.1 Introduction 38-1
38.2 University of Washington μ_{AUT} Values 38-1
38.2.1 μ_{AUT} Estimated from Low F/M SBR Method 38-2
38.2.2 μ_{AUT} Estimated from High F/M Method 38-2
38.2.3 μ_{AUT} Estimated from Washout Method 38-3
38.3 McMaster University μ_{AUT} Values 38-3
38.3.1 μ_{AUT} Estimated from Low F/M SBR Method 38-4
38.3.2 μ_{AUT} Estimated from High F/M Method 38-4
38.3.3 μ_{AUT} Estimated from Washout Method 38-4
38.4 City of Phoenix μ_{AUT} Values 38-5
38.4.1 μ_{AUT} Estimated from Low F/M SBR Method 38-5
38.4.2 μ_{AUT} Estimated from High F/M Method 38-5
38.4.3 μ_{AUT} Estimated from Washout Method 38-5

39.0 BACKGROUND ON MEASURING RBCOD 39-1
39.1 Background 39-1
39.2 OUR-Based Respirometric Methods 39-2
39.3 Physicochemical Methods 39-3
39.4 Constitution of the Soluble Effluent COD 39-4
39.4.1 Influent COD 39-4
39.4.2 Intermediates and End Products 39-5
39.4.3 Cell Lysis Products 39-5
39.4.4 Quantifying the "Internal" Generation of Soluble Effluent COD 39-6
39.4.5 Summary 39-8
39.5 Selection of Filtration Medium 39-8
39.6 References 39-9

40.0 READILY BIODEGRADABLE COD: UNIVERSITY OF WASHINGTON INVESTIGATIONS 40-1
40.1 Background 40-1
40.2 Experimental Methods 40-1
40.2.1 Flocculation/Filtration COD Method 40-1
40.2.2 Ultrafiltration COD Method 40-2
40.2.3 Respirometric Test Method 40-2
40.3 Results and Discussion 40-5
40.3.1 Test Reproducibility for Flocculation/Filtration and Ultrafiltration Methods 40-5
40.3.2 Recovery of Soluble COD 40-5
40.3.3 Comparison of Flocculation/Filtration and Ultrafiltration Methods 40-6
40.3.4 Comparison of Flocculation/Filtration and Respirometric Methods 40-8
40.3.5 Flocculation/Filtration After Ultrafiltration 40-9
40.4 Ease of Use 40-10
40.5 References 40-10
40.6 Appendix: Experimental Data 40-11

APPENDIX A – NOMENCLATURE A-1
A.1 Discussion A-1

APPENDIX B - BOD CALCULATIONS B-1
B.1 Background B-1
B.2 Basis for BOD Calculations B-2
B.3 BOD Associated with Soluble Biodegradable COD (S_B): B-3
B.3.1 Method 1 .... B-3
B.3.2 Method 2 .... B-4
B.4 BOD Associated with Particulate Slowly Biodegradable COD (X_S): B-5
B.5 BOD Associated with Active Biomass B-7
B.6 Example B-7
B.7 COD versus BOD as a Modeling Parameter B-9
B.8 References B-11

APPENDIX C - OPERATING A BENCH-SCALE CONTINUOUS FLOW SYSTEM ... C-1
C.1 Background C-1
C.2 Operation C-1
C.2.1 COD Balance C-3
C.2.2 Nitrogen Balance C-3
C.2.3 System Sludge Age (SRT) C-4
C.2.4 Note on Estimating f_{XI} C-4
C.3 Example of Continuous Flow System C-4
C.3.1 COD Balance C-5
C.3.2 Nitrogen Balance C-5
C.3.3 Unbiodegradable Soluble COD (f_{SI}) C-5
C.3.4 Ammonia Nitrogen (f_{SNH}) C-6
C.3.5 Unbiodegradable Soluble Organic Nitrogen (f_{SNI}) C-6
C.3.6 Division of Biodegradable Organic Nitrogen into Soluble and Particulate: f_{XNB} C-6

C.3.7 System Sludge Age (SRT) C-6
C.3.8 Particulate Unbiodegradable COD (f_{XI}) C-6
C.4 References C-7

APPENDIX D - OPERATING A BENCH-SCALE SBR SYSTEM D-1
D.1 Background D-1
D.2 SBR Operation D-1
D.2.1 Influent Wastewater D-3
D.2.2 Selection of SBR Sludge Age D-3
D.2.3 SBR Start-up Period D-4
D.2.4 Monitoring during 12 Day Intensive Testing D-5
D.3 SBR Data Validation: Mass Ballances D-8
D.3.1 COD Balance D-8
D.3.2 Nitrogen Balance D-8
D.4 Characteristics from Direct Measurements D-10
D.5 Characteristics from Modelling of SBR Operation D-12
D.5.1 Particulate Unbiodegradable COD Fraction (f_{XI}) D-13
D.5.2 Soluble Unbiodegradable TKN Fraction (f_{SNI}) D-13
D.5.3 Nitrification Rate (μ_{AUT}) D-14
D.6 Estimated Labor Requirements D-15
D.7 References D-16

APPENDIX E - DEVELOPMENT OF HIGH F/M METHOD E-1
E.1 Background E-1
E.2 Basis for the High F/M Method E-1
E.3 Nitrite/Nitrate Response in the High F/M Method E-2
E.4 Simulated Batch Test Response E-3
E.5 Antoniou et al. (1990) Analysis Method E-5
E.6 Sözen et al. (1996) Analysis Method E-6
E.7 Influence of Ammonia Concentration E-11
E.8 Influence of Experimental Error in Nitrite/Nitrate Analysis E-14
E.9 Modified High F/M Method E-15
E.10 References E-16

APPENDIX F - HIGH F/M METHOD: DETAILS F-1
F.1 Basic Equipment F-1
F.2 Initial Steps F-2
F.3 pH Control Method A – DO Control with Alkalinity Addition F-4
F.4 pH Control Method B – Continuous Aeration Supplemented with Carbon Dioxide F-5
F.5 Commencing the Test F-6
F.6 Sampling F-7
F.7 Estimated Labor Requirements F-8

APPENDIX G - WASHOUT METHOD: DETAILS G-1
G.1 Basic Equipment G-1
G.2 Initial Steps G-2
G.3 Commencing the Test G-7

G.4 Sampling G-7
G.5 Estimated Labor Requirements G-9

APPENDIX H - PARAMETER ESTIMATION METHOD FOR DETERMINING NITRIFICATION RATES H-1
H.1 Background H-1
H.2 Non-Linear Regression Analysis H-1
H.2.1 Example H-2
H.3 Confidence Intervals H-3
H.3.1 Example II-5
H.4 Diagnostics H-9
H.4.1 Analysis of Residuals H-9
H.4.2 Parameter Correlations H-10
H.4.3 Example H-10
H.5 Application of Methods to Washout Test Data H-12
H.6 References H-12

APPENDIX I - ESTIMATION OF SRT AT FULL-SCALE PLANTS I-1
I.1 Background I-1
I.2 Definitions I-1
I.3 Example I-2
I.4 Additional Considerations for SRT Estimation I-4
I.4.1 Sludge Mass in the Clarifier I-4
I.4.2 Additional Sampling Problems I-5

LIST OF TABLES

ES-1 Summary of Measurement Methods ES-3
ES-2 Advantages and Disadvantages of Bioassay Methods. ES-4
1-1 Guide to This Report 1-8
3-1 Activated Sludge Simulators Surveyed 3-2
3-2 Summary of Information Collected from Simulator Vendors. 3-3
4-1 COD of Unfiltered, Glass-Fiber Filtered and 0.45 μm Filtered Raw Municipal Influent Wastewater 4-4
4-2 Typical Municipal Wastewater Characteristic Parameter Values. 4-17
4-3 Summary of Municipal Wastewater Characteristics and Measurement Techniques. 4-18
6-1 Comparison of Observed and Predicted VSS, OUR and Effluent Nitrate for Different Estimates of Influent Particulate Unbiodegradable COD Fraction 6-3
6-2 Example Calculation of Particulate Unbiodegradable Portion of Influent VSS 6-6
8-1 Estimation of Colloidal/Particulate Fractionation of SBCOD 8-5
10-1 Gas Chromatograph Set-Up Example for VFA Analysis 10-2
12-1 Analysis of Unfiltered and Glass-Fiber Filtered Municipal Influent Wastewater 12-2
14-1 SBR Mixed Liquor VSS and TKN Data 14-3
14-2 Biological Phosphorus Removal SBR Steady State Results with Fixed Influent COD Load and Varying Phosphorus Load 14-3
15-1 Yield under Anoxic Conditions with Various Soluble Substrates. 15-3
16-1 Examples of Reported μ_{AUT} Temperature Sensitivity Factors 16-25
16-2 Comparison of Advantages and Disadvantages of the Three Nitrification Rate Measurement Methods 16-28
17-1 Temperature Dependency θ Factors Based on Change in (μ_{AUT} - b_{AUT}) or μ_{AUT} with Temperature 17-17
20-1 Parameter Categories in Sensitivity Analysis 20-3
20-2 Activated Sludge Plant Specifications 20-5
20-3 ASM1 Default Parameters Used in the Sensitivity Example 20-6
20-4 Default Influent (Driving Function) Values 20-6
20-5 Operational Parameters Analyzed 20-7
20-6 Results of the Two Base Cases 20-8
20-7 Model Parameter Sensitivity Table (3.3 d SRT) 20-10
20-8 Analysis of Influent and Operational Parameters (3.3 d SRT) 20-11
20-9 Model Parameter Sensitivity Functions (15 d SRT) 20-12
20-10 Analysis of Influent and Operational Parameters (15 d SRT) 20-12
21-1 General Data Gathering Guideline for Different Types of Municipal Activated Sludge Wastewater Treatment Plants 21-12
22-1 Design and Operating Parameters for the Bench Scale Activated Sludge System 22-2
22-2 Bench Scale Activated Sludge System Daily Monitoring Parameters 22-2
22-3 Summary of Monitoring Results 22-2
22-4 Initial Estimates of Model Parameters 22-4
22-5 Comparison of Observed and Predicted Values Using Initial Parameters 22-5
22-6 Comparison of Observed and Predicted VSS, OUR and Effluent Nitrate for Different Estimates of Influent Particulate Unbiodegradable COD Fraction 22-5
22-7 Calibrated Estimates of Model Parameters 22-6
22-8 Comparison of Observed and Predicted Values Using Calibrated Parameters 22-6
22-9 Design and Operating Parameters for the Anoxic-Aerobic Pilot Scale System 22-7
22-10 Calibrated Estimates of Model Parameters for the Pilot Scale Anoxic-Aerobic System 22-9
23-1 Dimensions of Unit Tankage at the Aerobic System Case Study Plant 23-2
23-2 Summary of Influent COD Fractions 23-5
23-3 "Rule of Thumb" Estimates of Loading and Sludge Production Rates 23-11
24-1 Dimensions of Unit Tankage at the 91st Avenue WWTP 24-2
24-2 Summary of RAS Characterization Data 24-4
24-3 Summary of Wastewater Characterization Data for Primary Effluent and Anoxic Effluent 24-5
24-4 Summary of Wastewater Characterization Data for OX6 and Secondary Effluent 24-6
24-5 Summary of Diurnal Wastewater Characterization Data for Primary Effluent and AT AX5 (Anoxic Effluent) 24-7
24-6 Summary of Diurnal Wastewater Characterization Data for OX6 and Secondary Effluent 24-8
24-7 Summary of Steady State Modeling 24-9

25-1 Treated Effluent Permit Limits for the Bonnybrook WWTP 25-2
25-2 Design Parameters for Secondary C at the Bonnybrook WWTP 25-4
25-3 List of Data Routinely Collected for Bonnybrook Secondary C and Associated Primary Clarification and Primary Sludge Fermentation Systems 25-6
25-4 Raw Sewage Carbon, Nitrogen and Phosphorus Fractions Used in the Initial Model Calibration 25-12
25-5 Raw Sewage Carbon, Nitrogen and Phosphorus Fractions Used in the Refined Model Calibration 25-16
27-1 List of Components in the Petroleum Refinery Activated Sludge Model 27-3
27-2 Comparison of Observed Values and Model Predictions for Steady State Operation of a Bench Scale Activated Sludge System Operated on Petroleum Refinery Wastewater at a 20-day SRT 27-11
28-1 Dimensions of Unit Tankage at the Petrochemical Wastewater Treatment Plant 28-2
28-2 Historical Averages for Routine Monitoring Parameters at the Petrochemical Wastewater Treatment Plant 28 3
28-3 List of Components in the Petrochemical Refinery Activated Sludge Model 28-3
29-1 Dimensions and Capacities of Unit Processes at the Wastewater Treatment Plant for a Single Train 29-2
29-2 The Historical Average Values of the Wastewater Characteristics 29-2
29-3 List of Components in the Petroleum Wastewater Treatment Activated Sludge Model 29-3
29-4 Estimated Parameters Used in the Final Model 29-10
29-5 Pilot Reactor Experimental Conditions 29-11
29-6 Comparison of Results and Measurement for Full-Scale, Long-Term Operation 29-14
30-1 Influent Wastewater Characteristics 30-2
30-2 Preliminary Steady-State Simulation of Effluent Using Temperature-Corrected Default ASM1 Parameters 30-4
30-3 Influent Wastewater COD Fractions 30-4
30-4 Calibrated Model Parameters 30-5
30-5 Steady State Simulation Solution Using Experimentally Determined Parameters and Manually Adjusted Nitrification Parameters 30-7
32-1 Ammonia, Nitrate and OUR Profile Data for Day 3 32-4
32-2 Ammonia, Nitrate and OUR Profile Data for Day 5 32-6
32-3 Daily Results from SBR Intensive Testing Period 32-7
32-4 COD and Nitrogen Mass Balances on Data from SBR Intensive Testing Period 32-9
32-5 Data Derived from Results of SBR Intensive Testing Period 32-11
33-1 Ammonia, Nitrate and Nitrate Profile Data for Day 2 33-5
33-2 Ammonia and Nitrate Profile Data for Day 3 33-6
33-3 Daily Results from SBR Intensive Testing Period 33-7
33-4 Nitrogen Mass Balances on Data from SBR Intensive Testing Period 33-8
33-5 Data Derived from Results of SBR Intensive Testing Period 33-10
34-1 Low F/M SBR Operating Conditions at the Renton Facility 34-2
34-2 Average Influent and SBR Performance Data for Different SBR Feeding Regimes 34-5
34-3 NH_4-N and NO_3-N Profile Data during Nitrification Rate Tests in Phases I and II 34-6
34-4 NH_4-N and NO_3-N Profile Data during Nitrification Rate Tests in Phase III 34-7
34-5 Summary of Estimated Nitrate Production Rates 34-8
34-6 Summary of Test Results and Maximum Specific Nitrifier Growth Rates 34-8
34-7 Low F/M SBR Operating Conditions at the University of Washington Laboratory 34-9
34-8 Average Influent and Performance Data for Different Temperature SBRs 34-10
34-9 NH_4-N and NO_3-N Data during Nitrification Rate Tests in Phase III 34-12
34-10 Summary of Estimated Nitrate Production Rates 34-13
34-11 Summary of Test Results and Maximum Specific Nitrifier Growth Rates 34-13
35-1 Comparison of Nitrate Production Rates for Evaluation of the Effect of Alkalinity and Carbon Dioxide/Air Ratios 35-4
35-2 Comparison of $\mu_{AUT\ 20\ deg\ C}$ Values for Different Wastewaters 35-4
35-3 Ammonia and Nitrate Concentrations Response to Changing Initial Ammonia Concentration in High F/M Test 35-5
35-4 Effect of Initial Ammonia Concentration on μ_{AUT} 20°C in High F/M Tests 35-7
35-5 Effect of Different Wastewaters on μ_{AUT} 20°C in High F/M Tests 35-7
36-1 Washout Test Operating Conditions at the University of Washington Laboratory 36-3
36-2 Washout Test Data Collection Program 36-3
36-3 Summary of Washout Test μ_{AUT} 20°C Values 36-5
37-1 Summary of Nitrifier Decay Rate Test Results 37-4
37-2 Estimation of Nitrifier Decay Rate for 12°C Test Data 37-4

37-3 Comparison of Calculated Average μ_{AUT} 20°C Values Using Default and Experimentally Determined Nitrifier Decay Rates 37-6
37-4 Predicted Effluent Ammonia Concentrations versus SRT for a CMAS Process as a Function of b_{AUT} and μ_{AUT} Values at 20°C 37-7
38-1 Summary of μ_{AUT} Estimates Obtained from University of Washington Low F/M SBR Experiments 38-2
38-2 Summary of μ_{AUT} Estimates Obtained from University of Washington High F/M Experiments 38-3
38-3 Summary of μ_{AUT} Estimates Obtained from University of Washington Washout Experiments 38-3
38-4 Summary of μ_{AUT} Estimates Obtained from McMaster University High F/M Experiments 38-4
38-5 Summary of μ_{AUT} Estimates Obtained from McMaster University Washout Experiments 38-4
38-6 Summary of μ_{AUT} Estimates Obtained from Phoenix Washout Experiments 38-5
39-1 Example of Breakdown of RBCOD Components in European Raw Municipal Wastewater 39-1
39-2 Example of Molecular Weight Fractionation of Effluent COD 39-7
40-1 Influent and Effluent Filtrate COD for Renton WWTP Wastewater 40-5
40-2 Acetate Recovery in the Flocculation/Filtration and Ultrafiltration Methods 40-6
40-3 Acetate COD Recovery in the Respirometric Method 40-6
40-4 RBCOD by Flocculation/Filtration versus Ultrafiltration 40-7
40-5 COD Removal after Ultrafiltration Followed by Flocculation/Filtration 40-9

LIST OF FIGURES

1-1 Essential Requirements for Wastewater Treatment Process Simulation 1-3
1-2 Comparison of Plant-Derived Nitrifier Growth Rates with U.S. EPA (1975) Nitrifier Growth Rate Equation 1-5
2-1 Schematic Illustration of Characteristic EBPR Behavior in an Anaerobic/Aerobic Sequenced Bioreactor 2-2
4-1 Influent Total (Unfiltered) COD, Glass-Fiber Filtrate COD and 0.45 μm Membrane Filtrate COD of Daily Raw Municipal Wastewater Grab Samples 4-4
4-2 Quantile-Quantile Probability Plot of COD Fraction Passing through Glass-Fiber Filter 4-5
4-3 Quantile-Quantile Probability Plot of COD Fraction Passing through 0.45 μm Membrane Filter 4-5
4-4 Division of Municipal Wastewater COD into Components 4-7
4-5 Schematic Representation of COD Components for Municipal Wastewater 4-7
4-6 Division of Municipal Wastewater TKN into Components 4-12
4-7 Division of Municipal Wastewater Phosphorus into Constituent Fractions. 4-14
4-8 Glass-Fiber Filtrate COD and 0.45 μm Membrane Filtrate COD: Effluent Samples from a 10-day SRT System 4-21
4-9 Glass-Fiber Filtrate COD and 0.45 μm Membrane Filtrate COD: Effluent Samples from a 15-day SRT System Initially Exhibiting Unstable Operation 4-21
4-10 0.45 μm Membrane Filtrate COD and ffCOD: Effluent Samples from a 15-day SRT SBR System 4-22
4-11 0.45 μm Membrane Filtrate COD and ffCOD: Effluent Samples from two 15-day SRT SBR Systems Operated during this Study 4-23
6-1 Relationship of Bioreactor VSS Concentration to SRT and f_{XI}. 6-4
7-1 Idealized Oxygen Utilization Rate Response Observed in an Aerobic Batch Test (or in a Sequencing Batch Reactor during the React Phase) 7-2
7-2 Example of an Oxygen Utilization Rate Response Observed during the React Period in an SBR Receiving a High Concentration of RBCOD 7-4
7-3 Example of an Oxygen Utilization Rate Response Observed during the React Period in an SBR with a Low Nitrifier Growth Rate 7-5
7-4 Monod Relationship between Organism Growth Rate and Readily Biodegradable COD Concentration (S_s) 7-6
7-5 Components Contributing to the Measured OUR in the Batch Test 7-8
8-1 Diagram Depicting the Retention/Passage of Influent Wastewater COD Components through Sequential 1.2 μm Glass-Fiber Filtration, Flocculation and 0.42 μm Membrane Filtration 8-4
8-2 Example of a Fitted BOD Curve for Determination of Total Biodegradable COD (BCOD) Concentration 8-7
9-1 Example of OUR Response for an Aerobic Batch Test on Raw Municipal Wastewater 9-3
9-2 Plot of ln (OUR) versus Time for Initial Period of Increasing OUR in Fig. 9-1 9-3
12-1 Influent COD and TSS Concentrations (Daily Grab Samples) 12-3
12-2 Influent ISS (X_{ISS}) Concentration (Daily Grab Samples) 12-3
12-3 Influent ISS/TSS Ratio (Daily Grab Samples). 12-4
12-4 Influent Solids COD/VSS Ratio (Daily Grab Samples) 12-6
12-5 Quantile-Quantile Probability Plot of Influent Solids COD/VSS Ratio (Daily Grab Samples) 12-7
12-6 Influent Solids TKN/VSS Ratio (Daily Grab Samples) 12-8
12-7 Quantile-Quantile Probability Plot of Influent Solids TKN/VSS Ratio (Daily Grab Samples) 12-8
13-1 Mixed Liquor COD/VSS Ratio (Daily Grab Samples) 13-3
13-2 Quantile-Quantile Probability Plot of Mixed Liquor COD/VSS Ratio (Daily Grab Samples) 13-4
16-1 Impact of the Arrhenius Coefficient (θ) on the Calculated Minimum SRT (SRT_{min}) Required for Nitrification 16-3
16-2 Example of Nitrite/Nitrate and Ammonia Response in a Low F/M Batch Test for Estimating μ_{AUT} 16-6
16-3 Nitrite/Nitrate Production Rate in a Low F/M Batch Test for Estimating μ_{AUT} 16-7
16-4 Nitrite/Nitrate and Ammonia Profiles in a High F/M Batch Test for Estimating μ_{AUT} 16-12
16-5 A High F/M Test Conducted with Inadequate pH Control Likely Will Exhibit Data Irregularities. 16-14
16-6 Nitrite/Nitrate Profiles in Side-by-Side High F/M Batch Test with Either Plant Influent Wastewater or Secondary Effluent as Diluent 16-15
16-7 Nitrate Response in a Washout Test for Estimating μ_{AUT} 16-17
16-8 Influence of Initial Conditions on Washout Test Response 16-19
16-9 High F/M Residuals Plot Exhibiting a Characteristic Sinusoidal Wave Shape 16-20

16-10 Example of Fitting Simulated Ammonia Response Data to Observed Data over 24 Hours in a Full-Scale Treatment Plant 16-24
16-11 Examples of Parallel μ_{AUT} Estimates from High F/M Tests, Washout Tests and the Low F/M SBR with Mixed Liquor from Processes All with the Same Influent Wastewater 16-27
17-1 Nitrifier Mass as a Function of SRT for Different Nitrifier Decay Rates 17-5
17-2 Estimating Maximum Specific Growth Rate Based on a Low Decay Rate (0.04 /d^{-1}) 17-6
17-3 Nitrate Concentration Response in the Batch Reactor (Eq. 17.10-6) 17-10
17-4 Nitrifier Concentration Response in the Batch Reactor (Linear Scale) 17-11
17-5 Semi-Logarithmic Plot of NPR Response 17-11
17-6 Configuration for Simulating Nitrifier Decay Rate Measurement Test 17-13
17-7 Simulated Nitrate Concentration Response in the Batch Reactor (cf. Fig. 17-3) 17-13
17-8 Simulated Nitrifier Concentration Response in the Batch Reactor (cf. Fig. 17-4) 17-14
17-9 Least Squares Estimation of Nitrifier Decay Rate from Limited NPR Data 17-15
17-10 Least Squares Estimation of Nitrifier Decay Rate from Extensive NPR Data 17-16
17-11 Impact of the Temperature Dependency Coefficient (θ) on Minimum SRT for Nitrification 17-18
18-1 Nitrate Profile in an Anoxic Batch Test Combining Mixed Liquor and Municipal Wastewater 18-6
18-2 Configuration for Simulating Nitrifier Decay Rate Measurement Test 18-7
18-3 Simulation of Anoxic Batch Test Nitrate Response 18-7
19-1 Essential Requirements for Wastewater Treatment Process Simulation 19-2
19-2 Generalized Methodology for Using a Simulator 19-4
19-3 Example of Multi-Component Influent Stream in a Simulator Configuration 19-5
20-1 Normalized Diurnal Pattern Used for Dynamic Analysis 20-13
20-2 Diurnal Ammonia Profile 20-14
20-3 Diurnal Sensitivity Functions 20-15
21-1 Example of Daily Composite MLSS Sampling 21-19
21-2 Example of Non-Representative Bi-Hourly Grab MLSS Sampling 21-20
21-3 Example of Diurnal Treated Effluent Ammonia-N Variations 21-21
21-4 Example of Diurnal Treated Effluent Nitrate-N Variations 21-21
21-5 Example of Cumulative ISS Balance 21-22
22-1 Simulator Configuration for the Bench Scale Aerobic System 22-4
22-2 Simulator Configuration for the Pilot - Scale Anoxic-Aerobic System 22-7
22-3 Sensitivity of the Bioreactor Nitrate Profile to the Value of f_{ss} 22-8
22-4 Simulated and Observed Reactor VSS Profile Using Calibrated Parameter Estimates 22-8
22-5 Simulated and Observed Reactor OUR Profile Using Calibrated Parameter Estimates 22-9
22-6 Simulated and Observed Reactor Soluble COD Profile Using Calibrated Parameter Estimates 22-10
22-7 Simulated and Observed Reactor Nitrate-N Profile Using Calibrated Parameter Estimates 22-10
22-8 Simulated and Observed Reactor Soluble TKN Profile Using Calibrated Parameter Estimates 22-10
23-1 Simulator Configuration of the Aerobic System Case Study Plant 23-3
23-2 Calibration Period Measured Primary Influent Flow 23-4
23-3 Calibration Period Simulated and Measured Primary Influent $CBOD_5$ 23-4
23-4 Calibration Period Measured Primary Influent TKN and Ammonia (-N) 23-4
23-5 Calibration Period Measured Process Temperature 23-5
23-6 Calibration Period Measured and Simulated Primary Influent TSS 23-6
23-7 Calibration Period Calculated Primary Influent COD Concentration 23-6
23-8 Measured and Simulated Primary Effluent TSS 23-7
23-9 Measured and Simulated Primary Effluent $CBOD_5$ 23-7
23-10 Measured and Simulated MLSS and Simulated MLVSS as Monitored in the Last Aeration Tank 23-8
23-11 Measured and Simulated Secondary Effluent $CBOD_5$ 23-8
23-12 Measured and Simulated Secondary Effluent TSS 23-8
23-13 Measured and Simulated Secondary Effluent TKN 23-9
23-14 Measured and Simulated Secondary Effluent Ammonia-N 23-9
23-15 Measured and Simulated Secondary Effluent Total Phosphorus (TP) 23-9
23-16 Gravity Thickener Underflow Measured and Simulated TSS and VSS 23-10
23-17 Actual and Adjusted WAS Flows and WAS TSS Concentrations 23-12
24-1 Simulator Configuration of the 91st Avenue WWTP 24-2
24-2 Sampling Locations in the 91st Avenue WWTP 24-4
24-3 Diurnal Variability in Plant Nitrate Concentration 24-11
24-4 Simulation of Nitrate Concentration and Ammonia Concentration vs. Observed Concentrations 24-12
25-1 Flowsheet for Bonnybrook WWTP 25-2

25-2 Bonnybrook Bioreactor No. 10 as Configured in the Simulator 25-5
25-3 Wastewater Temperature for the Bonnybrook WWTP – 1999 25-7
25-4 Primary Effluent, NMLR, RAS and WAS Flowrates in Bioreactor No.10 at the Bonnybrook WWTP – 1999 25-8
25-5 Primary Sludge Feed, Recycle and Waste Sludge Flows for Fermenter No. 2 at the Bonnybrook WWTP – 1999 25-8
25-6 Raw Sewage COD Concentration in Bioreactor No.10 at the Bonnybrook WWTP – 1999 25-9
25-7 Raw Sewage TKN Concentrations in Bioreactor No.10 at the Bonnybrook WWTP – 1999 25-9
25-8 Raw Sewage Total Phosphorus Concentration for Bioreactor No.10 at the Bonnybrook WWTP – 1999 25-10
25-9 Primary Effluent Alkalinity to Bioreactor No.10 at the Bonnybrook WWTP – 1999 25-10
25-10 Raw Sewage ISS Concentration (Estimated) for Bioreactor No.10 at the Bonnybrook WWTP – 1999 25-11
25-11 Initial Model Predictions for Mixed Liquor, Autotrophic Organism and Ammonia-Nitrogen Concentrations 25-13
25-12 MLSS Concentration in Bioreactor No.10 at the Bonnybrook WWTP – 1999 25-14
25-13 WAS Flowrate in Bioreactor No.10 at the Bonnybrook WWTP – 1999 25-15
25-14 Predicted TSS and Autotrophic Organism Concentrations in the Mixed Liquor for the Refined Calibration 25-17
25-15 Predicted CODsol and TSS Concentrations in the Secondary Effluent for the Refined Calibration 25-17
25-16 Predicted Ammonia, Nitrate and TKN Concentrations in the Secondary Effluent for the Refined Calibration 25-18
25-17 Predicted TSS Concentration in the Fermenter Complete Mix Tank and VFA Concentration in the Gravity Sludge Thickener 25-18
27-1 Division of Petroleum Refinery Wastewater COD into Constituent Fractions 27-3
27-2 Division of Petroleum Refinery Wastewater TKN into Constituent Fractions 27-4
27-3 Division of Petroleum Refinery Wastewater Sulfur into Constituent Fractions 27-4
27-4 Example of Petroleum Refinery Model Calibration for Estimation of the Nitrifier Maximum Specific Growth Rate by Matching Measured TKN to Model Predictions 27-10
27-5 Comparison of Petroleum Refinery Predictions and Measured Values of OUR during a Cyclic Square Wave Feed Pattern on the 20-day SRT Bench Scale System 27-12
28-1 Typical Configuration of the Petrochemical Wastewater Treatment Plant 28-2
28-2 Petrochemical Plant Wastewater Characterization of the Total Influent COD, TKN and Sulfur 28-4
28-3 A Simulation of a Batch Test for Stream 2, Illustrating the Different Regions of the OUR Response Curve 28-8
28-4 Impact of the Inhibition Constant, K_I, Associated with Growth of Organisms on Phenolic Compounds on the Simulated OUR Response of a Stream 4 Batch Test (Default K_I =60 Mg/L) 28-9
28-5 Impact of the Half-Saturation Coefficient, K_{SI}, Associated with Growth of Organisms on Phenolic Compounds on the Simulated OUR Response of a Stream 4 Batch Test (Default K_{SPH} = 1 Mg/L) 28-9
28-6 Impact on Bioreactor Soluble COD in Response to a 14-Fold Increase in Tertiary Butyl Alcohol Concentration for a Period of 2.4 Hours 28-10
29-1 Graphical Representation of Model Components and Their Relationships 29-4
29-2 Schematic Diagram Demonstrates the Procedures Used in Model Calibration and Verification 29-6
29-3 Laboratory Batch Experimental Procedure 29-7
29-4 Model Parameters Estimation Procedure 29-8
29-5 A Simulation of a Batch Test for the V-353 Waste, with OUR and COD Responses 29-9
29-6 Shocking Loading Conditions Throughout the Pilot Plant Test Period Showing Influent and Effluent Soluble COD Concentration 29-12
29-7 Simulated versus Measured Effluent Soluble COD Concentration 29-12
29-8 Test Results for Ammonia Shock Load Study 29-13
29-9 The Measured and Simulated OUR for the Pilot – Scale Reactor 29-14
30-1 Flow Schematic of Chemical Plant Wastewater Treatment Plant 30-2
30-2 A Typical Respirogram for Determining the Heterotrophic Yield Coefficient 30-6
30-3 Plant Influent Flow, COD and TKN Data Used for Dynamic Calibration 30-8
30-4 Results of the Optimized Simulation for Determination of the Nitrification Parameters 30-9
32-1 Plot of Influent COD, SBR TSS and SBR VSS Prior to Intensive Period 32-2
32-2 Ammonia, Nitrate and OUR Rresponse over the Initial Period on Day 3 32-5
32-3 Ammonia, Nitrate and OUR Response over the Initial Period on Day 5 32-5
32-4 Configuration for Simulating the SBR 32-12

32-5 Simulated and Observed VSS in the SBR over 7 days 32-13
32-6 Simulated and Observed TSS in the SBR over 7 days 32-13
32-7 Simulated and Observed Nitrite/Nitrate in the SBR over 7 days 32-14
32-8 Simulated and Observed Ammonia and Nitrite/Nitrate Response in the SBR over Days 3 and 5 32-15
32-9 Simulated and Observed OUR Response in the SBR over Days 3 and 5 32-15
33-1 Plot of Influent COD, SBR TSS and SBR VSS Prior to Intensive Period 33-2
33-2 Ammonia, Nitrite and Nitrate Response over the Initial Period on Day 2 33-4
33-3 Ammonia, Nitrite and Nitrate Response over the Initial Period on Day 3 33-5
33-4 Configuration for Simulating the SBR 33-11
33-5 Simulated and Observed VSS in the SBR over 9 days 33-12
33-6 Simulated and Observed TSS in the SBR over 9 days 33-12
33-7 Simulated and Observed Nitrite/Nitrate in the SBR over 9 days 33-13
33-8 Simulated and Observed Ammonia and Nitrite/Nitrate Response in the SBR over Days 2 and 3 33-13
33-9 Simulated and Observed OUR Response in the SBR over Day 3 33-14
34-1 Consecutive Stages of a Daily Cycle Followed in Phase II 34-3
34-2 Ammonia and Nitrogen Profiles during Test No. 2 in Phase I Batch Fed SBR Experiment 34-7
34-3 Ammonia Concentration and OUR Profiles during a 22°C Nitrification Rate Test 34-11
34-4 Example of Ammonia, Nitrite and Nitrate Nitrogen Concentration Profiles during a 22°C Nitrification Test 34-11
35-1 Overview of Parallel High F/M Reactors in the University of Washington Laboratory 35-2
35-2 Exponential Nitrate Concentration Response in High F/M Test on 11/20/00 with Initial Ammonia Concentration of 30 mg N/L 35-6
35-3 Exponential Nitrate Concentration Response in High F/M Test on 11/20/00 with Initial Ammonia Concentration of 120 mg N/L 35-6
35-4 Exponential Nitrite and Nitrate Concentration Response in High F/M Test on 6/28/01 with Puyallup WWTP Mixed Liquor Seed and Puyallup Secondary Effluent 35-8
35-5 Exponential Nitrite/Nitrate Concentration Response in High F/M Test #1 on 6/04/01 with McMaster University SBR Mixed Liquor Seed and McMaster University SBR Decant 35-9
35-6 Exponential Nitrite/Nitrate Concentration Response in High F/M Test #2 on 6/04/01 with McMaster University SBR Mixed Liquor Seed and McMaster University SBR Decant 35-10
35-7 pH versus Time in High F/M Test #1 on 6/04/01 with McMaster University SBR Mixed Liquor Seed and McMaster University SBR Decant 35-10
35-8 Exponential Nitrite/Nitrate Concentration Response in High F/M Test #1 on 6/18/01 with McMaster University SBR Mixed Liquor Seed and Dundas, Ontario WWTP Raw Influent 35-11
35-9 Exponential Nitrite/Nitrate Concentration Response in High F/M Test #2 on 6/18/01 with McMaster University SBR Mixed Liquor Seed and McMaster University SBR Decant 35-12
35-10 Exponential Nitrite/Nitrate Concentration Response in High F/M Test #1 on 8/20/01 with Dundas Ontario WWTP Mixed Liquor Seed and Dundas, Ontario WWTP Raw Influent 35-13
35-11 Exponential Nitrite/Nitrate Concentration Response in High F/M Test #2 on 8/20/01 with Dundas Ontario WWTP Mixed Liquor Seed and Dundas, Ontario WWTP Secondary Efffluent 35-13
35-12 Exponential Nitrite/Nitrate Concentration Response in High F/M Test #1 on 9/17/01 with Dundas Ontario WWTP Mixed Liquor Seed and Dundas, Ontario WWTP Secondary Efffluent 35-14
35-13 Exponential Nitrite/Nitrate Concentration Response in High F/M Test #2 on 9/17/01 with Dundas Ontario WWTP Mixed Liquor Seed and Dundas, Ontario WWTP Secondary Efffluent 35-15
35-14 Exponential Nitrite/Nitrate Concentration Response in High F/M Test #1 on 10/01/01 with Dundas Ontario WWTP Mixed Liquor Seed and Dundas, Ontario WWTP Secondary Efffluent 35-16
35-15 Exponential Nitrite/Nitrate Concentration Response in High F/M Test #2 on 10/0120/01 with Dundas Ontario WWTP Mixed Liquor Seed and Dundas, Ontario WWTP Secondary Efffluent 35-17
35-16 pH versus Time in High F/M Test #1 with Dundas, Ontario WWTP Mixed Liquor Seed and Dundas, Ontario WWTP Secondary Effluent 35-18
35-17 Summary of μ_{AUT} Values Estimated for the McMaster University Investigations 35-19
35-18 Exponential Nitrite/Nitrate Concentration Response in Phoenix High F/M Test on 4/19/01 with Phoenix SBR Mixed Liquor Seed and Phoenix 91st Avenue Secondary Effluent 35-20
35-19 Exponential Nitrite/Nitrate Concentration Response in Phoenix High F/M Test on 5/14/01 with Phoenix 91st Avenue Mixed Liquor and Secondary Effluent 35-21
35-20 Exponential Nitrite/Nitrate Concentration Response in Phoenix High F/M Test on 5/26/01 with Phoenix 91st Avenue Mixed Liquor and Secondary Effluent 35-22
36-1 Overview of the Washout Apparatus at the University of Washington Laboratory 36-2
36-2 Washout Test Data for Test No. 4 with Puyallup WWTP Mixed Liquor and Secondary Effluent 36-4

36-3 Nitrite/Nitrate – Time Response for McMaster University Washout Test #1 Commenced on 9/10/01 36-6
36-4 Nitrite/Nitrate – Time Response for McMaster University Washout Test #2 Commenced on 6/25/01 36-7
36-5 Nitrite/Nitrate – Time Response for Phoenix Washout Test Commenced on 5/26/01 36-8
36-6 Nitrite/Nitrate – Time Response for Phoenix Washout Test Commenced on 9/17/01 36-9
36-7 Nitrite/Nitrate – Time Response for Phoenix Washout Test Commenced on 9/24/01 36-10
36-8 Nitrite/Nitrate – Time Response for Phoenix Washout Test Commenced on 10/01/01 36-11
37-1 Semi-Log Plot of Specific Endogenous Nitrifier Decay Rate at 22°C 37-5
37-2 Semi-Log Plot of Specific Endogenous Nitrifier Decay Rate at 12°C.. 37-5
40-1 Example of OUR Response Observed during the SBR Aeration Period 40-4
40-2 Flocculation/Filtration RBCOD versus 100K MWCO Ultrafiltration RBCOD 40-7
40-3 Respirometric RBCOD versu Flocculation/Filtration RBCOD 40-8

LIST OF ACRONYMS

The following are definitions of acronyms commonly used in this document. All other symbols are defined in Appendix A, Nomenclature.

ADR	Ammonia Disappearance Rate
APP	Aquifer Protection Permit
$CBOD_5$	5-day Carbonaceous Biological Oxygen Demand
CMAS	Complete-Mixed Activated Sludge (system)
CR	Consumptive Ratio
CSTR	Continuous Stirred Tank Reactors
DO	Dissolved Oxygen
DOC	Dissolved Organic Carbon
EPA	U. S. Environmental Protection Agency
ffCOD	Flocculated and Filtered COD
F/M	Food to Microorganism Ratio
GAO	Glycogen-Accumulating Organisms
GC	Gas Chromatography
HRT	Hydraulic Retention Time
IC	Ion Chromatography
LC	Liquid Chromatography
MGD	Million Gallons per Day
MW	Molecular Weight
MWCO	Molecular Weight Cut Off
MWD	Molecular Weight Distribution
NDEBPR	Nitrification, Denitrification and Excess Biological Phosphorus Removal
NPR	Nitrate Production Rate
PAO	Phosphorus-Accumulating Organisms
PHA	Polyhydroxyalkanoates
PHB	Poly-β-hydroxybutyrate
PHV	Poly-β-hydroxyvalerate
PST	Primary Settling Tank
RAS	Return Activated Sludge
SCFA	Short-Chain Fatty Acids
SDNR	Specific Denitrification Rate
STOWA	Dutch Foundation for Applied Water Research
TMDL	Total Maximum Daily Load
TOC	Total Organic Carbon
UCT	University of Cape Town
VOC	Volatile Organic Compound
VFA	Volatile Fatty Acids
WAS	Waste Activated Sludge
WWTP	Wastewater Treatment Plant

EXECUTIVE SUMMARY

This report is a state-of-the-art manual on peer-reviewed methods for measuring key influent wastewater characteristics and model parameters in activated sludge systems. It offers guidance and examples on the use of these methods and on model calibration.

Background

The introduction of the Activated Sludge Model Number 1 (ASM1) in 1987 was *one of the most important advancements in the wastewater treatment field in the past 25 years*. ASM1 was the first structured model that received widespread attention and acclaim in both academic and consulting/designing communities. It was the result of a major effort by researchers from Europe, Japan, South Africa and the USA. Through their efforts, and those of many numerous university scientists and "early adopting" engineering consultants worldwide, the model was refined and extended to account for carbonaceous oxidation, nitrification, denitrification and enhanced biological phosphorus removal. The availability of such tools is timely because of the *increasing need for plants that can nitrify and can implement biological nutrient removal (BNR),* a need spawned by state-mandated Total Maximum Daily Load (TMDL) regulations and the U.S. Environmental Protection Agency's new national nutrient criteria policy.

Rapid advancement in computer technology has improved the engineering community's ability to utilize these structured activated sludge models and has led to wide acceptance of computer-based simulation tools. *The reliability of wastewater treatment plant designs based on these computer-aided simulations depends in large part upon well-calibrated models*; that is, upon good estimates of influent wastewater characteristics and activated sludge model kinetic and stoichiometric parameters. Unfortunately, variability in reported values of model parameters often leads designers to make conservative estimates resulting in significant over-design. It has been estimated that if design engineers continue to use conservative values for the maximum specific nitrifier growth rate parameter alone, *over-design of nitrification facilities will cost agencies approximately $2 billion in excess expenditures* over the next 20 years.

The Water Environment Research Foundation (WERF), recognizing these concerns, sponsored this project on methods for measuring activated sludge model parameters. Historically, such methods have been perceived to be relatively expensive and labor intensive and have not been used extensively by the industry. *There is a need for methods that would be readily embraced by municipalities and consulting engineering companies*; that is, they should be simple to execute, use non-sophisticated equipment, give reproducible results and be relatively inexpensive. Another project objective was *to provide guidance on the most appropriate methods to use for estimating these parameters and on the use of activated sludge model simulators*. This report summarizes the results of this project; it consists of 40 chapters, divided into eight sections, and nine appendices (a guide to the report is given in Chapter 1.0).

Research into parameter measurement methods development was conducted at two locations: at McMaster University, Hamilton, Ontario, Canada and at the University of Washington, Seattle, Wash. The methods were tested in the field at King County's South Plant in Renton, Wash., and the City of Phoenix's 91st Avenue Multi-Cities Plant in Phoenix, Ariz. Many municipalities and industries participating in this project made data available to enable the development of case histories illustrating model calibration and the utilization of process simulators.

Key Wastewater Characteristics and Model Parameters

Good model calibration requires knowledge of not only model parameters but also influent wastewater characteristics, which can significantly influence plant performance, especially BNR systems. Probably the most important of these are readily biodegradable COD (RBCOD), because of its impact on biological phosphorus removal, and unbiodegradable particulate COD, because of its impact on sludge production and oxygen demand. *If a model is to provide reasonable predictions of system behavior, adequate knowledge of wastewater characteristics is crucial.*

An important issue with respect to measurement of wastewater characteristics relates to the pore size of the filter used to measure filtered or "soluble" components because of its impact on the amount of colloidal material that is retained on the filter in the separation process. A portion of the colloidal material can pass through the standard 0.45 μm membrane filter. In contrast, the majority of the colloidal material passes through glass-fiber filters used to measure influent solids as these have a pore size of approximately 1.2 μm. *Therefore, care is required in interpreting filtered data according to the type of filter used.* Some models do not differentiate between colloidal and particulate COD whereas others do. This has an impact on, for example, the modeling of primary clarifiers where correctly quantifying COD removal based on percent solids removal requires distinguishing between colloidal (non-settleable) and particulate components.

Table ES-1 lists the wastewater characteristics and model parameters for which guidance on measurement methods is provided and where these are located in this report.

Maximum Specific Nitrifier Growth Rate

A workshop, sponsored by WERF at the start of the project, was attended by leading practitioners of activated sludge model development and application. *The key recommendation was that research should focus on methods for the accurate determination of the maximum specific nitrifier growth rate parameter (μ_{AUT}) because of its enormous impact on activated sludge design.* Modeling and measurement of μ_{AUT} is reviewed in detail in Chapters 16.0 and 17.0.

Table ES-1. Summary of Measurement Methods

Wastewater characteristics	Chapter
Overview of methods	4.0
Unbiodegradable (inert) soluble COD	5.0
Unbiodegradable (inert) particulate COD	6.0
Readily biodegradable COD	7.0
Slowly biodegradable COD	8.0
Active biomass COD	9.0
VFA COD	10.0
Total Kjeldahl nitrogen & ammonia	11.0
Soluble biodegradable & unbiodegradable organic nitrogen	11.0
Influent solids – TSS, VSS, ISS, N/P content	12.0
Model Stoichiometric Parameters	**Chapter**
Biosolids COD/VSS	13.0
Biosolids nitrogen & phosphorus content	14.0
Anoxic heterotroph yield	15.0
Model Kinetic Parameters	**Chapter**
Maximum specific nitrifier growth rate	16.0
Nitrifier decay rate	17.0
Effective specific anoxic growth rate	18.0
Effective anoxic hydrolysis rate	18.0

There are two approaches for measuring the μ_{AUT} parameter: with bioassay methods in bench-scale tests or through simulation, in which modeled activated sludge system dynamic response is matched to field measurements. The bioassay methods are based on combining nitrifier biomass with ammonia at a concentration in excess of the half saturation coefficient, K_{NH}, so that the nitrification rate is at the maximum during the test.

Three bioassay methods were evaluated; the first is a well-established method, and the other two were developed and/or modified from methods in the literature in a search for shorter, less expensive methods:

1. *Low F/M test,* where the change in nitrifier mass is small, and there is a linear response in the measured parameter, usually oxidized nitrogen (nitrite and nitrate concentrations). F/M denotes ammonia/nitrifier concentration ratio in this context.

2. *High F/M test,* where the initial nitrifier mass is small relative to the mass produced during the test, and there is an exponential-type response to the measured parameter.

3. *Washout test*, where an intermediate F/M is selected allowing nitrifier washout to occur, and there is a non-linear, die-away response in the measured parameter.

Experimental technique is extremely important to obtain reproducible results. Variables such as alkalinity, pH, dissolved oxygen concentration and temperature must be carefully controlled to levels that will not skew results. Extensive experimental work was conducted at both McMaster University and the University of Washington to evaluate the impact of these variables and to define appropriate values to maintain during the tests. These investigations are summarized in Chapters 31.0 through 38.0. Detailed procedures for the three tests are given in Appendices D, F and G, respectively. Table ES-2 summarizes the advantages and disadvantages of the three bioassay methods.

Table ES-2. Advantages and Disadvantages of Bioassay Methods

Test Method	Advantages	Disadvantages
Low F/M SBR	- Simple operation - Provides other useful process data: RBCOD, non-biodegradable COD fractions. - Provides longer-term information on wastewater characteristics and possible nitrification inhibition. - Can be operated at the same SRT as full-scale plant to minimize the error from nitrifier decay rate assessment. - Can be used to measure nitrifier decay rate, after being operated to steady state. - Simple pH control with aid of DO controller.	- Requires long-term reactor operation (6 weeks). - Requires significant daily data collection to estimate amount of nitrifier biomass and to maintain SRT. - Requires intensive sampling at the end of the test for batch nitrification rate tests.
High F/M	- Simplest procedure. - Test is completed within 5-7 days. - No need for long-term wastewater and reactor sampling and analyses.	- More complex pH control. - Requires careful selection of seed size. - Evaluates only one-time wastewater samples. Need to repeat at least 2 times. - Population limited by initial seed selection.
Washout	- Test is completed within 5-6 days. - No need for long-term wastewater and reactor sampling and analysis. - Simple pH control. - Performed with mixed liquor from an actual full-scale system. - Relatively simple to ensure non-ammonia-limited growth.	- More complex in terms of reactor operation, maintenance, and HRT/SRT control. - More equipment - pumps and controller. - Evaluates only one-time wastewater samples. Need to repeat at least 2 times. - Difficult to estimate initial SRT/HRT correctly. - Population limited by initial seed selection.

It was demonstrated that all three methods, when conducted in parallel, provided the same estimate of μ_{AUT}. The least preferred, experimentally, is the washout test because of its complexity and set up of initial conditions. The high F/M test is the simplest of the three and requires minimal time, thereby minimizing cost. Although the low F/M method requires a longer period, it does provide additional information on wastewater characteristics. Also, once the system is set up, it is relatively simple to repeat the test several times to assess whether the wastewater is likely to exhibit nitrifier toxicity or inhibition effects. *Both the high F/M and low F/M tests are recommended, with selection dependent upon specific test program objectives.*

Estimating μ_{AUT} by simulation of a well-characterized activated sludge system is a valid method, conceptually. For example, simulation of the high F/M systems operated in this study (with accurate SRT control and accurate knowledge of influent and effluent loading) yielded very similar estimates to those derived from the other methods. Simulation of full-scale plant behavior was evaluated at the 91st Avenue treatment plant in Phoenix, Ariz. (Chapter 24.0). A major constraint of this approach is that extensive plant data must be collected to apply this approach to full-scale systems successfully. In many cases it is likely that parameters impacting estimation of nitrification rate will not be known, measured or well-controlled. These uncertainties are compounded in complex plants with multiple recycles, return streams and multiple flow splits. In the Phoenix case study a lower value of μ_{AUT} was obtained compared to those in the bioassay tests. *Simulation of full-scale plant response is not regarded as a preferred method for estimating μ_{AUT} accurately.*

During the project, μ_{AUT} measurements were conducted, using the three bioassay methods, at three locations across North America. *In all cases the μ_{AUT} estimates, corrected to 20°C, were all very close, in the range of 0.90 to 0.95 d^{-1}.* In addition, when data from several low F/M studies outside of this project were re-analyzed, μ_{AUT} estimates, corrected to 20°C, also were in the range of 0.90 to 0.95 d^{-1}. This finding suggests that there may well be consistency in nitrifier growth kinetics from plant to plant. *This hypothesis questions the widely held perception that μ_{AUT} varies significantly from plant to plant. Although the amount of testing was limited in this project, it is suggested that the variability in rate reported previously, aside from situations with toxicity or inhibition problems, is perhaps an artifact of the incorrect assumption of a low nitrifier decay rate (see below).* Nevertheless, it should be recognized that nitrifiers are susceptible to inhibition and toxicity, particularly with industrial waste treatment systems. Therefore, *it is strongly recommended that μ_{AUT} testing be conducted for all nitrification and BNR systems*. This will identify any problem situations, and provide a database of information to test the hypothesis on the consistency in μ_{AUT}. Approaches for identifying toxicity/inhibition problems are proposed in Chapter 16.0.

Nitrifier Decay Rate

This research has identified and demonstrated several important aspects concerning nitrifier decay rate (b_{AUT}), and the interaction between decay rate and estimation of μ_{AUT} (Chapter 17.0). A correct estimate for b_{AUT} is important in estimating μ_{AUT}. In particular, in the low F/M method for measuring μ_{AUT} (at high SRT), assuming too low a b_{AUT} value results in a significant underestimate of μ_{AUT}, perhaps by a factor of two. The traditional method for determining b_{AUT} from the slope of a semi-logarithmic nitrate production rate versus time plot underestimates b_{AUT} significantly. That approach overlooks regeneration of nitrifiers on nitrogen lysis from heterotroph decay. A new method for data analysis and estimating b_{AUT} has been presented. *Data* *from this study indicate a value for b_{AUT} of 0.17 d^{-1} (at 20°C) and an associated temperature dependency factor (θ) of 1.029.*

Nitrifier Growth Rate Temperature Dependency

There is considerable uncertainty in the literature on the impact of temperature on μ_{AUT}. An important finding in this work is that higher, more conservative growth rate temperature dependency factors reported in the literature may well be a mathematical artifact of calculating the factor from either (1) changes in (μ_{AUT} - b_{AUT}) with temperature rather than changes in μ_{AUT}, and/or (2) incorporating the assumption that b_{AUT} is "negligible." Limited parallel measurements of μ_{AUT} were conducted at 12°C and 22°C and yielded a value of 1.072 for θ. This is at the low end of the range reported in the literature. From a design or plant capacity-rating standpoint, the crucial factor is the value of μ_{AUT} at the minimum operating temperature. Therefore, if facilities are available to conduct tests at the minimum temperature using sludge taken from the plant when operating at the minimum temperature, this approach is recommended.

Readily Biodegradable COD

Bioassay tests based on oxygen utilization conceptually are best for measuring RBCOD concentration for process modeling because they directly measure the impact of the material on the process. However, these require more sophisticated equipment and expertise than physicochemical methods, which are simpler to perform. Of these, the flocculation/filtration, or ffCOD test, is easier to use than the ultrafiltration method. There has not been sufficient rigorous comparison between methods to conclude that the ffCOD, ultrafiltration and bioassay tests provide identical results.

Model Calibration

Chapters 19.0 through 30.0 address activated sludge model calibration. Chapter 19.0 provides guidance on the use of process simulators, Chapter 20.0 illustrates the sensitivity of simulator output based on changes to parameter values and Chapter 21.0 provides guidance on model calibration. Examples of calibration with municipal wastewater treatment systems are given for carbonaceous oxidation only (Chapter 23.0), for biological nitrogen removal (Chapter 24.0), for biological nitrogen and phosphorus removal (Chapter 25.0) and for the case of laboratory- and pilot-scale applications (Chapter 22.0). Examples are presented for industrial applications such as petroleum refineries (Chapters 27.0 and 29.0), a petrochemical plant (Chapter 28.0) and a chemical plant (Chapter 30.0).

Concluding Remarks

1. A comprehensive manual on methods for measuring key influent wastewater characteristics and activated sludge model parameters is presented with guidance on the use of these methods.

2. Significant advancements on measuring μ_{AUT} are presented and two cost effective methods are recommended. The high F/M method is the simplest; equipment requirements are less complex than for the washout method, but pH control demands specific attention. The low F/M (SBR) method involves longer-term operation with extensive sample analysis, but also provides estimates of important wastewater characteristics. Agencies that need to implement nitrification and/or BNR in their facilities are strongly recommended to use these peer-reviewed methods to improve their designs and to minimize capital expenditures.

3. Experiments have demonstrated that the nitrifier decay rate is more significant than reported in many earlier studies. A higher value for nitrifier decay rate (0.17 d^{-1} at 20°C) than is commonly accepted was derived from this work. This finding has a large impact on estimating μ_{AUT}.

4. Higher, more conservative growth rate temperature dependency factors reported in the literature may be a mathematical artifact of calculating the factor from changes in (μ_{AUT} - b_{AUT}) with temperature rather than changes in μ_{AUT}, and/or incorporating the assumption that b_{AUT} is "negligible."

5. Examples of activated sludge model calibration are presented for municipal and industrial wastewater plants. Guidance is offered on model calibration procedures.

6. The report is supplemented with the results of experimental investigations that support the above recommendations and detailed mathematical derivations of the expressions developed for interpreting results from these methods.

CHAPTER 1.0

INTRODUCTION

1.1 Background

Mathematical modeling of system behavior implemented in various simulators has had a significant impact on the development of design procedures for activated sludge systems in recent years. For full-scale plant operation it has also found application as an optimization tool in assessing the effects of changes in waste flows and loads, operational modifications (e.g., changes in recycles) and proposed modifications to plant configuration. Modeling has also proved valuable in operator training; through simulation exercises using the model, the operator acquires "instant" experience in the behavior to be expected with changes in inputs, system configuration and operational strategies.

The utility of simulation models becomes almost essential for biological nutrient removal (BNR) systems incorporating nitrification, denitrification and excess biological phosphorus removal, in addition to carbonaceous material removal. The behavior of these systems involves a large number of compounds and biological reactions, many of which interact. At least three separate groups of microorganisms (polyP or phosphorus-accumulating heterotrophs, non-polyP heterotrophs, nitrifying autotrophs) occur in BNR systems, operating on a large number of chemical components in three distinct environmental regimes (aerobic zones, anoxic zones where nitrate but not oxygen is present, and anaerobic zones where both nitrate and oxygen are excluded as far as possible). These features make for complex behavior that has increased the level of difficulty in design, operation and control.

There are thousands of nitrification plants and a few hundred BNR applications in the United States, but the need for future systems is growing rapidly owing to the current round of state-mandated Total Maximum Daily Load (TMDL) investigations as well as the U.S. Environmental Protection Agency (EPA)'s new national nutrient criteria policy, which will spur states to set more restrictive water quality criteria. As a result, a significant increase in the application of both nitrifying and BNR systems is expected in the coming years.

Because of the variability in reported values of model parameters (most notably the maximum specific nitrifier growth rate), conservative estimates often are used, resulting in significant over-design. If designers continue to use these low default values without on-site characterization of model parameters, it has been estimated that over-design of nitrification facilities will cost about $100 million annually in excess capital expenditures. This is based on planned expenditures for nitrification from the U.S. EPA *Needs Survey of 1997*, and overall, represents $2 billion during a twenty year period (Parker, 1999; Parker and Goehring, 2002). In any individual mid- to large-sized treatment plant, site-specific determination of the nitrifier growth rate by the methods given in this manual can be expected to save millions of dollars in capital expenditures.

1.2 Activated Sludge Models

Currently, the primary approach to modeling activated sludge systems is to use mechanistic models. These mechanistic models incorporate mathematical expressions that represent the biological interactions, based on hypotheses proposed for the biological processes occurring within the system. Most are based on the standardizing work of the International Water Association (IWA). The historical development and status of the current generation of models are summarized in the subsequent two chapters.

Secondary clarifiers form an integral part of the activated sludge process, but are not discussed in this guidance manual. Interested users should refer to other IWA-sponsored work by Ekama et al. (1997).

1.3 Models versus Simulators

The biological models should not be confused with simulation programs (simulators). A simulator is a computer program that incorporates the biological model with models for other unit operations, and must link the units of a specific system according to the particular flow scheme. Figure 1-1 highlights the requirements for setting up a simulation of a typical wastewater treatment facility that incorporates the activated sludge process. Depending upon the intended use of the simulator, information is needed on the inputs to the process, its configuration and its operating conditions (Wilson and Dold, 1998).

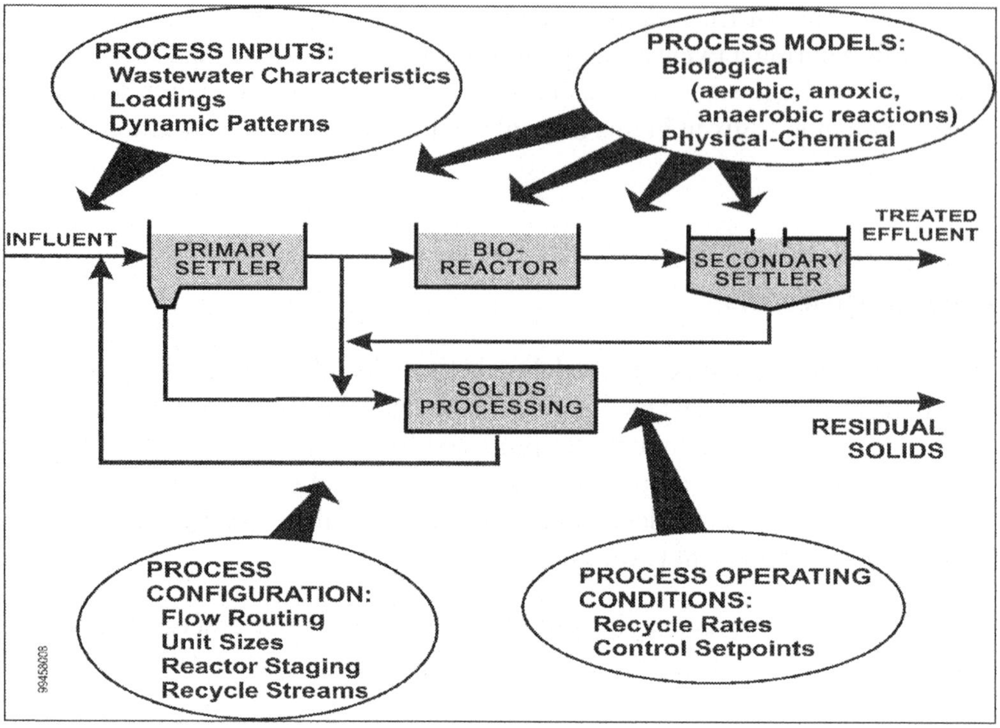

Figure 1-1. Essential Requirements for Wastewater Treatment Process Simulation (Wilson and Dold, 1998)

1.4 Simulator Calibration

Some degree of calibration usually is required for a simulator to accurately predict effluent characteristics, sludge production, oxygen requirements and other system characteristics. The importance of this step in the simulation process cannot be over-emphasized, because inappropriate calibration can lead to erroneous interpretation of simulator output as well as incorrect or non-optimal designs.

In terms of Figure 1-1, to simulate an activated sludge system, it is necessary to obtain details of the physical configuration (such as reactor volumes and clarifier dimensions), operating conditions (such as wastage rate, recycle rates and dissolved oxygen (DO) concentrations) and the influent loading pattern (including flow rate, COD and TKN). All of this information essentially is "measurable," and it should be possible to establish these data without difficulty. In practice, problems often are encountered with determining sludge wasting rate. This is crucial as it determines the system sludge age or solids retention time (SRT). Also, influent concentration measurements often are unreliable as a result of sampling problems. These and other potential sources of error and the way to manage them are discussed in Chapter 21.0.

In addition to the "measurable" parameters, the current generation of biological models requires information concerning:

- Influent wastewater composition, and
- Model kinetic and stoichiometric parameters.

1.4.1 Influent Wastewater Characteristics

The term "wastewater characteristics" refers to the partitioning of influent organic material into biodegradable and inert portions, the influent total nitrogen into ammonia and organic nitrogen and so on. Influent wastewater characteristics may vary, often appreciably, from one municipal waste to another. Wastewater characteristics have a very significant impact on system performance, particularly for nutrient removal systems. A single characteristic, such as the readily biodegradable COD fraction, can determine whether or not a system designed for excess phosphorus removal will in fact remove phosphorus. Therefore, if the model is to provide reasonable predictions of system behavior, adequate knowledge of wastewater characteristics is extremely important.

1.4.2 Activated Sludge Model Kinetic and Stoichiometric Parameters

The biological models contain many *stoichiometric and kinetic parameters*, including yields and growth rates. Several studies have shown that the stoichiometric and kinetic parameters (with one notable exception discussed below) do not change appreciably for different systems *treating municipal wastewaters*. For example, the decay rate estimated from the decline in oxygen utilization rate in a batch aerobic sludge digestion test is very similar for activated sludge drawn from aerobic systems in different parts of the world (when standardized to 20°C). The uniformity of parameters likely reflects uniformity in the composition of municipal wastewaters, and a resultant similarity in the diversity of the microorganism populations in different systems.

It is proposed that the values for the stoichiometric and kinetic model parameters tabulated for a particular model (presumably derived from calibration studies on a range of treatment systems) should be acceptable initial estimates, and should not require substantial adjustment in model calibration exercises.

Note: If large changes in default kinetic and stoichiometric parameters are necessary to match simulated and measured results in municipal plants, this usually indicates that the system or the wastewater was not properly characterized. The most common problem is poor information on sludge wastage. Default values of kinetic and stoichiometric parameters are likely to be unrepresentative for industrial wastewater and for municipal plants with substantial industrial inputs.

The exception referred to above is the maximum specific growth rate of the nitrifiers, μ_{AUT}. This parameter shows marked variations between systems treating different wastewaters as shown in Figure 1-2. The design curve in Figure 1-2 is from the U.S. EPA 1975 *Nitrogen Control Manual* (also included in the 1993 edition). Figure 1-2 was created from a summary of data, which supported that curve (shown as circles), as well as more recently collected data, which showed considerable scatter from the original design curve (Parker and Richards, 1994).

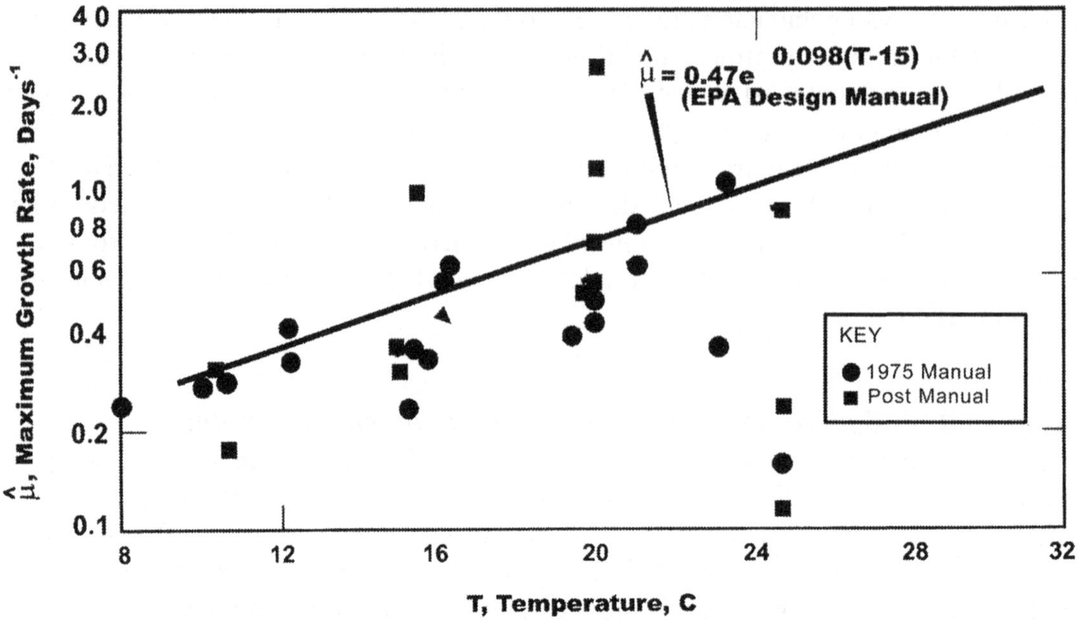

Figure 1-2. Comparison of Plant-Derived Nitrifier Growth Rates with U.S. EPA (1975) Nitrifier Growth Rate Equation

In the models, nitrification usually is considered a single-step process mediated by one group of nitrifying autotrophic bacteria. The growth rate parameter for the nitrifiers (μ_{AUT}) defines the maximum specific growth rate of this surrogate organism mass. Values for μ_{AUT} (at 20°C) reported in the literature range from 0.2 to 1.0 d^{-1} for activated sludge systems treating different municipal wastewaters. It has been suggested that the variation appears to correlate with the extent of the industrial component in the wastewater; with increased industrial input there likely is an increased possibility of inhibition of nitrifying organisms. However, many other factors influence nitrification behavior, such as pH and alkalinity. Better measurement methods also recently have reduced variation in this parameter. The project described in this report presents information that accounts for this variability and the hypothesis that it may be an artifact of the incorrect assumption of a low nitrifier decay rate (Chapter 17.0). Since μ_{AUT} has a major impact on nutrient removal system design and performance, and on model predictions as well, special attention should be paid to its measurement or estimation. In a sense, μ_{AUT} can be regarded as a wastewater characteristic.

1.5 Industrial Wastewaters

Activated sludge systems treating organic-containing industrial wastewaters embrace many industrial sectors ranging from petroleum refineries to food processing and from pulp and paper plants to tanneries. The IWA-type models have been developed specifically for municipal treatment systems. It is likely that these models can be applied directly (with appropriate wastewater characterization) to certain industrial wastewater treatment systems such as food processing. However, this is not the case for wastewater from all industrial sectors. In certain cases, the biological and physicochemical processes impacting system performance, and the kinetics and stoichiometry of the processes, may differ substantially from municipal treatment systems. For example, in the treatment of petroleum refinery wastes, it would be appropriate to consider the following:

- Biomass yield coefficients for the un-oxygenated hydrocarbons are lower than for organics in municipal wastewaters.
- Biological processes such as autotrophic oxidation of sulfides and other reduced sulfur compounds may impact oxygen demand significantly.
- Volatilization and stripping can be important "removal" mechanisms.

In such a case, the municipal models such as ASM1 and ASM2 cannot be applied directly to industrial systems, and it may be necessary to develop an industry-specific model (e.g., Baker, 1994; Jones et al., 1995). However, the IWA-type models provide a useful starting point for model development. This is because many of the basic processes used to model the treatment of municipal wastewater will apply in an activated sludge process used to treat the industrial wastewaters.

1.6 Objectives

The objectives of this project are:

- To summarize the state of the art in activated sludge modeling and the state of practice when applying these models.
- To identify key parameters in IWA-based models that display significant sensitivity to changing environmental conditions, wastewater characteristics and process type.
- To identify the most appropriate methods for measuring the key parameters and the frequency of measurement required.
- To evaluate the performance of the measurement methods and the level of effort required to implement them.
- Describe the basis for partitioning the organic and nitrogenous material into various wastewater characteristic components.
- To compile the findings of these investigations and to create a manual that provides guidance on the use of protocols for measuring essential parameters in IWA-based models and on model and simulator calibration.

1.7 Summary of Peer Review Workshop

A team of peer reviewers and the Project Subcommittee provided critical review, in a workshop environment, of a preliminary compendium of methods of measuring activated sludge model parameters that had been prepared at that time. The workshop was held in New Orleans, LA., October 8-9, 1999; invitees included leading practitioners in the development and application of activated sludge models. The reviewers were asked to concur with the recommendations in the preliminary manual of methods, and to identify a short list of critical parameters that needed improved or new procedures.

Many useful suggestions were made by the reviewers, which greatly contributed to the comprehensiveness of this report. Resolution of key issues that provided guidance for the project included the following:

- Accurate determination of maximum specific nitrifier growth rate was regarded of paramount importance in the calibration of activated sludge models. Experimental work was to focus on this parameter and measurement of nitrifier growth rate was to reflect actual operating conditions.
- Also of major importance in the calibration of activated sludge models was the determination of readily biodegradable COD (RBCOD). Experimental work was also to address methods of measuring RBCOD, but to a lesser extent than nitrification rate.
- Physical-chemical methods for wastewater characterization were to be utilized/developed whenever possible, especially for RBCOD.
- Procedures for all model parameters were to relate to existing models only. Investigating model changes was considered to be beyond the scope of this project.

1.8 Report Outline

This report is a manual of practice on activated sludge model calibration directed to practitioners who are/have been actively using activated sludge models. The manual provides guidance on the use of protocols for measuring essential parameters in IWA-based models and on model and simulator calibration. It consists of forty chapters and nine appendices as summarized in Table 1-1. The nine appendices provide additional detail on nomenclature, on specific procedures described in the main body of the report and on mathematical derivations of equations.

Table 1-1. Guide to This Report

Topic	Description	Chapter Nos.
Project overview	Synopsis of project background and objectives and of current activated sludge models and simulators	1.0-3.0
Wastewater characterization	Identification and discussion of parameter measurement/estimation methods. Describes background and methods of measuring six components of COD, TKN components and influent solids characterization	4.0-12.0
Stoichiometric model parameters	Description of background and methods of measuring COD, N and P content of biosolids	13.0-15.0
Nitrification, denitrification	Description of background and methods of measuring nitrification rate, nitrifier decay rate and denitrification rate	16.0-19.0
Model calibration	Guidance on the use of activated sludge model simulators and model calibration	19.0-21.0
Examples of model calibration – municipal applications	Demonstration of utilizing bench and pilot scale data, and full-scale applications for carbonaceous oxidation, biological nitrogen control and biological nitrogen and phosphorus control	22.0-25.0
Examples of model calibration – industrial applications	Model calibration in petroleum refineries, a petrochemical plant and a chemical plant	26.0-30.0
Compilation of experimental work in support of method development	Summary of experimental data for testing the low F/M, high F/M and washout methods of measuring nitrification rate and the flocculation/filtration, ultrafiltration and bioassay methods of measuring RBCOD	31.0-40.0

1.9 References

Baker, A.J. 1994. *Modeling activated sludge treatment of petroleum and petrochemical wastes.* Ph.D. Thesis, McMaster University, Hamilton, Ontario, Canada.

Ekama, G.A., J.L. Barnard, F.W. Günthert, P. Krebs, J.A. McCorquodale, D.S. Parker, and E.J. Wahlberg. 1997. *Secondary settling tanks: theory, modelling, design and operation*, IAWQ Scientific and Technical Report No. 6. London, UK: International Water Association.

Grady C.P.L., G.T. Daigger, and H.C. Lim. 1999. *Biological wastewater treatment* 2nd Edition. New York, NY: Marcel Dekker, Inc.

Jones, R.M., P. L. Dold, A. J. Baker, and T. Briggs. 1995. Optimization of a biological wastewater treatment process at a petrochemical plant using process simulation. In *Proc. 68th Annual Conference of the Water Environment Federation*, Miami, FL. Alexandria, VA: Water Environment Federation.

Parker, D.S., and J.T. Richards. 1974. Discussion of kinetic and process analysis of nitrification in coupled trickling filter-activated sludge systems. *Water Env. Res.* 66(7):934-935.

Parker, D. S. 1999. A proposed WERF research program for debottlenecking, optimizing and re-rating existing wastewater treatment plants. *WERF Final Report, Project 99-WWF-1: Research priorities for debottlenecking, optimizing and re-rating wastewater treatment plants*, 2.1-2.10. Alexandria, VA: Water Environment Research Foundation.

Parker, D., and C. Goehring. 2002. The conversation the consulting engineer and the utility manager ought to have: what does getting more for less mean? In *Proc. 75th Annual Conference of the Water Environment Federation* Chicago, IL, CD-ROM. Alexandria, VA: Water Environment Federation.

U.S. EPA. 1975. *Process design manual for nitrogen control.* Report prepared for U.S. EPA by Brown and Caldwell, EPA/625/1-77/007. Washington DC: Office of Technology Transfer, U.S. Environmental Protection Agency.

Water Research Commission. 1984. *Theory, design and operation of nutrient removal activated sludge processes.* P. O. Box 824, Pretoria, South Africa: Water Research Commission.

Wilson A.W., and P.L. Dold. 1998. General methodology for applying process simulators to wastewater treatment plants. In *Proc. 71st Annual Conference of the Water Environment Federation*, Orlando, FL. Alexandria, VA: Water Environment Federation.

RECOMMENDATIONS FOR FUTURE RESEARCH

Based on the work reported here, the following topics are recommended for further research:

- **Nitrifier growth rate temperature dependency.** Results from this study indicated a lower nitrifier temperature dependency factor than reported in many cases previously, and a basis was provided for this finding. Nevertheless, further research on temperature dependency of nitrifier growth and decay rates is needed. In the interim, if temperature-controlled facilities are available, it is suggested that testing should be conducted at, or close to, the minimum annual temperature experienced by a full-scale plant (using sludge drawn from the plant at that low temperature). The rate can then be back-calculated to the reference temperature of 20°C. This would remove the uncertainty over predicting plant performance at the minimum temperature.

- **Nitrifier growth rate measurement – impact of sudden temperature changes.** A question remains as to whether nitrification rate data will be estimated correctly if mixed liquor withdrawn from a plant at, say, 12°C is tested at 20°C (a usual laboratory temperature) using the methods presented in this manual. The reverse situation is also a concern: namely, can sludge be withdrawn from a plant at 20°C, and a test be conducted at 12°C to correctly predict nitrifier growth rate at 12°C? There are a number of issues here. For example, does the rapid temperature shock have an impact? Also, is it possible that population shifts with slow seasonal changes in temperature confound the results? These issues should be addressed in research.

- **Modeling nitrification as a single- versus a two-step process.** The conclusions from this work should be viewed in the context of the assumption that it is sufficient to model nitrification as a single-step process. It is suggested that research be directed to determine if there is reason to consider a two-population approach, distinguishing between the behavior of ammonia- and nitrite-oxidizers.

- **Effect of redox potential on nitrifier specific decay rates.** The laboratory work in this study considered aerobic-only systems. Further research should be undertaken on nitrifier decay under anoxic/anaerobic conditions in light of the findings of a reduced anoxic decay rate by other researchers. The work also should evaluate the effect of the anoxic time and possibly anaerobic contact time (in biological nutrient removal (BNR) systems) on the decay rate. The work should include using molecular biology methods to assess any nitrifier population selection effects under aerobic versus anoxic/aerobic growth.

- **Effect of nitrifier acclimation to low dissolved oxygen (DO) concentration conditions**. The activated sludge models incorporate DO dependency factors (i.e., switching functions) to reduce nitrification rates with decreasing DO concentration. This aspect was not addressed in this study. Most research on the effect of DO has involved testing where nitrifying sludge grown at appreciable DO is subjected to low DO for short periods. There are some claims more recently, with regard to simultaneous nitrification-denitrification systems operated at low DO, that nitrifiers acclimate to low DO concentrations such that their rate is not greatly depressed. Long-term studies on DO effects including molecular biology methods to evaluate population characteristics are needed.

- **Readily biodegradable COD.** The limited testing conducted in this project suggested that RBCOD values measured using a DO probe respirometric method were consistently lower than those measured by the flocculated-filtered COD method. In other work, a different respirometric method yielded similar RBCOD values as the flocculated-filtered COD method. Further research work should be directed to understand the differences between these techniques; this should include comparison with the ultrafiltration technique.

- **Colloidal COD.** Existing activated sludge models do not account for the possibility that part of the influent colloidal COD may be unbiodegradable. (In fact, many models do not identify a colloidal COD fraction at all.) This raises a number of issues for wastewater characterization. It would appear that information is limited on this topic and it is recommended that research be directed to resolve methods for measuring colloidal COD and assessing the biodegradability and the fate of this material in the system.

CHAPTER 2.0

Review of IWA Activated Sludge Models

2.1 Historical Background

In 1983, the International Association on Water Pollution Research and Control [IAWPRC, subsequently IAWQ (International Association of Water Quality) and now IWA (International Water Association)] appointed a task group to review modeling of activated sludge systems incorporating *carbonaceous energy removal, nitrification and denitrification*. The initial deliberations of the group resulted in a preliminary version of the "IAWPRC model" (Grady et al., 1986). Dold and Marais (1986) conducted a comprehensive evaluation of the preliminary model. It was proposed that certain changes be made, in particular with respect to the way in which the fate of organic nitrogen was modeled. These changes were subsequently adopted in the final version of the IAWPRC Activated Sludge Model No. 1 [ASM1] (Henze et al., 1987a, 1987b).

The task group drew on a wide range of information in formulating the ASM1 model. One research initiative that had a major influence on the model was the dynamic activated sludge model developed by Marais and co-workers at the University of Cape Town (Dold et al., 1980; van Haandel et al., 1981). This dynamic model evolved out of the steady state model of Marais and Ekama (1976). The steady state model, in turn, constituted a development from a number of previous models for carbonaceous and nitrogenous material conversion and removal (McKinney, 1962; McKinney and Ooten, 1969; Lawrence and McCarty, 1970; Downing et al., 1964).

Note: ASM1 was the first structured model that received widespread attention and publicity in both academic and consulting/designer circles. It is possibly the most significant step towards the high level and sophistication the wastewater process modeling field has attained over the past twenty years. A great benefit of ASM1 was to establish a common framework of understanding and agreement to serve as a starting point for further research and application. There are numerous university scientists as well as "early adopting" consultants who contributed to the research and investigations that led to improvements in the initial model and influenced the development of ASM type models. The efforts of these unnamed hundreds are recognized, even though the many valuable contributions are not cited in this short review.

The ASM1 model did not include a major mechanism in the activated sludge process, the phenomenon of excess biological phosphorus removal (EBPR). (Note: the following paragraphs draw from Copp's (1998) succinct summary of EBPR behavior.) EBPR is a microbial process that involves the cultivation, within the mixed community, of microorganisms that have the ability to take up phosphorus in excess of that required for growth. The net effect of this uptake is a reduced total effluent phosphorus concentration that can be less than 1mg/L in a well-operated system. The removal of phosphorus is achieved through the microbially mediated storage of polyphosphate granules, which can occupy up to 60 percent of the cell volume with a phosphorus mass as much as 38 percent of the volatile suspended solids (Lotter et al., 1986; Wentzel et al., 1989a, 1989b). (These high figures were for acetate-fed enhanced cultures, not for municipal systems.) Polyphosphate storage is encouraged in these systems by exposing the biomass to anaerobic and aerobic stages sequentially, favoring the growth of polyphosphate-accumulating facultative anaerobes.

Within an EBPR system, characteristic behavior is observed in each stage of the process (Figure 2-1). During the anaerobic stage, substrate is sequestered by the biomass and converted internally to polyhydroxyalkanoates (PHAs) while phosphorus is released from the biomass causing an increase in soluble phosphate. During the aerobic stage, PHA degradation occurs and soluble phosphate is accumulated as polyphosphate within the biomass, leaving a reduced level of phosphorus in the effluent.

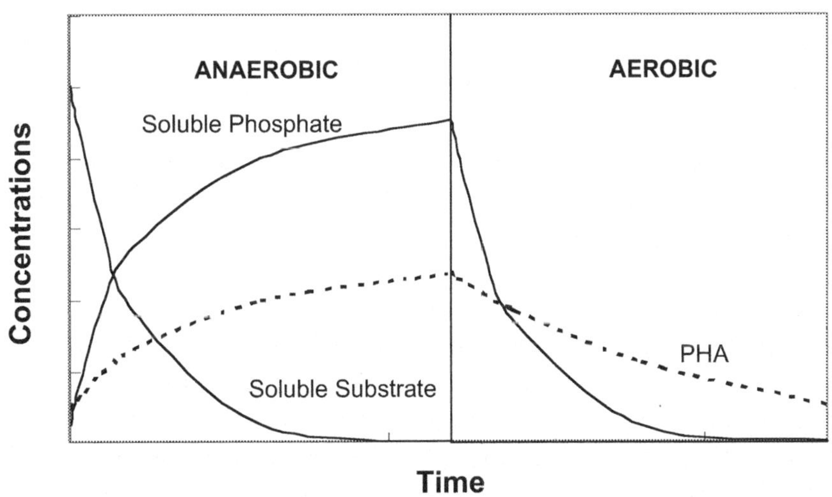

Figure 2-1. Schematic Illustration of Characteristic EBPR Behavior in an Anaerobic/Aerobic Sequenced Bioreactor (after Copp, 1998)

Proposed explanations of the biochemical behavioral patterns associated with P release and uptake (and net P removal) have been presented in a number of models; for example, that of Comeau et al. (1986), extended and modified by Wentzel et al. (1986), and that of Mino et al. (1987). The biochemical models are largely in agreement regarding the biochemical control mechanisms and have provided an explanation for the essential requirements for attaining EBPR, namely, an alternating anaerobic/aerobic sequence with the provision of short-chain fatty acids (SCFA, also referred to as volatile fatty acids (VFA)) during the anaerobic phase. These SCFAs are taken up by the polyP organisms and stored as organic polymers, generally as either poly-β-hydroxybutyrate (PHB) or poly-β-hydroxyvalerate (PHV) (referred to collectively as poly-β-hydroxyalkanoates (PHA)).

Since Comeau et al. (1986) and Wentzel et al. (1986) first introduced their biochemical models of excess biological phosphorus removal (EBPR), a substantial amount of research has resulted in variations to the original models (Mino et al., 1987; Arun et al., 1988; Wentzel et al., 1991; Satoh et al., 1994; Smolders et al., 1994a, b, c). However, the essence of the original models remains. More recently, biochemical models have been proposed as extensions of these, but which include both glycogen-accumulating organisms (GAOs) and phosphorus-accumulating organisms (PAOs) that co-exist in these systems (Satoh et al., 1992; Mino et al., 1995, 1998; Pereira et al., 1996). It is proposed that observed behavior in EBPR systems is the net result of the combined behavior of both GAOs and PAOs and that the proportion of GAOs and PAOs in a single culture can significantly affect the behavior of the mixed microbial community. It is proposed that PAOs possess a selective advantage in the presence of sufficient phosphorus, but limiting PAO growth by limiting the available phosphorus allows for the proliferation of GAOs. Both PAOs and GAOs anaerobically sequester substrate that is stored internally as PHA. Under aerobic conditions, the breakdown of PHA provides the carbon and energy for growth utilizing oxygen as a terminal electron acceptor. In PAOs, excess energy generation is stored as polyphosphate, whereas GAOs store excess energy as carbohydrate (glycogen).

In "normal" municipal wastewater, the SCFA content usually is minimal. In EBPR systems the readily biodegradable COD component is transformed to SCFA by the non-polyP organism mass, thereby making SCFA available to the polyP organisms (Meganck et al., 1985; Brodisch, 1985; Wentzel et al., 1985). Aside from this linkage, the polyP and non-polyP organisms in EBPR systems have been shown to act essentially independently of one another. For this reason Wentzel et al. (1988) adopted the approach of developing "enhanced" cultures of polyP organisms as the basis for studying the kinetics and stoichiometry of EBPR without the behavior being masked by that of the non-polyP organisms.

Enhanced cultures of polyP organisms were developed by Wentzel et al. (1988) in continuous flow activated sludge systems (modified Bardenpho and UCT process configurations), with acetate as the only organic substrate. The sludge mass in these systems was shown to comprise the polyP organism *Acinetobacter spp.* in excess of 90 percent. Based on observations of the continuous flow systems and batch experiments using mixed liquor drawn from these systems, Wentzel et al. (1989a, 1989b) developed a kinetic mechanistic model for the enhanced culture EBPR system. The model provided a very reasonable description of the response observed in a number of continuous flow enhanced culture systems and the batch experiments with *a single set of kinetic and stoichiometric parameters*. The enhanced culture model constituted a most significant step towards the development of a general activated sludge model capable of modeling the biological processes of carbonaceous energy removal, as well as nitrification, denitrification *and excess biological phosphorus removal* (NDEBPR).

Dold (1990, 1991) merged the ASM1 model for non-polyP heterotrophic organisms and autotrophic organisms (Henze et al., 1987a, 1987b) and the Wentzel et al. (1989b) model for polyP organisms. In combining the models, certain extensions and modifications were incorporated in both the ASM1 and the Wentzel model components. Since the initial development, this NDEBPR model has been evaluated extensively against experimental data from laboratory-scale and full-scale treatment plants. This has led to further model refinement (Barker and Dold, 1997a, 1997b).

The task group also extended ASM1 to include simulation of combined NDEBPR processes. The initial ASM2 model introduced PAOs as a new group of organisms and was proposed as "a compromise between complexity and simplicity, and between the many viewpoints on how the correct model should look like; it should be used as a conceptual platform for further model development" (Henze et al., 1994, 1995). Ekama and Wentzel (1999) discussed in detail the limitations of ASM2 with respect to the microbiology and biochemistry of the process, and with respect to the anoxic process, fermentation and anaerobic hydrolysis. The model has undergone many refinements, and the current version is denoted as ASM2d (Henze et al., 1999).

In parallel work, the task group developed ASM3, a new model member of the ASM family, which simulates carbonaceous energy removal, nitrification and denitrification (Gujer et al, 1999). ASM3 modifies the conceptual model of ASM1. It introduces storage of organic substrates as a new process, and replaces the death-decay process for heterotrophic organisms with an endogenous respiration process. As a result, hydrolysis is now independent of the electron donor. All processes (except for hydrolysis) run at a reduced rate under anoxic conditions compared to aerobic conditions. ASM3 also uses lower anoxic yield coefficients. A recent development has been to add an EBPR component to ASM3 (Reiger et al., 2001).

In 2001, the task group published an overview of the family of ASM models, which summarizes the development of ASM1, ASM2, ASM2d and ASM3 (Henze et al., 2001).

2.2 The Basis of IWA Activated Sludge Models

The IWA task group initially began the process of modeling activated sludge systems by addressing systems in which carbon oxidation, nitrification and denitrification are accomplished in a single sludge. The task group followed several distinct steps. First, the fundamental processes occurring within such a system were identified, where the term "process" is used to describe a "micro" event such as cell growth or maintenance, rather than a series of "macro" operations as in an activated sludge process. Next, the kinetics and stoichiometry of those processes were characterized. The process rate expressions were then incorporated into mass balance equations depicting the physical configuration of the system. These equations then need to be solved. This sequence of developments is described in detail in several publications: Grady et al. (1986), Gujer and Henze (1991) and Henze et al. (1987a, 1987b).

The task group adopted the notation developed by a previous task group (Grau et al., 1982). The task group concluded that the matrix format offered the best method of presenting the large amount of information in this complex system; the above four publications describe the matrix format with examples. The matrix format facilitates a continuity check on any component and allows rapid and easy recognition of the fate of each component, which aids in the preparation of mass balance equations.

The task group employed the concept of switching functions to turn process rate equations on and off as environmental conditions are changed. It also regarded COD to be the best measure of the concentration of organic matter in wastewater because it alone provides a link between electron equivalents in the organic substrate, the biomass and the oxygen utilized. Therefore, the concentration of all organic materials, including biomass, is expressed in COD units in IWA-based models. The organic matter (as well as the nitrogenous matter) in wastewater is divided into a number of categories based upon biodegradability and on physical state (soluble or particulate). The biodegradable fraction is further sub-divided into readily biodegradable and slowly degradable components.

With these assumptions, the ASM1 matrix included thirteen components and eight processes. Wastewater was characterized in terms of seven dissolved and six particulate components that are used to describe two forms of biomass, seven fractions of COD (organic material) and four fractions of nitrogen. Dissolved oxygen concentration and alkalinity were incorporated as two further wastewater characteristics. The eight processes included three related to the growth of heterotrophic and autotrophic organisms, two representing biomass decay and three hydrolysis processes.

The following were recognized as limitations in applying ASM1:

- The model was developed for municipal wastewaters.
- Temperature and pH were assumed to be constant.
- Parameter values in the rate expressions were assumed to be constant.
- The effects of nutrient limitations were not considered.
- The correction factors for denitrification (η_g and η_h) were assumed to be constant.
- The coefficients for nitrification were assumed to be constant.
- The biomass was assumed to remain homogeneous.
- The entrapment of particulate organic matter in the biomass was assumed to be instantaneous.
- Hydrolysis of organic matter and organic nitrogen were assumed to occur simultaneously and with equal rates.
- The type of electron acceptor was assumed not to affect loss of biomass by decay.
- The user must ensure that sludge settleability is good. Therefore, the activated sludge system to be modeled must be maintained within SRT constraints of 3 to 30 days, and organic loadings must not be high enough to cause poor settling conditions.

2.3 Developments in IWA Activated Sludge Models

As described in the Historical Background section above, the IWA task group developed ASM2 to incorporate EBPR so as to facilitate the modeling of EBPR as well as NDEBPR systems. ASM2d was subsequently developed to address some of the limitations in ASM2. In particular, it includes two additional processes to account for the fact that phosphorus-accumulating organisms (PAOs) can use cell internal organic storage products for denitrification, whereas ASM2 assumes PAOs to grow only under aerobic conditions. In addition to the constraints outlined for the application of ASM1, the following limitations apply to ASM2d (Henze et al., 1999):

- The model assumes sufficient magnesium and potassium content in the wastewater to maintain EBPR.
- It is known that nitrite and nitrogen monoxide (NO) have inhibitory effects on EBPR and these effects are not considered in the model.
- ASM2d is valid for temperatures ranging from 10°C to 25°C.

ASM3 was developed by the task group to address many of the limitations described above for ASM1 and ASM2. Only a limited amount of calibration work has been undertaken for ASM3 to date (Wichern et al., 2001), so it remains to be seen if these limitations have been successfully eliminated. The task group identified the following specific limitations for ASM3 (Gujer et al. in Henze et al., 2001):

- ASM3 is valid for temperatures ranging from 8°C to 23°C and for pH in the range of 6.5 to 7.5.
- ASM3 (and ASM1) do not include any processes that describe biomass behavior in an anaerobic environment. Simulation of systems with large fractions of anaerobic reactor volumes may lead to errors.
- ASM3 does not address nitrite inhibition.
- ASM3 (and ASM1) cannot be used for activated sludge systems with very high organic loads or with low SRTs (1 day), where flocculation or adsorption of particulate biodegradable organics and storage may become limiting.
- ASM3 contains a reduced heterotrophic yield for anoxic conditions to improve predictions of solids production and denitrification.

Another member of the ASM type models, the General Model (Barker and Dold, 1997a) has a different structure and several variations from the models listed above. Most importantly, it incorporates a detailed anaerobic model, based on Jones et al. (1992) that, among others, provides a more detailed description of the crucial VFA generation processes in the anaerobic and aerobic zones of BNR plants.

2.4 References

Arun, V., T. Mino, and T. Matsuo. 1988. Biological mechanism of acetate uptake mediated by carbohydrate consumption in excess biological phosphorus removal systems. *Water Res.* 22(5):565-570.

Barker, P.S., and P.L. Dold. 1997a. General model for biological nutrient removal activated sludge systems: model presentation. *Water Environ. Res.* 69(5):969-984.

Barker, P.S., and P.L. Dold. 1997b. General model for biological nutrient removal activated sludge systems: model application. *Water Environ. Res.* 69(5):985-999.

Brodisch, K.E.U. 1985. Interaction of different groups of microorganisms in biological phosphate removal. *Water Sci. Tech.* 17(11-12):139-146.

Comeau, Y., K.J. Hall, R.E.W. Hancock, and W.K. Oldham. 1986. Biochemical model for enhanced biological phosphorus removal. *Water Res.* 20:1511-1521.

Copp, J.B. 1998. *COD balances in biological nutrient removal (nitrogen and phosphorus) activated sludge systems*, Ph. D. Thesis, McMaster University, Hamilton, Ontario, Canada, pp. 82-83.

Dold, P.L. 1990. A general activated sludge model incorporating biological excess phosphorus removal. In *Proc. Annual Conf C.S.C.E.,* Hamilton, Ontario, Canada.

Dold, P.L. 1991. Modification du modèle général des boues activées pour tenir compte de la déphosphatation. *Sciences et Techniques de L'eau* 24:229.

Dold, P.L., G.A. Ekama, and G.v.R. Marais. 1980. A general model for the activated sludge process. *Prog. Water Technol.* 12:47-77.

Dold, P.L., and G.v.R. Marais. 1986. Evaluation of the general activated sludge model proposed by the IAWPRC task group. *Water Sci. Tech.* 18(6):63-89.

Downing, A.L., H.A. Painter, and G. Knowles. 1964. Nitrification in the activated sludge process. *J. Proc. Inst. Sew. Purif.* 64:130-158.

Ekama, G.A., and M.C. Wentzel. 1999. Difficulties and developments in biological nutrient removal technology and modeling. *Water Sci. Tech.* 39(6):1-11.

Grady, C.P.L., W. Gujer, M. Henze, G.v.R. Marais, and T. Matsuo. 1986. A model for single-sludge wastewater treatment systems. *Water Sci. Tech.* 18(6):47-61.

Grau, P., P.M. Sutton, M. Henze, S. Elmaleh, C.P.L. Grady, W. Gujer, and J. Koller. 1982. Recommended notation for use in the description of biological wastewater treatment processes. *Water Res.* 16:1501-1505.

Gujer, W., and M. Henze. 1991. Activated sludge modeling and simulation. *Water Sci. Tech.* 23(4-6):1011-1023.

Gujer, W., M. Henze, W., T.Mino, and M.C.M. van Loosdrecht. 1999. Activated Sludge Model No. 3. *Water Sci. Tech.* 39(1):183-193.

Henze, M., C.P.L. Grady, W. Gujer, G.v.R. Marais, and T. Matsuo. 1987a. Activated Sludge Model No. 1, *IAWPRC Scientific and Technical Report No. 1*, London, UK: International Water Association.

Henze, M., C.P.L Grady, W. Gujer, G.v.R Marais, and T. Matsuo. 1987b. A general model for single sludge wastewater treatment systems. *Water Res.* 21(5):505-515.

Henze, M., W. Gujer, T. Mino, T. Matsuo, M. Wentzel, and G.v.R. Marais. 1995. Activated Sludge Model No. 2. *IAWQ Scientific and Technical Report No. 3*, London, UK: International Water Association.

Henze, M., W. Gujer, T.Mino, T. Matsuo, M. Wentzel, and G.v.R. Marais. 1995. Activated Sludge Model No. 2: biological phosphorus removal. *Water Sci. Tech.* 31(2):1-11.

Henze, M., W. Gujer, T.Mino, T. Matsuo, M. Wentzel, G.v.R. Marais, and M.C.M. van Loostrecht. 1999. Activated Sludge Model No. 2d, ASM2D. *Water Sci. Tech.* 39(1):165-182.

Henze, M., W. Gujer, T. Mino and M.C.M. van Loosdrecht. 2000. Activated Sludge Models ASM1, ASM2, ASM2d and ASM3. *IWA Scientific and Technical Report No. 9*, London, UK: International Water Association.

Jones, R.M., J.F. MacGregor, K.L. Murphy, and E.R. Hall. 1992. Towards a useful dynamic model of the anaerobic digestion process. *Water Sci. Technol.* 25 (7), 61-72.

Lawrence, A.W., and P.L. McCarty. 1970. Unified basis for biological treatment design and operation. *J. San. Eng. Div., ASCE* 96:757-778.

Lotter, L.H., M.C. Wentzel, R.E. Loewenthal, G.A. Ekama, and G.v.R. Marais. 1986. A study of selected characteristics of *Acetinobacter spp.* isolated from activated sludge in anaerobic/anoxic/aerobic and aerobic systems. *Water SA* 12(4):203-208.

Marais, G.v.R., and G.A. Ekama. 1976. The activated sludge process: part 1–steady state behavior. *Water SA* 2:163-199.

McKinney, R.E. 1962. Mathematics of complete mixing activated sludge. *J. San. Eng. Div., ASCE* 88, SA3, Proc Paper 3133:87-113.

McKinncy, R.E., and R.J. Ooten. 1969. Concepts of complete mixing activated sludge. In *Trans. 19th San. Eng. Conf.* (pp. 32-59). University of Kansas.

Meganck, M., D. Malnou, P. Le Flohic, G.M. Faup, and J.M. Rovel. 1985. The importance of acidogenic microflora in biological phosphorus removal. *Water Sci. Tech.* 17(11-12):199-212.

Mino, T., V. Arun, Y. Tsuzuki, and T. Matsuo. 1987. Effect of phosphorus accumulation on acetate metabolism in the biological phosphorus removal process. In: (R Ramadori: ed.) *Advances in Water Pollution Control - Biological Phosphate Removal from Wastewaters* (pp.27-38). Oxford, UK: Pergamon Press.

Mino, T., W.T. Liu, F. Kurisu, and T. Matsuo. 1995. Modeling glycogen storage and denitrification capability of microorganisms in enhanced biological phosphate removal processes. *Water Sci. Tech.* 31(2):25-34.

Mino, T., M.C.M. van Loosdrecht, and J.J. Heijnen. 1998. Microbiology and biochemistry of the enhanced biological phosphate removal process. *Water Res.* 32(11):3193-3207.

Pereira, H., P.C. Lemos, M.A. Reis, J.P.S.G. Crespo, M.J.T. Carrondo, and H. Santos. 1996. Model for carbon metabolism in biological phosphorus removal processes based on *in vivo* ^{13}C-NMR labeling experiments. *Water Res.* 30(9):2128-2138.

Reiger, L., G. Koch, M. Kuhni, W. Gujer, and H. Seigrist. 2001. The EAWAG Bio-P module for Activated Sludge Model No. 3. *Water Res.* 35(16):3887-3903.

Satoh, H., T. Mino, and T. Matsuo. 1992. Uptake of organic substrates and accumulation of polyhydroxyalkanoates linked with the glycolysis of intracellular carbohydrates under anaerobic conditions in the biological excess phosphate removal process. *Water Sci. Tech.* 26(5/6):933-942.

Satoh, H., T. Mino, and T. Matsuo. 1994. Deterioration of enhanced biological phosphorus removal by the domination of microorganisms without polyphosphate accumulation. *Water Sci. Tech.* 30(11): 203-211.

Satoh, H., W.D. Ramey, F.A. Koch, W.K. Oldham, T. Mino, and T. Matsuo. 1996. Anaerobic substrate uptake by the enhanced biological phosphorus removal activated sludge treating real sewage. *Water. Sci. Tech.* 34(1-2):8-15.

Smolders, G.J.F., J. van der Meij, M.C.M. van Loosdrecht and J.J. Heijnen (1994a) Model of the anaerobic metabolism of the biological phosphorus removal process: stoichiometry and pH influence. *Biotech. Bioeng.* 43:461-470.

Smolders, G.J.F., M.C.M. van Loosdrecht, and J.J. Heinen. 1994b. A metabolic model for the biological phosphorus removal process. *Water Sci. Tech.* 31:79-93.

Smolders, G.J.F., J. van der Meij, M.C.M. van Loosdrecht, and J.J. Heijnen. 1994c. Stoichiometric model of the aerobic metabolism of the biological phosphorus removal process. *Biotech. Bioeng.* 43:837-848.

Van Haandel, A.C., G.A. Ekama, and G.v.R. Marais. 1981. The activated sludge process: part 3 - single sludge denitrification. *Water Res.* 15(10):1135-1152.

Wentzel, M.C., P.L. Dold, G.A. Ekama, and G.v.R. Marais. 1985. Kinetics of biological phosphorus removal. *Water Sci. Tech.* 17(11-12):57-71.

Wentzel, M.C., L.H. Lotter, R.E. Loewenthal, and G.v.R. Marais. 1986. Metabolic behavior of *Acinetobacter spp.* in enhanced biological phosphorus removal–a biochemical model. *Water SA* 12(4):209-224.

Wentzel, M.C., R.E. Loewenthal, G.A. Ekama, and G.v.R. Marais. 1988. Enhanced polyphosphate organism cultures in activated sludge systems. Part I: enhanced culture development. *Water SA* 14(2):81-92.

Wentzel, M.C., G.A. Ekama, R.E. Loewenthal, P.L. Dold, and G.v.R. Marais. 1989a. Enhanced polyphosphate organism cultures in activated sludge systems. Part II: experimental behavior. *Water SA* 15(2):71-88.

Wentzel, M.C., P.L. Dold, G.A. Ekama, and G.v.R. Marais. 1989b. Enhanced polyphosphate organism cultures in activated sludge systems. Part III: kinetic model. *Water SA* 15(2):89-102.

Wentzel, M.C., L.H. Lotter, G.A. Ekama, R.E. Loewenthal, and G.v.R. Marais. 1991. Evaluation of biochemical models for biological excess phosphorus removal. *Water Sci. Tech.* 23(4-6):567-576.

Wichern, M., F. Obenaus, P. Wulf, and K.H. Rosenwinkel. 2001. Modeling of full-scale wastewater treatment plants with different treatment processes using the Activated Sludge Model No. 3. *Water Sci. Tech.* 44(1):49-56.

CHAPTER 3.0

ACTIVATED SLUDGE PLANT SIMULATORS

3.1 Introduction

The synthesis of fundamental knowledge about activated sludge systems into models as described in Chapter 2.0 is taken further in simulation where solutions to modeling problems are developed using numerical methods and computer hardware. A simulator is a computer program that allows the user to link various unit processes together (such as bioreactors and clarifiers) according to the flow scheme of a particular treatment plant, and then to mimic the performance of the plant for specified operational and influent loading conditions. The simulator may also incorporate models describing unit processes beyond the activated sludge system such as chemical precipitation, anaerobic digesters and sludge handling processes. Each unit process incorporates one or more mathematical models of which the bioreactor model usually is the most complex.

Because of the complexity of many wastewater treatment plants, a simulator should offer a high level of flexibility with minimal user training. With the advent of powerful graphics software, the use of object-oriented modeling concepts can effectively be used to achieve these goals. With this approach, the user does not require any knowledge of programming; instead, an icon-based modeling approach is used to specify a problem, that is, to specify components of a system, physical and kinetic properties and initial conditions. Once the system is specified graphically, and the compositions of inputs to the system are defined, the simulation can be executed to generate results. Graphical output often is a useful means for the user to view the state of the system as the simulation progresses.

The potential benefit of simulators in the analysis, design and operation of wastewater treatment systems is tremendous. Such tools can provide a better understanding of the mechanistic basis of models, in that they allow users to appreciate system responses to changes in control variables. Simulators also may be used effectively for the calibration of complex models and training of plant operators.

3.2 Survey of Activated Sludge Model Simulators

To complete the review of the state of the art of activated sludge modeling, a survey of commercially available activated sludge simulators was carried out. At the time of the survey, there were seven such major simulators available worldwide. Table 3-1 summarizes information on the simulators surveyed and pertinent data on their availability.

Table 3-1. Activated Sludge Simulators Surveyed

Simulator	Vendor	Location	Website
ASIM	EAWAG (Swiss Federal Institute for Environmental Science &Technology)	Switzerland	www.eawag.ch
BioWin	EnviroSim Associates Limited	Canada	www.envirosim.com
EFOR	DHI, Inc.	Denmark	www.dhi.com
GPS-X	Hydromantis, Inc.	Canada	www.hydromantis.com
SIMBA	IFAK-System GmbH	Germany	www.ifak-system.com
STOAT	WRc Group	United Kingdom	www.wrcplc.co.uk
WEST	Hemmis N.V.	Belgium	www.hemmis.com

All the simulators allow the user to select more than one bioreactor model. For example, GPS-X offers all the IWA models (ASM1-3), and models attributed to Dold, among others, that have been modified by Hydromantis, Inc.; EFOR offers ASM1 and versions of ASM2d and ASM3 that have been modified by DHI, Inc.; STOAT offers modified versions of ASM1, ASM2d, unmodified ASM3 and models that are based on BOD (Stokes et al., 2000) rather than COD; BioWin is based on an extension of the Barker and Dold (1997) general model (which can be substituted with the IWA models (ASM1-3)).

The assumptions made and values ascribed to specific parameters in the bioreactor models will produce differences in simulator outputs. Researchers have tried to evaluate the impact of such model differences (O'Shaughnessy et al., 1998; Copp, 2001), but such an exercise requires an in-depth knowledge of each model. Model details change frequently as developers modify and improve their models. It should be recognized that the IWA-type models incorporate a large number of stoichiometric and kinetic parameters and that the different variations of the models imply differences in model structure. Therefore, it is important to recognize that parameters in one model cannot necessarily be transferred directly to other models *(even where the parameters sometimes have the same name)*. Nevertheless, the true test (and utility) of any particular model should be whether or not the model can, using a single set of model kinetic and stoichiometric parameters:

- Satisfactorily track the changes with time for a range of key parameters (e.g., soluble phosphorus, ammonia and nitrate concentrations; volatile suspended solids concentrations and oxygen uptake rate),
- Be used in different types of systems (e.g., aerobic systems (with and without nitrification), nitrogen removal systems, biological phosphorus removal systems, and biological nitrogen and phosphorus removal systems (under steady state and dynamic conditions)), and
- Be operated over a range of conditions of operating variables such as sludge retention time and recycle rates.

A questionnaire seeking information on the status of commercial wastewater treatment plant simulators that are using IWA-based activated sludge models was distributed to vendors during 2001. All seven vendors responded to the survey. Table 3-2 summarizes the information supplied by respondents.

Table 3-2. Summary of Information Collected from Simulator Vendors

	Simulator						
	ASIM[1]	BioWin	EFOR	GPS-X	SIMBA	STOAT	WEST
Year Introduced							
	1988	1990	1990	1991	1994	1994	1998
Current Version							
	ASIM v. 3.00	BioWin32 v. 1.2.1	EFOR 2001	GPS-X v. 4.0	SIMBA v. 4.0	STOAT v. 4.2	WEST v. 3.0
User Groups (Percent of Total)							
Consultants	20	66	60	40	24	50	40
Owners/ Operators		12	30	20	23	48	10
Academic	80	20	10	30	30	2	45
Others		2		10	23		5
Users by Region							
	Worldwide	Worldwide	Worldwide (mainly Europe)	Worldwide	Worldwide (mainly Germany, Holland)	Worldwide (mainly UK, US)	Worldwide
Computer Platform/Operating System							
	MS DOS (MS Windows in 2002)	MS Windows 95/NT 4.0 or later	MS Windows 95/NT3.51 or later	MS Windows 95/NT 4.0 or later	MS Windows 98/NT 4.0 or later	MS Windows 95 or later	MS Windows 95/NT 4.0 or later
Can models other than the base model be used?							
	Yes	Yes	Yes	Yes	Yes	Yes	Yes
Approximate number of licenses sold/leased							
	>300	>300	100	Hundreds	>170	>150	>150

[1] *EAWAG also sells a simulation environment called AQUASIM, which can operate on several platforms and is used primarily by academics.*

All the simulators were developed during the 1990s, building on the development of ASM1 during the 1980s. All simulators are configured within the Windows environment, which allows the use of graphics software. All of the simulators have reached worldwide use, although the use of some simulators such as EFOR and SIMBA is focused upon the location of the developers. Within the user distribution data sets provided by the vendors, the largest group of users for the BioWin, EFOR, GPS-X and STOAT simulators is consultants. Conversely, the largest group of users for the ASIM, SIMBA and WEST simulators is academic; in the case of ASIM, this was developed primarily as a teaching and research tool (Gujer and Larsen, 1995), whereas WEST has a transparent environment that can accommodate any model, an attribute that is favored by academics.

The survey clearly shows that plant-wide simulators have extended the use of IWA-based activated sludge models to a worldwide extent.

3.3 References

Barker, P.S., and P. L. Dold. 1997. General model for biological nutrient removal activated sludge systems: model presentation. *Water Environ. Res.* 69(5):969-984.

Copp, J.B. 2001. *The COST simulation benchmark: description and simulator manual.* Office for Official Publications of the European Community, Luxembourg. p. 154.

Gujer, W., and T. A. Larsen. 1995. The implementation of biokinetics and conservation principles in ASIM. *Water Sci. Tech.* 31(2):257-266.

O'Shaughnessy, M. M., G. V. Crawford, G. T. Daigger, and M. D. Elliott. 1998. Biological process models – comparative predictive performance evaluations. In *Proc. 71st Annual Conference of the Water Environment Federation*, Orlando, FL. Alexandria, VA: Water Environment Federation.

Stokes, A. J., J. R. West, C. F. Forster, and W. J. Davies. 2000. Understanding some of the differences between the COD- and BOD- based models offered in STOAT. *Water Res.* 34(4):1296-1306.

CHAPTER 4.0

WASTEWATER CHARACTERISTICS

4.1 Background

Activated sludge models distinguish between the mechanisms acting on different components in the influent wastewater stream. The term "wastewater characteristics" refers to the partitioning of influent organic material into biodegradable and unbiodegradable (inert) portions, the ammonia portion of the total nitrogen and so on. The influent wastewater characteristics may vary, often appreciably, from one municipal wastewater to another.

Wastewater characteristics have a very significant impact on system performance, particularly for nutrient removal systems. A single characteristic, such as the readily biodegradable COD fraction, can determine whether or not a system designed for excess phosphorus removal will in fact remove phosphorus. A second example is the portion of unbiodegradable particulate COD in the influent wastewater; this impacts sludge production and oxygen demand significantly. Therefore, if the model is to provide reasonable predictions of system behavior, adequate knowledge of wastewater characteristics is extremely important.

Note: Arguably the most significant municipal wastewater characteristics are the amounts of RBCOD and unbiodegradable particulate COD.

In the context of treatment plant simulation, wastewater characteristics not only are important for activated sludge system modeling; simulating the performance of other unit operations also is dependent on appropriate wastewater characterization. For example, the division of slowly biodegradable influent COD into colloidal and particulate portions significantly impacts the modeling of primary sedimentation.

Differences in influent wastewater characteristics usually are a function of a combination of the following factors: (1) socio-economic factors; (2) water usage; (3) degree and pattern of inflow and infiltration; (4) use of garbage grinders; (5) presence, magnitude and nature of any industrial content; (6) size, retention time and storage capacity within the sewage collection system; and (7) presence of a phosphate detergent ban.

This chapter presents information on how the influent wastewater COD, TKN and TP are divided into sub-components. An overview of measurement methods for the different components is provided. Details of the methods are presented in the chapters that follow.

Note: The structure of the wastewater characterization presented here is applicable to municipal wastewaters and certain industrial wastewaters such as food processing.

At the outset, a comment on nomenclature is appropriate. In certain treatment plant simulators the approach has been to define all components as concentrations. For example, the concentration of the influent COD and the concentration of COD components such as unbiodegradable soluble COD, readily biodegradable COD, etc., are defined in mg/L. Other simulators have followed the "totals and fractions" presentation approach. With this approach, component concentrations are presented as a fraction of the total. For example, if a wastewater with a total COD of 500 mg/L includes an *unbiodegradable soluble* COD concentration of 50 mg/L this is presented as having an *unbiodegradable soluble* COD fraction of 0.10. This difference in notation is purely semantic; the model equations obviously will be based on concentrations. An advantage of the "totals and fractions" approach is that it provides a logical basis for comparing the characteristics of different wastewaters.

4.2 Nature of Wastewaters

This section provides a brief discussion on the actual nature of wastewaters and how they are approximated for modeling purposes. Wastewater comprises a complex mixture of organic and inorganic material. The term "wastewater characteristics" refers to the partitioning of the various materials into groups. For modeling, the groups are defined according to how these materials behave in the activated sludge process. For example, the first level of division for organic material is related to whether the material is biodegradable or unbiodegradable. Within the biodegradable fraction, the material is further divided into readily biodegradable and slowly biodegradable substrate. The readily biodegradable substrate is presumed to consist of relatively small molecules (such as volatile fatty acids and low molecular weight carbohydrates), which can be easily transported into the cell resulting in an immediate response in the use of electron acceptor (oxygen or nitrate). The slowly biodegradable COD is assumed to consist of larger and more complex material, which requires extracellular breakdown prior to uptake and utilization (Dold et al., 1980). To facilitate modeling and measurement of the components, they have been related to a physical definition, i.e., readily biodegradable material is represented as soluble material while slowly biodegradable substrate is assumed to be particulate. (The latter assumption, that slowly biodegradable material is particulate, is discussed in the following section – a significant portion may be colloidal.) The unbiodegradable (inert) fraction represents material that does not react or reacts so slowly that any losses in the treatment plant are negligible. A portion of this material is soluble and passes directly through the system in the liquid phase, so that the effluent concentration is the same as the influent concentration. Unbiodegradable material that is particulate is entrapped in the sludge mass and removed with the excess sludge.

Methods for measurement of the different wastewater components can be related either to the impact of the component on the process or by the physical definition. For example, one method for measurement of the biodegradable soluble material is related to the oxygen uptake rate (OUR) observed during degradation of this material. These methods can be called bioassay methods, in that they derive the characteristics from observations of a biological system. A second technique for measurement of biodegradable soluble components utilizes a filtration method that is related to the assumption that the readily biodegradable material is soluble and the slowly degradable material is particulate. Such a method is referred to as a "direct method." Direct methods may require an arbitrary division, such as a filter size, to define the wastewater fraction.

In some cases, measurement methods are related to assumptions about a model structure. One example is the measurement of unbiodegradable soluble material. If one assumes that this is represented by the effluent soluble COD concentration, one must first assume that there is no generation of this material in the process. Some studies have explored the inclusion of a process for generation of unbiodegradable soluble microbial products. The modeling of soluble microbial product generation is discussed in Chapter 39. However, the IWA-type models do not include this process.

The wastewater characterization methods presented in this manual are concerned with the concentrations of organic materials, and the macronutrients nitrogen and phosphorus. Characterization of the organic material is important for a model to successfully predict oxygen demand, nitrogen removal and mixed liquor suspended solids concentrations. The characterization of the nitrogenous material also has a significant impact on prediction of the oxygen demand, as well as on biological nitrogen and phosphorus removal performance. Division of phosphorus into constituent fractions may also be important when low effluent phosphorus concentrations are required, but this has received limited attention to date.

4.3 Important Note on Filter Type and the Term "Soluble"

Analyses for concentration often are conducted on both unfiltered and filtered wastewater samples. In this regard it is very important to draw attention to confusion or a lack of clarity when reporting "soluble" concentrations. In many instances the term "soluble," in the context of wastewater treatment, generally refers to material that passes through a 0.45 μm membrane filter (Grady, 1989). However, in many cases other filter types are used. For example, for TSS measurements, glass-fiber filters with a nominal pore size of 1.2 to 1.5 μm are used. An analysis of the filtrate from this filtration may differ substantially from the 0.45 μm filtrate. Therefore, it is important to note the type of filter used when reporting "soluble" concentration analyses. Discrepancies can be significant for COD analyses on influent wastewater samples. This is because a significant portion of the COD is colloidal, and the fraction of this material passing through the filter is very dependent on filter type.

The significance of filter type when describing influent wastewater "soluble" components is best illustrated by example. Grab samples were collected once a day over a period of 60 days at a 35 ML/day treatment plant receiving a municipal flow. The samples were filtered through both standard glass-fiber (GF) filters and through 0.45 μm membrane filters (MF), and the filtrates were analyzed for COD concentration. Figure 4-1 shows a time-series plot of the daily total (unfiltered) COD and the corresponding filtrate ("soluble") CODs. Averages and standard deviations for the three sets of data are listed in Table 4-1. The data show that significantly more COD passes through the glass-fiber filter than through the 0.45 μm filter: 40% versus 26% on average. Presumably most of the colloidal COD passes through the glass-fiber filter, but the 0.45 μm filter retains a significant portion.

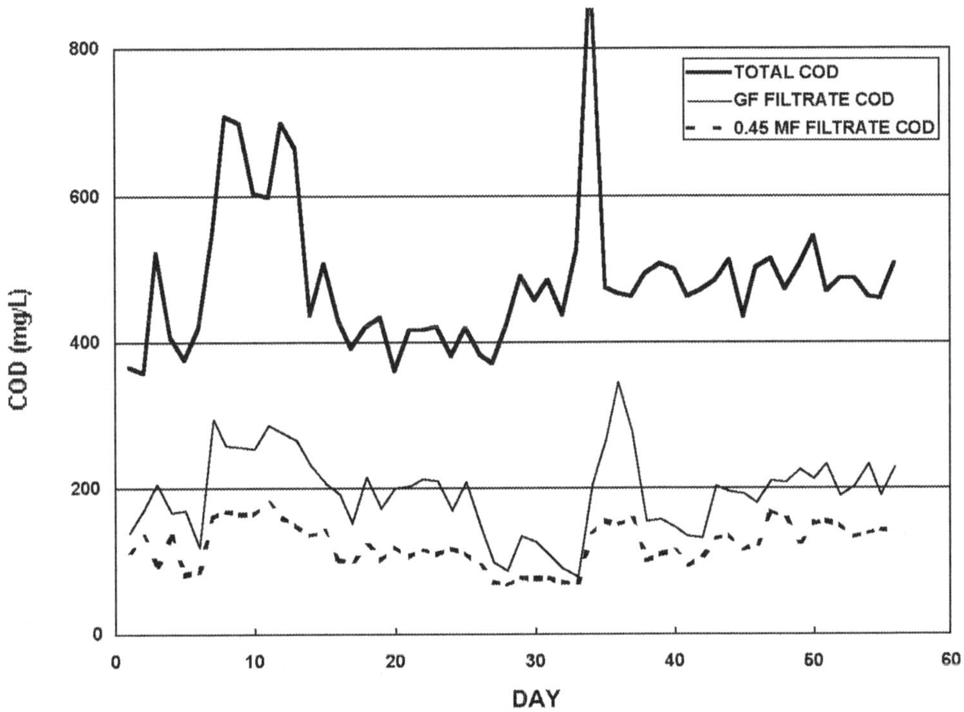

Figure 4-1. Influent Total (Unfiltered) COD, Glass-Fiber Filtrate COD and 0.45 μm Membrane Filtrate COD of Daily Raw Municipal Wastewater Grab Samples

Table 4-1. COD of Unfiltered, Glass-Fiber (GF) Filtered and 0.45 μm (MF) Filtered Raw Municipal Influent Wastewater (all as mg COD/L)

Parameter	Mean	Standard deviation
Total (unfiltered) COD	514	169
COD (GF filtrate)	209	89
COD (0.45 μm MF filtrate)	136	73

The values differ quite substantially from day to day. It is interesting to compare the fraction of the total COD that passes through each filter type. The daily fraction values are plotted in quantile-quantile probability plots (Chambers et al., 1983) in Figures 4-2 and 4-3 for glass-fiber filtrate and 0.45 μm membrane filtrate, respectively. In each case, the plotted points closely approximate a straight line, indicating that both fractions essentially exhibit a normal distribution (GF: mean = 0.407; standard deviation = 0.109; MF: mean = 0.261; standard deviation = 0.062).

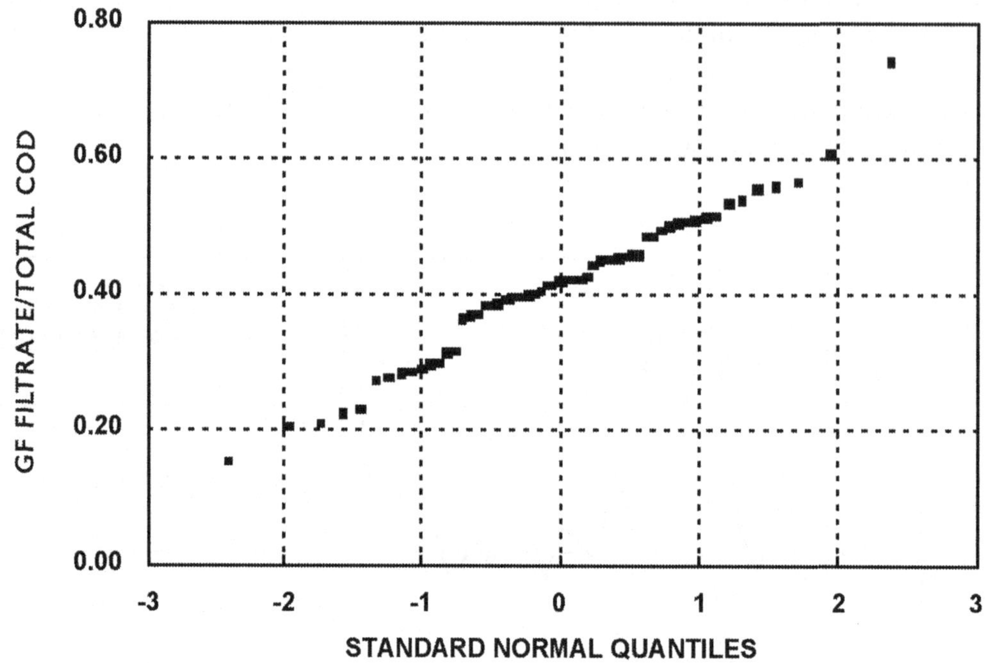

Figure 4-2. Quantile-Quantile Probability Plot of COD Fraction Passing through Glass-Fiber Filter (GF) (Daily Grab Samples)

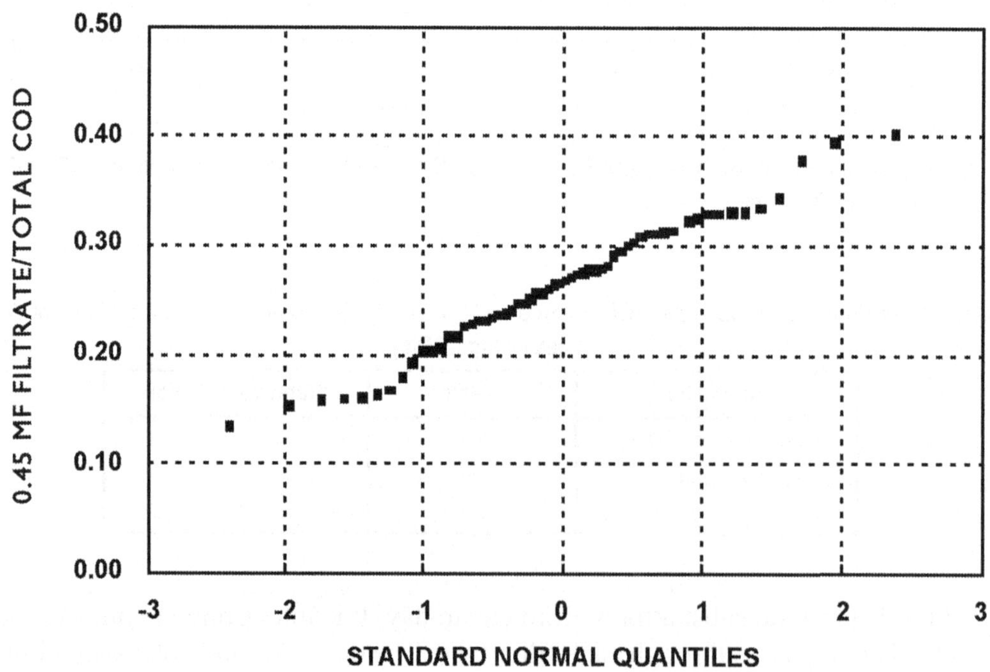

Figure 4-3. Quantile-Quantile Probability Plot of COD Fraction Passing through 0.45 μm Membrane Filter (MF) (Daily Grab Samples)

In most situations, colloidal material entering an activated sludge system largely appears to be adsorbed onto the biomass, and remains with the solids when effluent (or mixed liquor) samples are filtered. Therefore, less difference in filtrate concentrations from glass-fiber and 0.45 μm membrane filtration of effluent samples is evident compared to influent samples. There are a number of related issues that are important regarding wastewater characterization, for example:

- Whether all of the influent colloidal COD is biodegradable, or whether a portion is unbiodegradable;
- The validity of the assumption that adsorption is nearly complete in most activated sludge systems;
- Whether generation of colloidal unbiodegradable material within the activated sludge process is significant; and
- The types of treatment system where adsorption of colloidal material likely is incomplete (e.g. contact stabilization, trickling filters, extremely low SRT systems).

The authors are of the opinion that research information on these topics is not necessarily conclusive, or requires further investigation. A discussion of these topics is deferred until the end of the chapter.

4.4 Organic Material

Characterization of the carbonaceous material in municipal wastewater for modeling purposes usually is in terms of the chemical oxygen demand (COD). The advantage of selecting COD as the parameter for quantifying the "strength" of organic material in the influent, as opposed to BOD or TOC, is that it provides a consistent basis for description of the activated sludge process (see Appendix B). The division of the total influent COD ($COD_{T, INF}$) into the various fractions used in nutrient removal system design and modeling is shown in Figures 4-4 and 4-5.

4.4.1 Biodegradable and Unbiodegradable Portions

The first division of the influent COD is into biodegradable COD and unbiodegradable COD. Each of these fractions is divided further into two sub-portions. (Any COD contribution from active biomass in the influent wastewater is discussed separately below.)

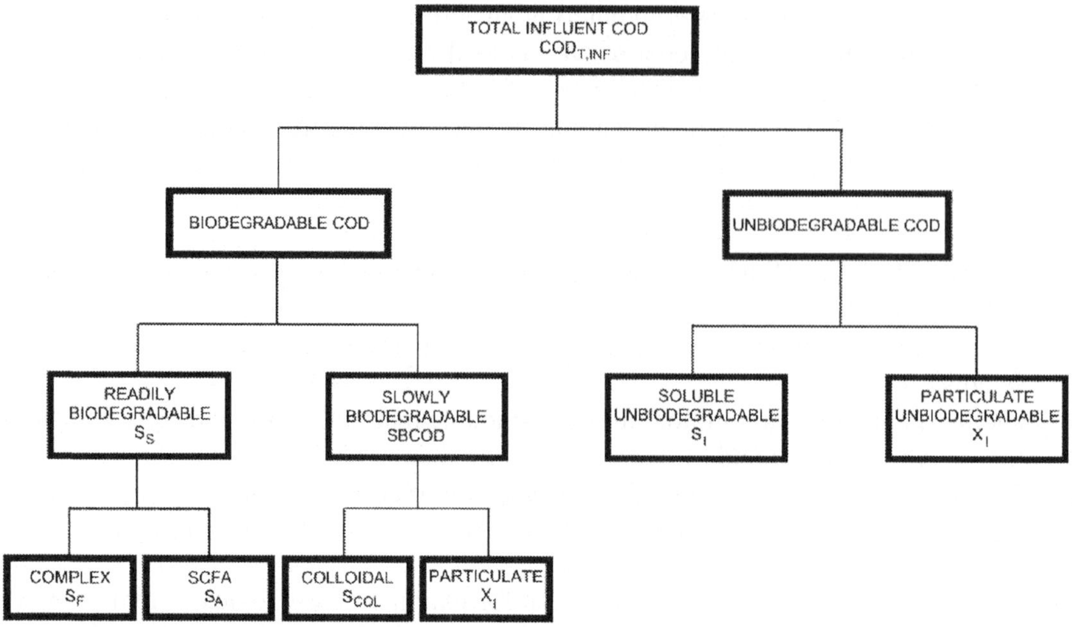

Figure 4-4. Division of Municipal Wastewater COD into Components

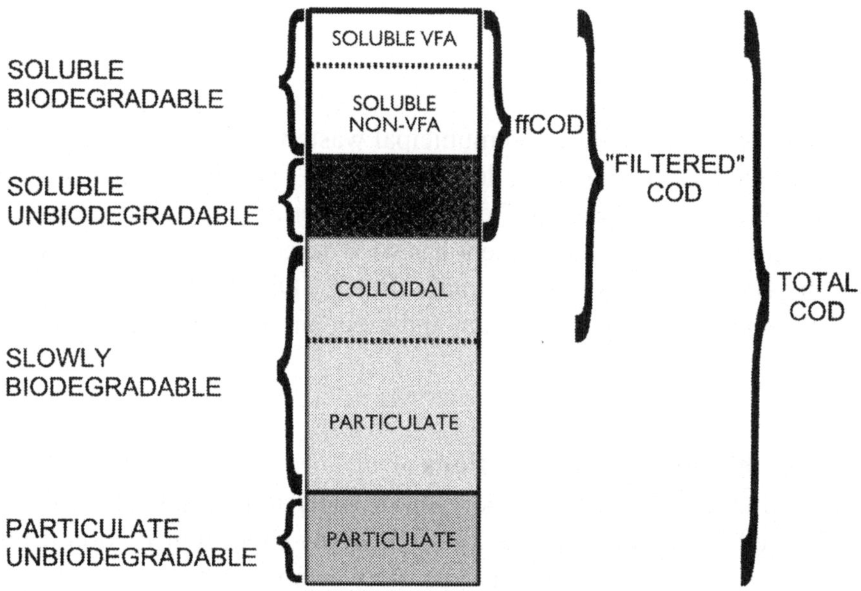

Figure 4-5. Schematic Representation of COD Components for Municipal Wastewater

Note: The term "ffCOD" refers to flocculated and filtered COD, and is discussed in Chapter 7.0, "Readily Biodegradable COD."

4.4.1.1 Biodegradable Sub-Portions

The biodegradable material is divided into a readily biodegradable portion (S_S, also termed RBCOD) and a slowly biodegradable portion (SBCOD). The readily biodegradable portion is hypothesized to consist of material that can be absorbed readily by the organism and metabolized for energy and synthesis. The slowly biodegradable portion is made up of particulate/colloidal material and complex organic molecules that require extracellular enzymatic breakdown prior to absorption and utilization. The apportionment between readily and slowly biodegradable COD has a major influence on process dynamic behavior (for example, oxygen demand) and is of crucial importance in the design of biological nitrogen and phosphorus removal systems. For purposes of modeling excess biological phosphorus removal systems it is necessary to distinguish a further division of the readily biodegradable material (S_S). The readily biodegradable portion consists of complex (fermentable) readily biodegradable COD (S_F) and readily biodegradable COD in the form of short chain volatile fatty acids (S_A).

Note: The distribution of slowly biodegradable COD between colloidal and particulate is important for modeling primary settling tank performance. The division is less significant for the activated sludge system because colloidal material is adsorbed rapidly onto the sludge matrix.

4.4.1.2 Unbiodegradable Sub-Portions

The unbiodegradable material is divided into a soluble unbiodegradable portion (S_I) and a particulate unbiodegradable portion (X_I). Both components are hypothesized to be unaffected by biological action in the system. In an activated sludge system, the S_I leaves via the secondary clarifier overflow at a concentration equal to that in the influent. The X_I is enmeshed in the sludge mass and accumulates in the system. At steady state, the mass of X_I entering the system in the influent will be balanced by the mass leaving via the sludge wastage stream. From a mass balance standpoint, the mass of X_I in the system will equal the influent mass per day multiplied by the system sludge age. In systems operated at a long sludge age, the X_I can constitute a significant portion of the mixed liquor settleable solids concentration.

4.4.2 Quantifying the Division

Division of the influent wastewater COD into the four sub-components can be defined by three fractional parameters (f_{SI} , f_{XI} and f_{SS}):

f_{SI} = fraction of the total influent COD which is unbiodegradable soluble

f_{XI} = fraction of the total influent COD which is unbiodegradable particulate

f_{SS} = fraction of the total influent COD which is readily biodegradable

Using these fractions, the division of the total influent COD into concentrations of the four sub-components may be expressed as follows:

$$S_I = f_{SI} \cdot COD_{T,INF} \quad (4.4.2\text{-}1)$$

$$X_I = f_{XI} \cdot COD_{T,INF} \quad (4.4.2\text{-}2)$$

$$S_S = f_{SS} \cdot COD_{T,INF} \quad (4.4.2\text{-}3)$$

$$SBCOD = (1 - f_{SS} - f_{SI} - f_{XI}) \cdot COD_{T,INF} \quad (4.4.2\text{-}4)$$

where:

S_I	= soluble unbiodegradable COD concentration
X_I	= particulate unbiodegradable COD concentration
S_S	= soluble readily biodegradable COD concentration
SBCOD	= slowly biodegradable COD concentration
$COD_{T,\ INF}$	= total influent COD concentration

and with:

$$COD_{T,INF} = S_S + S_I + X_I + SBCOD \qquad (4.4.2\text{-}5)$$

Note that many models (such as ASM1 and ASM2d) assume that the slowly biodegradable COD (SBCOD) is in a particulate form (i.e., X_S). However, for modeling the organics removal in a primary settling tank, and the COD composition and load of the stream passing to the activated sludge system, it is important to distinguish between the portions of the SBCOD that are colloidal and particulate. In other words, only a portion of the slowly biodegradable material is settleable. This division of SBCOD into colloidal and particulate portions can be defined by another fraction:

f_{XS} = fraction of the slowly biodegradable COD that is particulate (not colloidal)

$$X_S = f_{XS} \cdot SBCOD \qquad (4.4.2\text{-}6)$$

$$S_{COL} = (1 - f_{XS}) \cdot SBCOD \qquad (4.4.2\text{-}7)$$

where:

S_{COL}	= slowly biodegradable colloidal COD concentration
X_S	= slowly biodegradable particulate COD concentration

so that:

$$SBCOD = X_S + S_{COL} = (1 - f_{SS} - f_{SI} - f_{XI}) \cdot COD_{T,INF} \qquad (4.4.2\text{-}8)$$

and equation 4.4.2-5 becomes:

$$COD_{T,INF} = S_S + S_I + X_I + X_S + S_{COL} \qquad (4.4.2\text{-}9)$$

For excess phosphorus removal systems division of the readily biodegradable COD (S_S) into a complex (fermentable) portion (S_F) and a VFA portion (S_A) can be defined by another fraction (f_{SA}):

f_{SA} = fraction of the RBCOD in the form of VFA

$$S_A = f_{SA} \cdot S_S \qquad (4.4.2\text{-}10)$$

$$S_F = (1 - f_{SA}) \cdot S_S \qquad (4.4.2\text{-}11)$$

where:

S_A = soluble readily biodegradable volatile fatty acid COD concentration

S_F = soluble readily biodegradable complex COD concentration

with:

$$S_S = S_A + S_F \qquad (4.4.2\text{-}12)$$

4.4.3 Biomass COD in the Influent

Heterotrophic biomass may constitute a significant component of the influent wastewater COD. Reported values of influent heterotrophic biomass range from 7 percent to 25 percent of the total influent COD (Orhon and Çokgör, 1997). If modeling processes downstream of primary clarifiers, much of this biomass may originate from excess sludge recycled to the influent of the plant (see, for example, Kappeler & Gujer, 1992). Taking into account the active organisms in the influent, the total influent COD would be the sum of the biodegradable COD, the unbiodegradable COD, and the active organism mass COD.

$$COD_{T,INF} = S_S + S_I + X_I + S_{COL} + X_S + X_{HET} \qquad (4.4.3\text{-}1)$$

where:

X_{HET} = active heterotrophic organism concentration

If the biomass component is not measured separately, this material will be reflected in the calibrated influent values of X_I and X_S. In most cases, this will not significantly impact model predictions of plant performance. In very high rate systems, however, the determination of the biomass component may be more important.

4.4.4 Alternative Approach for Quantifying the COD Components

Guidelines for wastewater characterization recently have been developed by the Dutch Foundation for Applied Water Research (STOWA). Reference is made to these guidelines throughout the manual in the specific relevant chapters for each wastewater component. However, the overall structure of the characterization procedure is mentioned here (Roeleveld and van Loosdrecht, 2002):

1. S_I is determined based on soluble COD in the effluent of a wastewater treatment plant.
2. S_S is determined based on soluble COD in the influent less soluble COD in the effluent.

3. SBCOD is determined by measuring the total concentration of biodegradable material in a bioassay test, and subtracting the RBCOD from total.
4. X_I is determined by difference, i.e.:

$$X_I = COD_{T,INF} - S_S - S_I - SBCOD \qquad (4.4.4\text{-}1)$$

As a result of this sequence for characterization, all of the measurement errors propagate through to X_I. In other procedures presented in this manual, measurement errors are reflected in the SBCOD component. Because SBCOD typically is a much larger portion of the total influent COD, the relative error in this component will be smaller. Inaccuracies in the X_I component are reflected directly in important model predictions such as sludge production and the mixed liquor suspended solids concentration.

4.5 Nitrogenous Material

Characterization of the nitrogenous material in the influent (N_T) is in terms of total Kjeldahl nitrogen (TKN). It should be noted that in a limited number of cases, appreciable influent concentrations of nitrate/nitrite have been encountered in municipal wastewaters (often originating from plant recycles). This has a significant impact on nutrient removal system design and performance. For these cases, the influent nitrate/nitrite concentration must be included as an additional wastewater characteristic, separate from the TKN. For design and modeling purposes, it is necessary to specify the magnitudes of the various portions of the influent TKN according to the division shown in Figure 4-6.

4.5.1 Free and Saline Ammonia and Organically Bound Portions

The first division of the influent TKN is into free and saline ammonia (S_{NH}) and organically bound TKN. The organically bound TKN is divided further into two portions, biodegradable and unbiodegradable, each with soluble and particulate sub-portions.

Note: The soluble portion of the influent TKN (mainly ammonia) is very significant for modeling TKN removal in primary settling tanks. The division is less significant if only modeling an activated sludge system because particulate biodegradable organic nitrogen is hydrolyzed to ammonia within the system.

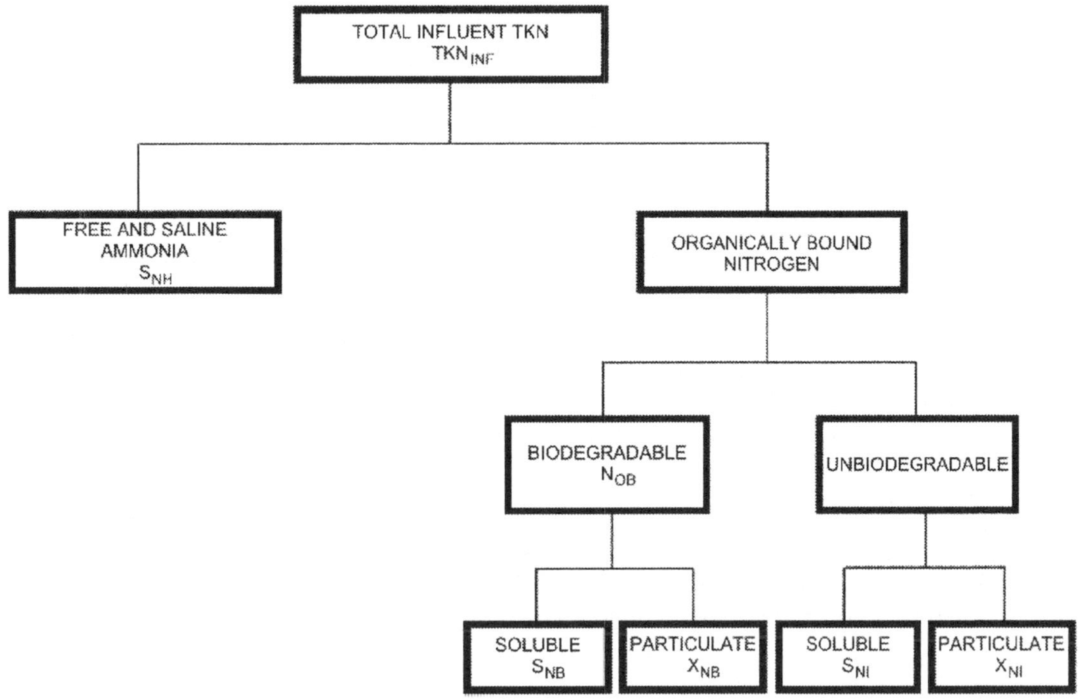

Figure 4-6. Division of Municipal Wastewater TKN into Components

4.5.1.1 Biodegradable Organically Bound TKN

The biodegradable organic nitrogen is divided into a soluble portion (S_{NB}) and a particulate portion (X_{NB}).

4.5.1.2 Unbiodegradable Organically Bound TKN

This component of the organic nitrogen is divided into a soluble unbiodegradable portion (S_{NI}) and a particulate unbiodegradable portion (X_{NI}). The unbiodegradable portions are hypothesized to be unaffected by biological action in the system. In an activated sludge system the soluble unbiodegradable material leaves via the secondary clarifier overflow (and waste stream) at a concentration equal to that in the influent. The particulate unbiodegradable portion is material associated with the unbiodegradable particulate COD in the influent and leaves the system via the sludge wastage stream.

4.5.2 Quantifying the Division

The influent wastewater TKN consists of five sub-components based on the division shown in Figure 4.6.

$$TKN_{INF} = S_{NH} + S_{NI} + X_{NI} + S_{NB} + X_{NB} \qquad (4.5.2\text{-}1)$$

The division into sub-components is not expressed as readily by defining fractional parameters as in the case for the division of COD. This is for a number of reasons, including (1) an inability to separately measure soluble biodegradable and unbiodegradable components in the effluent and (2) inconsistencies in the models in terms of the linkages between COD and *organic* N. However, it should be recognized that the major portion of the TKN is ammonia (often as much as 75 percent). Therefore, division of the remainder into four sub-components perhaps is not as critical. A practical approach is to identify two fractions: f_{SNH} and f_{SNI}:

f_{SNH} = fraction of the influent TKN which is free and saline ammonia;

f_{SNI} = fraction of the influent TKN which is soluble unbiodegradable.

In terms of concentrations:

$$S_{NH} = f_{SNH} \cdot TKN_{INF} \qquad (4.5.2\text{-}2)$$

$$S_{NI} = f_{SNI} \cdot TKN_{INF} \qquad (4.5.2\text{-}3)$$

The particuladegrte unbiodegradable influent TKN, X_{NI}, presumably is associated with the particulate unbioadable COD in the influent. Therefore, one means for quantifying this TKN portion is to assume a nitrogen content for the particulate unbiodegradable influent COD:

$$f_{XNI} = \frac{X_{NI}}{X_I} \; mgN/mgCOD \qquad (4.5.2\text{-}4)$$

f_{XNI} = N content of the influent particulate unbiodegradable COD

In terms of concentrations, noting that f_{XNI} is not defined as a fraction of the influent TKN (other "f" factors are defined as fractions of the total influent COD or TKN):

$$X_{NI} = f_{XNI} \cdot X_I = f_{XNI} \cdot f_{XI} \cdot COD_{T,INF} \qquad (4.5.2\text{-}5)$$

The biodegradable organic nitrogen (N_{OB}) is given by difference, i.e., the total influent TKN less the sum of S_{NH}, S_{NI} and X_{NI}. The division into soluble (S_{NB}) and particulate (X_{NB}) influent biodegradable organic nitrogen can be expressed through a fraction:

f_{XNB} = fraction of the biodegradable organic TKN which is particulate.

In terms of concentrations:

$$N_{OB} = X_{NB} + S_{NB} = TKN_{INF} - S_{NH} - S_{NI} - X_{NI} \qquad (4.5.2\text{-}6)$$

$$S_{NB} = (1 - f_{XNB}) \cdot N_{OB} \qquad (4.5.2\text{-}7)$$

$$X_{NB} = f_{XNB} \cdot N_{OB} \qquad (4.5.2\text{-}8)$$

4.6 Phosphorus Content

Characterization of the phosphorus content of the influent is in terms of the total phosphorus concentration (TP). In the context of nutrient removal, division of the TP into sub-components has not received as much attention as the COD and TKN. The approach in modeling and design has been to assume that a portion of the influent TP is associated with the unbiodegradable influent COD (approximately 10 to 15 percent of TP), and that the remainder can be considered as orthophosphate. (This assumption would not be appropriate for primary settling tank (PST) modeling.) Limited data on influent soluble reactive phosphorus (SRP, generally considered to be orthophosphate) would indicate this approach to be reasonable; that is, measurements of influent SRP are similar to TP less P associated with the unbiodegradable influent COD (assuming a P content of approximately 0.02 mg P per mg unbiodegradable particulate COD).

Several factors support the need for a more rigorous accounting of phosphorus fractions, considering the increasing emphasis on low effluent phosphorus concentrations (< 0.3 mg P/L). For example, up to 90 percent of the total soluble phosphorus (TSP) in aquatic systems can be comprised of soluble unreactive phosphorus (SUP) (Minear, 1972). If SUP were an appreciable fraction of the influent TP, this likely would impact the attainable minimum effluent P concentration for biological nutrient removal systems. A suitable method for accounting for the various influent phosphorus fractions is analogous to the division of influent TKN, and is shown in Figure 4-7. This division into components is presented below for completeness.

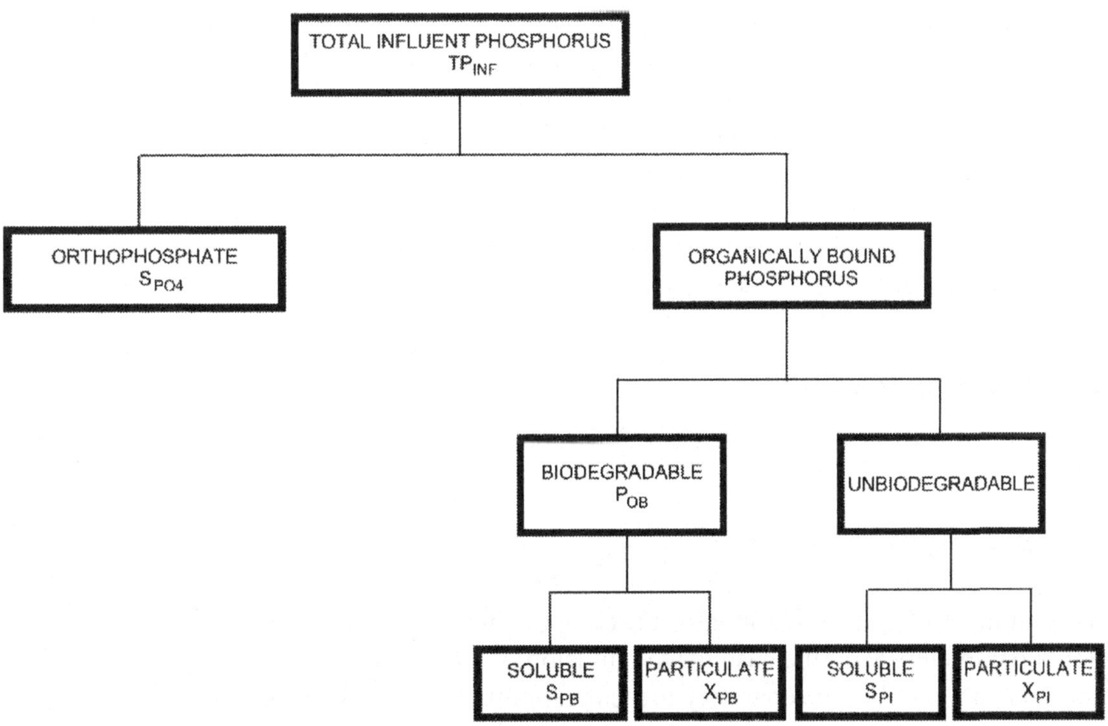

Figure 4-7. Division of Municipal Wastewater Phosphorus into Constituent Fractions

4.6.1 Orthophosphate and Organically Bound Portions

The first division of the influent TP is into orthophosphate (S_{PO4}) and organically bound TP. The organically bound TP is divided further into two portions (biodegradable and unbiodegradable), each with soluble and particulate sub-portions.

Note: The soluble portion of the influent TP (mainly orthophosphate) is very significant for modeling TP removal in primary settling tanks. In addition, there may be cases where a significant concentration of particulate inorganic phosphorus may be formed in the influent (e.g., when an industrial input contains chemicals that will precipitate phosphorus) and this also would impact primary settling tank performance. The division is less significant if only modeling an activated sludge system because particulate biodegradable organic phosphorus is hydrolyzed to orthophosphate within the system.

4.6.1.1 Biodegradable Organically Bound TP

The biodegradable organic phosphorus is divided into a soluble portion (S_{PB}) and a particulate portion (X_{PB}).

4.6.1.2 Unbiodegradable Organically Bound TP

This component of the organic phosphorus is divided into a soluble unbiodegradable portion (S_{PI}) and a particulate unbiodegradable portion (X_{PI}). These unbiodegradable portions are hypothesized to be unaffected by biological action in the system. In an activated sludge system, the soluble unbiodegradable material leaves via the secondary clarifier overflow (and waste stream) at a concentration equal to that in the influent. The particulate unbiodegradable portion is material associated with the particulate unbiodegradable COD in the influent and leaves the system via the sludge wastage stream.

4.6.2 Quantifying the Division

The influent wastewater TP consists of five sub-components based on the division shown in Figure 4-7.

$$TP_{INF} = S_{PO4} + S_{PI} + X_{PI} + S_{PB} + X_{PB} \qquad (4.6.2\text{-}1)$$

As with the TKN, the division into sub-components is not expressed as readily by defining fractional parameters as in the case for the division of COD. This is for a number of reasons, including (1) an inability to separately measure soluble biodegradable and unbiodegradable components in the effluent and (2) inconsistencies in the models in terms of the linkages between COD and *organic* P. However, it should be recognized that the major portion of the TP is orthophosphate (often as much as 75 percent). Therefore, division of the remainder into four sub-components perhaps is not as critical. A practical approach is to identify two fractions: f_{SPO4} and f_{SPI}:

f_{SPO4} = fraction of the influent TP which is orthophosphate;

f_{SPI} = fraction of the influent TP which is soluble unbiodegradable.

In terms of concentrations:

$$S_{PO4} = f_{SPO4} \cdot TP_{INF} \quad (4.6.2\text{-}2)$$

$$S_{PI} = f_{SPI} \cdot TP_{INF} \quad (4.6.2\text{-}3)$$

The particulate unbiodegradable influent TP, X_{PI}, presumably is associated with the particulate unbiodegradable COD in the influent. Therefore, one means for quantifying this TP portion is to assume a phosphorus content for the particulate unbiodegradable influent COD:

$$f_{XPI} = \frac{X_{PI}}{X_I} \ mgP/mgCOD \quad (4.6.2\text{-}4)$$

f_{XPI} = P content of the influent particulate unbiodegradable COD.

In terms of concentrations, noting that f_{XPI} is not defined as a fraction of the influent TP (other "f" factors are defined as fractions of the total influent COD or TP):

$$X_{PI} = f_{XPI} \cdot X_I = f_{XPI} \cdot f_{XI} \cdot COD_{T,INF} \quad (4.6.2\text{-}5)$$

The biodegradable organic phosphorus (P_{OB}) is given by difference, i.e., the total influent TP less the sum of S_{PO4}, S_{PI} and X_{PI}. The division into soluble (S_{PB}) and particulate (X_{PB}) influent biodegradable organic phosphorus can be expressed through a fraction:

f_{XPB} = fraction of the biodegradable organic TP which is particulate.

In terms of concentrations:

$$P_{OB} = X_{PB} + S_{PB} = TP_{INF} - S_{PO4} - S_{PI} - X_{PI} \quad (4.6.2\text{-}6)$$

$$S_{PB} = (1 - f_{XPB}) \cdot P_{OB} \quad (4.6.2\text{-}7)$$

$$X_{PB} = f_{XPB} \cdot P_{OB} \quad (4.6.2\text{-}8)$$

4.7 Typical Magnitudes for Wastewater Fractions

Table 4-2 lists magnitudes for the various characteristic fractions typically encountered with municipal wastewaters. This listing should be treated with caution. In many instances wastewaters exhibit characteristics very different from the "norm." Situations are encountered where wastewater composition differs substantially from "normal" ranges, and this likely will hold significant consequences.

Table 4-2. Typical Municipal Wastewater Characteristic Parameter Values

Wastewater Characteristic	Concentrations	Concentration Units	Fractions	Fraction Units
Organic Material				
Total COD	250 – 700	g COD m^{-3}	–	–
Readily biodegradable COD	25 – 125	g COD m^{-3}	0.05 – 0.25	g COD / g of total COD
Soluble unbiodegradable COD	20 – 50	g COD m^{-3}	0.04 – 0.16	g COD / g of total COD
Particulate unbiodegradable COD	35 – 110	g COD m^{-3}	0.07 – 0.22	g COD / g of total COD
Slowly biodegradable COD	200 – 400	g COD m^{-3}	0.4 – 0.80	g COD / g of total COD
Nitrogenous Material				
TKN	25 – 70	g N m^{-3}	–	–
Free and saline ammonia	20 – 30	g N m^{-3}	0.50 – 0.75	g N / g TKN
Soluble unbiodegradable TKN	0 – 5	g N m^{-3}	0 – 0.07	g N / g TKN
Biodegradable organically bound TKN	0 – 10	g N m^{-3}	0 – 0.25	g N / g TKN
Particulate unbiodegradable TKN	2 – 8	g N m^{-3}	0.03 – 0.07	g N / g particulate unbiodegradable COD
Phosphorus Material				
TP	4 – 15	g P m^{-3}	–	–
Orthophosphate	2 – 12	g P m^{-3}	0.50 – 0.85	g P / g TP
Soluble unbiodegradable TP	0 – ?	g P m^{-3}	0 – ?	g P / g TP
Biodegradable organically bound TP	0 – 10	g P m^{-3}	0 – 0.25	g P / g TP
Particulate unbiodegradable TP	1 – 4	g P m^{-3}	0.02 – 0.03	g P / g particulate unbiodegradable COD

4.8 Measurement Techniques

Methods for determining influent wastewater characteristics can be divided into two main categories: direct measurements using "standard method" analytical techniques and bioassay methods, where information is derived from the observation of a biological system (e.g., batch, continuous flow or sequencing batch reactor (SBR) activated sludge systems). Table 4-3 provides a summary of measurement or estimation techniques for the various wastewater characteristics.

Table 4-3. Summary of Municipal Wastewater Characteristics and Measurement Techniques

Parameter		Direct	Bioassay
COD	Total	✓	
	Filtered	✓	
	ffCOD	✓	
	f_{SS} (RBCOD)	✓	✓
	f_{SI}	✓	✓
	f_{XI}	✓	✓
TKN	Total	✓	
	Filtered	✓	
	Ammonia-N	✓	
	f_{SNH}	✓	
	f_{SNI}		✓
	f_{XNI}		✓
	f_{XNB}		✓
Nitrate/Nitrite N		✓	
Phosphorus	Total	✓	
	Filtered	✓	
	PO4-P	✓	
	f_{SPO4}	✓	
	f_{SPI}		✓
	f_{XPI}		✓
	f_{XPB}		✓
Solids	VSS	✓	
	TSS	✓	
	ISS	✓	
Alkalinity		✓	
Active Biomass			✓

4.8.1 Direct Methods

"Direct" methods may also include characteristics that can be inferred from direct measurements, such as the estimation of inert suspended solids (ISS) from the difference between the measured total suspended solids (TSS) and volatile suspended solids (VSS).

4.8.2 Bioassay Methods

A number of wastewater characteristics can be derived from observing the response in tests where wastewater is combined with activated sludge. These could be conducted in a number of different ways, and at different scales, for example, batch, continuous-flow or sequencing batch reactor (SBR) systems. Typically, batch tests would be conducted at laboratory-scale, but continuous flow systems could be either laboratory- or full-scale.

An example of a parameter that can be estimated from monitoring of a continuous flow system is the influent unbiodegradable particulate COD fraction, f_{XI}. This parameter is best estimated through measurements of VSS and oxygen utilization rate (OUR) on a system operated at a known sludge age, and with accurate information on influent COD loading. Very often the information available on a full-scale plant (flow rates, solids wastage, etc.) may be inaccurate or incomplete, and hence there will be some uncertainty in the accuracy of estimates. Also, the time-varying influent loading complicates estimation. For these reasons it likely will be necessary to operate a laboratory-scale system which allows for more accurate control of operating parameters (such as sludge age and flows).

In terms of laboratory-scale systems, a fill-and-draw or SBR system has a number of advantages over a continuous flow system: (1) less equipment is required (no pumps, tubing and the like); (2) no specialized equipment is needed (such as glassware or settling tank); (3) the process is simpler to operate (e.g., no pumping rates have to be adjusted); (4) problems such as unintentional denitrification in the clarifier causing loss of solids in the effluent are avoided; and (5) it is possible to control operating sludge age more accurately.

Chapters that follow outline procedures for measuring or estimating the various wastewater characteristics. In a number of cases, the methodology is based on the analysis of data from laboratory-scale experimental systems. For example, the unbiodegradable soluble COD fraction, f_{SI}, could be calculated from results gathered from either a fill-and-draw SBR or a continuous flow system. Procedures for operating continuous-flow and fill-and-draw SBR laboratory-scale systems, as well as the data collection protocols, are outlined in appendices.

4.9 Closing Comments on Colloidal COD

The IWA models (ASM1, ASM2d) do not include a distinction between colloidal and particulate COD. The models assume that:

- Influent slowly biodegradable COD (SBCOD) likely consists of both complex colloidal and fine particulate material.
- Colloidal influent material will be adsorbed onto the sludge matrix, and removed from the liquid phase, essentially instantaneously. Therefore, no distinction is made between colloidal and particulate SBCOD.

Extending from, and related to these assumptions, the models assume no colloidal matter appears in the secondary effluent.

Certain other models make a distinction between colloidal and particulate SBCOD in the structured influent COD characterization. A kinetic expression describing the rate of adsorption of the colloidal material, and rate of removal from the liquid phase, is included in the model. This approach is more complex as it requires specification of one more influent characteristic (as well as kinetic parameters related to the adsorption process). However, it offers two specific modeling advantages:

- The potential of unadsorbed influent colloidal material appearing in the effluent. This would be particularly important in certain types of treatment systems where adsorption of colloidal material likely is incomplete (e.g., contact stabilization, trickling filters, extremely low SRT systems).

- The distinction between colloidal and particulate material lends itself to modeling performance of separation units, particularly those upstream of activated sludge units prior to adsorption occurring. For example, correctly quantifying the COD removal in a primary settling tank based on a percent solids removal demands a distinction between colloidal (non-settleable) and particulate portions.

An aspect that modelers appear not to have considered is whether influent colloidal COD is all biodegradable (i.e., SBCOD), or whether a portion is unbiodegradable. This possibility raises a number of issues for wastewater characterization. These issues are not addressed directly in this document. It would appear that information is incomplete, and it would not be appropriate to propose new approaches or concepts in this pragmatic manual on the current status of modeling. However, the issue is identified as important for further investigation.

There is some merit, or at least interest, in presenting a few samples of data that identify some of the issues. In all cases, the data were from systems treating typical municipal wastewater. Also, in all cases, the data were insufficient for rigorous statistical comparison, and therefore the observations are somewhat speculative. (Also, note that there was considerable day-to-day variability in influent COD concentration.)

Figures 4-8 and 4-9 show daily values of effluent (decant) COD after either glass-fiber filtration (GF) or 0.45 µm membrane filtration (MF). In the first case (Figure 4-8 – a 10-day SRT system), it could be contended that there is not a significant difference between the two sets. Earlier observations in this chapter on influent wastewater filtration clearly demonstrated that glass-fiber filters allow passage of more colloidal material than 0.45 µm membrane filters. Therefore, the data in Figure 4-8 could be interpreted as indicating that the effluent contained little or no colloidal COD; otherwise, one would anticipate a difference. This observation falls within the paradigm of the current models.

The second case (Figure 4-9 – a 15-day SRT system) shows data from start up of a laboratory SBR system where the initial biomass concentration was low and mixed liquor solids increased with time as the system equilibrated. Over the initial period, there is a significant difference between GF and MF effluent COD. An obvious interpretation would be that initially, when solids concentration was low and the system was somewhat unstable, a portion of the influent colloidal COD was not being adsorbed, and appeared in the effluent. Once the system stabilized it appears that the two sets of filtration values are very similar.

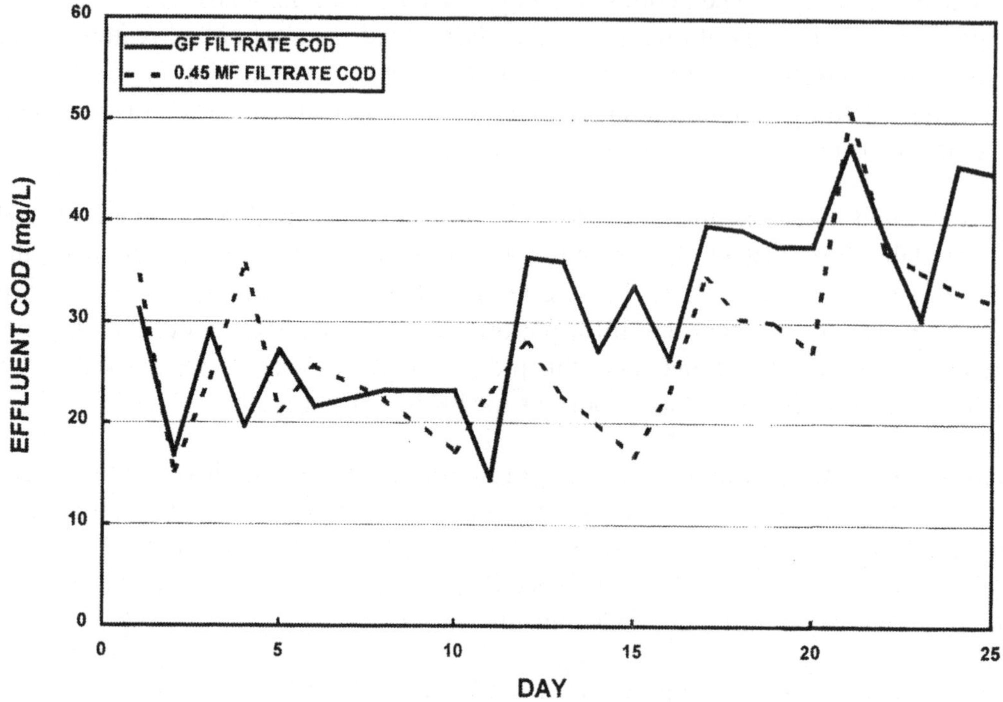

Figure 4-8. Glass-Fiber Filtrate COD, and 0.45 μm Membrane Filtrate COD: Effluent Samples from a 10-day SRT System

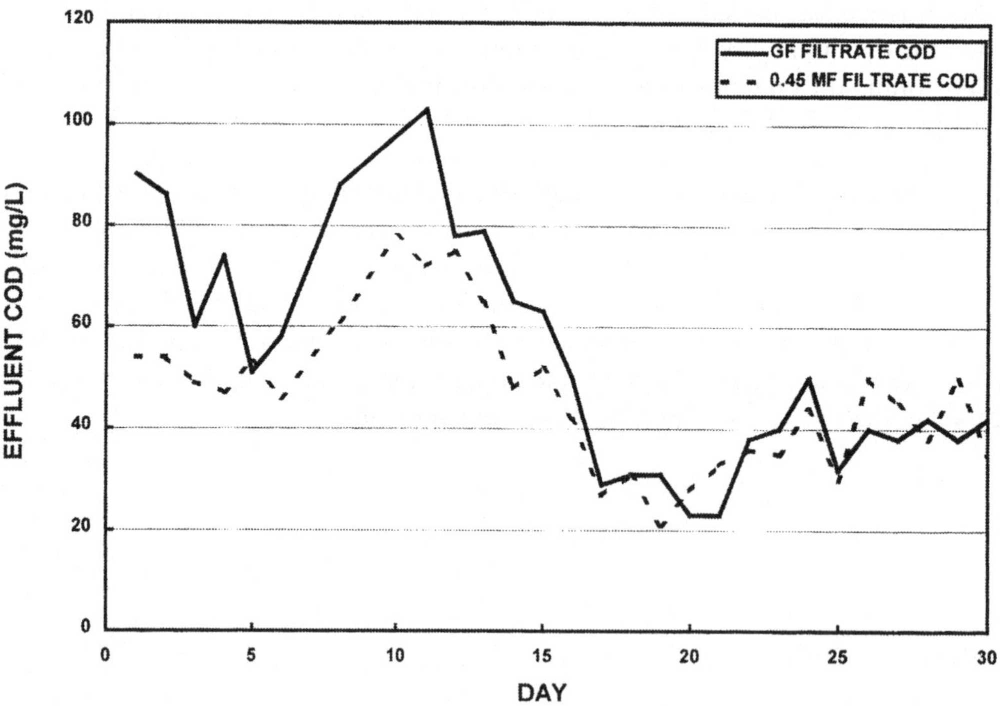

Figure 4-9. Glass-Fiber Filtrate COD, and 0.45 μm Membrane Filtrate COD: Effluent Samples from a 15-day SRT System Initially Exhibiting Unstable Operation

One of the test protocols to be introduced later (ffCOD) involves flocculation of effluent (and influent) samples prior to filtering through 0.45 μm membrane filters. The objective is to remove any residual colloidal matter to determine the concentration of true "soluble" COD. Figure 4-10 shows a comparison of effluent 0.45 μm membrane filtered COD (with and without pre-flocculation). The flocculation step perhaps appears to remove some COD. However, this is not very conclusive from these data, and if so, the amount removed is small. Again this indicates very little effluent colloidal matter.

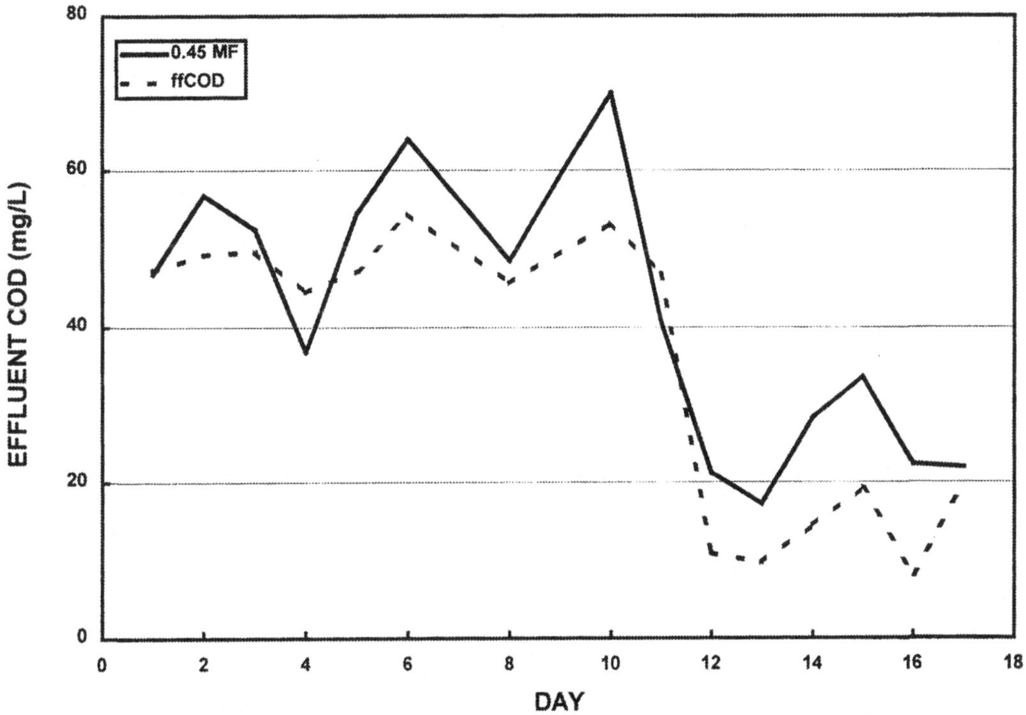

Figure 4-10. 0.45 μm Membrane Filtrate COD and ffCOD: Effluent Samples from a 15-day SRT SBR System

Figure 4-11 presents (very limited) data from analysis of effluents from two SBR systems operated during this study. In both cases, the pre-flocculation step reduces the COD by approximately 40 percent, indicating that the effluents contained significant colloidal COD. This conflicts with data presented in Figures 4-8 to 4-10.

The small amount of data here do not provide any meaningful conclusions. The objective has merely been to illustrate that considerations regarding the fate (and measurement) of colloidal content should be investigated further.

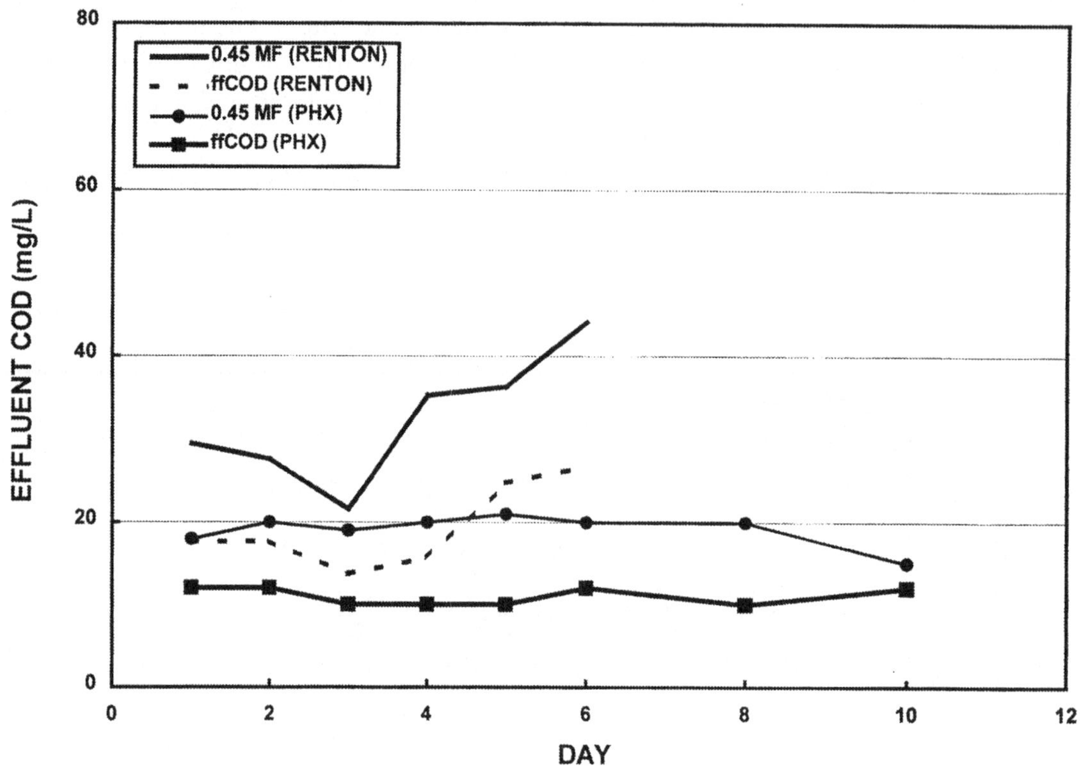

Figure 4-11. 0.45 μm Membrane Filtrate COD and ffCOD: Effluent Samples from Two 15-day SRT SBR Systems Operated during This Study

4.10 References

Barker, P.S., and P. L. Dold. 1997. General model for biological nutrient removal activated sludge systems: model presentation. *Water Environ. Res.* 69(5):969-984.

Chambers, J.M., W.S. Cleveland, B. Kleiner, and P.A. Tukey. 1983. *Graphical methods for data analysis*. Boston, MA: Duxbury Press. 395 pp.

Dold, P.L., G.A. Ekama, and G.v.R. Marais. 1980. A general model for the activated sludge process. *Prog. Water Technol.* 12:47-77.

Dold, P.L., and G.v.R. Marais. 1986. Evaluation of the general activated sludge model proposed by the IAWPRC task group. *Water Sci. Tech.* 18(6):63-89.

Grady, C.P.L. 1989. Dynamic modeling of suspended growth biological wastewater treatment processes. In G. Patry and D. Chapman (eds.) *Dynamic Modeling and Expert Systems in Wastewater Engineering* (pp. 1-38). Chelsea, Michigan: Lewis Publishers. 395 pp.

Kappeler, J., and W. Gujer. 1992. Estimation of kinetic parameters of heterotrophic biomass under aerobic conditions and characterization of wastewater for activated sludge modelling. *Water Sci. Tech.* 25(6):125-139.

Minear, R.A. 1972. Characterization of naturally occurring dissolved organophosphorus compounds. *Envir. Sci. Technol.* 6(5):431-437.

Orhon, D., and E.U. Çokgör. 1997. COD fractionation in wastewater characterization – the state of the art. *J. Chem. Tech. Biotechnol.* 68:283-293.

Roeleveld, P.J., and M.C.M. van Loosdrecht. 2002. Experience with guidelines for wastewater characterization in the Netherlands. *Water Sci. Tech.* 45(6):77-87.

Standard Methods. 1989. *Standard methods for the examination of water and wastewater* (17th ed.). Washington DC: American Public Health Association.

Water Research Commission. 1984. *Theory, design and operation of nutrient removal activated sludge processes*. P.O. Box 824, Pretoria, South Africa: Water Research Commission.

CHAPTER 5.0

SOLUBLE UNBIODEGRADABLE (INERT) COD

5.1 Background

The soluble unbiodegradable (inert) COD in influent wastewater is soluble organic material unaffected by biological action in the system. The magnitude of this component is important in the description of activated sludge system behavior, particularly with respect to effluent COD quality, volatile solids production and oxygen demand. The soluble unbiodegradable COD leaves the system in the effluent and wastage streams. Therefore the influent unbiodegradable COD concentration can be equated to the filtered effluent COD. The assumptions here are that:

- The residual soluble biodegradable material in the effluent is negligible compared to the unbiodegradable material. For example, in systems treating municipal wastewater operated at sludge ages exceeding about 3 days, the amount of biodegradable soluble COD in the effluent is very small (< 2 mg/L).
- There is no generation of soluble unbiodegradable COD within the system.

5.2 Method

The soluble unbiodegradable COD concentration (and fraction) is obtained directly from measured plant data as follows:

$$S_I = Filtered \cdot Effluent\ COD$$

$$f_{SI} = \frac{Filtered \cdot Effluent\ COD}{Unfiltered \cdot Effluent\ COD} \qquad (5.2\text{-}1)$$

5.3 Example

A bench-scale single reactor activated sludge system was operated at a nominal sludge age of 10.6 days with unsettled (raw) municipal wastewater as influent. A typical data set was as follows:

Unfiltered influent COD = 310 *mg/L*
Filtered effluent COD = 41 *mg/L*

$$S_I = 41 \cdot mg/L$$

$$f_{SI} = \frac{41 \cdot mg/L}{310 \cdot mg/L} = 0.13$$

5.4 Considerations

- In most municipal applications, very little colloidal matter is likely to appear in the effluent because it is adsorbed on mixed liquor solids. However, there can be significant colloidal matter in the effluent of some industrial wastewaters. For this reason, and to maintain consistency with the ffCOD method for determining readily biodegradable COD (S_S – see Chapter 7), it is recommended that the effluent be pre-flocculated before filtration when determining S_I. Note, however, that experience shows in most municipal cases that the difference between the effluent 0.45 μm membrane-filtered COD and the effluent ffCOD is less than 5 mg/L.
- The STOWA wastewater characterization guidelines recommend the use of membrane filtration (0.1 μm) to retain all colloidal material on the filter without a pre-flocculation step (Roeleveld and van Loosdrecht, 2002). These guidelines also apply correction factors to account for residual biodegradable organics in the effluent. The S_I concentration is calculated for two cases as follows:

  - (for low loaded WWTPs)

$$S_I = 0.9 \cdot COD_{EFF,FILT}$$

  - (for high loaded WWTPs)

$$S_I = 0.9 \cdot COD_{EFF,FILT} - 1.5 \cdot BOD_{5,EFF}$$

- The f_{SI} fraction will be higher for primary effluent compared to raw influent wastewater because of the disproportionate removal of particulate matter in the primary clarifier.
- The influent COD to a municipal treatment plant usually exhibits a characteristic diurnal variation in concentration. If the soluble unbiodegradable COD is a fixed proportion of the total influent COD, then the effluent soluble unbiodegradable COD concentration also should exhibit a diurnal variation.
- For treatment of municipal wastewaters, it usually is reasonable to assume that there is no generation of unbiodegradable soluble microbial products within the process. With higher-strength industrial wastewaters, residual microbial products may constitute a significant part of the effluent soluble COD. Germirli et al. (1991) describes an experimental procedure for differentiating between soluble COD that originates from the influent and the soluble microbial products. However, it should be noted that none of the ASM-series models include residual soluble microbial product generation.

5.5 References

Roeleveld, P.J., and M.C.M. van Loosdrecht. 2002. Experience with guidelines for wastewater characterization in the Netherlands. *Water Sci. Tech.* 45(6):77-87.

Germili, F., D. Orhon, and N. Artan. 1991. Assessment of the initial inert soluble COD in industrial wastewaters. *Water Sci. Tech.* 23(4/6):1077.

CHAPTER 6.0

PARTICULATE UNBIODEGRADABLE (INERT) COD

6.1 Background

Particulate unbiodegradable COD (X_I) from the influent is enmeshed in sludge mass and accumulates in the system. The magnitude of this component is important in the description of activated sludge system behavior, particularly with respect to volatile solids production and oxygen demand. At steady state, the mass of X_I entering the system in the influent will be balanced by the mass leaving via the sludge wastage stream and in the effluent. From a mass balance, the mass of X_I in the system will equal the influent mass per day multiplied by the system sludge age. Therefore, X_I accounts for an increasing proportion of the volatile solids with increasing sludge age. It is also important to note that X_I does not register as BOD in the influent. As a result, an increasing proportion of X_I as a component of influent COD will increase the COD to BOD ratio in the influent.

6.2 Method

The magnitude of the f_{XI} fraction is most easily estimated using an activated sludge simulation model by matching the observed and predicted mixed liquor volatile suspended solids (VSS) concentrations. An initial value for f_{XI} is assumed, and predictions are compared to observed data. The f_{XI} value is adjusted by trial until good correspondence is obtained between the observed and predicted values. It should be noted that changing the f_{XI} value also impacts predictions of oxygen utilization rate (OUR) and other process variables such as nitrate (in nitrifying plants), and so comparing observed and predicted values for these two parameters provides a useful cross-check on the f_{XI} estimate. For example, an increase in f_{XI} (1) increases VSS due to additional accumulation; (2) decreases carbonaceous OUR because the amount of biodegradable COD is decreased; and (3) increases the amount of nitrogen in the mixed liquor solids (for a given f_N), which decreases the amount of nitrogen available to be nitrified, and in turn decreases the OUR for nitrification. An advantage of the simulation approach is that the OUR and effluent nitrate concentration predictions can readily be compared to observed data together with the VSS prediction; OUR data are not always available for full-scale systems.

Although it is important to use the same model in estimating f_{XI} as will be used for the simulation applications, the steady state Water Research Commission (1984) model is useful for illustration. Assuming a uniform VSS concentration in an activated sludge system (i.e., no step feed), the VSS concentration can be estimated using the following equation:

$$X_{VSS} = \frac{QCOD_{T,INF}\theta_X}{f_{CV}V}\left\{\frac{(1-f_{SI}-f_{XI})Y_{HET}}{1+b_H\theta_X}(1+fb_H\theta_X)+f_{XI}\right\} \qquad (6.2\text{-}1)$$

where:

X_{VSS} = mixed liquor VSS concentration (mg VSS/L)

Q = influent flow rate (L/d)

$COD_{T,\ INF}$ = influent COD concentration (mg/L)

θ_X = sludge age (adjusted for solids loss in the effluent) (d)

V = reactor volume (L)

Y_{HET} = heterotroph yield coefficient
= 0.666 mg COD/mg COD

b_H = steady state theory endogenous decay rate
= $0.24\ (1.029)^{T-20}$ /d

f = steady state theory endogenous residue fraction
= 0.2 mg VSS/mg VSS

f_{CV} = mixed liquor solids COD/VSS ratio
= 1.48 mg COD/mg VSS

f_{XI} = fraction of total influent COD which is particulate unbiodegradable

f_{SI} = fraction of total influent COD which is soluble unbiodegradable

6.3 Example

Unsettled (raw) wastewater from the influent to a municipal WPCP was used to operate a 6.3L single-reactor activated sludge system. The system was operated at a nominal sludge age of 10.6 days, established by wasting a volume of 0.60 L of mixed liquor from the reactor each day. Temperature in the laboratory system was maintained at 20°C. The system was operated for a period of 50 days at a near-constant daily loading ($Q = 15$ L/d, influent COD $\cong 310$ mg COD/L), using five different batches of wastewater. Differences in the wastewater composition from batch to batch (TKN/COD ratio, readily biodegradable COD and so on) were relatively small. Once a steady state was attained, values of the parameters monitored daily were averaged for operation with each batch of wastewater. A typical data set was as follows:

Unfiltered influent COD	= 310 mg/L
Unfiltered influent TKN	= 36.0 mgN/L
Unfiltered effluent COD	= 71 mg/L
Filtered effluent COD	= 41 mg/L
Unfiltered effluent TKN	= 4.2 mgN/L
Filtered effluent TKN	= 2.2 mgN/L
Effluent NO_3-N	= 25.2 mgN/L
Mixed liquor VSS	= 1784 mg/L
Reactor OUR	= 24.1 mg/L/hr

Wastewater characteristics other than the unbiodegradable particulate COD fraction were determined according to procedures outlined in this report; this included the unbiodegradable soluble COD fraction (f_{SI} = 0.13). COD and nitrogen mass balances on the experimental system gave recoveries of 98.8 and 101.5 percent, respectively.

Before attempting to estimate the magnitude of f_{XI} by trial, the exact sludge age of the system was recalculated to account for the small loss of solids in the effluent. This yielded a value of 8.24 days (see Appendix C).

With this sludge age, together with the other operating conditions and measured influent parameters as inputs, a simulator was used to estimate f_{XI} iteratively. No changes were made to the default values of the kinetic and stoichiometric parameters in the model. Table 6-1 below shows a comparison of the observed VSS, OUR and effluent nitrate concentration to the predicted values for different estimates of f_{XI}. The strong influence of f_{XI} on volatile solids production even at this relatively low sludge age is evident. From the simulation results, a value of 0.22 gives a good compromise between the observed and predicted values. It should be noted that this value is higher than that usually encountered for a raw municipal wastewater; a value of 0.11 to 0.13 would be more typical.

Table 6-1. Comparison of Observed and Predicted VSS, OUR and Effluent Nitrate for Different Estimates of Influent Particulate Unbiodegradable COD Fraction

Parameter	Observed Value	Predicted Value for Different f_{XI}			
		0.10	0.15	0.20	0.25
VSS (mg/L)	1,784	1,478	1,613	1,748	1,883
OUR (mgO_2/L/h)	24.1	28.5	27.2	25.8	24.4
NO_3 (mgN/L)	25.2	26.3	25.6	25.0	24.3

Application of equation 6.2-1 (the simplified steady state equation) using an f_{XI} value of 0.22 yields:

$$X_V = \frac{15 \cdot 310 \cdot 8.24}{1.48 \cdot 6.3}\left\{\frac{(1-0.13-0.22)0.666}{1+0.24 \cdot 8.24}(1+0.2 \cdot 0.24 \cdot 8.24)+0.22\right\} \quad (6.3\text{-}1)$$
$$= 1733\ mg/L$$

Figure 6-1 shows the VSS estimated by equation 6.2-1 using the given input conditions over a range of SRTs and at different values of f_{XI}. It is clear from this plot that the influence of f_{XI} on the predicted value of the VSS increases with increasing sludge age. The importance of an accurate measurement of the SRT is also evident. (By inspection of equation 6.2-1, it can also be seen that it is crucial to accurately measure the influent flow rate and the influent COD concentration). The plot suggests that a least squares estimator might be a suitable method for finding f_{XI}. However, it is rarely practical to operate a system to different steady states over a range of different SRTs. Fortunately, the VSS estimate is quite sensitive to the value of f_{XI} at any given SRT. With a simple steady state model such as equation 6.2-1, the value of f_{XI} can be found by solving for it explicitly. A more sophisticated simulator will require an iterative search for the best value of f_{XI}.

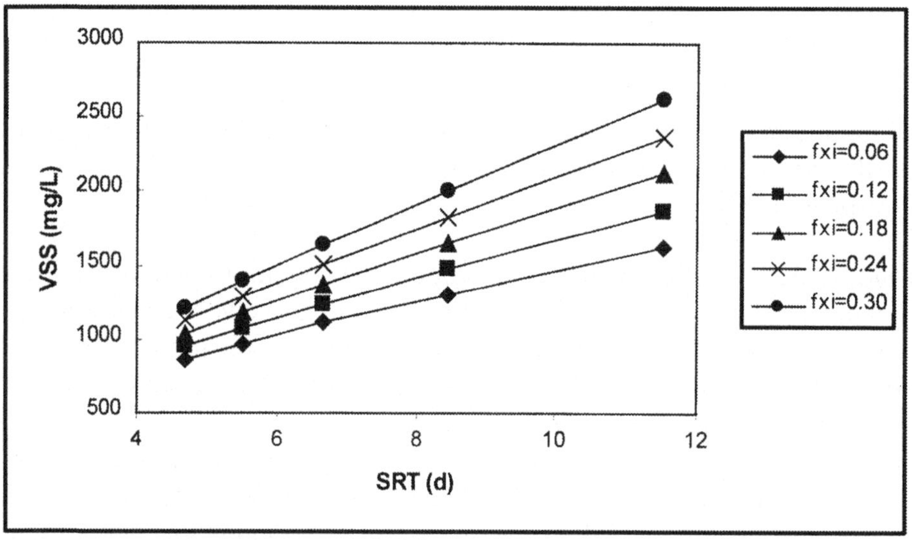

Figure 6-1. Relationship of Bioreactor VSS Concentration to SRT and f_{XI}

6.4 Considerations

- The most important factor in estimating the unbiodegradable particulate COD concentration by the method outlined above is the SRT of the system.
- The method can be applied directly to full-scale plant data. However, the SRT often is not known accurately (Appendix I) and may not be controlled accurately. For these reasons bench-scale systems provide more reliable data for estimating f_{XI} (see Appendices C and D).

- Although both of the above approaches require values for certain kinetic and stoichiometric parameters, these values generally can be considered as relatively constant from site to site for municipal systems.
- The terms in equation 6.2-1 are written for a continuous flow system. If the f_{XI} value is estimated from data for an SBR system, X_{VSS} is the VSS concentration during the mixed phase, Q is the feed volume per day and V is the SBR volume during the mixed phase.
- The f_{XI} fraction will be lower for primary effluent compared to raw influent wastewater because of the disproportionate removal of particulate matter in the primary clarifier.
- The STOWA wastewater characterization guidelines measure or estimate all of the other COD constituents, and then calculate the unbiodegradable particulate portion by difference (Roeleveld and van Loosdrecht, 2002). This is in contrast to the method presented in this manual where the slowly biodegradable COD (SBCOD) portion is calculated by difference (see Chapter 8 for a description of the STOWA method for estimating SBCOD). As discussed in the cited reference for this method, the reliance on a value extrapolated from a BOD measurement is "intrinsically inaccurate," and all of the errors and inaccuracies are reflected in the fraction f_{XI}, which can have a significant impact on modeling results.

When plant data are not available for estimating the influent unbiodegradable particulate COD concentration by the method above, several other parameters may provide an indication of the f_{XI} fraction, or at least an indication whether the value may be non-typical. Also, if data are available on influent wastewater COD, $cBOD_5$ and VSS, the f_{XI} fraction can be estimated using an empirical BOD_5 calculation. These are considered below.

6.5 Influent COD/cBOD$_5$ Ratio as an Indicator for f_{XI}

If data are available for both influent wastewater COD and carbonaceous BOD_5, this allows calculation of the ratio of COD to $cBOD_5$:

$$Influent\ \ COD:BOD\ \ ratio = \frac{Influent\ \ COD}{cBOD_5} \qquad (6.5\text{-}1)$$

For municipal raw influent and primary effluent the ratio typically is in the ranges listed below:

Raw municipal influent	= 2.0 – 2.2 mg COD/mg BOD
Municipal primary effluent	= 1.9 – 2.0 mg COD/mg BOD

The values for the unsettled influent are higher than those for the settled influent. This is to be expected as settling will remove a portion of the unbiodegradable particulate COD.

In municipal wastewater treatment, cases where the ratio is significantly higher than the typical values often are an indication of an unusually high f_{XI} fraction.

6.6 Mixed Liquor VSS/TSS Ratio as an Indicator for f_{XI}

For non-phosphorus removal full-scale systems (i.e., without EBPR or chemical precipitation), if there are accurate data on the influent inorganic suspended solids (ISS) loading, the ratio of mixed liquor VSS to TSS may provide some indication of the influent f_{XI} value.

$$Mixed\ \ liquor\ \ VSS:TSS\ \ ratio = \frac{VSS}{TSS} \qquad (6.6\text{-}1)$$

For systems treating municipal raw influent or primary effluent, the ratio typically is in the region of the values listed below:

Treating raw municipal influent = 0.75 mg VSS/mg TSS

Treating municipal primary effluent = 0.83 mg VSS/mg TSS

In cases where the ratio is significantly higher than the typical values, it often is an indication of an unusually high f_{XI} fraction.

6.7 Influent VSS Concentration as an Indicator of f_{XI}

For systems treating raw municipal influent, the unbiodegradable particulate portion of the influent VSS is typically in the range of 20 percent to 40 percent. This range can be used to cross-check the value of f_{XI} estimated by the methods presented in this chapter against the observed values of the VSS concentration. Table 6-2 illustrates the calculation of the VSS from some typical influent fractions, and the subsequent calculation of the particulate unbiodegradable portion of the VSS.

Table 6-2. Example Calculation of Particulate Unbiodegradable Portion of Influent VSS

Parameter	Concentration (mg/L)	Fraction	Fraction Units
Total COD	400		
Soluble unbiodegradable COD	20	0.05	(mg COD/mg total COD)
Soluble readily biodegradable COD	80	0.20	(mg COD/mg total COD)
Slowly biodegradable COD	248	0.62	(mg COD/mg total COD)
Particulate	186	0.75	(mg COD/mg SBCOD)
Colloidal	62	0.25	(mg COD/mg SBCOD)
Particulate unbiodegradable COD	52	0.13	(mg COD/mg total COD)
VSS*	149		
Particulate unbiodegradable portion of VSS**		0.22	

* VSS = (particulate slowly biodegradable COD + particulate unbiodegradable COD)/1.6

**Particulate unbiodegradable portion of VSS = Particulate unbiodegradable COD/1.6 /VSS

Note: These calculations assume a COD/VSS ratio of 1.6

6.8 Estimation of f_{XI} from Influent COD, $cBOD_5$ and VSS Concentrations

In many cases, historical plant data on organic composition are available only for influent COD, $cBOD_5$ and VSS. Usually it is relatively simple to estimate the soluble COD components (S_S and S_I). The unbiodegradable particulate COD content (i.e., X_I or f_{XI}) can be estimated using these data, based on assuming a value for the COD/VSS ratio of the influent solids. The step-wise iterative calculation procedure is outlined below.

Note: The level of confidence in the estimation approach outlined below can be increased by making a limited number of measurements of the influent particulate material COD/VSS ratio using the methods outlined in Chapter 12.0.

Also, if only influent BOD_5 and VSS data are available, then a limited number of analyses of both BOD_5 and COD (say, 20 on each) on split samples of influent wastewater should provide a reasonable estimate of the COD/BOD_5 ratio and this can be used to approximate COD concentrations from historical BOD_5 data.

Step 1: Assume a COD/VSS ratio for the influent *particulate* matter. Typically this is in the range of 1.6 to 1.8 mg COD / mg VSS. Note that this is not the total COD; it is the COD of the influent solids (see Chapter 12.0). Then:

$$(X_I + X_S) = \left(\frac{COD}{VSS}\right)_{ASSUMED} \cdot VSS \qquad (6.8\text{-}1)$$

Step 2: Calculate the influent colloidal SBCOD concentration, S_{COL}:

$$S_{COL} = COD_T - S_S - S_I - (X_I + X_S) \qquad (6.8\text{-}2)$$

Step 3: Assume a value for X_I. Then:

$$X_S = (X_I + X_S) - X_I \qquad (6.8\text{-}3)$$

At this point, all influent COD component concentrations are "known" – S_S, S_I, S_{COL}, X_S and X_I.

Step 4: Apply the empirical method outlined in Appendix B for calculating $cBOD_5$ based on these influent COD component concentrations.

Step 5: If the calculated $cBOD_5$ does not match the observed value, return to Step 3 and adjust the X_I estimate. Continue until the calculated $cBOD_5$ matches the observed value.

This procedure also allows calculation of the particulate/colloidal partition of the SBCOD (i.e., f_{XS}):

$$f_{XS} = \frac{X_S}{X_S + S_{COL}} \qquad (6.8\text{-}4)$$

Note that:

- It is assumed that the COD/VSS ratio for unbiodegradable particulate COD (X_I) and particulate SBCOD (X_S) are equal. A small adjustment to the equations would be necessary if different COD/VSS values are assumed for these two components.

- Some adjustment to the equations would be required in cases where the influent wastewater contains a substantial biomass COD contribution. The COD/VSS ratio for biomass (e.g., 1.42 mg COD/mg VSS) likely is lower than the ratio for X_S and X_I.

- The method does not account for the possibility that the influent wastewater contains unbiodegradable colloidal COD.

Note: An alternative approach for estimating the X_I content, as well as other influent fractions, is to examine data on raw influent and primary settling tank (PST) overflow composition. Specifying the raw influent COD composition (allowing calculation of all settleable and non-settleable COD concentrations, as well as $cBOD_5$ and VSS) and the percent solids removal in the PST enables calculation of COD component concentrations, $cBOD_5$ and VSS in the PST overflow. The raw wastewater composition can be adjusted until the estimated input and overflow total COD, $cBOD_5$ and VSS concentrations match the observed values reasonably.

6.9 References

Roeleveld, P.J., and M.C.M. van Loosdrecht. 2002. Experience with guidelines for wastewater characterization in the Netherlands. *Water Sci. Tech.* 45(6):77-87.

Water Research Commission. 1984. *Theory, design and operation of nutrient removal activated sludge processes*. P. O. Box 824, Pretoria, South Africa: Water Research Commission.

CHAPTER 7.0

READILY BIODEGRADABLE COD

7.1 Background

Ekama et al. (1986) outlined three experimental methods for estimating the readily biodegradable COD (S_S) fraction in an influent wastewater. Of these, the aerobic batch method probably is the most practical as it has the least requirement for analytical/laboratory facilities and lends itself to on-site application at wastewater treatment plants. A similar test also can be conducted using an SBR. Mamais et al. (1993) proposed a physical/chemical method for determining the S_S that involves a flocculation step prior to filtration to remove colloidal COD. Dold et al. (1986) developed and tested a physical method based on ultrafiltration.

This chapter outlines the bioassay or oxygen uptake rate (OUR)-based respirometric procedures and the physical/chemical methods for determining the RBCOD fraction in an influent wastewater.

7.2 OUR-Based Respirometric Methods

7.2.1 Aerobic Batch Method

The following equipment is required:

- a vessel such as a 4L glass flask
- either a mechanical mixer or a magnetic stirrer for maintaining the reactor contents uniformly mixed
- an air supply (compressed air cylinder, small air compressor or a fish-tank aerator)
- a dissolved oxygen (DO) meter with rapid response probe
- a means of recording the meter output (e.g., a strip chart recorder)

In the test, a measured volume of wastewater (V_{WW}) of known total COD (COD_T) is combined with a measured volume of activated sludge mixed liquor (V_{SL}) in the batch vessel. Mixing intensity in the reactor when the air supply is off should be such that the contents are maintained well-mixed and the DO probe gives a stable reading, but not so intense that the liquid surface is turbulent; this would allow significant entrainment of air which would cause errors in oxygen utilization rate (OUR) measurements. From immediately after the start of the test OUR measurements are made at intervals of a few minutes; measurement continues for a period of approximately 1 to 2 hours, as appropriate. Each OUR is measured as follows:

- The contents of the reactor are aerated rapidly to raise the oxygen level to approximately 6 to 8 mgO_2/L. Aeration is then terminated and the ensuing decrease in DO concentration with time is recorded until the DO has reduced to approximately 3 mgO_2/L (or until the trace of DO versus time allows the linear decrease in DO with time to be measured accurately). The OUR is obtained from the slope of the linear section DO response – that is, the decrease in DO over a measured time interval. It is important that the plot of DO show a clear linear decrease to allow accurate measurement of the OUR. Curvature in the plot indicates either transfer of oxygen at the liquid surface or very poor calibration of the DO meter.

An idealized example of an OUR response observed in an aerobic batch test with a municipal wastewater (or in an SBR during the react phase – see later) is shown in Figure 7-1. The OUR response is the sum of a number of distinct phenomena (see discussion at end of this chapter).

The area under the OUR curve is divided into four sections: area 1 indicates the mass of oxygen utilized for oxidation of RBCOD; area 2 indicates the mass utilized for nitrification; area 3 denotes the mass consumed for oxidation of slowly biodegradable COD (rate limited by hydrolysis of the enmeshed COD). The remaining area under the OUR curve indicates the oxygen associated with endogenous respiration.

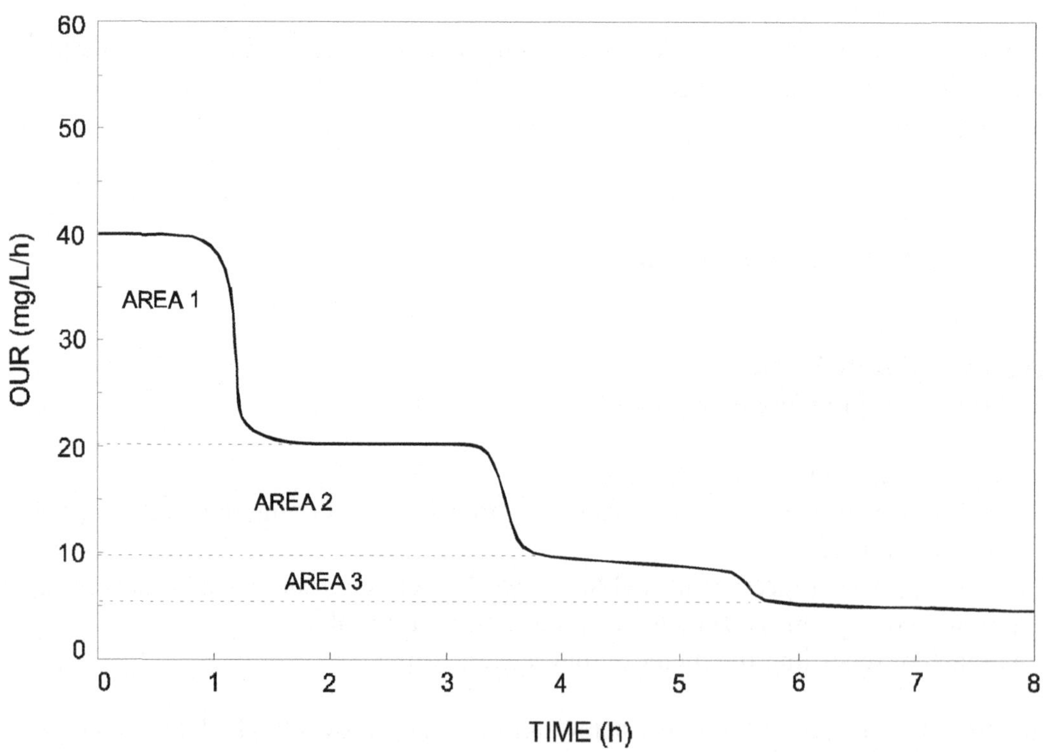

Figure 7-1. Idealized Oxygen Utilization Rate Response Observed in an Aerobic Batch Test (or in a Sequencing Batch Reactor during the React Phase)

The RBCOD concentration is estimated based on measuring the area under the OUR curve (Area 1 shown in Figure 7-1), and applying equation 7.2.1-1 below:

$$S_S = \frac{M_O}{1-Y_{HET}} \cdot \frac{V_{SL}+V_{WW}}{V_{WW}}$$
$$f_{SS} = \frac{S_S}{COD_T} \qquad (7.2.2\text{-}1)$$

where:

M_O = mass of oxygen consumed per liter batch test volume in utilization of S_S (mgO_2/L) (Area 1 in Figure 7-1)

Y_{HET} = heterotrophic yield coefficient (0.666 mg cell COD/mg substrate COD)

V_{SL} = volume of activated sludge mixed liquor used in test (L)

V_{WW} = volume of wastewater of COD concentration S_T used in test (L)

The ratio of wastewater to mixed liquor used in the batch test can be interpreted as a food to microorganism ratio (F/M). Changing the F/M should not change the estimate of f_{SS}; area 1 is determined by the mass of readily biodegradable COD in the wastewater. However, changing F/M will influence the shape of the OUR response. Selection of the quantities of wastewater and mixed liquor combined in the test should be such that the response enables accurate estimation of the mass of oxygen consumed for utilization of the readily biodegradable COD. If F/M is too low, then the rate of utilization will be high and the S_S will be utilized very rapidly; this will result in a tall, narrow response. On the other hand, if F/M is too high, the response will be very flat, making it difficult to identify accurately the drop in OUR. In either case, estimation of the correct area becomes difficult. Selection of a suitable F/M is influenced by a number of factors, including: (1) the active organism concentration in the mixed liquor used in the test is a function of sludge age (and other factors), and will vary from one situation to another and (2) the maximum specific growth rate of the heterotrophs can differ substantially between sludges. Therefore there is no definitive guideline as to the correct F/M for all situations; the appropriate ratio of wastewater to mixed liquor should be determined by trial. Perhaps a ratio which gives an F/M of approximately 0.6 mg COD/mg VSS can be used as a starting point.

7.2.2 SBR Method

In the aerobic batch method described above, a measured volume of wastewater (V_{WW}) is combined with a measured volume of activated sludge mixed liquor (V_{SL}) in an aerated batch reactor. In the bench-scale SBR operation (see Appendix D) the same procedure is followed at the start of the react period; that is, the instantaneous addition of the V_{WW} volume of influent wastewater to the SBR containing a volume of (V_P - V_{WW}) mixed liquor. Estimation of the readily biodegradable COD concentration (S_S) can be achieved through analysis of the OUR response measured in the reactor over the first part of the react period. From the start of the react period, the OUR in the reactor should follow the type of pattern shown in Figure 7-1 (described in the previous section). The readily biodegradable COD concentration in the influent wastewater can be calculated using the same method as for the aerobic batch method:

$$S_S = \frac{M_O}{1-Y_H} \cdot \frac{V_P}{V_{WW}} \qquad (7.2.1\text{-}2)$$

where:

V_P = volume of filled SBR (L)

V_{WW} = volume of wastewater added to the SBR during the feed phase (L)

Two examples of typical OUR responses obtained from SBR studies are shown below. Figure 7-2 shows the OUR response observed in an SBR receiving a significant concentration of RBCOD. Figure 7-3 shows the response of a system receiving a moderate concentration of RBCOD, but with a low nitrifier growth rate. In the second case nitrification is complete only after 10 hours.

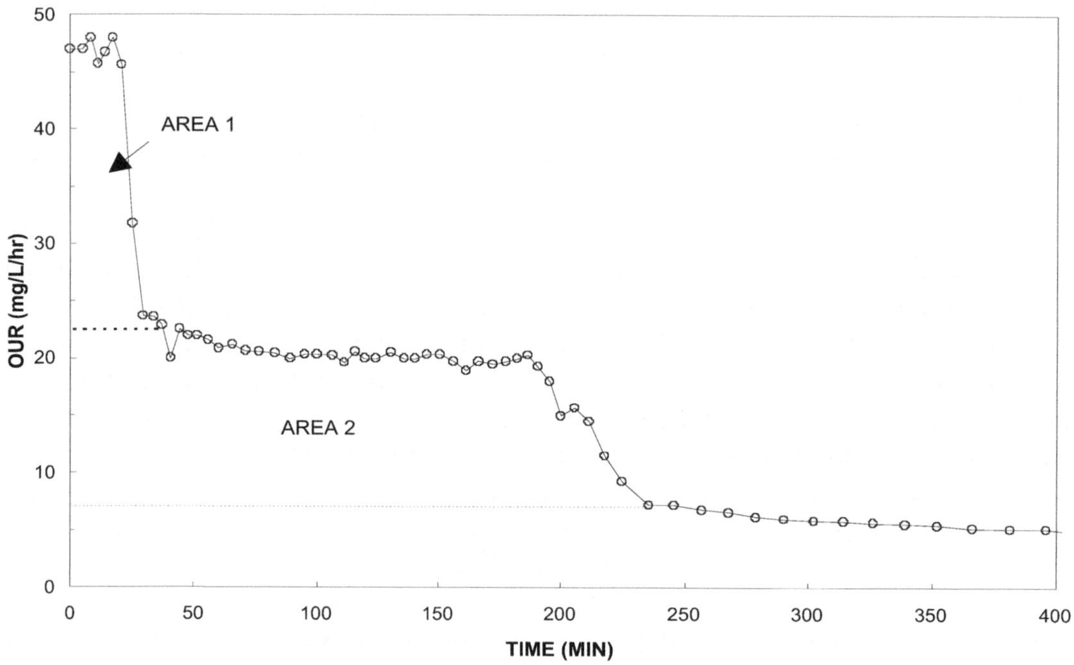

Figure 7-2. Example of an Oxygen Utilization Rate Response Observed during the React Period in an SBR Receiving a High Concentration of RBCOD (Garman et al., 1996)

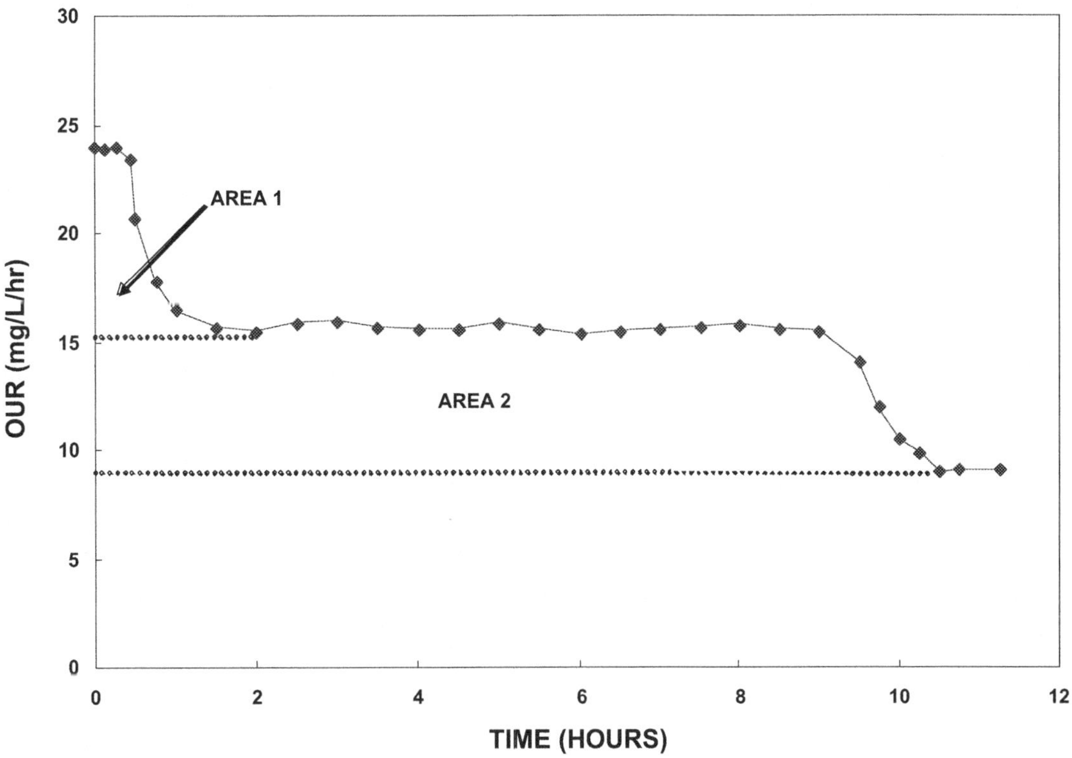

Figure 7-3. Example of an Oxygen Utilization Rate Response Observed during the React Period in an SBR with a Low Nitrifier Growth Rate

7.2.3 Automated Respirometric Methods

All of the bioassay methods mentioned above utilize respirometry, defined as "the measurement and interpretation of the biological oxygen consumption rate under well-defined experimental conditions" (Spanjers et al., 1998). A number of different respirometric techniques are available (Spanjers et al., 1998), including some that have been automated and applied online (Gernaey et al., 1999). The reader is referred to the quoted references for detailed descriptions.

7.2.4 Batch Test Without (External) Seed

Wentzel et al. (1999) describe a batch test procedure for estimation of S_S that does not require an activated sludge seed source because it relies on the heterotrophic organisms present in the influent. The method for determining the influent active organism concentration is summarized in Chapter 9.0. However, determination of S_S is dependent on the use of a specific model for the activated sludge process, and the method will not be described here. The reader is referred to the references for this procedure (Wentzel et al., 1999; Wentzel et al., 1995).

7.3 OUR Response in Batch Test

The basis of the batch method for estimating readily biodegradable COD concentration (S_S) is that the oxygen consumption associated with utilization of all of the S_S can be isolated and measured. Accepting that there is a fixed oxygen requirement per unit COD utilized ($1 - Y_{HET}$), it is possible to estimate the amount of S_S at the start of the test from the associated amount of oxygen consumed. The response of OUR under the conditions of the test and the means for isolating the oxygen consumption associated with S_S uptake can be anticipated from consideration of the model structure with respect to kinetics of substrate removal (and other oxygen-consuming processes). Over the course of the test a number of components contribute to the observed total OUR (i.e., the measured OUR).

7.3.1 OUR for Readily Biodegradable COD Utilization

Removal of this material is governed by a saturation-type (Monod) relation between growth rate and readily biodegradable COD concentration of the form shown in Figure 7-4. At the start of the test, the concentration of S_S will be high (say, at point A), sufficient to ensure that the rate of utilization is at the maximum. Utilization will continue at the maximum (*constant*) rate until the concentration has been reduced to that approximately corresponding to point B. Thereafter there will be a decline in rate as the remaining small amount of S_S is consumed. The OUR associated with utilization of the S_S is directly related to the rate of utilization. Therefore this component of the OUR will be constant for a period from the start of the test (A to B), after which the rate will drop to zero rapidly. The test is designed to continue for a period after all of the S_S has been utilized.

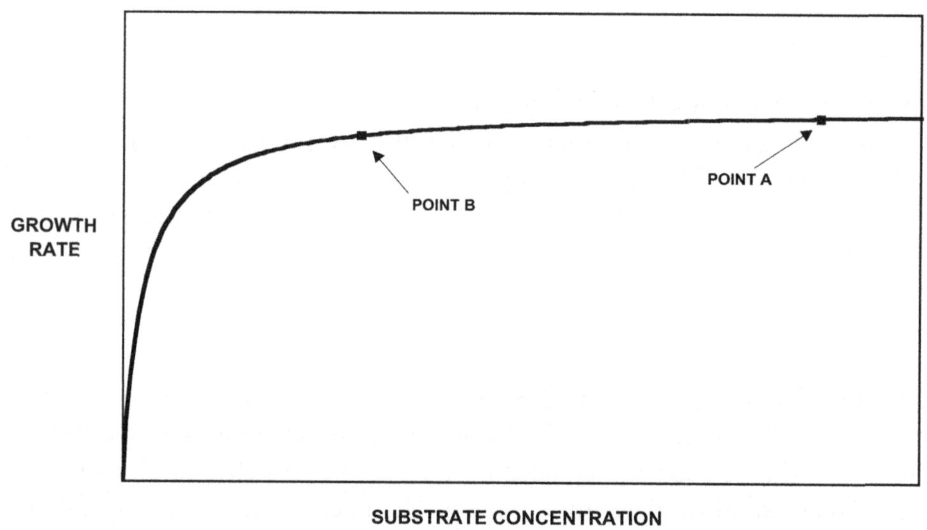

Figure 7-4. Monod Relationship between Organism Growth Rate and Readily Biodegradable COD Concentration (S_S)

7.3.2 OUR Associated with Slowly Biodegradable COD Removal

The rate of removal of the slowly biodegradable COD (X_S and S_{COL}) is governed by the relatively slow rate of hydrolysis; this process is also modeled by a saturation function. At the high sludge loading, the rate of hydrolysis will be at the maximum at the start of the test. Because the rate of utilization of S_S is rapid, and because the amount of S_S usually is significantly less than the amount of SBCOD at the start of the test, the rate of hydrolysis will continue at the maximum (*constant*) rate even after all of the S_S has been removed. Therefore, for the duration of the test the OUR associated with utilization of the products of hydrolysis will be constant.

7.3.3 OUR for Endogenous Respiration

The rate of organism decay is modeled as a first order process with respect to the concentration of active organisms. Over the duration of the batch test, the change in active organism concentration will be small. Therefore the rate of oxygen utilization associated with decay (endogenous respiration) essentially will be constant.

7.3.4 OUR for Nitrification

If nitrification occurs during the test there will be an associated OUR. The kinetics of this process also are described by a Monod-type function with the rate dependent on the ammonia concentration. For the conditions of the test (initial ammonia concentration and nitrifier organism mass) it is likely that nitrification will continue at the maximum (*constant*) rate after the S_S has been utilized. Therefore the OUR for nitrification will be constant during the test. However, if the ammonia concentration is reduced to less than about 2 mgN/L before the removal of S_S is complete, then the associated drop in OUR may be exaggerated or masked by a decrease in OUR for nitrification. In this circumstance nitrification should be suppressed in the test. This can be achieved through the addition of allyl thiourea (20 mg/L) to the mixed liquor prior to commencing the test.

The OUR observed in the batch test reflects the sum of the components described above, as shown in Figure 7-5. Adding the components it is evident that the OUR will be near constant initially. After a period there will be a drop in OUR corresponding to complete utilization of the S_S. Thereafter, the OUR will continue at a second constant rate.

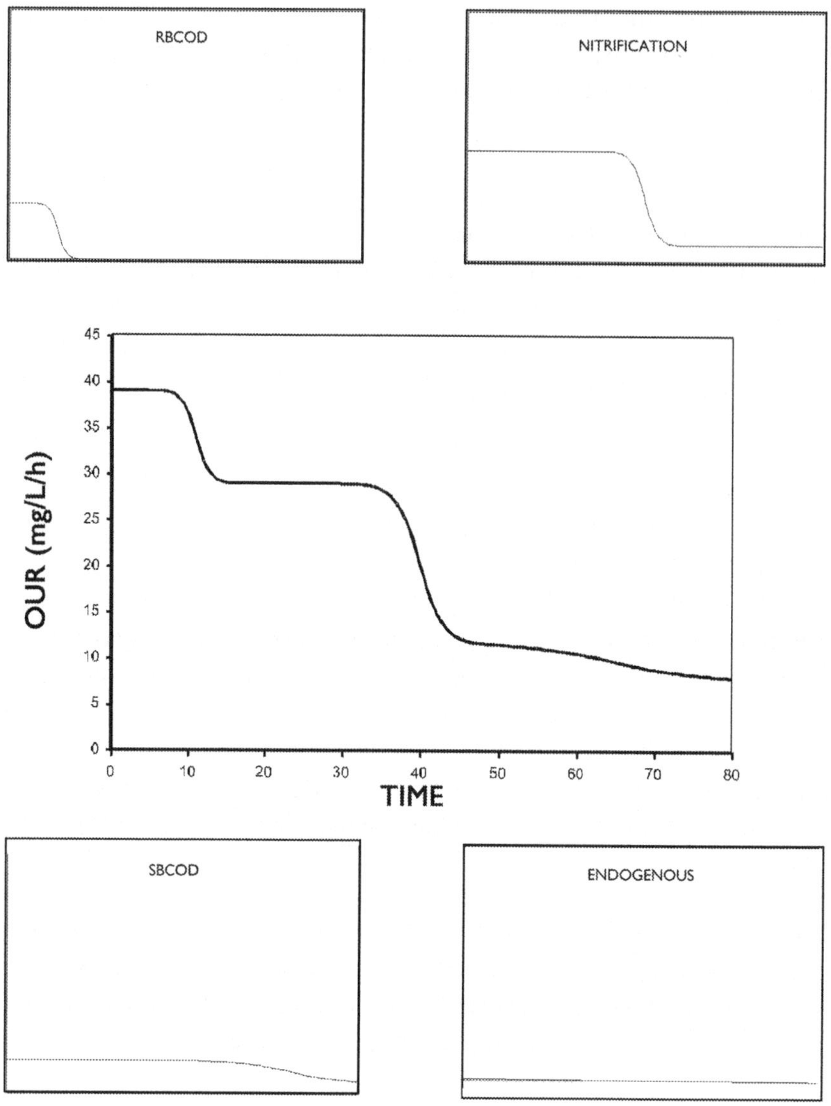

Figure 7-5. Components Contributing to the Measured OUR in the Batch Test

7.4 Physicochemical Methods

The readily biodegradable COD (S_S) in municipal wastewaters is presumed to consist of relatively small molecules (such as volatile fatty acids and low molecular weight carbohydrates) which can be easily transported into the cell, while the slowly biodegradable COD is assumed to consist of larger and more complex colloidal and particulate material which requires extracellular breakdown prior to uptake and utilization (Dold et al., 1980). The inferred parallel between the biokinetic division and the division based on physical characteristics appears to offer a basis for measuring the S_S without having to operate a biological system - namely, by physical separation and COD measurement. Note that this parallel will not necessarily be true for industrial wastewaters that may have a large number of soluble compounds with a wide range of biodegradation rates. In these cases, the aerobic batch reactor and the SBR methods described in this chapter are the most useful.

The obvious separation method is filtration. However, standard filtration applied in wastewater treatment (i.e., 0.45μm membrane filter or 1.5μm glass-fiber filter used for solids) does not remove all of the colloidal material from the influent wastewater. That is, the "soluble" COD in municipal wastewater consists of (see Figure 4-2):

- Readily biodegradable COD (S_S)
- Unbiodegradable soluble COD (S_I)
- A portion of the colloidal (slowly biodegradable) COD

Two approaches have been proposed for isolating the RBCOD (S_S) by physicochemical means:

- ffCOD (or flocCODsol) method of Mamais et al. (1993): The concept with this method is that the colloidal material can be removed from the liquid phase by flocculation prior to filtration.
- Ultrafiltration method of Dold et al. (1986): The concept with this method is that the S_S comprises low molecular weight (MW) compounds, and can be separated using a low MW cut-off ultrafiltration membrane.

A physical method for S_S has been applied in Europe based on guidelines developed by the Dutch Foundation for Applied Water Research (STOWA). This approach simplifies the ffCOD method by eliminating the flocculation step and using membrane filtration with a smaller pore size (i.e., 0.1 μm) (Roeleveld and van Loosdrecht, 2002).

Each of these methods, when applied to an influent wastewater, also measures the soluble unbiodegradable COD (S_I). Therefore, to isolate the S_S, the method must be applied to the secondary effluent, and S_S is determined by difference:

$$\begin{aligned} S_S &= Influent\ physchem.\ COD\ (S_S + S_I) \\ &\quad - Effluent\ physchem.\ COD\ (S_I) \end{aligned} \qquad (7.4\text{-}1)$$

Potential problems are apparent in the application of the physicochemical methods for measuring the S_S portion of a municipal wastewater:

1. The amount of the readily biodegradable COD remaining in the effluent should be negligible. Therefore, the methods should only be applied to systems with sludge ages in excess of about 3 days so that this assumption is valid.
2. The assertion that the effluent filtrate COD (representing the unbiodegradable soluble portion of the effluent) is near equal to the influent S_I assumes that the generation of inert soluble material within the process is negligible.
3. The soluble/colloidal material in a municipal wastewater must span a wide range of molecular sizes and weights. A problem with the ultrafiltration method is to select the appropriate MW cut-off membrane that will allow passage of small-molecule readily biodegradable COD but which will retain colloidal molecules.

7.4.1 Flocculated and Filtered COD (ffCOD) Method

Mamais et al. (1993) proposed a direct method for estimating the influent readily biodegradable COD concentration (S_S) as an alternative to the more time-consuming biological assay methods described above. The method is based on a physical separation, which involves pre-flocculation of the sample followed by filtration (referred to as the flocCODsol test or "ffCOD"). It is assumed that the flocculation step will remove the colloidal material, resulting in a filtrate that contains only "truly soluble" material. The procedure is outlined briefly below:

- 1 mL of 100 g/L zinc sulfate solution is added to 100 mL of wastewater;
- The sample is then mixed vigorously for approximately 1 minute;
- The sample pH is adjusted to approximately 10.5 using 6 M sodium hydroxide solution;
- The sample then is allowed to settle, and a sample of the supernatant is withdrawn;
- The supernatant sample is filtered using a 0.45 μm membrane filter, and the filtrate COD is analyzed.

The unbiodegradable soluble COD must be subtracted to obtain the influent readily biodegradable COD, i.e., the RBCOD is estimated by determining the flocCODsol of the influent, and subtracting from it the flocCODsol of the effluent. For a more detailed outline of the procedure the reader is referred to Mamais et al. (1993). Note that it is important to use zinc sulfate as the flocculant as outlined in this procedure. In the studies by Narkis et al. (1980), where lime and ferric chloride were used as flocculants, removals of volatile acids up to 20 percent from raw wastewater samples were observed. However, zinc sulfate is less adsorptive than these other flocculants, so the removals of readily degradable material in the ffCOD test should be much less significant.

$$S_S = Influent\ ffCOD - Effluent\ ffCOD$$

$$f_{SS} = \frac{S_S}{Influent\ Unfiltered\ COD} \qquad (7.4.1\text{-}1)$$

7.4.2 Ultrafiltration Method

If the readily biodegradable material is in fact made up of small molecular units which are transported into the cell rapidly, then separation (by ultrafiltration) of the low molecular weight COD from an influent wastewater sample should enable direct measurement of the S_S fraction.

Ultrafiltration membranes are susceptible to blinding by colloidal and particulate matter. Therefore, prior to filtering wastewater samples it is necessary to follow a pre-treatment procedure to remove as much of this material as possible. Dold and Bagg (1986) proposed the following procedure.

Note: This step may not be necessary; particulate material (say, > 1μm) should be removed by filtering the sample through a glass microfiber filter. However, while this method can be used, it was found that these filters blocked very rapidly when filtering raw wastewater samples and several 47 mm diameter filters were required to process 100mL of sample. Centrifugation thus reduced the cost of the analysis.

---

1. Centrifuge the sample to remove a large fraction of the particulate matter, and decant the supernatant. In this study samples were centrifuged at 6500 rpm for 30 minutes.
2. Both influent and effluent samples are subjected to two stages of pre-filtration after centrifugation. First, the samples are filtered under vacuum through two successive glass

microfiber (e.g., Whatman GF/C) membranes; then the samples are passed through a 0.45μm membrane filter. This step ensures removal of (1) residual particulate matter and (2) colloidal matter, the presence of which often is indicated by a "milky" color in certain wastewater samples (see comments on filter blinding below).

3. Finally the filtrates from step 2 are passed through an ultrafiltration membrane in a stirred, static cell (e.g., Amicon Corp. 50 mL cell) under pressure of nitrogen gas.

The RBCOD concentration is calculated as follows:

$$S_S = \textit{Influent Ultrafiltrate COD} - \textit{Effluent Ultrafiltrate COD}$$

$$f_{SS} = \frac{S_S}{\textit{Influent Unfiltered COD}} \qquad (7.4.2\text{-}1)$$

An important consideration with the filtration method for RBCOD determination is selection of the appropriate filter. Although information is limited, it appears that there is considerable flexibility in the choice of membrane. For example, a molecular mass cut-off of say, 1,000, 10,000 or even 100,000 probably would be equally suitable for isolating the readily biodegradable influent COD fraction in this wastewater. As proposed, filtration of influent and effluent samples with the same cut-off membrane should yield a close estimate of RBCOD by difference (see Chapter 39.0 for more detail).

7.5 Considerations

- Because they directly measure the impact of the material on the process, bioassay tests conceptually are the best for measuring the readily biodegradable COD concentration for process modeling. However, bioassay methods require more sophisticated equipment and expertise than physicochemical methods. The physicochemical methods have the advantage of being simpler to perform.
- Of the physicochemical methods, the ffCOD test is the easiest to use. The ultrafiltration method is more labor intensive and time consuming due to the repetitive treatment of the sample via centrifugation and double filtration before membrane filtration at high pressure.
- There has not been sufficient rigorous comparison between methods to conclude that the ffCOD test and the bioassay tests for S_S provide equivalent results. The original reference for the ffCOD test (Mamais et al.,1993) provided a limited amount of data to demonstrate that the results were equivalent to a bioassay test. Preliminary testing conducted in this study and in others showed S_S values measured with a bioassay method were consistently lower than those measured by the flocculation/filtration method (see Chapter 40.0).
- In most municipal treatment plants operated at sludge ages greater than about 3 days, the colloidal material in the effluent is low because it tends to adsorb to the mixed liquor solids. If 0.45μm membrane filtration is applied to an effluent without pre-flocculation, the effluent COD typically will be only slightly higher (i.e., < 5 mg/L difference) than if pre-flocculation is applied. This may not be the case with industrial effluents such as meat processing wastes where the concentrations of effluent colloidal material can be significant. In these cases, flocculation of the effluent is essential to estimate S_S.

- The principal experimental problem encountered by Dold et al. (1986) in the investigation of the ultrafiltration procedure was one of filter blocking and blinding encountered with the ultrafiltration membranes and, at times, with the 0.45μm membranes even when the major portion of the particulate material was removed by centrifugation. Three indications of filter blinding were apparent: (1) rapidly declining filtration rates; (2) the formation of a brown slime layer (gel) on the filtration membrane; and (3) unexpectedly low filtrate COD values indicating retention of a portion of the soluble COD by the membranes. The pre-filtration procedure was developed by trial to overcome this problem.

7.6 References

Dold, P.L., G.A. Ekama, and G.v.R. Marais. 1980. A general model for the activated sludge process. *Prog. Water Technol.* 12:47-77.

Dold, P.L., W.K. Bagg, and G.v.R. Marais. 1986. *Measurement of the readily biodegradable COD fraction (S_{bs}) in municipal wastewater by ultrafiltration*. Research Report No.W57, Cape Town, South Africa: University of Cape Town.

Ekama, G.A., P.L. Dold, and G.v.R. Marais. 1986. Procedures for determining influent COD fractions and the maximum specific growth rate of heterotrophs in activated sludge systems. *Water Sci. Tech.* 18(6):91-114.

Garman, K.R., M.J. Tetreault, P.L. Dold, D.S. Parker, and H.R. Finley. 1996. Evaluation of critical activated sludge parameters for dynamic process models. In *Proc. 69th Annual Water Environment Federation Conference,* Dallas, TX. Arlington VA: Water Environment Federation.

Gernaey, K., B. Petersen, J.-P. Ottoy, and P. Vanrolleghem. 1999. Biosensing activated sludge. *Water Quality International*, June, 16-21.

Mamais, D., D. Jenkins, and P. Pitt. 1993. A rapid physical-chemical method for the determination of readily biodegradable soluble COD in municipal wastewater. *Water Res.* 27(1):195-197.

Narkis, N., S. Henefeld-Fourrier, and M. Rebhon. 1980. Volatile organic acids in raw wastewater and in physico-chemical treatment. *Water Res.* 14:1215-1224.

Roeleveld, P.J., and M.C.M. van Loosdrecht. 2002. Experience with guidelines for wastewater characterization in the Netherlands. *Water Sci. Tech.* 45(6):77-87.

Spanjers, H., P.Vanrolleghem, G. Olsson, and P.L. Dold. 1998. Respirometry in control of the activated sludge process: principles. *IAWQ Scientific and Technical Report No. 7*, London, UK: International Water Association.

Ubisi, M.F., T.W. Jood, M.C. Wentzel, and G.A. Ekama. 1997. Activated sludge mixed liquor heterotrophic active biomass. *Water SA* 23(3):239-248.

Wentzel, M.C., A. Mbewe, and G.A. Ekama. 1995. Batch test for measurement of readily biodegradable COD and active organism concentrations in municipal Wastewaters. *Water SA* 21(2):117-124.

Wentzel, M.C., A. Mbewe, M.T. Lakay, and G.A. Ekama. 1999. A test for characterization of the carbonaceous materials in municipal wastewaters. *Water SA* 25(3):327-335.

CHAPTER 8.0

SLOWLY BIODEGRADABLE COD

8.1 Background

The slowly biodegradable COD (SBCOD) is made up of complex organic material that requires extracellular enzymatic breakdown prior to absorption and utilization by the organisms. On a physical basis, this material likely comprises fine particulate matter, colloidal matter and perhaps some large complex organic molecules that can be considered soluble. However, for the purposes of modeling the treatment of municipal wastewaters, the slowly degradable material is considered to be particulate and colloidal. Note that industrial wastewaters may contain a number of soluble slowly biodegradable components. This issue is discussed in the specific chapters dedicated to characterization and modeling of industrial wastewaters.

When colloidal slowly biodegradable material is combined with activated sludge, the colloidal matter is likely adsorbed onto the floc rapidly, and removed from the liquid phase. Therefore, within the process, the fate of the colloidal matter essentially is associated with the particulate phase. For modeling the bioreactor it is, therefore, sufficient to group all the colloidal and particulate biodegradable COD in one category ("slowly biodegradable," or SBCOD), which is considered as particulate in separators (e.g., secondary clarifiers) downstream of bioreactors.

The division of SBCOD between particulate and colloidal portions is particularly significant for modeling separation processes prior to combining the wastewater with activated sludge (where the colloidal material is rapidly adsorbed onto the particulate phase). For example, model predictions of organics removal (both COD and BOD removal) in a primary settling tank is directly impacted by the colloidal/particulate SBCOD fractionation because the colloidal material is not settleable. Where modeling of primary settling tanks is included in simulations, this division should be specified to correctly predict the COD composition and load of the stream passing to the activated sludge system.

8.2 Method

The slowly biodegradable COD (SBCOD) concentration is obtained by difference from measured plant data and previously estimated influent fractions as follows:

$$SBCOD = X_S + S_{COL} = (1 - f_{SS} - f_{SI} - f_{XI}) \cdot COD_T \quad (8.2\text{-}1)$$

where:

f_{SS}	= readily biodegradable COD fraction
f_{SI}	= soluble unbiodegradable COD fraction
f_{XI}	= particulate unbiodegradable COD fraction
COD_T	= influent COD concentration (mg/L)
X_S	= particulate slowly biodegradable COD concentration
S_{COL}	= colloidal slowly biodegradable COD concentration

8.3 Example

Although the wastewater characterization procedures presented in this manual do not require an estimate of the slowly biodegradable COD influent fraction, this example illustrates how the slowly biodegradable influent COD concentration is calculated by difference after other COD fractions are estimated.

Unsettled (raw) wastewater from the influent to a municipal WPCP was used to operate a 6.3L single-reactor activated sludge system. The system was operated at a nominal sludge age of 10.6 days. The following typical data set from this system has been used to characterize the wastewater as illustrated in previous chapters of this report ("filtered" values are from the analysis of 0.45 μm filtrate samples):

Unfiltered influent COD	= 310 mg/L
Unfiltered influent TKN	= 36.0 mgN/L
Unfiltered effluent COD	= 71 mg/L
Filtered effluent COD	= 41 mg/L
Unfiltered effluent TKN	= 4.2 mgN/L
Filtered effluent TKN	= 2.2 mgN/L
Effluent NO_3-N	= 25.2 mgN/L
Mixed liquor VSS	= 1,784 mg/L
Reactor OUR	= 24.1 mg/L/hr

The following influent COD fractions were determined using the procedures outlined in this report:

f_{SS} = 0.05 (estimated using the aerobic batch method – see Chapter 7.0)

f_{SI} = 41/310 = 0.13 (assumes negligible colloidal material in effluent – see Chapter 7.0)

f_{XI} = 0.22 (estimated through simulation – see Chapter 6.0)

Based on these wastewater characteristics, the slowly biodegradable COD concentration can be calculated from Eq. 8.2-1 as follows:

$$\begin{aligned} SBCOD &= (1 - 0.05 - 0.13 - 0.22) \cdot COD_T \\ &= 0.6 \cdot COD_T \\ &= 186\ mg/L \end{aligned}$$

8.4 Particulate and Colloidal SBCOD

A distinction can be made between the portions of the influent wastewater SBCOD that are colloidal and particulate. Because the colloidal portion is non-settleable, this fractionation is particularly significant for modeling the organics removal in a primary settling tank, and the COD composition and load of the stream passing to the activated sludge system. The division of SBCOD into colloidal (S_{COL}) and particulate (X_S) portions can be defined by a fraction (f_{XS}):

f_{XS} = fraction of the SBCOD that is particulate (not colloidal)

$$X_S = f_{XS} \cdot SBCOD \qquad (8.4\text{-}1)$$

$$S_{COL} = (1 - f_{XS}) \cdot SBCOD \qquad (8.4\text{-}2)$$

The distribution between colloidal and particulate portions cannot be derived by any *direct* method such as filtration because the unbiodegradable particulate COD (X_I) and biodegradable particulate COD (X_S) cannot be separated. This is evident from Figure 8-1, which depicts the retention or passage of the influent wastewater COD components through sequential 1.2μm glass-fiber filtration, flocculation (to remove colloidal matter from the liquid phase) and 0.45μm membrane filtration.

The division of SBCOD between particulate and colloidal portions (i.e., f_{XS}) can be estimated using the approach outlined below if data are available on:

- Influent total COD (COD_T);
- COD of the *filtrate from the glass-fiber filtration for TSS/VSS* measurement; and
- estimates of the RBCOD and unbiodegradable soluble and particulate COD concentrations (i.e., S_S, S_I and X_I, or the associated fractions f_{SS}, f_{SI} and f_{XI}) [If ffCOD data on the influent are available, note that ffCOD = $S_S + S_I$].

The concentrations of RBCOD and unbiodegradable soluble and particulate COD are:

$$S_S = f_{SS} \cdot COD_T \qquad (8.4\text{-}3)$$

$$S_I = f_{SI} \cdot COD_T \qquad (8.4\text{-}4)$$

$$X_I = f_{XI} \cdot COD_T \qquad (8.4\text{-}5)$$

From Figure 8-1 it is evident that:

$$(X_S + X_I) = COD_T - \textit{Glassfiber filtrate COD} \qquad (8.4\text{-}6)$$

Hence, with X_I known:

$$X_S = COD_T - \textit{Glassfiber filtrate COD} - X_I \qquad (8.4\text{-}7)$$

Also:

$$S_{COL} = \textit{Glassfiber filtrate COD} - S_S - S_I = \textit{Glassfiber filtrate COD} - \textit{ffCOD} \qquad (8.4\text{-}8)$$

Knowing X_S and S_{COL}:

$$F_{XS} = \frac{X_S}{X_S + S_{COL}} \qquad (8.4\text{-}9)$$

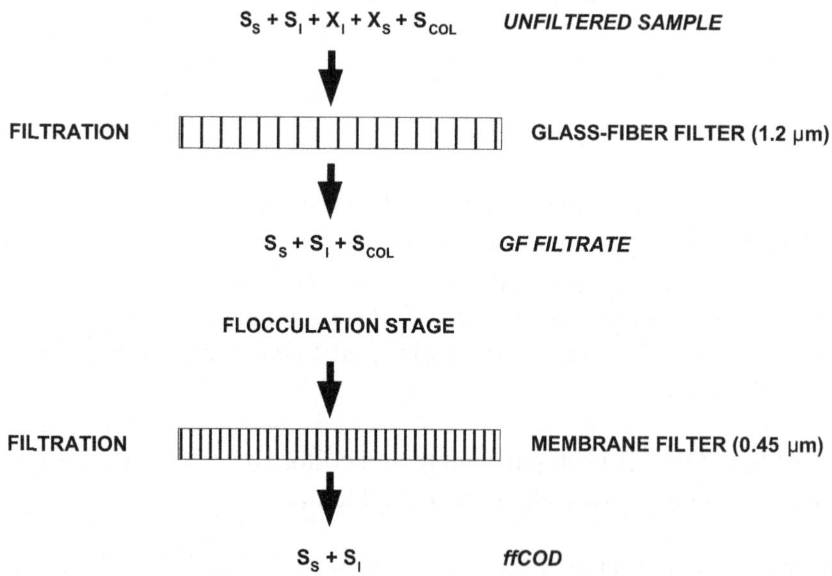

Figure 8-1. Diagram Depicting the Retention/Passage of Influent Wastewater COD Components through Sequential 1.2 μm Glass-Fiber Filtration, Flocculation and 0.42 μm Membrane Filtration

An example calculation is presented below, with reference to the data in Table 8-1. The data are averages from the analysis of grab samples of raw influent wastewater collected once a day over a period of 60 days at a 35 ML/day treatment plant receiving a typical municipal flow.

(See Chapter 12.0 for details.) Measured parameters are shown in bold text in the table. The following influent COD fractions were determined using procedures outlined in this report:

$f_{SS} = 0.17$; hence $S_S = 87$ mg/L

$f_{SI} = 0.06$; hence $S_I = 31$ mg/L

$f_{XI} = 0.06$; hence $X_I = 31$ mg/L

Hence:

$X_S = 514 - 209 - 31 = 274$ mg/L

$S_{COL} = 209 - 118 = 91$ mg/L

$f_{XS} = 274 / (274 + 91) = 0.75$

In many situations data are available for influent total COD, BOD_5 and VSS, but glass-fiber filtrate COD data are not available. If X_I has not been estimated, the f_{XS} parameter can be estimated together with estimating f_{XI} (i.e., X_I) by the method presented in Chapter 6.0.

Table 8-1. Estimation of Colloidal/Particulate Fractionation of SBCOD (all units mg/L) (Directly Measured Parameters in Bold Text)

Parameter	Value
COD (unfiltered)	**514**
COD (GF filtrate)	**209**
COD (0.45 _m filtrate)	**136**
VSS	**169**
TSS	**204**
S_S (RBCOD)	87
S_I (unbiodeg. sol. COD)	31
(ffCOD)	(87 + 31 = 118)
X_I (unbiodeg. part. COD)	31
X_S	274
S_{COL}	31
f_{XS}	0.75

A number of observations are pertinent regarding the data in Table 8-1, and underlying assumptions with this method:

- The f_{XS} value of 0.75 is typical for raw municipal wastewaters.
- The measured 0.45 μm filtrate COD value (without flocculation) is significantly less than the glass-fiber filtrate COD, but significantly higher than the ffCOD value (see Chapter 7.0). This demonstrates that at least a portion of the colloidal SBCOD passes through a 0.45 μm filter.

- Caution should be exercised when applying this approach with certain industrial wastewaters. For example, meat processing wastewaters may contain substantial amounts of colloidal COD (possibly both biodegradable and unbiodegradable) that is not adsorbed completely in the activated sludge process. This will result in an incorrect estimate of the unbiodegradable influent COD concentration based on a 0.45 μm filtration of the effluent without pre-flocculation because at least part of the colloidal matter will pass through the filter, and the approach outlined above would not be valid.
- Some adjustment to the equations would be required in cases where the influent wastewater contains a substantial biomass COD contribution.

8.5 Alternate Method for Determining SBCOD

The Dutch Foundation for Applied Water Research (STOWA) guidelines for wastewater characterization present a method for measurement of the SBCOD concentration (Roeleveld and van Loosdrecht, 2002). The total concentration of biodegradable COD in the influent is the sum of the readily biodegradable COD (RBCOD) and the SBCOD. In the STOWA guidelines, the total influent concentration of biodegradable COD is estimated from a BOD test. Subtracting the RBCOD concentration (determined separately from a physical test – see Chapter 7.0) from the total biodegradable COD provides an estimate of the SBCOD concentration. A brief summary of the method is provided below.

A BOD test is conducted on a wastewater, and the BOD is measured as a function of time. An example is shown in Figure 8.2 where the BOD was measured on days 1, 2, 4, 6 and 8 during a test on a municipal wastewater. The data are used to estimate a BOD first order rate constant, k_{BOD} by fitting a BOD curve to the measured data as shown in the figure. When the k_{BOD} constant is determined for a given wastewater, the ultimate BOD concentration (BOD_{tot}) can be estimated from a given BOD_t test using the following equation:

$$BOD_{tot} = \frac{1}{1 - e^{-k_{BOD} \cdot t}} BOD_t \qquad (8.5\text{-}1)$$

For example, the BOD_{tot} can be estimated from the influent BOD_5 concentration measured during routine plant monitoring. To estimate the total biodegradable COD (BCOD) concentration, a correction factor (f_{BOD}) must be applied to account for the unbiodegradable endogenous residue created through the decay of biomass during the test, so that the BCOD concentration is estimated as follows:

$$BCOD = \frac{1}{1 - f_{BOD}} BOD_{tot} \qquad (8.5\text{-}2)$$

Some example values for this test are shown in Figure 8-2. The slowly biodegradable COD concentration is calculated by subtracting the RBCOD concentration (determined separately) from the BCOD concentration:

$$SBCOD = BCOD - RBCOD \qquad (8.5\text{-}3)$$

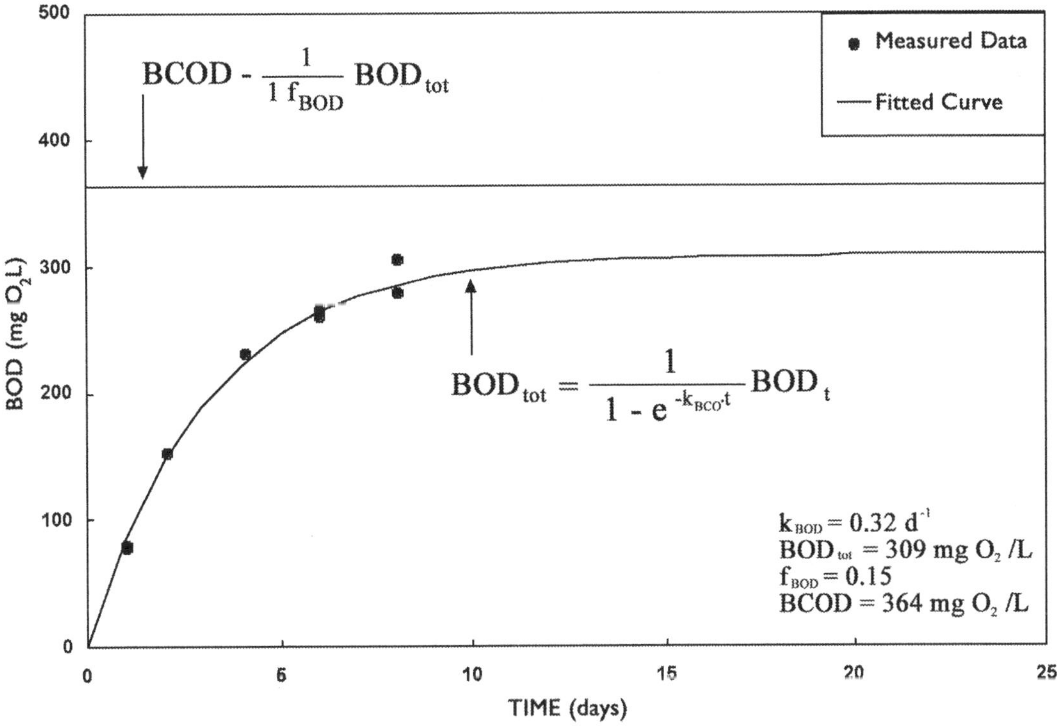

Figure 8-2. Example of a Fitted BOD Curve for Determination of Total Biodegradable COD (BCOD) Concentration (after Roeleveld and van Loosdrecht, 2002)

8.6 Considerations

- When determining the characteristics of influent wastewater, it is important to understand that it is only the particulate fraction of the slowly degradable COD that will be removed in a primary clarifier. Failure to recognize this can result in an underestimate of the COD concentration to the bioreactor when the influent characteristics are based on samples collected upstream of primary settling tanks.
- Some researchers have suggested that the slowly biodegradable COD can be determined from the OUR curve in a batch test, similar to the manner in which the readily biodegradable COD fraction can be determined. However, in many cases, the region of the OUR curve associated with the slowly biodegradable COD is not defined clearly enough to be useful in this estimate.
- In the STOWA guidelines for wastewater characterization, the unbiodegradable particulate COD (X_I) concentration is found by difference after the SBCOD concentration is determined as shown in this chapter, and the other COD components (S_I and S_S) are determined using physical methods. However, the inaccuracies in a BOD test often are significant, and all of the characterization inaccuracies are absorbed by the estimate of X_I (see Chapter 6.0).

8.7 References

Roeleveld, P.J., and M.C.M. van Loosdrecht. 2002. Experience with guidelines for wastewater characterization in the Netherlands. *Water Sci. Tech.* 45(6):77-87.

CHAPTER 9.0

Active Biomass COD

9.1 Background

Heterotrophic biomass may constitute a significant component of the influent wastewater COD. Reported values of influent heterotrophic biomass range from 7 percent to 25 percent of the total influent COD (Orhon and Çokgör, 1997). If modeling processes downstream of primary clarifiers, much of this biomass may originate from excess sludge recycled to the influent of the plant (see, for example, Kappeler & Gujer, 1992). Since the continuous seeding of biomass from the influent of an activated sludge system can significantly impact the behavior of the process (particularly the solids balance and organism wash-out solids retention time), in some cases it can be important to quantify this fraction of the wastewater.

9.2 Method

A batch method for estimating the mass of active heterotrophic organisms in an activated sludge sample was proposed by Kappeler and Gujer (1992). The sample of activated sludge is combined with a sample of centrifuged wastewater supernatant. The combined sample is mixed, and the OUR is then monitored over a period of time (until there is a significant drop). This method was modified by Wentzel et al. (1995), and applied to determining the mass of active organisms in a sample of influent wastewater. In this test, a volume (typically 3 or 4 L) of influent wastewater of known COD concentration (COD_T) is mixed and aerated, and the OUR is measured semi-continuously for an extended period of approximately 10 to 20 hours as appropriate. The basis of the test is that the influent wastewater sample will contain at least a small (seed) concentration of heterotrophic bacteria ($X_{HET,0}$), the characteristic to be determined. At the start of the test, there should be an appreciable RBCOD concentration (well in excess of the half-saturation coefficient for growth), and therefore, the first part of the test should reflect a logarithmic growth phase. That is, the OUR should increase exponentially from an initial low value until the RBCOD is nearly completely consumed. Thereafter, there should be a precipitous drop in OUR.

Based on the OUR response, the mass of organisms can be estimated using an appropriate activated sludge model (such as an ASM-type model). Wentzel et al. (1995) used a simplified version of the UCT model (Dold et al., 1980), and obtained an expression relating the observed OUR at time *t* to the initial heterotroph concentration, $X_{HET,0}$. In terms of the IWA model, this expression is:

$$ln(OUR_t) = ln\left[\frac{\mu_{HET} \cdot (1-Y_{HET}) \cdot X_{HET,0}}{24 \cdot Y_{HET}}\right] + (\mu_{HET} - b_{HET}) \cdot t/24 \qquad (9.2\text{-}1)$$

where:

OUR_t = OUR at time t hours (mg/L/h)

μ_{HET} = maximum specific heterotrophic organism growth rate (/d)

Y_{HET} = yield coefficient (COD basis)

= 0.666 mg COD/mg COD

b_{HET} = death-regeneration organism decay rate

= $0.62\ (1.029)^{T-20}$ (/d)

From the above equation, a plot of *ln* (OUR_t) *versus* time (t) should yield a straight line with:

$$slope = (\mu_{HET} - b_{HET})/24$$

and

$$y\text{-}intercept = ln\left[\frac{\mu_{HET} \cdot (1-Y_{HET}) \cdot X_{HET,0}}{24 \cdot Y_{HET}}\right]$$

Therefore the active heterotrophic organism concentration in the influent wastewater sample can be estimated as follows:

$$X_{HET,0} = \frac{e^{(y\text{-}intercept)} \cdot 24}{(slope \cdot 24 + b_{HET})(1-Y_{HET})/Y_{HET}} \qquad (9.2\text{-}2)$$

9.3 Example

An example of an OUR response from an aerobic batch test on raw influent wastewater is shown in Figure 9-1. From this data, a plot of *ln* (OUR_t) versus time (t) yields a straight line as shown in Figure 9-2, where the line slope and intercept are also shown.

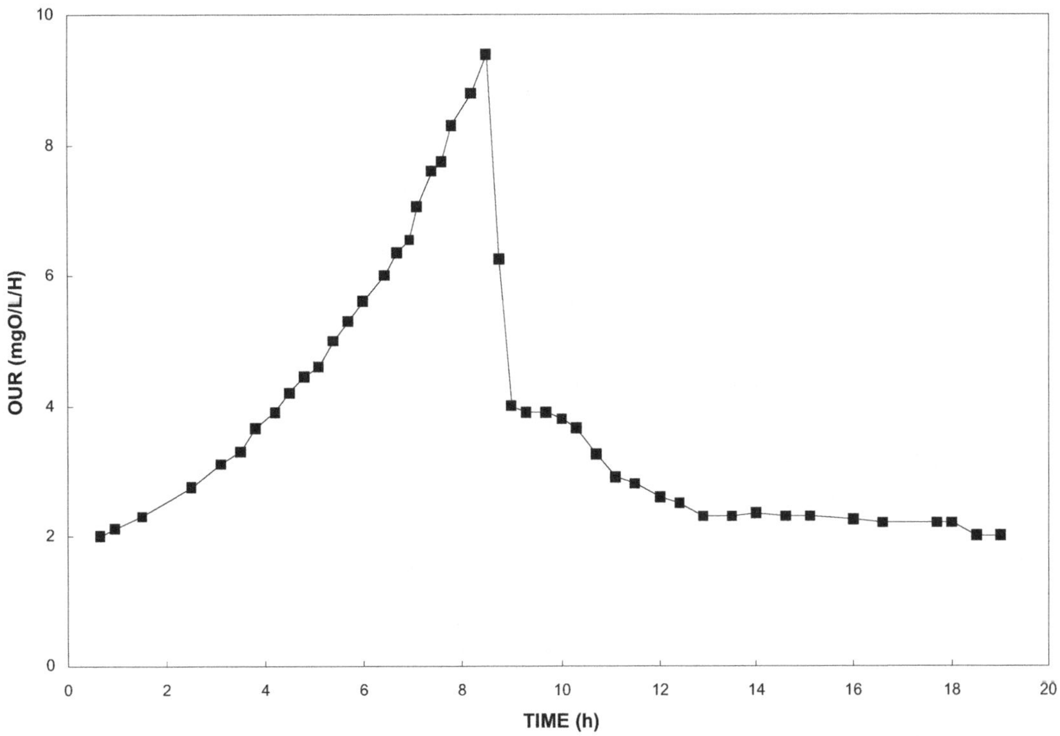

Figure 9-1. Example of OUR Response for an Aerobic Batch Test on Raw Municipal Wastewater (after Ubisi et al., 1997)

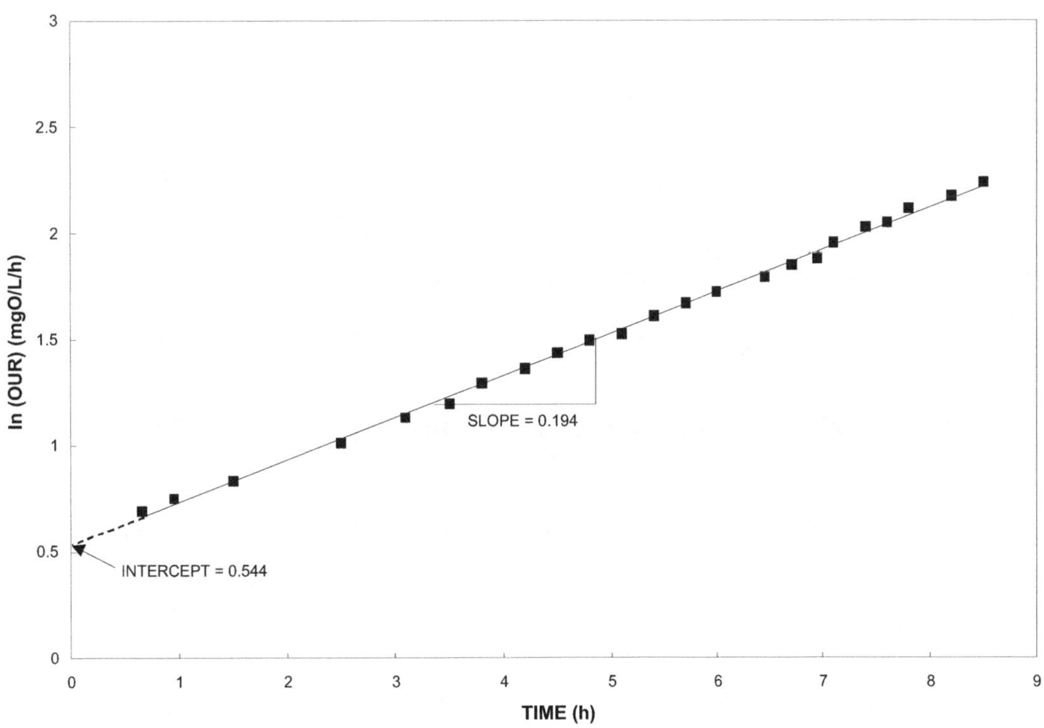

Figure 9-2. Plot of ln (OUR) versus Time for Initial Period of Increasing OUR in Figure 9-1

Using these data in equation 9.2-2:

$$X_{BH,0} = \frac{e^{0.544} \cdot 24}{(0.194 \cdot 24 + 0.62)(1-0.666)/0.666} = 15.6\ mg\ COD/L$$

This accounted for approximately 4 percent of the COD of this influent wastewater sample.

9.4 Considerations

- In the test, the OUR, particularly at the start, is low. Therefore steps should be taken to avoid any transfer of oxygen into the sample through the surface as this would cause errors in the OUR measurements. This can be achieved in several ways; for example, covering the surface with floating balls, or conducting the test in a near-full Erlenmeyer flask so as to minimize the surface exposed to the air.
- To determine if nitrification is exerting an oxygen demand during the test, samples can be analysed for nitrate and nitrite at the start and at the end of the test. However, it is unlikely that any significant nitrification will occur due to the short duration of the test and the very low concentration of nitrifiers relative to the heterotrophs.
- If the biomass component is not measured separately, this material will be reflected in the calibrated influent values of X_I and X_S. In most cases, this will not significantly impact model predictions of plant performance. However, in very high rate systems, the determination of the biomass component may be more important.

9.5 References

Dold, P.L., G.A. Ekama, and G.v.R. Marais. 1980. A general model for the activated sludge process. *Prog. Water Technol.* 12:47-77.

Kappeler, J., and W. Gujer. 1992. Estimation of kinetic parameters of heterotrophic biomass under aerobic conditions and characterization of wastewater for activated sludge modelling. *Water Sci. Tech.* 25(6):125-139.

Ubisi, M. F., T.W. Jood, M.C. Wentzel, and G.A. Ekama. 1997. Activated sludge mixed liquor heterotrophic active biomass. *Water SA* 23(3):239-248.

Wentzel, M.C., A. Mbewe, and G.A. Ekama. 1995. Batch test for measurement of readily biodegradable COD and active organism concentrations in municipal wastewaters. *Water SA* 21(2):117-124.

CHAPTER 10.0

VFA COD

10.1 Background

For the purposes of modeling excess biological phosphorus removal (EBPR) systems, it is necessary to distinguish between "complex" (fermentable) readily biodegradable COD and readily biodegradable COD comprising short-chain volatile fatty acids (VFAs).

VFAs are essential for the biochemical mechanism resulting in EBPR (see Chapter 2.0). For EBPR to occur in the system, VFAs either must be present in the influent or must be produced in the system. The concentration of VFAs in the influent can dramatically impact the performance of an EBPR system.

Note: It is only necessary to distinguish between "complex" readily biodegradable COD and readily biodegradable COD comprising short-chain volatile fatty acids when modeling biological excess phosphorus removal systems.

The concentrations of individual VFAs in a raw municipal wastewater generally are on the order of a few milligrams per liter, typically in the range of 5 mg/L to 15 mg/L as acetate, with the major contributors being acetate and propionate. The influent concentration depends on factors such as sewer temperature, residence time, gradient, the amount of organic sediment deposited in the sewers and amount of infiltration.

The most effective means of analyzing for these compounds is by gas chromatography (GC) or ion chromatography (IC). Titration methods have also been used to monitor VFA levels in anaerobic digestors. However, these methods are not sensitive enough for the low VFA levels observed in most municipal wastewaters. This chapter provides an overview of GC and IC methods for determining VFA concentration in wastewaters.

10.2 Method (1) – Gas Chromatograph

Gas chromatography is a chromatographic technique that can be used to separate volatile organic compounds. A gas chromatograph consists of a flowing mobile phase, an injection port, a separation column containing the stationary phase and a detector.

Mobile phases are generally inert gases such as helium or nitrogen. The injection port consists of a rubber septum through which a syringe needle is inserted to inject the sample. The injection port is maintained at a higher temperature than the boiling point of the least volatile component in the sample mixture. Since the partitioning behavior is dependent on temperature, the separation column usually is contained in a thermostatically controlled oven. Most columns contain a liquid stationary phase on a solid support. Specific methodologies for VFAs are available from the literature. See, for example, Lie and Welander (1997). Table 10-1 provides an example of a gas chromatograph set-up for VFA analysis.

Table 10-1. Gas Chromatograph Set-Up Example for VFA Analysis

Standards	
Compounds	Acetic, propionic, iso-butyric, butyric acids
Concentrations	5 to 100 mg/L
Detector	
Type	Flame ionization detector (FID)
Temperature	300 °C
Column	
Type	Wide bore capillary (e.g., Hewlett Packard FFAP (free fatty acid phase) column)
Dimensions	30 m long 0.32 mm ID 0.5 _m film thickness
Sample Volume	0.5 _L
Inlet	
Pressure	8 psi
Temperature	240 °C
Oven Temperature Program	
0 to 1 min.	80 °C
1 to 3 min.	Ramp from 80 °C to 120 °C
3 to 11.5 min.	Ramp from 120 °C to 205 °C
11.5 to 13.5 min.	205 °C

10.3 Method (2) – Ion Chromatograph

Ion chromatography is a type of liquid chromatography (LC) where ion exchange resins are used to separate atomic or molecular ions based on their interaction with the resin. For the measurement of VFAs, the anions of the various volatile acids will react with a cation in the resin. The differences in the elution times of the VFAs are related to the difference in equilibrium constants of these reactions. The solvent or eluent for VFAs is usually NaOH. Detection is by measuring the conductivity of the solution leaving the column. To eliminate background interference caused by the anions in the eluent, the instrument will include an eluent suppressor that supplies H^+ to neutralize the OH^-, and a membrane to selectively remove the Na^+. This method has been found to work well in an application for resolving C6 and shorter acids as well as lactate, and problems with deactivation of the column from higher acids were found to be minimal. In addition, the use of an aqueous carrier in IC eliminates problems with water and salts from aqueous samples that can occur with a GC (Lalman and Bagley, 2000).

10.4 Considerations

Influent VFAs can come from several sources:

- Fermentation within the collection system,
- Fermentation within the primary clarifier,
- On-site fermenters operated specifically for VFA generation, and
- Supplemental VFA addition.

When VFAs originate from uncontrolled sources (e.g., collection system, primary clarifiers) the influent concentration can vary considerably. For instance, low temperatures and/or high flows could reduce VFA generation in the collection system.

10.5 References

Lie, E., and T. Welander. 1997. A method for the determination of the readily fermentable organic fraction in municipal wastewater. *Water Res.* 31(6):1269-1274.

Lalman, J.A., and D.M Bagley. 2000. Anaerobic degradation and inhibitory effects of linoleic acid. *Water Res.* 34(17), 4220-4228.

CHAPTER 11.0

INFLUENT TKN FRACTIONS

11.1 Background

Characterization of the nitrogenous material in the influent is in terms of the total Kjeldahl nitrogen (TKN). The division of the total influent TKN into components is shown in Figure 4-3. It is important to note that 60 percent to 75 percent of the influent TKN is in the ammonia form, while the remainder of the nitrogen is bound in organic material which may be soluble or particulate. Particulate organic material in the influent is composed of both biodegradable (X_S) and unbiodegradable (X_I) COD (and possibly some biomass). Methods for measuring the influent solids do not distinguish between the components such as X_S, and X_I. This creates difficulties for assessing how the nitrogen is distributed. Nevertheless, it is important to estimate this distribution because it impacts the amount of nitrogen that is available for nitrification, and therefore the final effluent nitrogen concentrations.

It also should be noted that, in a limited number of cases, appreciable influent concentrations of nitrate/nitrite have been encountered in municipal wastewaters (often originating from plant recycles). This has a significant impact on nutrient removal system design and performance. For these cases, the influent nitrate/nitrite concentration must be included as an additional wastewater characteristic, separate from the TKN.

11.2 Ammonia Fraction of TKN: f_{SNH}

Note: Typically, ammonia accounts for the major portion of the influent TKN in municipal wastewaters. Therefore, the fractionation of the remaining organic portions is less critical for modeling purposes.

The fraction of the total influent TKN that is free and saline ammonia (f_{SNH}) can be determined by direct analysis. A sample of the wastewater is taken and a portion filtered through a 0.45 μm membrane filter to remove particulate material. The unfiltered sample is analyzed using standard techniques to determine the concentration of TKN, and the filtrate is analyzed for ammonia concentration. The ammonia concentration will be the same in the filtered and unfiltered sample. The fraction of the total influent TKN that is free and saline ammonia is:

$$f_{SNH} = \frac{Influent\ NH_3\text{-}N}{Influent\ Unfiltered\ TKN} \quad (11.2\text{-}1)$$

In raw municipal wastewaters, the f_{SNH} fraction typically ranges from 0.65 to 0.75 mg N/mg N. However, values as low as 0.45 have been measured at North American plants.

11.3 Soluble Unbiodegradable Organic Nitrogen Fraction: f_{NSI}

Note: Biodegradable and unbiodegradable soluble organic nitrogen in the effluent are not distinguished in the TKN test.

The soluble unbiodegradable fraction of the influent TKN can only be estimated based on the filtered effluent TKN and ammonia concentrations from a fully-nitrifying activated sludge system. The filtered effluent TKN comprises residual ammonia, residual (yet-to-be-converted-to-ammonia) soluble biodegradable organic nitrogen (S_{NB}) and any soluble unbiodegradable organic nitrogen (S_{NI}) from the influent:

$$Filtered\ effluent\ TKN = S_{NH} + S_{NB} + S_{NI} \quad (11.3\text{-}1)$$

The difference between the filtered TKN and ammonia (S_{NH}) concentrations will be the sum of soluble biodegradable and unbiodegradable organic nitrogen (S_{NB} and S_{NI}, respectively). For a fully-nitrifying system, usually the filtered effluent TKN will be in the range of 0.5 mgN/L, and the ammonia concentration will be low, say 0.1 mgN/L – that is, a difference of about 0.4 mgN/L. Model applications indicate that the residual concentration of *biodegradable* organic nitrogen (i.e., material that has not been converted to ammonia) in the effluent typically is about 0.4 mgN/L. That is, usually there is no unbiodegradable organic nitrogen in the effluent (or influent). However, in certain instances, the filtered effluent TKN from a nitrifying system may be high, in the range of 3 to 5 mgN/L, even though the ammonia concentration is low (and presumably the soluble biodegradable organic nitrogen is still 0.4 mgN/L). In order to simulate such a case, it is necessary to assume that there is a portion of soluble unbiodegradable TKN from the influent passing out in the effluent. Based on the suppositions above, this could be estimated as follows:

$$S_{NI} \approx Filtered\ effluent\ TKN\ -\ S_{NH} - 0.4\ mgN/L$$

$$f_{SNI} = \frac{S_{NI}}{TKN} \quad (11.3\text{-}2)$$

This method for estimating f_{SNI} is not particularly satisfactory. However, there is no means for distinguishing between biodegradable and unbiodegradable soluble TKN. Fortunately, the amount of unbiodegradable soluble TKN usually is only a small fraction of the total influent TKN (< 3 percent), so errors in the estimate are not of particular significance.

11.4 Nitrogen Content of the Particulate Unbiodegradable Influent COD: f_{NXI}

This material usually accounts for approximately 10 percent of the influent TKN. As with f_{SNI}, there is no means to distinguish between particulate biodegradable and unbiodegradable influent TKN. For municipal wastewaters, usually it appears reasonable to assume that the nitrogen content of the particulate unbiodegradable influent COD is the same as the nitrogen content of the mixed liquor solids. (For industrial wastes this assumption may not be reasonable.) Hence, on a COD basis, from equation 11.3-2:

$$\begin{aligned} f_{NXI} &= \frac{X_{NI}}{X_I} \\ &\approx f_N = \frac{TKN\ of\ mixed\ liquor\ solids}{COD\ of\ mixed\ liquor\ solids} \quad (\approx 0.07\ mgN/mgCOD) \end{aligned} \qquad (11.4\text{-}1)$$

A possible check on the validity of this estimate would be to assume that the particulate biodegradable and unbiodegradable TKN is equally distributed between the particulate biodegradable and unbiodegradable COD. If this is the case, then:

$$\begin{aligned} f_{NXI} &\approx \frac{Particulate\ Influent\ TKN}{Particulate\ Influent\ COD} \\ &\approx \frac{TKN - TKN_{FILT}}{COD_T - COD_{FILT}} \end{aligned} \qquad (11.4\text{-}2)$$

11.5 Division of Biodegradable Organic Nitrogen into Soluble and Particulate: f_{XNB}

The division of biodegradable organic TKN into soluble (S_{NB}) and particulate (X_{NB}) portions also can only be estimated. A sample of the wastewater is taken and a portion filtered through a 0.45 μm membrane filter to remove particulate material. The unfiltered sample and the filtrate are then analyzed using standard techniques to determine the concentrations of COD (COD_T) and TKN in the unfiltered sample, as well as the TKN and ammonia in the filtrate (TKN_{FILT} and S_{NH}, respectively).

For the *unfiltered* sample:

$$\begin{aligned} TKN = {} & ammonia\text{-}N\ (f_{SNH} \cdot TKN) \\ & + soluble\ biodegradable\ organic\ N\ \ (S_{NB}) \\ & + soluble\ unbiodegradable\ organic\ N\ \ (f_{NSI} \cdot TKN) \\ & + particulate\ biodegradable\ organic\ N\ \ (X_{NB}) \\ & + particulate\ unbiodegradable\ organic\ N\ \ (f_{NXI} \cdot f_{XI} \cdot COD_T) \end{aligned} \qquad (11.5\text{-}1)$$

And for the *filtered* sample:

$$\begin{aligned} TKN_{FILT} = {} & ammonia\text{-}N\ (f_{SNH} \cdot TKN) \\ & + soluble\ biodegradable\ organic\ N\ \ S_{NB}) \\ & + soluble\ unbiodegradable\ organic\ N\ \ (f_{NSI} \cdot TKN) \end{aligned} \qquad (11.5\text{-}2)$$

Hence:

$$f_{XNB} = \frac{X_{NB}}{S_{NB} + X_{NB}} = \frac{TKN - TKN_{FILT} - f_{NXI} \cdot f_{XI} \cdot COD_T}{(1 - f_{SNH} - f_{NSI})\, TKN - f_{NXI} \cdot f_{XI} \cdot COD_T} \quad (11.5\text{-}3)$$

By assuming a typical value of 0.07 mgN/mgCOD (≅ 0.1 mgN/mgVSS) for f_{NXI}, then calculation of f_{XNB} requires knowledge of two other wastewater parameters: f_{SNH}, f_{NSI} and f_{XI}. Noting that f_{NSI} usually is very small (< 0.03) and small compared to f_{SNH}, setting f_{NSI} to zero in equation 11.5-3 will not result in an undue error in the estimate of f_{XNB}.

11.6 Considerations

- The assumption that the nitrogen content of the particulate unbiodegradable influent COD is the same as the nitrogen content of the mixed liquor solids is reasonable because a significant portion of the mixed liquor solids are comprised of unbiodegradable particulate COD from the influent. If the N content of the mixed liquor solids is significantly different from the typical organism value of 0.07mgN/mgCOD, then this may indicate a different value for f_{NXI}.

CHAPTER 12.0

Influent Solids Characterization

12.1 Background

The foregoing chapters have focused mainly on the organic content of influent wastewater and the fractionation of the COD into components. When simulating the performance of a particular treatment plant, these characteristics are important in determining factors such as effluent quality and aeration requirements, as well as volatile solids production. However, influent wastewater also contains inorganic solids that contribute to the total solids production. The inorganic suspended solids (X_{ISS}) accumulate in the mixed liquor and can have a substantial impact on the sizing of aeration tanks and secondary clarifiers, as well as solids handling facilities.

Determination of influent inorganic suspended solids (X_{ISS}) requires measurement of influent total and volatile suspended solids concentrations (TSS and VSS, respectively); X_{ISS} is the difference between TSS and VSS. The analysis involves filtering a sample of influent wastewater through a standard glass-fiber filter to capture the suspended solids. The pre-weighed filter is dried, weighed again (for TSS), ignited and weighed again (for VSS).

Usually the TSS and VSS analyses are conducted in parallel with analyses on the unfiltered influent for total COD, TKN, ammonia, TP, etc. Three additional analyses – COD, TKN and TP – on the *filtrate from the glass-fiber filtration for TSS/VSS* provide information on the composition of the influent solids which is important for model calibration; namely:

- COD/VSS ratio of the influent solids;
- N/VSS or N/COD content of the influent solids; and
- P/VSS or P/COD content of the influent solids.

This chapter presents part of the results of a wastewater characterization study, namely, data specific to the influent solids characteristics. The data are used to calculate the three parameters listed above, as well as influent X_{ISS}. The significance of influent inorganic suspended solids concentration on the mixed liquor solids in a typical activated sludge system is also demonstrated.

12.2 Example of Influent Solids Data Collection

Grab samples of raw influent wastewater were collected once a day over a period of 60 days at a 35 ML/day treatment plant receiving a typical municipal flow (TKN data were only collected for days 19 to 60). The samples were filtered through standard glass-fiber filters and analyzed for:

- TSS and VSS;
- Unfiltered COD and TKN; and
- Glass-fiber filtrate COD and TKN.

Note: In this case no measurements were conducted for TP. Typically this parameter should be included in the analysis of unfiltered and glass-fiber filtered samples.

Table 12-1 lists the averages of the measured and calculated parameters; these are separated by a solid line in the table. It should be noted that grab samples were collected each day (at approximately the same time) rather than daily composites. Therefore the concentrations differed quite substantially from day to day. As an example, Figure 12-1 shows a plot of the daily COD and TSS concentrations. Figure 12-2 shows the corresponding plot for the influent inorganic suspended solids (ISS or X_{ISS}). Aside from the fact that these were grab samples, some of the variation in concentration was due to rainfall events.

Note: It is important to note that the objective is to measure the COD, TKN and TP of the solids retained on the glass-fiber filter. Therefore, calculation of the COD/VSS, TKN/VSS and TP/VSS of the influent solids requires measurement of the glass-fiber filtrate COD, TKN and TP as opposed to filtrate from a different filtration such as a 0.45 μm filter. Standard glass-fiber filters have an effective pore size of approximately 1.2 μm and allow passage of colloidal material that is largely retained on a 0.45 μm filter. For example, the data in Table 12-1 show glass-fiber and 0.45 μm filtrate CODs of 209 and 136 mg/L, respectively.

Table 12-1. Analysis of Unfiltered and Glass-Fiber (GF) Filtered Municipal Influent Wastewater

Parameter	Value	Units
COD (unfiltered)	514	mg/L
COD (GF filtrate)	209	mg/L
COD (0.45 μm filtrate)	136	mg/L
TKN (unfiltered)	39.3	mg/L
TKN (GF filtrate)	30.6	mg/L
TSS	204	mg/L
VSS	169	mg/L
ISS	35	mg/L
Solids COD/VSS	1.83 (std. dev. = 0.27)	mg COD / mg VSS
Solids TKN/VSS	0.054 (std. dev. = 0.017)	mg N / mg VSS

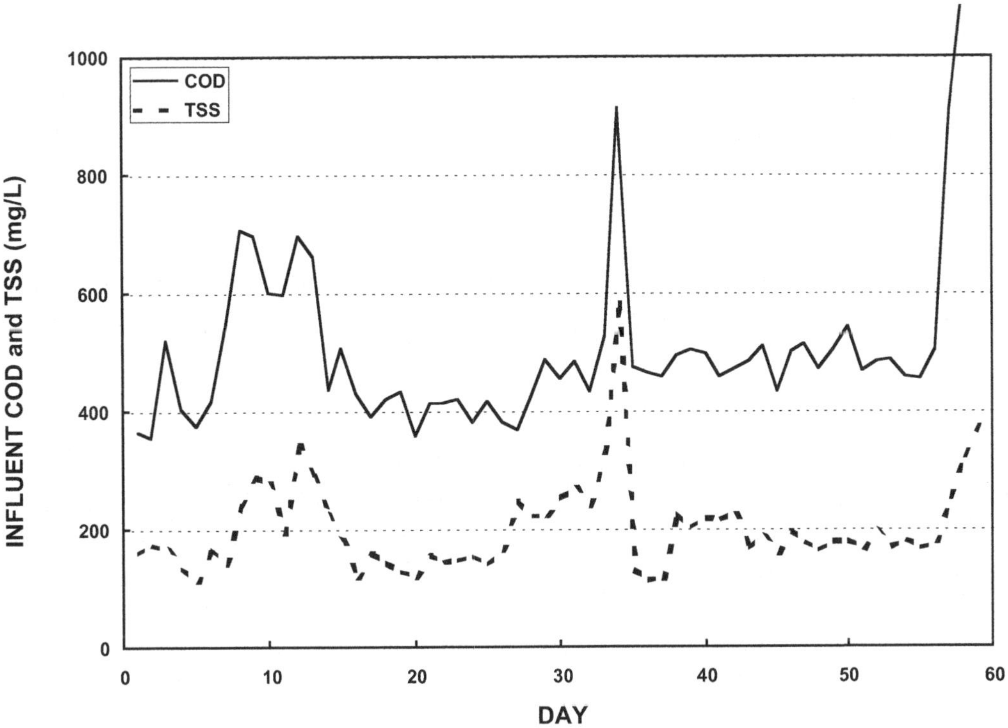

Figure 12-1. Influent COD and TSS Concentrations (Daily Grab Samples)

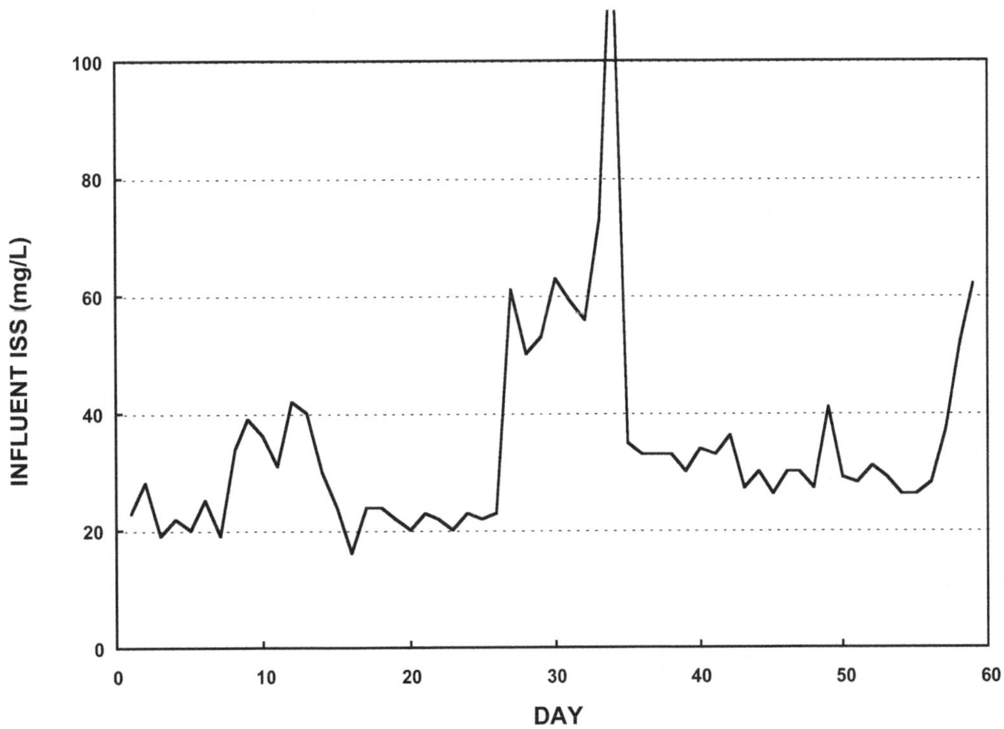

Figure 12-2. Influent ISS (X_{ISS}) Concentration (Daily Grab Samples)

12.3 Inorganic Suspended Solids (X_{ISS})

The concentration of X_{ISS} in the influent wastewater is simply the difference between the total and volatile solids concentrations:

$$X_{ISS,INFL} = Influent\ TSS\ -\ Influent\ VSS \qquad (12.3\text{-}1)$$

In North American municipal wastewaters, the average daily flow-weighted composite concentrations of X_{ISS} in raw and primary settled municipal wastewater typically are in the ranges of 25-45 mg/L and 10-25 mg/L, respectively. The difference reflects the removal of solids with primary settling.

As noted, the grab sample X_{ISS} data shown in Figure 12-2 exhibited considerable variation over the period. This variation roughly follows the variations in COD and TSS, with the exception of a period from Days 27 to 37, where the concentration almost doubled. This is demonstrated more specifically in Figure 12-3 where the ratio of ISS to TSS is plotted. This situation is not usual; typically the ratio remains more uniform (as do the ratios for other parameters such as COD/VSS, VSS/TSS and TKN/COD). The reasons for the unusual response seen here were not established.

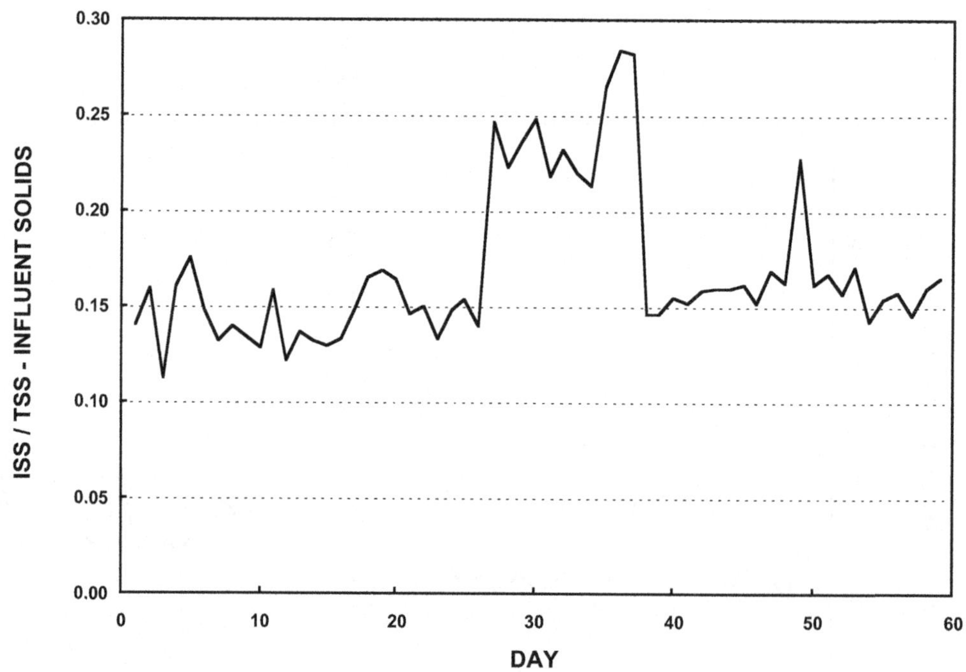

Figure 12-3. Influent ISS/TSS Ratio (Daily Grab Samples)

12.3.1 Significance of Inorganic Suspended Solids (X_{ISS})

The impact of influent X_{ISS} on activated sludge mixed liquor suspended solids concentration is best illustrated by the following example.

Consider the following (non-phosphorus removal) treatment plant:

$$Influent\ Total\ COD = 500\ mg/L$$
$$Influent\ X_{ISS} = 20\ mg/L$$
$$Influent\ Flowrate = 12000\ m^3/d$$
$$Aeration\ Tank\ Volume = 4000\ m^3$$
$$SRT = 10\ days$$
$$MLVSS = 3875\ mg/L$$

The X_{ISS} concentration expected in the aeration tank can be estimated directly from a mass balance as follows:

$$Aeration\ Tank\ X_{ISS} = \frac{Influent\ X_{ISS} \times Influent\ Flowrate \times SRT}{Aeration\ tank\ volume}$$
$$= \frac{20\ mg/L \times 12000\ m^3/d \times 10\ days}{4000\ m^3}$$
$$= 600\ mg/L$$

The resulting MLSS is:

$$MLSS = MLVSS + X_{ISS}$$
$$= 3875 + 600$$
$$= 4475\ mg/L$$
$$(VSS/TSS = 0.87)$$

Now, consider the same system if the influent X_{ISS} is equal to 40 mg/L:

$$Aeration\ \tan k\ X_{ISS} = \frac{40\ mg/L \times 12000\ m^3/d \times 10\ days}{4000\ m^3}$$
$$= 1200\ mg/L$$
$$MLSS = 3875 + 1200$$
$$= 5075\ mg/L$$
$$(VSS/TSS = 0.76)$$

It is clear from this example that the influent X_{ISS} concentration can have a significant impact on the system MLSS for a given SRT.

Note: In non-phosphorus removal systems (i.e., systems without chemical precipitation or EBPR), X_{ISS} data can be used as a check on the system solids balance. X_{ISS} is a conservative substance in biological treatment systems so there should be a 1:1 relation between input and output X_{ISS}.

12.4 Influent Solids COD/VSS

The activated sludge simulation models are based on COD as the parameter for quantifying organic material. The volatile suspended solids (VSS) concentration of, say, an influent stream or in the mixed liquor of a process is a derived parameter, calculated based on specifying a ratio for COD/VSS for the particulate COD components.

Note: Certain activated sludge models assume a single value for the ratio of COD/VSS for all particulate COD components (such as biomass and particulate unbiodegradable COD). Other models have the facility to assign different COD/VSS ratios for the different particulate COD components.

The ratio of COD to VSS for the influent solids is calculated from the monitoring data as follows:

$$COD/VSS = \frac{Unfiltered\ COD - Glassfiber\ filtrate\ COD}{VSS} \qquad (12.4\text{-}1)$$

Figure 12-4 shows a plot of the COD/VSS for the grab sample daily data. There appears to be considerable variation in the data from day to day. However, it should be recognized that the calculated COD/VSS is derived from a ratio of two differences. Also, the VSS concentration is quite small, and the relative error in a VSS measurement increases at lower concentrations. Therefore, some variation in the COD/VSS value would be anticipated even if the actual value were relatively constant. The daily data values are plotted in a quantile-quantile probability plot in Figure 12-5 (Chambers et al., 1983). The plotted points closely approximate a straight line, indicating that the data exhibit a normal distribution (mean = 1.83 mgCOD/mgVSS; standard deviation = 0.27).

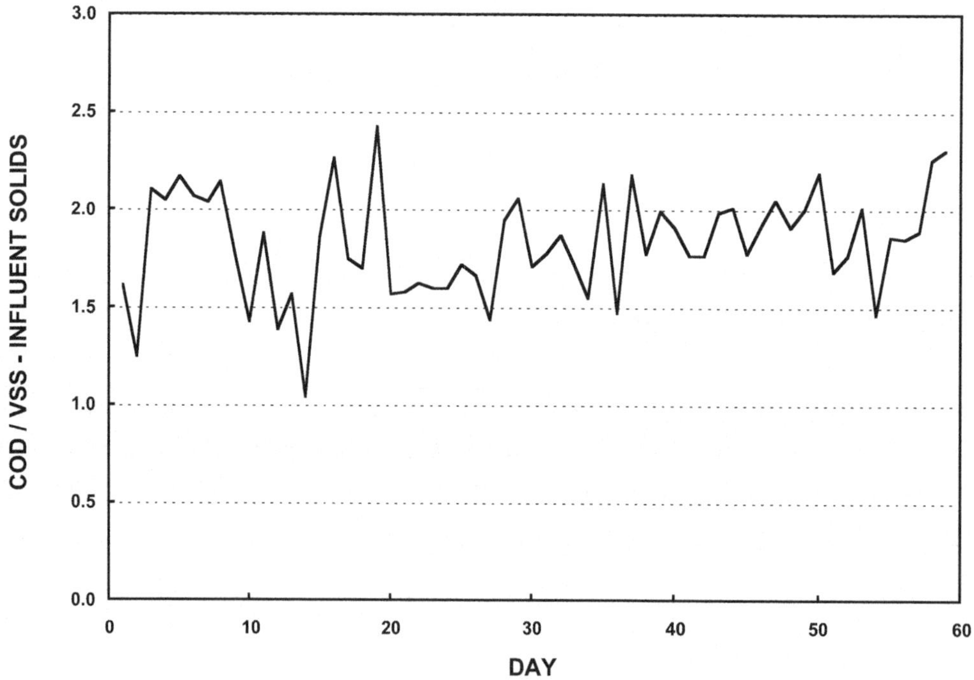

Figure 12-4. Influent Solids COD/VSS Ratio (Daily Grab Samples)

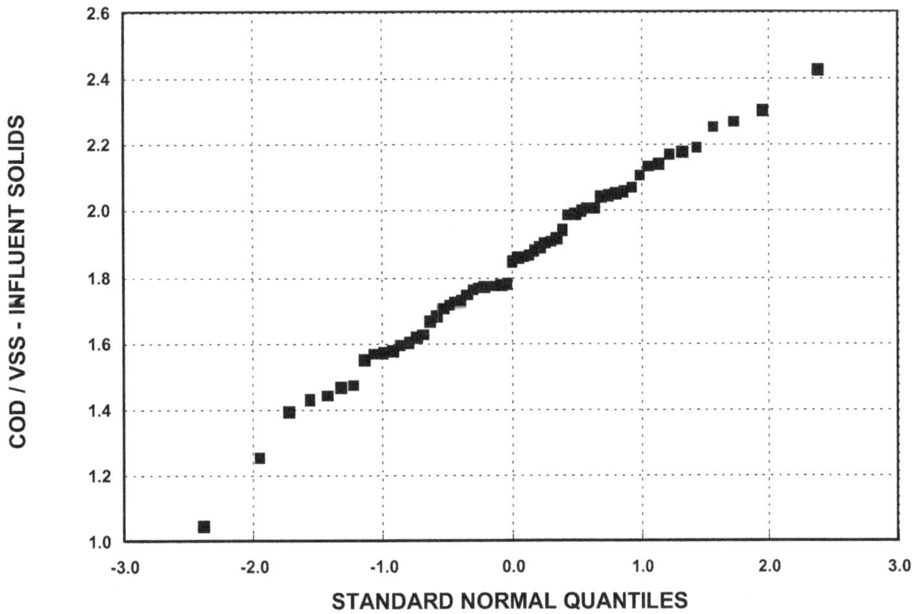

Figure 12-5. Quantile-Quantile Probability Plot of Influent Solids COD/VSS Ratio (Daily Grab Samples)

Particulate organic material in municipal influent wastewater generally includes both unbiodegradable (X_I) and biodegradable (X_S) components (and perhaps a biomass component). It is not possible to separate out the two components so as to establish whether the COD/VSS factors differ for each. From a modeling standpoint, the best first approximation is to assume that the COD/VSS ratios for both unbiodegradable COD (X_I) and biodegradable COD (X_S) are equal.

It is important to note that the COD/VSS ratio for influent solids observed here is significantly different from the ratio suggested for biomass (i.e., 1.42 mg COD/mg VSS – see Chapter 13.0). Also, within the activated sludge process, X_I accounts for a major portion of the VSS, particularly at longer SRT. Therefore, it would appear reasonable to assume that models with the facility to specify different COD/VSS ratios for different particulate COD components will better-predict VSS concentrations.

12.5 Influent Solids Nitrogen Content

In the activated sludge models a portion of the influent TKN is assumed to be associated with the particulate unbiodegradable influent COD (X_I) and is not available as a synthesis nitrogen source or for nitrification. Therefore, an overall accounting of the nitrogen balance for a particular system requires an estimate of the nitrogen associated with X_I.

The nitrogen content of the influent solids is calculated from the monitoring data as follows:

$$N/VSS = \frac{Unfiltered\ TKN\ -\ Glassfiber\ filtrate\ TKN}{VSS} \qquad (12.5\text{-}1)$$

Figure 12-6 shows a plot of the influent solids TKN/VSS for the grab sample daily data. As with the COD/VSS data, there is considerable variation in the data from day to day. Again, it should be recognized that the calculated TKN/VSS is derived from a ratio of two differences and that the relative error in the VSS measurement may be appreciable. The daily data values are plotted in a quantile-quantile probability plot in Figure 12-7. The plotted points also approximate a straight line (except for a few apparent outliers), indicating that the data exhibit a normal distribution (mean = 0.054 mg N/mg VSS; standard deviation = 0.017).

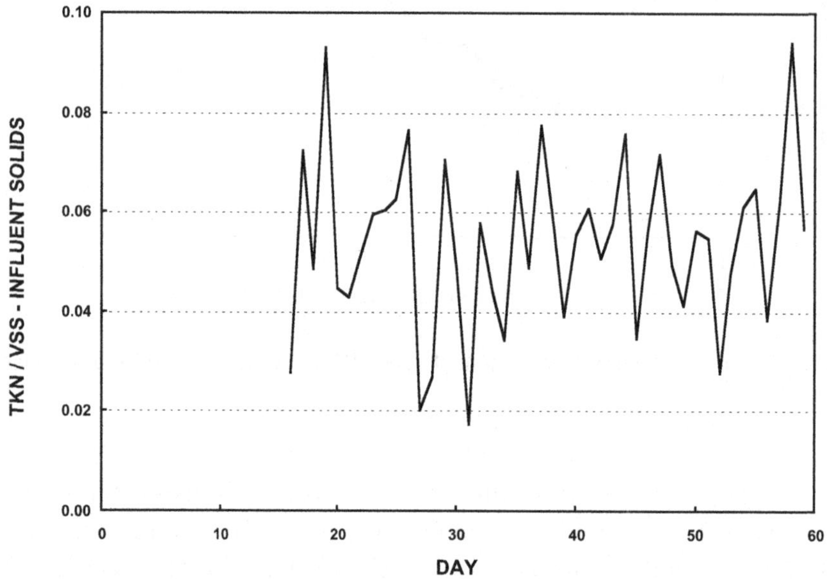

Figure 12-6. Influent Solids TKN/VSS Ratio (Daily Grab Samples)

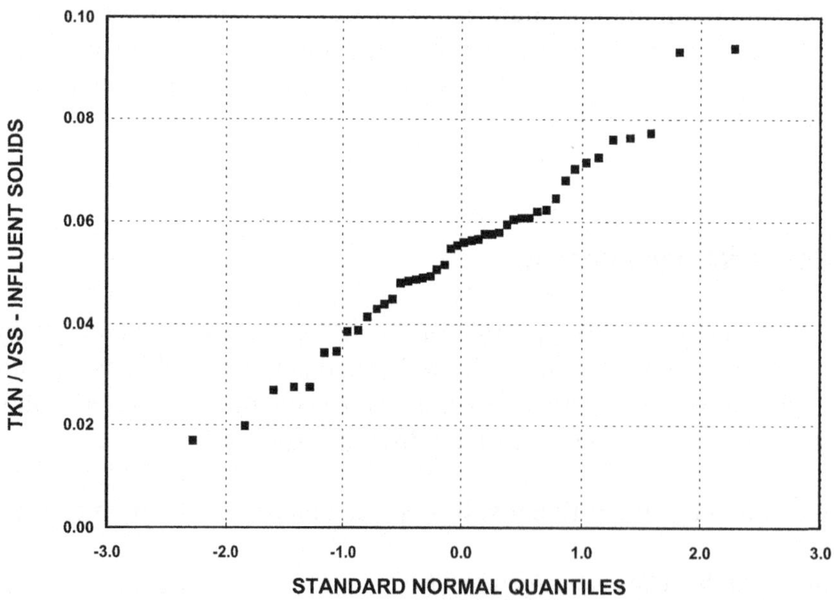

Figure 12-7. Quantile-Quantile Probability Plot of Influent Solids TKN/VSS Ratio (Daily Grab Samples)

Analogous to the comments on COD/VSS, it should be recognized that the VSS generally includes both unbiodegradable (X_I) and biodegradable (X_S) components (and perhaps a biomass component). It is not possible to separate out the two components so as to establish whether the TKN/VSS factors differ for each. From a modeling standpoint, the best first approximation is to assume that the TKN/VSS content for both unbiodegradable COD (X_I) and biodegradable COD (X_S) are equal.

It is important to note that the TKN/VSS ratio for influent solids observed here is significantly different from the ratio suggested for mixed liquor VSS (i.e., 0.1 mg N/mg VSS – see Chapter 14.0). Therefore, it would appear reasonable to assume models with the facility to specify different TKN/VSS ratios for different particulate COD components will better-predict VSS concentrations.

12.6 Influent Solids Phosphorus Content

The phosphorus content the influent solids can be calculated in the same way as the nitrogen content as follows:

$$P/VSS = \frac{Unfiltered\ TP\text{ - }\ Glassfiber\ filtrate\ TP}{VSS} \qquad (12.6\text{-}1)$$

12.7 Considerations

Particulate organic material in municipal influent wastewater generally includes both unbiodegradable (X_I) and biodegradable (X_S) components with associated N and P (and perhaps a biomass component). It is not possible to separate out the two components so as to establish whether the COD/VSS, N/VSS or P/VSS factors differ for each. From a modeling standpoint the best first approximation is to assume that:

- The COD/VSS ratios for both particulate unbiodegradable COD (X_I) and particulate biodegradable COD (X_S) are equal.
- N and P associated with influent solids are evenly distributed between particulate unbiodegradable COD (X_I) and particulate biodegradable COD (X_S).

12.8 References

Chambers, J.M., W.S. Cleveland, B. Kleiner, and P.A. Tukey. 1983. *Graphical methods for data analysis*. Boston, MA: Duxbury Press. 395 pp.

CHAPTER 13.0

MIXED LIQUOR COD/VSS

13.1 Background

The activated sludge simulation models are based on COD as the parameter for quantifying organic material. The volatile suspended solids (VSS) concentration of the mixed liquor in a process is a derived parameter, calculated based on specifying a ratio for COD/VSS for the particulate COD components. Certain activated sludge models assume a single value for the ratio of COD/VSS for all the particulate COD components. Other models have the facility to assign different COD/VSS ratios for the different particulate components. In Chapter 12.0 it was noted that the COD/VSS ratio for influent solids likely is significantly different from the ratio suggested for biomass (1.42 mg COD/mg VSS – see below). Therefore it would appear reasonable to assume that models with the facility to specify different COD/VSS ratios for different particulate COD components will better-predict VSS concentrations.

Particulate organic material that contributes to the measured COD and VSS in the mixed liquor generally includes several components:

- Active biomass and endogenous residue;
- Particulate unbiodegradable COD (X_I) derived from the influent; and
- Residual undegraded particulate biodegradable COD.

The relative amounts of the different components in the mixed liquor are a function of a number of factors – principally, the system SRT and the concentration of particulate unbiodegradable COD (X_I) in the influent wastewater. Because measures of mixed liquor COD and VSS do not distinguish between the different components, the observed COD/VSS ratio reflects the composite value.

When gathering data for an activated sludge process model calibration exercise it is useful to monitor the COD/VSS ratio of the composite mixed liquor solids even though this is not a specific model stoichiometric parameter per se. The reason for monitoring mixed liquor COD/VSS is that it provides an additional crosscheck on simulation results through comparing the observed and model-predicted values. Typically values in the range 1.45 to 1.55 mg COD/mg VSS have been observed for mixed liquor from systems over a range of SRTs and wastewater characteristics.

This section presents the experimental method for determining the COD/VSS ratio, presents example results and demonstrates the basis for the widely applied value of 1.42 mg COD/mg VSS for biomass.

13.2 Biomass COD/VSS

The "formula" $C_5H_7O_2N$ has been suggested as representing the chemical composition of biomass. This can be used to calculate a theoretical biomass COD/VSS ratio by writing the balanced equation for the complete oxidation to carbon dioxide and water (without oxidizing the nitrogen component):

$$\begin{aligned}
&C_5H_7O_2N + 5\,O_2 \rightarrow 5\,CO_2 + NH_3 + 2\,H_2O \\
&Mass\ O_2 = 5 \cdot 32 = 160\ g \\
&Mass\ C_5H_7O_2N = 113\ g \\
&\frac{COD}{Mass} = \frac{160}{113} = 1.42\ gCOD/g
\end{aligned} \qquad (13.2\text{-}1)$$

13.3 Method for Measuring Mixed Liquor COD/VSS

- A mixed liquor sample is removed from the full-scale process or from a bench-scale process operated for wastewater characterization.
- The mixed liquor sample is split into two aliquots, each of which is analyzed as follows:

1. Aliquot (1): A measured volume of sample is filtered through a standard glass-fiber filter (i.e., 1.2 μm nominal pore size) for analysis of mixed liquor VSS content of the sample using the standard method. The filtrate from this sample is analyzed for mixed liquor filtered COD.
2. Aliquot (2): The mixed liquor floccs in this sample are broken up by mixing the sample in a food blender or by inserting an ultrasonic probe. This sample is then analyzed for mixed liquor total COD.

- Using the experimental values derived above, the COD/VSS ratio is determined as follows:

$$COD/VSS = \frac{ML\ Unfiltered\ COD - GF\ Filtrate\ COD}{Mixed\ Liquor\ VSS} \qquad (13.3\text{-}1)$$

Note that the mixed liquor filtered COD should be closely equal to the effluent filtered COD, so that the ratio can also be calculated as follows:

$$COD/VSS = \frac{ML\ Unfiltered\ COD - Effluent\ Filtered\ COD}{Mixed\ Liquor\ VSS} \qquad (13.3\text{-}2)$$

13.4 Examples

A laboratory-scale aerobic activated sludge system receiving raw municipal wastewater as influent was operated at an SRT of approximately 10 days at 20°C. Over a period of 37 days, sampling was conducted to measure the mixed liquor COD/VSS ratio each day. The daily values are plotted in a quantile-quantile probability plot in Figure 13-2 (mean = 1.47 mg COD/mg VSS; standard deviation = 0.07). In this case, the plotted points do not approximate a straight line (normal distribution) as closely as the data in Figure 12-5. This could be for a number of reasons: (1) the smaller data set; (2) inconsistent sampling; or (3) the system not being at steady state, and the relative amounts of the contributing components (such as biomass and influent solids) changing over the sampling period.

It is interesting to note that the average COD/VSS ratio for the mixed liquor (1.47 mg COD/mg VSS) differs significantly from the average COD/VSS ratio for the influent solids to this system (1.83 mg COD/mg VSS). This supports the contention that the mixed liquor COD/VSS is a composite measure, assuming that the 1.42 mg COD/mg VSS value is correct for biomass.

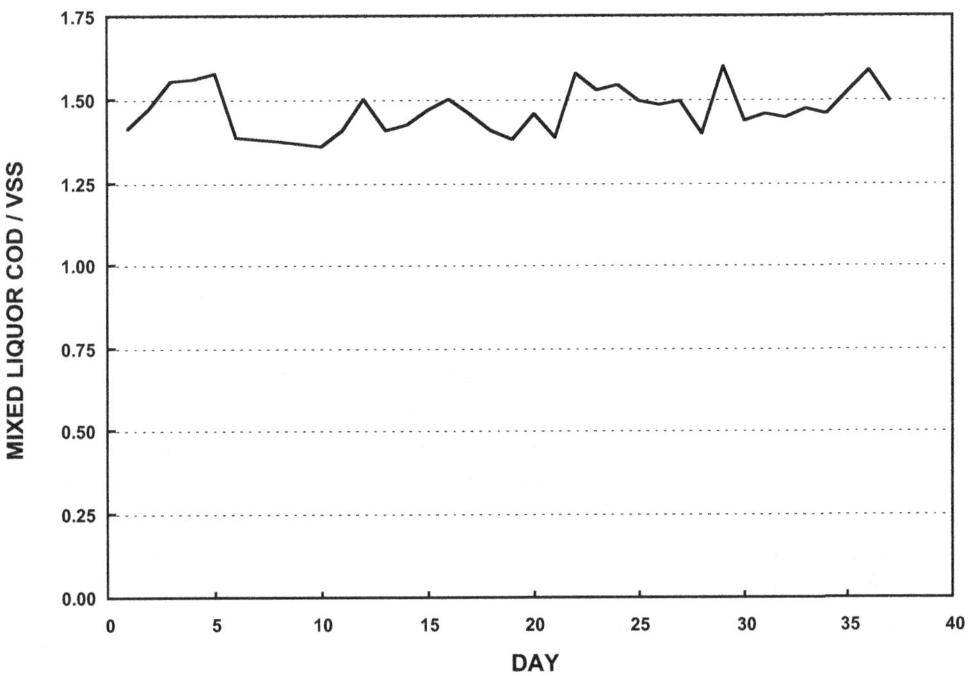

Figure 13-1. Mixed Liquor COD/VSS Ratio (Daily Grab Samples)

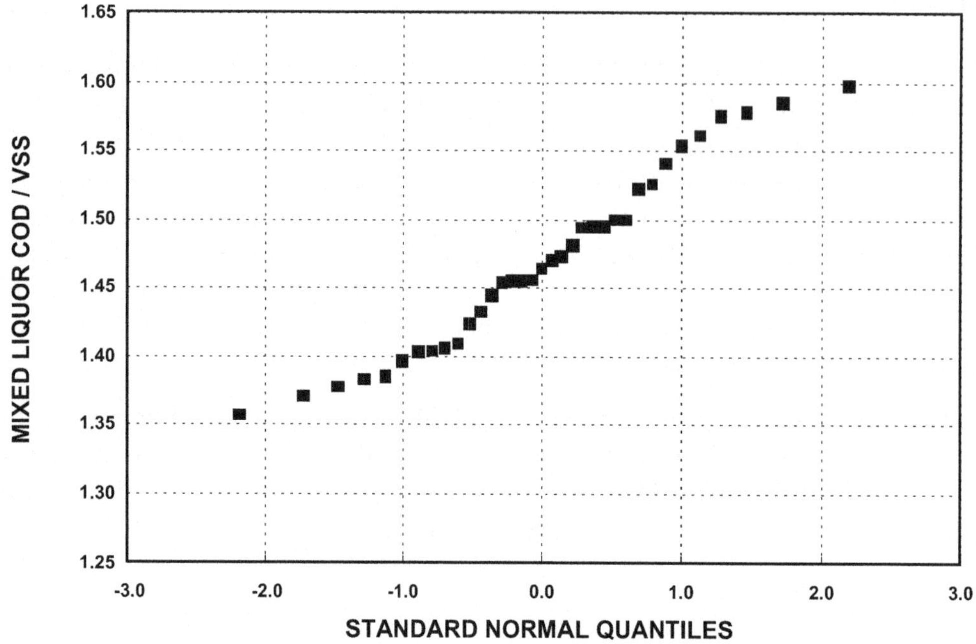

Figure 13-2. Quantile-Quantile Probability Plot of Mixed Liquor COD/VSS Ratio (Daily Grab Samples)

13.5 Considerations

- The evaluation of COD/VSS requires computation from a ratio of experimental data. Small random errors in these data are magnified when the ratio is calculated, potentially leading to a wide variability in experimentally determined values.
- Caution is advised in using limited experimental mixed liquor COD/VSS data to justify adjustment to any activated sludge model parameters such as the biomass COD/VSS ratio. It is difficult to obtain a homogenous, representative sample (of small volume) for the COD analysis. Therefore, deviations in measured mixed liquor COD/VSS values from the usual range (1.45 to 1.55 mg COD/mg VSS) typically are the result of experimental error.

CHAPTER 14.0

Mixed Liquor Nitrogen and Phosphorus Content

14.1 Background

When gathering data for an activated sludge process model calibration exercise it is useful to monitor the nitrogen and phosphorus content of the composite mixed liquor solids even though these are not specific model stoichiometric parameters per se. Nitrogen (N) and phosphorus (P) associated with solids in the mixed liquor includes several components:

- Synthesis N and P in active biomass and endogenous residue
- N and P associated with particulate unbiodegradable COD (X_I) derived from the influent
- Residual N and P associated with particulates from the influent, which have not been lysed and converted to ammonia or phosphate, respectively.
- Stored polyphosphate (in biological P removal systems).

The relative amounts of the different components in the mixed liquor are a function of a number of factors – principally, the system SRT and the amount of N or P associated with particulate unbiodegradable COD (X_I) in the influent wastewater. Because measures of mixed liquor VSS and N or P content do not distinguish between the different components, the observed COD/VSS ratio reflect the composite value.

Typically, ranges of values for mixed liquor from systems over a range of SRTs and wastewater characteristics are:

- 0.08 to 0.12 mg N/mg VSS (equivalent to 0.054 to 0.081 mg N/mg COD for a mixed liquor COD/VSS of 1.48 mg COD/mg VSS);
- 0.025 to 0.035 mg P/mg VSS in systems without biological P removal (equivalent to 0.017 to 0.024 mg P/mg COD for a mixed liquor COD/VSS of 1.48 mg COD/mg VSS).

This section presents the experimental method for determining the mixed liquor N/VSS and P/VSS ratios, and presents example results.

14.2 Method for Measuring Mixed Liquor N/VSS and P/VSS

- A mixed liquor sample is removed from the full-scale process or from a bench-scale process operated for wastewater characterization.
- The mixed liquor sample is split into two different aliquots, each of which is analyzed as follows:

1. Aliquot (1): A measured volume of sample is filtered through standard glass-fiber filter paper (i.e., 1.2 μm nominal pore size) for analysis of mixed liquor VSS content of the sample using a standard method. The filtrate from this sample is analyzed for mixed liquor filtered TKN and filtered TP.
2. Aliquot (2): The mixed liquor flocs in this sample are broken up by mixing the sample in a food blender or by inserting an ultrasonic probe into the sample. This sample is then analyzed for mixed liquor total TKN and TP.

- Using the experimental values derived above, the N/VSS and P/VSS ratios are determined as follows:

$$N/VSS = \frac{ML\ Unfiltered\ TKN - GF\ Filtrate\ TKN}{Mixed\ Liquor\ VSS} \qquad (14.2\text{-}1)$$

$$P/VSS = \frac{ML\ Unfiltered\ TP - GF\ Filtrate\ TP}{Mixed\ Liquor\ VSS} \qquad (14.2\text{-}2)$$

Alternatively, the mixed liquor filtered TKN and TP should be closely equal to the effluent filtered TKN and TP, respectively, so that the ratios can also be calculated as follows:

$$N/VSS = \frac{ML\ Unfiltered\ TKN - Effluent\ Filtered\ TKN}{Mixed\ Liquor\ VSS} \qquad (14.2\text{-}3)$$

$$P/VSS = \frac{ML\ Unfiltered\ TP - Effluent\ Filtered\ TP}{Mixed\ Liquor\ VSS} \qquad (14.2\text{-}4)$$

14.3 Example

Two SBRs were operated at SRTs of 12 days for 44 days to determine wastewater characteristics. SBR1 was operated with unsettled primary influent. SBR2 was operated with primary effluent. The data in Table 14-1 are averages for daily analyses conducted over the last 15 days of SBR operation. No TP data were collected in this study, so a P/VSS value was not calculated.

Table 14-1. SBR Mixed Liquor VSS and TKN Data

Unit	Parameter		
	Mixed Liquor VSS (mg/L)	Mixed Liquor Unfiltered TKN (mg/L)	Effluent (Decant) Filtered TKN (mg/L)
SBR1	1,206	96	2.7
SBR2	686	68	2.7

From these data, the following values for the N/VSS can be calculated:

- SBR1:

$$N/VSS = \frac{96 - 2.7}{1206} = 0.077\ mgN/mgVSS$$

- SBR2:

$$N/VSS = \frac{68 - 2.7}{686} = 0.095\ mgN/mgVSS$$

14.4 P/VSS in Biological Phosphorus Removal Systems

Monitoring phosphorus content of the mixed liquor solids is very important in the analysis and modeling of biological phosphorus removal activated sludge systems. The data provide a basis for assessing the overall P mass balance and system P removal performance. As an example, Copp (1998) operated laboratory scale anaerobic/aerobic SBR systems with a synthetic influent wastewater at an approximately uniform SRT of 8 to 10 days. Each system received a uniform influent COD load of acetate (440 mg COD/L). The influent phosphate concentration ranged from 4.5 mg/L to 55.5 mg/L. At the low end, most of the 4.5 mg P/L was required for biomass synthcsis, and very little P was available for excess biological removal. At the high end, with 55.5 mg P/L, the system still reduced effluent soluble P to less than 1 mg P/L. Table 14-2 summarizes results pertinent to the P content of the mixed liquor solids for each system. The VSS concentrations were reasonably uniform because the loading was fixed and the SRTs only varied over a limited range. However, as the influent P concentration increased (and essentially all the excess P was removed biologically), so the P/VSS ratio increased.

Table 14-2. Biological Phosphorus Removal SBR Steady State Results with Fixed Influent COD Load and Varying Phosphorus Load

Influent P (mg/L)	SRT (days)	VSS (mg/L)	TSS (mg/L)	VSS:TSS	P Content (mgP/mgVSS)
4.5	10.2	2,041	2,448	0.83	0.051
10.6	8.2	1,456	2,062	0.71	0.107
12.5	10.0	2,068	2,953	0.70	0.120
25.0	9.3	1,892	3,347	0.57	0.188
55.5	9.1	2,146	5,241	0.41	0.416

14.5 Considerations

- The reason for monitoring mixed liquor N/VSS and P/VSS is that these provide additional crosschecks on simulation results through comparing the observed and model-predicted values.
- For non-biological P removal systems, N/VSS is the most important parameter to monitor, in conjunction with the influent solids N/VSS. These are important as a crosscheck on the system N balance and the amount of N available for nitrification.
- The activated sludge models include a stoichiometric parameter for the amount of synthesis N taken up per unit biomass generated in growth. Typically the value is say 0.07 mg N/mg biomass COD (0.10 mg N/mg VSS at a biomass COD/VSS of 1.42 mg COD/mg VSS). (The models may allow different factors for different biomass populations.) The models also may allow specification of a different factor for the N content of endogenous residue. Whether or not the active biomass and endogenous residue have the same N content is an issue for further research. However, it has been suggested that the endogenous residue N content is lower. This has implications for nitrification modeling because the consequence of a lower endogenous residue N content is more N available for nitrification.
- The above issue also applies to the P content.

14.6 References

Copp, J.B. 1998. *COD balances in biological nutrient removal (nitrogen and phosphorus) activated sludge systems*, Ph. D. Thesis, McMaster University, Hamilton, Ontario, Canada.

Water Research Commission. 1984. *Theory, design and operation of nutrient removal activated sludge processes*. P.O. Box 824, Pretoria, South Africa: Water Research Commission.

CHAPTER 15.0

DENITRIFICATION STOICHIOMETRY

15.1 Background

In activated sludge systems incorporating unaerated zones (or phases), heterotrophic organisms utilize nitrate in place of oxygen as the terminal electron acceptor in the utilization of organic substrate. Modeling of denitrification involves consideration of both the stoichiometry and the kinetics. This chapter deals with the stoichiometry of denitrification. Chapter 18.0 considers the kinetics.

Note: The extent of denitrification in systems receiving only municipal wastewater is strongly governed by the amount of readily biodegradable COD (RBCOD) in the influent. This endorses the importance of measuring RBCOD content for accurate modeling of nitrogen removal.

15.2 Heterotroph Yield Coefficient in Anoxic Growth

Microbiological studies of the similarities and differences between aerobic and anoxic respiration in pure cultures indicate that essentially the same pathways are followed, but with lower ATP generation when nitrate is the electron acceptor versus oxygen. A direct relationship exists between ATP production and biomass production, so sludge production should be lower under anoxic conditions.

In aerobic growth, utilization of 1 unit of substrate COD requires ($1 - Y_{HET}$) units of oxygen, where Y_{HET} is the yield coefficient (mg biomass (as COD) formed per unit substrate COD utilized). In reverse, the amount of substrate COD stabilized by 1 mg of oxygen is:

$$COD = \frac{1}{1 - Y_{HET}} \qquad (15.2\text{-}1)$$

The value of Y_{HET} in activated sludge models typically is approximately 0.666 mg biomass COD per mg COD utilized. That is, the models predict that 3 mg of COD is utilized when 1 mg of oxygen is consumed.

A 2.86 conversion factor is incorporated in this expression when modeling denitrification. This is merely a factor to account for the amount of oxygen or nitrate required to accept the same number of electrons, and has been confirmed experimentally (e.g., Copp and Dold, 1998a). The factor can be deduced from writing the half reactions for accepting a mole of electrons:

$$e^- + \frac{1}{4} O_2 + H^+ \rightarrow \frac{1}{2} H_2O \qquad (15.2\text{-}2)$$

$$e^- + \frac{1}{5} NO_3^- + \frac{6}{5} H^+ \rightarrow \frac{1}{10} N_2 + \frac{3}{5} H_2O \qquad (15.2\text{-}3)$$

Equating the masses of oxygen and nitrate:

$$\frac{1}{5}\ mol\ NO_3^- \equiv \frac{1}{4}\ mol\ O_2$$

$$\frac{14}{5}\ g\ NO_3^- - N \equiv \frac{32}{4}\ g\ O_2 \qquad (15.2\text{-}4)$$

$$1\ g\ NO_3^- - N \equiv 2.86\ g\ O_2$$

Therefore, in anoxic growth the amount of nitrate (as N) consumed in the utilization of 1 mg of substrate COD is:

$$1\ mg\ COD \xleftrightarrow{stabilized\ by} \left(1 - Y_H\right) mg\ O_2 \cdot \frac{1\ mg\ NO_3 - N}{2.86\ mg\ O_2} = \frac{\left(1 - Y_H\right)}{2.86}\ mg\ NO_3 - N \qquad (15.2\text{-}5)$$

For anoxic growth, the equivalent to Eq. 15.2-1, but for the amount of substrate COD required to denitrify 1 mg nitrate (as N) is:

$$COD = \frac{2.86}{1 - Y_{HET}} \qquad (15.2\text{-}6)$$

Assuming the value for $Y_{HET} = 0.666$ for aerobic growth, equation 15.2-6 predicts that 8.6 mg of substrate COD are required to denitrify 1 mg of nitrate (as N). However, a lower anoxic yield implies that less COD is required to denitrify a given amount of nitrate. Conversely, a lower anoxic yield implies that more nitrate is denitrified for a given amount of COD consumed.

Re-arranging equation 15.2-6:

$$Y_{HET} = 1 - \frac{2.86}{COD} \quad (15.2\text{-}7)$$

An indication of whether the yield is lower in anoxic versus aerobic growth can be obtained from equation 15.2-7, using literature information on substrate requirements for denitrification. For example, in methanol addition for denitrification the amount of methanol required typically is reported as 3.2 mg methanol per mg NO_3-N removed. This corresponds to 4.8 mg COD/mgNO_3-N because methanol has a COD/mass ratio of 1.5. The anoxic yield can be calculated from equation 15.2-7:

$$Y_{HET,ANOXIC} = 1 - \frac{2.86}{4.8} \quad (15.2\text{-}8)$$

$$= 0.404 \; mg \; biomass \; COD/mg \; substrate \; COD$$

Copp and Dold (1998b) conducted 32 batch tests to demonstrate a significant difference in biomass yield under anoxic and aerobic conditions using a variety of simple soluble substrates. Batch tests were performed under anoxic and aerobic conditions using a seed of either activated sludge from one of two municipal wastewater treatment facilities or a prepared sample of *Pseudomonas denitrificans*. Irrespective of the test organism, the calculated yields were consistent for each electron acceptor. Results for the anoxic tests are summarized in Table 15-1. For these tests, the average anoxic yields were 62 percent of the aerobic yield. COD balance calculations were performed and confirmed that all of the initial COD could be accounted for in terms of residual soluble COD, biomass production and electron acceptor consumption.

Table 15-1. Yield under Anoxic Conditions with Various Soluble Substrates (Copp & Dold, 1998b)

Seed Organism(s)	Soluble Substrate	# of Expts	Yield (g COD/g COD)	COD Balance
P. denitrificans	citrate	12	0.424	0.95
P. denitrificans	acetate	2	0.192	1.04
P. denitrificans	glucose	1	0.382	1.06
Mixed liquor	citrate	4	0.512	1.09
Mixed liquor	glucose	2	0.365	1.01
Mixed liquor	nutrient broth	2	0.366	0.98
Mixed liquor	Bacto-peptone	2	0.363	0.95
Mixed liquor	yeast extract	1	0.363	1.02

St. Amant and McCarty (1969) reported data on substrate requirements for denitrification on the basis of a "consumptive ratio" – the equivalent amount of organic carbon used per equivalent amount of nitrate reduced. The CR (consumptive ratio) values ranged from 1.3 to 1.47 to 1.69 for methanol, ethanol and sugar, respectively. Substrates with lower CR values have lower cell yields and are more efficient for nitrate reduction. Acetate has a CR value similar to that of methanol.

Much of the literature information on substrate requirements for denitrification has been for separate-sludge tertiary denitrification systems. Very limited information on substrate addition to anoxic-aerobic activated sludge systems is available, particularly with regard to accurately quantifying yield in anoxic growth. Spérandio et al. (1999) measured carbon dioxide evolution rates in batch tests and reported aerobic and anoxic yields for activated sludge samples sourced from a full-scale aerobic plant or an anoxic-aerobic pilot plant. Several different substrates were used in the batch tests. Excluding the results for acetate as sole substrate (which gave a low value) the results for a mixture of acetic acid and starch or glucose gave mean Y_{HET} values for aerobic growth of 0.66 to 0.67, and average anoxic Y_{HET} values of 0.54 to 0.57 (all as mg biomass COD per mg substrate COD). The latter compares favourably with the theoretical value of 0.50 recommended by Orhon et al. (1996) as a default Y_{HET} for anoxic growth based on bioenergetics for a municipal sewage.

In summary, from both theoretical considerations and experimentation, the heterotrophic yield coefficient applied in activated sludge models should be lower under anoxic conditions than aerobic conditions. A limited review of the literature appears to indicate that anoxic yields (and possibly aerobic yields) with very simple substrates such as acetate are lower than for the mix of soluble substrates in a typical municipal wastewater. The consensus appears to be that a value for anoxic Y_{HET} in the range of 0.52 to 0.54 mg biomass COD per mg substrate COD is appropriate for the mix of substrates in municipal wastewater.

15.3 Reduced Anoxic Yield: Modeling Implications

The issue of a reduced yield under anoxic *versus* aerobic conditions is extremely important in modeling denitrification behavior. Equation 15.2-6 quantifies the amount of substrate required to denitrify 1 mg of nitrate-N. Applying the aerobic Y_{HET} of 0.666 mg biomass COD per mg substrate COD predicts that 8.6 mg of substrate COD are required to denitrify 1 mg of nitrate. In contrast, a value of 0.54 predicts that only 6.2 mg COD is required. The consequences of applying the higher aerobic Y_{HET} will be to:

- Underpredict the amount of denitrification.
- Overpredict volatile solids production. (This aspect is made more complex by recent suggestions in the literature that heterotroph decay rate under anoxic conditions is lower than under aerobic conditions, presumably as a result of reduced predation in the absence of oxygen. This would slightly offset the reduced sludge production when applying a reduced anoxic yield coefficient.)

The concept of a separate reduced anoxic yield coefficient in the activated sludge models was first introduced in the Barker and Dold (1997) model. The ASM1 and ASM2d models do not incorporate a reduced Y_{HET} for anoxic growth.

Specific simple substrates such as methanol and acetate often are added to systems to enhance denitrification. These simple substrates exhibit an even lower Y_{HET} than the 0.52 to 0.54 value suggested for typical municipal wastewater. For correct modeling of denitrification behavior in systems receiving influent wastewater, and with supplemental carbon addition to anoxic zones, future model improvements should include:

- The specific supplemental carbon compound as a state variable; and
- Model kinetic process equations for the anoxic utilization of the specific compound based on a separate anoxic yield coefficient for the compound.

15.4 Biodegradable COD Available for Denitrification

With certain wastewaters (likely industrial wastewaters) it is possible that not all of the soluble aerobically degradable organic substrate can be utilized in anoxic growth. In cases where this is considered a possibility, parallel aerobic and anoxic batch tests can be performed, combining acclimated mixed liquor and influent, and monitoring the filtered COD change with time. If the anoxic test levels off at a higher soluble COD, this would indicate the presence of soluble organic substrate that cannot be utilized for denitrification under anoxic conditions. Useful modeling information can be derived if oxygen uptake rate (OUR) is monitored in the aerobic test, and nitrate concentration in the anoxic test.

Currently the models do not distinguish between soluble biodegradable substrate that can or cannot be used for denitrification. In cases where this phenomenon is evident, extension of the base model will be necessary.

15.5 References

Barker, P.S., and P.L. Dold. 1997. General model for biological nutrient removal activated sludge systems: model presentation. *Water Environ. Res.* 69(5):969-984.

Copp, J.B., and P.L. Dold. 1998a. Confirming the nitrate-to-oxygen conversion factor for denitrification. *Water Res.* 32:1296-1304.

Copp, J.B., and P.L. Dold. 1998b. Comparing sludge production under aerobic and anoxic conditions. *Water Sci. Tech.* 38:285-294.

St. Amant, P.P., and P.L. McCarty. 1969. Treatment of high nitrate wastewaters. *AWWA Journal* 61(12):659-662.

Orhon, D., S. Sözen, and N. Artan. 1996. The effect of heterotrophic yield on the assessment of the correction factor for anoxic growth. *Water Sci. Tech.* 34(5-6):67-74.

Spéranddio, M., V. Urbain, J.M. Audic, and E. Paul. 1999. Use of carbon dioxide evolution rate for determining heterotrophic yield and characterizing denitrifying biomass. *Water Sci. Tech.* 39(1):139-146.

CHAPTER 16.0

NITRIFICATION RATE

16.1 Background

Nitrifier growth rate arguably is the most important parameter in the design of activated sludge systems, particularly for nutrient removal systems. Therefore, estimation of the μ_{AUT} parameter is extremely important for design purposes. This is also the case if process simulation is to provide reliable results.

Existing methods for measuring nitrification rate are time-consuming and costly. Therefore, in this project substantial research attention has been directed at developing simple and rapid methods for determining this crucial parameter.

16.2 Modeling Nitrification

Nitrification is the oxidation of saline ammonia to nitrate. Traditionally, this has been ascribed to two specific genera of autotrophic bacteria, the *Nitrosomonas* and the *Nitrobacter*, where the *Nitrosomonas* oxidize free and saline ammonia to nitrite and the *Nitrobacter* oxidize nitrite to nitrate. Research in recent years has identified other bacteria involved in the nitrification process (Dionisi et al., 2002, Biesterfeld et al., 2001, Daims et al., 2000, 2001, Juretschko et al., 1998, Scharmm et al., 1998, Wagner et al., 1998). The first step (i.e., conversion of ammonia to nitrite) usually is the rate-limiting one in the two-step nitrification process. As a result, in the activated sludge models, nitrification usually is considered a single step process mediated by one group of surrogate nitrifying autotrophic bacteria. The growth rate parameter for the nitrifiers (μ_{AUT}) defines the maximum specific growth rate of this surrogate organism mass.

$$r_{XAUT} = \mu_{AUT} \frac{S_{NH}}{K_{NH} + S_{NH}} \cdot X_{AUT}$$
$$r_{SNO} = \left(\frac{\mu_{AUT}}{Y_{AUT}}\right) \frac{S_{NH}}{K_{NH} + S_{NH}} \cdot X_{AUT} \qquad (16.2\text{-}1)$$

where:

r_{XAUT} = nitrifier organism growth rate

r_{SNO} = nitrite/nitrate production rate

μ_{AUT} = maximum specific nitrifier growth rate

Y_{AUT} = nitrifier yield coefficient

K_{NH} = nitrification half-saturation coefficient

S_{NH} = ammonia concentration

Values for μ_{AUT} (at 20°C) ranging from 0.2 to 1.0 d^{-1} have been observed in activated sludge systems treating a number of different wastewaters. (In a sense, μ_{AUT} has been regarded as a wastewater characteristic.) The variation may correlate with the extent of the industrial component in the wastewater; it has been suggested that with increased industrial input there is an increased possibility of inhibition of nitrifying organisms. However, many other factors influence the observed nitrate production rate; for example:

- Dissolved oxygen (DO) concentration,
- Temperature,
- PH,
- Alkalinity (or perhaps the availability of HCO_3), and
- Mixing and floc size, and potential for pH/DO gradients within the floc.

Temperature impacts nitrification kinetics significantly. For example, the EPA *Design Manual for Nitrogen Control* (U.S. EPA, 1993) suggests the following temperature dependency:

$$\mu_{AUT,T} = 0.47 \cdot e^{0.098(T-15)} = 0.47 \cdot 1.103^{(T-15)} \qquad (16.2\text{-}2)$$

There is considerable uncertainty in the literature regarding the impact of temperature on nitrifier growth rate. Researchers have suggested temperature dependency coefficients (1.103 in above equation) in the range of 1.076 to 1.127 (a higher temperature dependency factor implies a larger temperature dependency). The value chosen is very significant in full-scale plant capacity rating and design because of the impact on the SRT required for nitrification at the winter temperature condition (Figure 16-1).

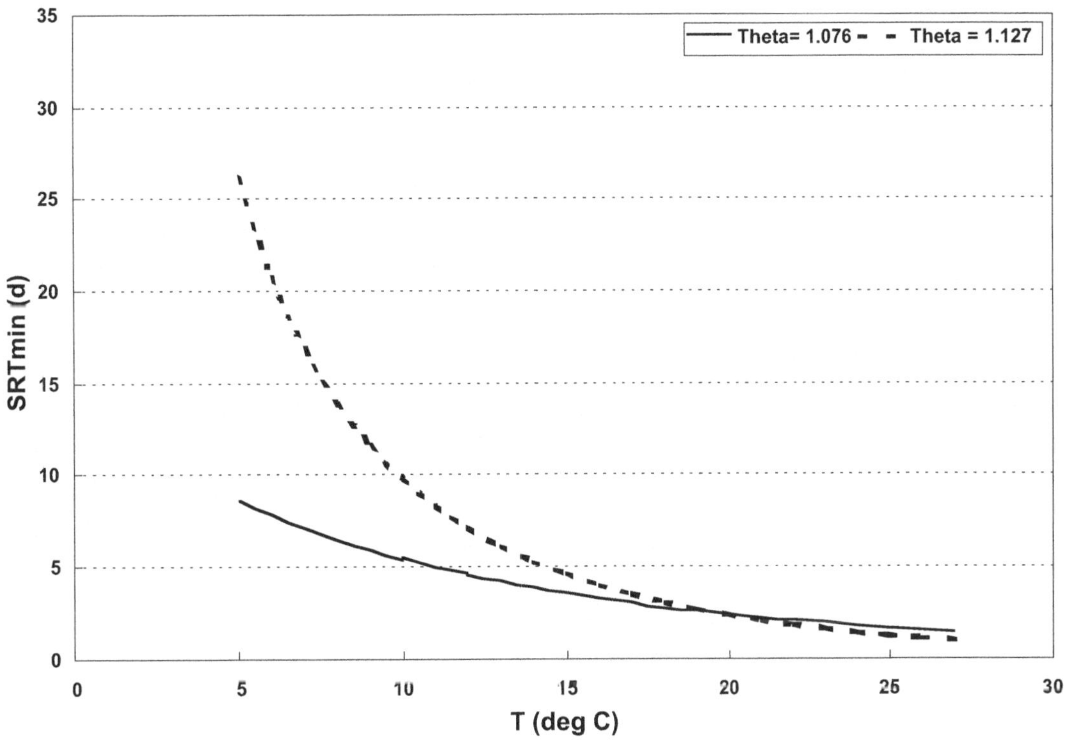

Figure 16-1. Impact of the Temperature Dependency Coefficient (θ) on the Calculated Minimum SRT (SRT_{min}) Required for Nitrification

In the activated sludge models, the impact of DO concentration on nitrification rate is incorporated by multiplying the process rate by a switching function that ranges from zero (at zero DO concentration) to 1 (at high DO concentrations). Accordingly, equation 16.2-1 is extended as follows:

$$r_{NO} = \left(\frac{\mu_{AUT}}{Y_{AUT}}\right)\frac{S_{NH}}{K_{NH}+S_{NH}} \cdot X_{AUT} \cdot \frac{S_O}{K_{SO}+S_O} \qquad (16.2\text{-}3)$$

where:

S_O = DO concentration (mg/L)

K_{SO} = switching function half-saturation coefficient

Note: Low DO concentration reduces nitrification rate. However, there may be situations where DO is implicated in a rate reduction, even though DO is relatively high. For example, in situations of low mixing intensity and large floc size, but high bulk liquid DO concentration, there may be some diffusional limitations that reduce nitrification rate. The K_{SO} parameter should be applied with caution.

16.3 Approaches to Measuring or Estimating μ_{AUT}

Essentially there are two approaches for measuring or estimating the μ_{AUT} parameter:

- Bioassay methods.
- By simulation, through matching modeled activated sludge system dynamic response behavior (ammonia and nitrate) to experimental observations. With this approach, difficulties often are encountered because many of the parameters impacting nitrification rate either are not known/measured or are not controlled. Also, extensive data gathering is required for input to the simulation model.

Conceptually, the bioassay methods involve two extremes of the measurement regimes, both of which are based on combining nitrifier biomass with ammonia (usually at a concentration well in excess of the half-saturation coefficient, K_{NH}, so that the rate is at the maximum):

1. Low F/M tests where the change in nitrifier mass during the test is small, and there is a linear response in the measured parameter (usually nitrate concentration).
2. High F/M tests where the initial nitrifier mass is small relative to the mass produced during the test (i.e., exponential growth), and there is an exponential-type response in the measured parameter (usually nitrate concentration, or perhaps oxygen utilization rate).

Note: F/M denotes the ammonia/nitrifier concentration ratio in this context.

An intermediate F/M condition is applied in the washout test procedure described later. In this test, a flow-through reactor is initiated with nitrifying sludge from the full-scale plant, and then operated at a low SRT (less than the washout SRT).

Note: Downing et al. (1964) used the concept of the minimum SRT as a means for estimating nitrifier growth rate in activated sludge systems; namely, operating continuous flow systems at low SRT to identify the SRT below which nitrifier growth could not be sustained. This was termed the "washout" method. This washout approach was used widely as a means for estimating nitrifier maximum specific growth rate in continuous flow activated sludge systems, particularly during the 1970s. (The Washout Method presented in this chapter is based on the same concept, but is implemented differently.)

In the low F/M tests (such as the SBR method – see Appendix D), it is necessary to know the concentration of nitrifiers in the test. This is a non-trivial requirement, and usually is not known with any degree of accuracy for mixed liquor drawn from a full-scale plant. This is because in order to estimate X_{AUT}, it is necessary to have detailed plant operating information on parameters such as influent loading, sludge age and effluent quality. Usually this detailed information is not available, and a full-scale plant is seldom close to a steady state. For this reason, the tests usually are performed in bench-scale systems where these factors can be controlled and measured accurately.

It is essential that *both* nitrite and nitrate concentrations are measured in all of the different bioassay methods (e.g., low F/M, high F/M and washout). This is necessary because in modeling nitrification as a single-step process (i.e., using equation 16.2-1), the nitrification rate is reflected by the sum of nitrite and nitrate production. Neglecting nitrite production from the analysis would lead to an erroneous (i.e., depressed) nitrification rate estimate. In the discussion of the bioassay methods below, equations are presented in terms of the total nitrite/nitrate concentration (S_{NO} = nitrite + nitrate).

Note: Before discussing the individual bioassay methods it should be stressed that experimental technique is extremely important. Care must be taken to ensure that factors such as alkalinity, pH and dissolved oxygen concentration are carefully controlled to levels that will not skew the results.

Note: In the examples presented in this chapter, the specific nitrifier decay rate, and its temperature dependency, are assumed as:

$$b_{AUT,T} = 0.17 \cdot 1.029^{(T-20)} \ d^{-1}$$

Selection of b_{AUT} is discussed in detail in Chapter 17.0.

16.4 Low F/M Bioassay

In the low F/M bioassay an appropriate amount of nitrifying mixed liquor is combined with influent wastewater. Ammonia may be added to the influent, or alternatively the test could be performed using secondary effluent spiked with ammonia. The DO concentration is maintained at a level greater than about 6 mg/L to ensure that a low DO concentration does not limit the nitrification rate. Parameters monitored during the test would include ammonia and nitrate concentrations, pH and perhaps oxygen utilization rate (OUR).

The basis of this method for estimating μ_{AUT} is that the rate of production of nitrite/nitrate over the first part of the test is constant, and can be directly linked to μ_{AUT}. An example of the response of nitrite/nitrate and ammonia concentrations, and OUR, in a batch test combining influent wastewater and mixed liquor is shown in Figure 16-2. Features to note are:

- The S_{NO} response essentially is linear until ammonia is depleted.
- The rates of change of ammonia and S_{NO} concentration are not the same. The ammonia response is impacted by hydrolysis and ammonification of organic nitrogen, ammonia requirements for synthesis and nitrification.
- In the OUR response, initially the OUR is high (for utilization of RBCOD), after which the OUR drops to a plateau corresponding to the maximum nitrification rate. OUR drops off when ammonia is depleted.

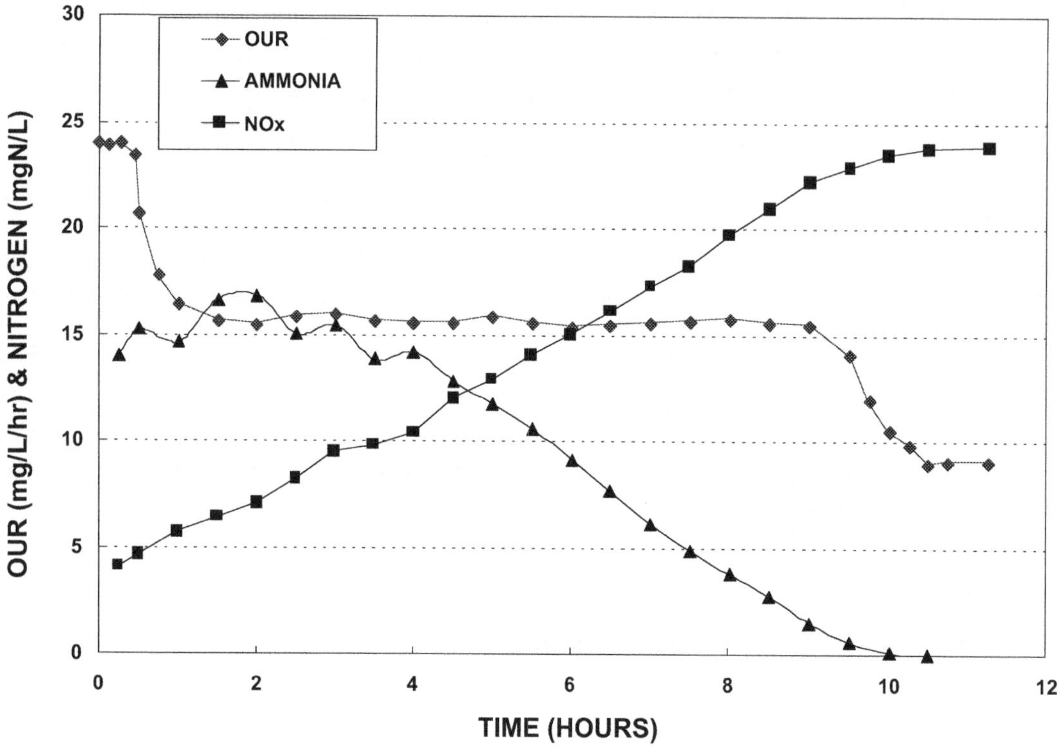

Figure 16-2. Example of Nitrite/Nitrate and Ammonia Response in a Low F/M Batch Test for Estimating μ_{AUT}

16.4.1 Estimating μ_{AUT} from Nitrite/Nitrate Response

The rate of generation of nitrite/nitrate is equal to the rate of disappearance of ammonia *utilized for nitrification*:

$$\frac{dS_{NO}}{dt} = -\frac{dS_{NH}}{dt} \qquad (16.4.1\text{-}1)$$

At the start of the test, the concentration of ammonia in the reactor is high enough (relative to K_{NH}) to ensure that the nitrification rate is at the maximum. That is, the specific growth rate ($\mu = \mu_{AUT} \cdot S_{NH}/(K_{NH} + S_{NH})$) essentially is equal to the maximum specific growth rate (μ_{AUT}). Hence, from equation 16.2-1:

$$\frac{dS_{NO}}{dt} = \mu_{AUT} \cdot \frac{X_{AUT}}{Y_{AUT}} \qquad (16.4.1\text{-}2)$$

Rearranging equation 16.4.1-2:

$$\mu_{AUT} = \frac{dS_{NO}}{dt} \Big/ \frac{X_{AUT}}{Y_{AUT}} \qquad (16.4.1\text{-}3)$$

The rate of change of nitrite/nitrate concentration (nitrification rate) is obtained from the slope of the linear increase in nitrite/nitrate concentration observed during the test period as shown in Figure 16-3, i.e., the increase in nitrite/nitrate over a measured time interval.

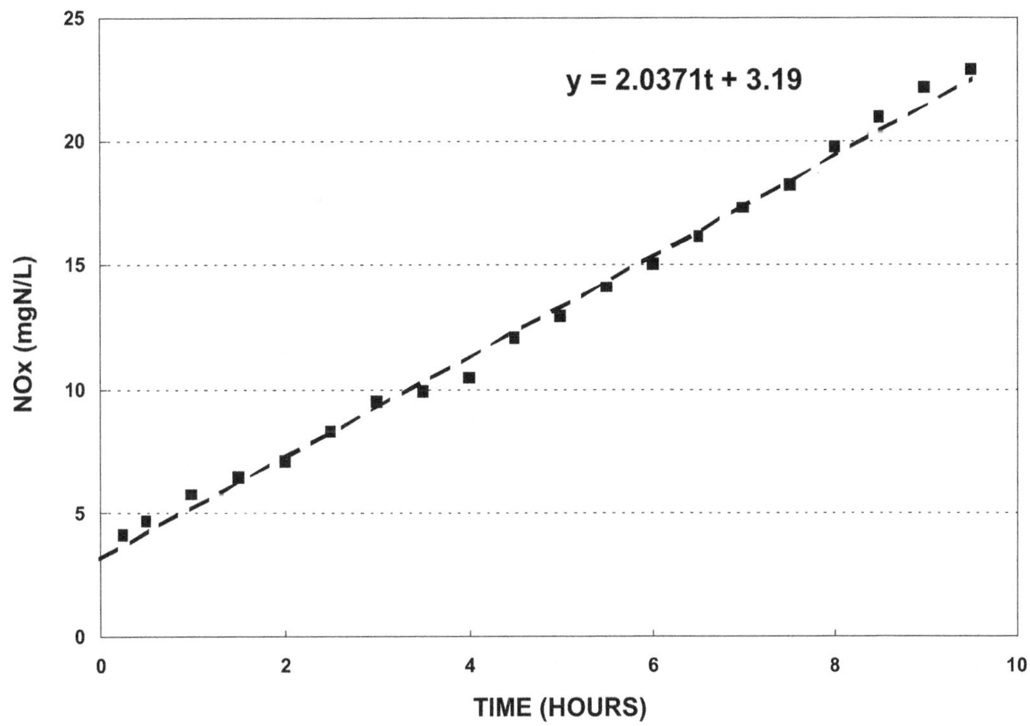

Figure 16-3. Nitrite/Nitrate Production Rate in a Low F/M Batch Test for Estimating μ_{AUT}

In this test, the nitrite/nitrate production rate was 2.0371 mgN/L/hour, and the quotient (X_{AUT}/Y_{AUT}) was estimated to be 51.1 mg/L (see below). Substituting in equation 16.4.1-3 yields:

$$\mu_{AUT} = 2.0371 \cdot {}^{24}\!/_{51.1} = 0.96\ d^{-1}$$

The test was performed at 23°C. The value for μ_{AUT} can be adjusted to a reference temperature (of 20°C here) using a temperature dependency correction factor of 1.072 in equation 16.4.1-4. This yielded an estimate for $\mu_{AUT,20}$ of 0.78 d^{-1} for this case.

$$\mu_{AUT,T} = \mu_{AUT,20} \cdot \theta^{(T-20)} \qquad (16.4.1\text{-}4)$$

In this test, the DO concentration was controlled between 4 mg/L and 6 mg/L (average of 5 mg/L). The μ_{AUT} value to apply in the model is derived by adjusting the observed rate for the DO concentration switching function according to equation 16.2-3:

$$\mu_{AUT,20} = \mu_{AUT,MEASURED} \Big/ \frac{S_O}{K_{SO} + S_O} = 0.78 \Big/ \frac{5}{0.5+5}$$

$$= 0.86\ d^{-1}$$

16.4.2 Calculation of Nitrifier Concentration

Estimating μ_{AUT} from the batch test results essentially is based on knowing the concentration of nitrifiers. Applying equation 16.4.1-3 actually requires calculation of the quotient (X_{AUT}/Y_{AUT}) rather than the nitrifier concentration itself (i.e., the estimate obtained for μ_{AUT} is independent of the value assumed for the nitrifier yield).

The quotient (X_{AUT}/Y_{AUT}) in a system can be estimated using an activated sludge model. For example, the ASM1 or ASM2 models could be used to simulate the system from which the sludge for the batch test is drawn, and thus generate the predicted value for X_{AUT}. The quotient (X_{AUT}/Y_{AUT}) should be calculated based on the Y_{AUT} value applied in the model.

Note: The second form of equation 16.4.2-1 is valid only for systems without denitrification.

An alternative approach would be to hand-calculate the quotient (X_{AUT}/Y_{AUT}) using a simple steady state model such as that in Water Research Commission (1984):

(The reader is cautioned that, strictly speaking, there is not a steady-state condition for an SBR process, and this model should be applied only once the SBR has reached a quasi-steady state.)

$$\frac{X_{AUT}}{Y_{AUT}} = \frac{Q \cdot (TKN_{INF} - TKN_{EFF} - N_S) \cdot \theta_X}{(1 + b_{AUT,T} \cdot \theta_X) \cdot V_P} = \frac{Q \cdot (S_{NO,EFF} - S_{NO,INF}) \cdot \theta_X}{(1 + b_{AUT,T} \cdot \theta_X) \cdot V_P} \quad \textit{[no denitrification]} \qquad (16.4.2\text{-}1)$$

where:

Q = influent flow (L/d)

V_P = system volume (L)

TKN_{INF} = influent TKN concentration (mgN/L)

TKN_{EFF} = filtered effluent TKN concentration (mgN/L)

$S_{NO,INF}$ = influent nitrite/nitrate concentration (mgN/L) (usually zero)

$S_{NO,EFF}$ = effluent nitrite/nitrate concentration (mgN/L)

N_S = nitrogen required for sludge production (mgN/L)

θ_X = sludge age (SRT) (days)

$b_{AUT,T}$ = nitrifier decay rate (/d) (as used in model) (e.g. $0.17\,(1.029)^{(T-20)}$)

N_S $\cong 0.041$ mg N/mg VSS

If the first form of equation 16.4.2-1 is applied (based on TKN), then it is necessary to calculate the concentration of nitrogen per unit influent flow that is directed into waste solids, N_S:

$$N_S = f_N \frac{V_P \cdot X_{VSS}}{Q \cdot \theta_X} \qquad (16.4.2\text{-}2)$$

where:

f_N = nitrogen content of the sludge

≅ 0.1 mg N/mg VSS

X_{VSS} = system VSS concentration (mg VSS/L)

The batch test data shown in Figure 16-2 was the response observed over part of one cycle in an SBR (see Appendix D). The system details were as follows:

Volume = 6.5 L

Influent volume = 5.0 L/d

SRT = 12.1 days

Effluent S_{NO} = 17.8 mg/L

Applying the second form of equation 16.4.2-1:

$$\frac{X_{AUT}}{Y_{AUT}} = \frac{Q \cdot (S_{NO,EFF} - S_{NO,INF}) \cdot \theta_X}{(1 + b_{AUT,T} \cdot \theta_X) \cdot V_P}$$

$$= \frac{5 \cdot (17.8 - 0) \cdot 12.1}{(1 + 0.17 \cdot 1.029^{(23-20)} \cdot 12.1) \cdot 6.5}$$

$$= 51.1 \; mg/L$$

The crucial part of this method for measuring μ_{AUT} is being able to estimate the quotient (X_{AUT}/Y_{AUT}) reliably. As noted earlier, the information for this calculation generally is not known accurately for full-scale systems, in particular, the exact SRT and TKN/COD loading. Hence, low F/M tests using mixed liquor drawn directly from a full-scale treatment plant are unlikely to give reliable results.

16.4.3 Implementation of Low F/M Bioassay in SBR

A bench-scale fill-and-draw SBR system (receiving the same influent as the full-scale plant) is ideal for generating the requisite information for measuring μ_{AUT}.

- The system is simple to operate compared to continuous-flow systems.
- The linear nitrate response can be monitored whenever influent is added. (In fact, the example response shown in Figure 16-2 was taken from an SBR system.)
- Influent and effluent data (for estimating X_{AUT}/Y_{AUT}) can be gathered readily.

- It is simple to control SRT accurately. The μ_{AUT} value can be calculated by calculating X_{AUT}/Y_{AUT} from the operating data, or through simulating the response (adjusting μ_{AUT} to fit the data).

A drawback of this method is that the SBR must be operated for an extended period (e.g., six weeks) before a steady state is attained and reliable data gathered. However, the system allows estimation of a number of important wastewater characteristics in addition to μ_{AUT}.

Details of operating and monitoring an SBR system for estimating μ_{AUT} are presented in Appendix D, and are not repeated here. However, it is worth noting a specific advantage of the SBR approach; namely, the nitrifier μ_{AUT} can be estimated by simulation of the system behavior. This is useful because:

- Simulating system behavior over the entire period of operation accounts for small changes in process conditions such as influent loading, temperature and DO concentration.
- Fitting the nitrate/ammonia response to estimate μ_{AUT} is crosschecked by simultaneously fitting the OUR response data. OUR reflects nitrification behavior quite sensitively in this system.
- Direct application of the model, and estimation of the other wastewater characteristics (such as f_{SI} and f_{XI}) to fit other system parameters such as VSS, provides an increased level of confidence in the μ_{AUT} estimate.
- The modeling approach obviates the need to calculate X_{AUT} from equation 16.4.2-1, and accounts for changes in X_{AUT} due to any changes in influent loading.

Note: It has been suggested that the once-a-day batch feeding in the low F/M SBR protocol may result in the growth of a nitrifier population different from that in the full-scale plant with continuous feed. In this study, three different feeding schedules for the low F/M SBR method were evaluated. These were: once-a-day feeding, six-times-per-day feeding and continuous feed for 23 hours over the react phase. In all cases, the measured μ_{AUT} values were the same.

16.5 Rapid Bioassay Methods

As part of this study, two new methods for measurement of μ_{AUT} have been developed. The principal advantage of these methods is that they result in an estimate of μ_{AUT} in approximately one week, whereas traditional methods (e.g., the low F/M SBR method) require a much longer analysis period (Low F/M SBRs must be run for a period of 3-4 SRTs (i.e., 40-60 days) in order to reach a steady state suitable for rigorous data collection). The following sections outline the theory and development of these methods:

- high F/M batch test method
- washout method

16.6 High F/M Batch Test

In the high F/M batch test, a relatively low concentration of nitrifying mixed liquor is spiked with ammonia, and the increase in nitrite/nitrate concentration with time is monitored for a period of, say, 4 days. The principal advantages of the high F/M method over the low F/M method are that: (1) Computation of μ_{AUT} does not depend on an estimate of the mass of nitrifiers in the test and (2) there is not a requirement to operate a laboratory-scale system for an extended period.

The basis of this method for estimating μ_{AUT} is that the initial nitrifier concentration is small, and the rate of growth of nitrifiers during the test is exponential. A typical response observed in a 3-liter high F/M test is shown in Figure 16-4. In the test, a small volume of nitrifying mixed liquor was added to a diluent (treatment plant secondary effluent in this case) to result in an initial seed concentration in the test of 30 to 35 mg VSS/L. At the start of the test, diluent and seed were added to the reactor, mixing and aeration were started, and reagents (1.38 g of ammonium chloride and 1 g of sodium bicarbonate) were added. Ammonium chloride was added in an amount that resulted in an initial ammonia-nitrogen concentration of about 120 mg/L. The pH of each test was checked with a pH probe about 3 times per day, and 0.5 g to 1.0 g of sodium bicarbonate was added if the pH dropped to 7.2. Aeration was controlled by a DO concentration controller, which cycled aquarium pumps on and off to maintain a DO concentration setpoint between 4 mg/L and 6 mg/L (for a detailed outline of the test method, see Appendix F). Features to note are:

- The nitrite/nitrate response is exponential, i.e., not linear as with the low F/M test.
- Test duration typically is 4 to 7 days.
- The test leads to an estimate of ($\mu_{AUT} - b_{AUT}$), not μ_{AUT} explicitly.
- The ($\mu_{AUT} - b_{AUT}$) value is estimated by fitting a model equation to the observed nitrite+nitrate response.

Note: In the high F/M test careful attention to pH control is required.

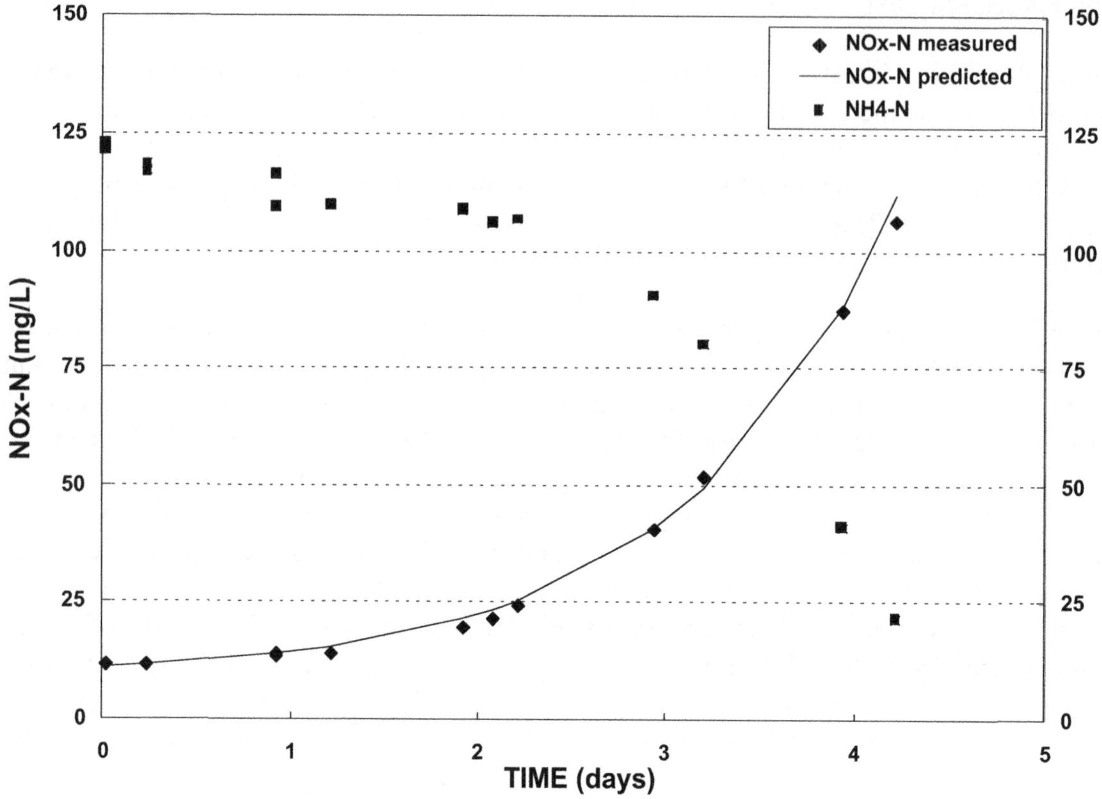

Figure 16-4. Nitrite/Nitrate and Ammonia Profiles in a High F/M Batch Test for Estimating μ_{AUT}

16.6.1 Brief Equation Formulation

At the start of the test, the concentration of ammonia in the reactor is high enough (relative to K_{NH}) to ensure that the nitrification rate is at the maximum. That is, the specific growth rate ($\mu = \mu_{AUT} \cdot S_{NH}/(K_{NH} + S_{NH})$) essentially is equal to the maximum specific growth rate (μ_{AUT}). Hence, as with equation 16.4.1-2:

$$\frac{dS_{NO}}{dt} = \frac{\mu_{AUT}}{Y_{AUT}} \cdot X_{AUT} \qquad (16.6.1\text{-}1)$$

The change in nitrifier biomass concentration, X_{AUT}, is determined by growth and decay:

$$\begin{aligned}\frac{dX_{AUT}}{dt} &= \mu_{AUT} \cdot X_{AUT} - b_{AUT} \cdot X_{AUT} \\ &= \left(\mu_{AUT} - b_{AUT}\right) \cdot X_{AUT}\end{aligned} \qquad (16.6.1\text{-}2)$$

Integrating equation 16.6.1-2 from time zero to time t:

$$X_{AUT,t} = X_{AUT,0} \cdot e^{\left(\mu_{AUT} - b_{AUT}\right)t} \qquad (16.6.1\text{-}3)$$

where:

$X_{AUT,t}$ = nitrifier concentration at time t

$X_{AUT,0}$ = nitrifier concentration at time zero

b_{AUT} = nitrifier decay rate

μ_{AUT} = maximum specific nitrifier growth rate

Substituting equation 16.6.1-3 in equation 16.6.1-1, and integrating from time zero to time t:

$$S_{NO,t} = S_{NO,0} + \frac{\mu_{AUT} \cdot X_{AUT,0}}{Y_{AUT} \cdot \left(\mu_{AUT} - b_{AUT}\right)} \cdot \left[e^{\left(\mu_{AUT} - b_{AUT}\right)t} - 1\right] \qquad (16.6.1\text{-}4)$$

where:

$S_{NO,t}$ = oxidized nitrogen concentration at time t

$S_{NO,0}$ = oxidized nitrogen concentration at time zero

Y_{AUT} = nitrifier yield coefficient

$X_{AUT,0}$ = nitrifier concentration at time zero

The development of these equations is discussed in more detail later in this chapter. For estimating ($\mu_{AUT} - b_{AUT}$), non-linear regression is used to fit equation 16.6.1-4 to test nitrite/nitrate data such as shown in Figure 16-4. The parameter estimation approach is discussed in detail in Appendix H.

16.6.2 High F/M Batch Test – pH Control Issue

A number of parameters require close monitoring throughout the course of a high F/M experiment. Methods for addressing these experimental issues are discussed in detail in Appendix F. Of these, pH control is the most important. Methods for controlling pH are also discussed in Appendix F. An example of a pH control problem is presented here to highlight the importance of this issue.

The pH in the high F/M test reactor should be maintained between 7.2 and 7.8. If the pH drops below 7.2 or rises above 7.8, nitrification is inhibited, leading to an underestimated value for μ_{AUT}. In addition, if the pH rises above 7.8, problems with ammonia stripping occur, complicating the analysis because a nitrogen mass balance cannot be performed to check the integrity of the experimental data.

Potential for changing pH in high F/M tests is the result of two factors:

- In the test approximately 100 mgN/L of nitrate is generated. This will consume approximately 700 mg/L of alkalinity as $CaCO_3$. Usually this will far exceed the amount available in the diluent, resulting in potential for pH decline unless alkalinity is supplemented.

- The test vessel obviously must be aerated. If aeration is continuous, pH will tend to rise due to stripping of carbon dioxide. For example, continuous aeration of high F/M tests with secondary effluent (pH = 7.4) from the Dundas WWTP as diluent raised the pH to above 9 within a few hours.

An example of a high F/M test with a pH control problem is shown in Figure 16-5. The ammonia response is nearly linear (rather than exponential) because stripping is the dominant ammonia removal mechanism. In this case, the high pH is a problem in itself because the nitrification rate obviously was limited as evidenced by the small increase in nitrate concentration.

Nitrogen balance: In the test, very little nitrogen is utilized for cell synthesis. Therefore, the total inorganic nitrogen concentration (TIN = ammonia-N + nitrite-N + nitrate-N) should remain near constant. Summing the three concentrations at each sample time, and scanning the TIN response provides a simple check on the validity of the experimental data. A decreasing trend in TIN values over the test often is indicative of ammonia stripping (as a result of a too-high pH).

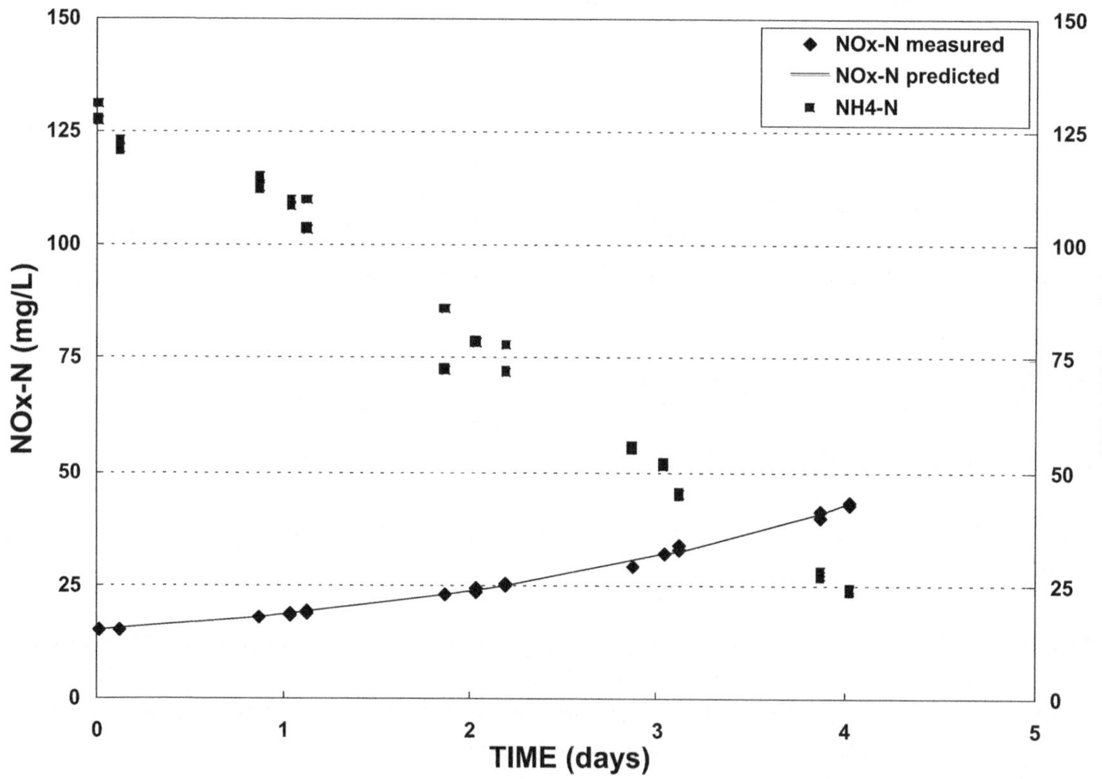

Figure 16-5. A High F/M Test Conducted with Inadequate pH Control Likely Will Exhibit Data Irregularities

---

Note: Ammonia concentration: At the outset of the test the ammonia concentration in the high F/M reactor is quite high (e.g., 120 mgN/L) and as the test progresses, it will decrease exponentially. As discussed in Appendix E, when the ammonia concentration reaches values of less than 20 mgN/L, it appears that the half-saturation effect becomes evident (i.e., K_{NH} value), and nitrification rate is not at the maximum. Therefore, fitting to the data should only include that portion of the test where ammonia concentration exceeds 20 mg/L.

16.6.3 Inhibition/Toxicity Evaluation

It is sometimes contended that influent wastewaters may contain biodegradable organics that are inhibitory to the nitrifying organisms. The consequence of this situation will be that nitrification may be inhibited at the front end of the plant, and that the effective nitrification rate will increase through the plant once the biodegradable inhibitory organic is removed. The high F/M method provides an ideal technique for testing such a hypothesis. Two high F/M tests can be conducted in parallel, one using plant influent wastewater as the diluent and the other, using secondary effluent as the diluent. Figure 16-6 shows an example of nitrite/nitrate response in a pair of such side-by-side tests. Note that:

- In the test with secondary effluent, the initial nitrate concentration is higher because the plant was nitrifying and the effluent contained nitrate; hence the offset in the responses.
- The turnover in response for the last pair of data points occurs because ammonia has been depleted.
- In this case, the responses were almost identical, and no inhibition was evident.

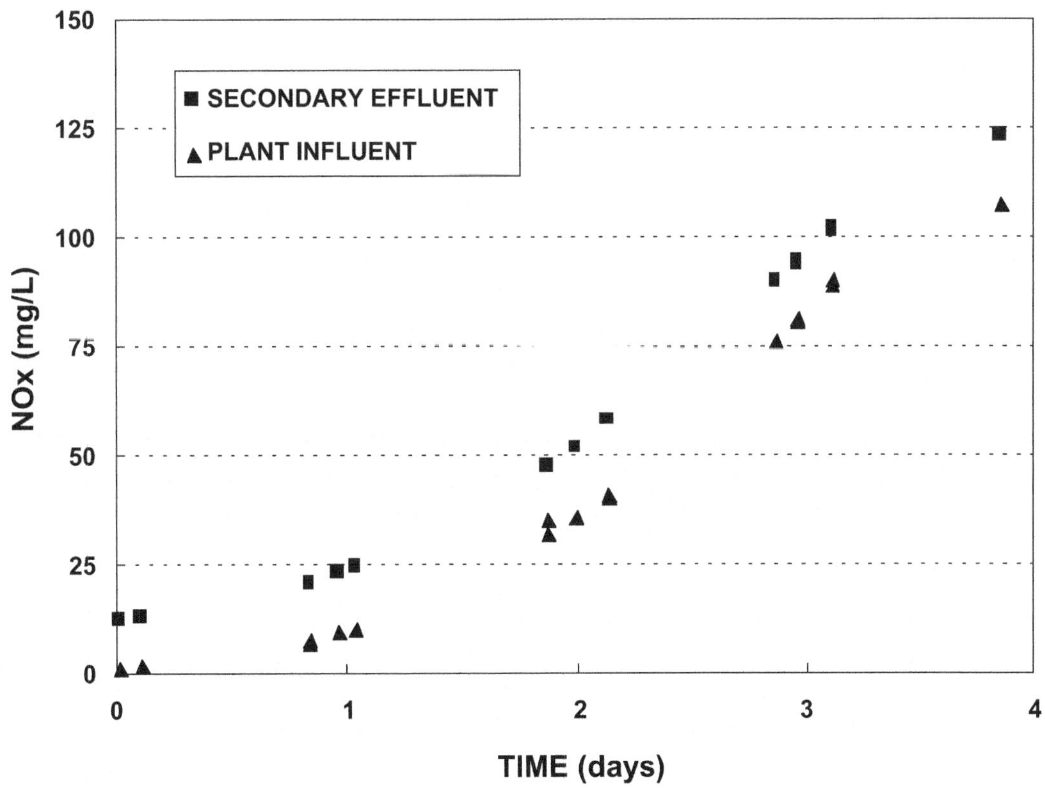

Figure 16-6. Nitrite/Nitrate Profiles in Side-By-Side High F/M Batch Test with Either Plant Influent Wastewater or Secondary Effluent as Diluent

Note: In situations where a non-nitrifying treatment plant is to be upgraded, it would be useful to know whether the influent is likely to cause nitrification toxicity problems. An indication of the potential may be evident from parallel testing along the lines of the approach above. Sludge from a nitrifying plant could be used as the seed in two parallel tests: diluent for one test would be influent from the nitrifying plant, and diluent for the other test would be influent from the non-nitrifying plant (i.e., the one to be evaluated.). Presumably a difference in response between the two tests would indicate possible inhibition or toxicity problems, although this would not be conclusive evidence. However, data from the tests would be useful for identifying the need for further evaluation.

16.7 Washout Bioassay

The washout method for estimating μ_{AUT} is based on monitoring the effluent nitrite/nitrate response over a period of several days in a flow-through reactor initially filled with nitrifying mixed liquor from the full-scale plant. The influent flow rate Q is selected such that the hydraulic retention time (HRT = V/Q (= SRT for a flow-through reactor)) will result in washout of nitrifiers. As with the high F/M method, the principal advantages of the washout method over the low F/M SBR method are that (1) computation of μ_{AUT} does not depend on an estimate of the mass of nitrifiers in the test, and (2) there is not a requirement to operate a laboratory-scale system for an extended period. Other perceived advantages are:

- There is very little nitrifier growth during the test. Essentially, the test considers the response of nitrifiers taken from the full-scale plant.
- The organisms are receiving the actual plant influent (with ammonia added).

A typical washout test response is shown in Figure 16-7. In the test, primary influent was spiked with sufficient ammonium chloride to result in a total influent ammonia-nitrogen concentration of approximately 50 mg/L. Sodium bicarbonate also was added to achieve an influent alkalinity concentration of approximately 350 mg/L as $CaCO_3$. The influent flow rate was set at 8 L/d., resulting in a SRT of 0.75 day. An effluent pump was set at the same rate to maintain a constant level in the reactor. In this test, ammonium chloride was added to the reactor at the start of the test to avoid initial substrate limitations. Effluent samples from the washout test were collected and filtered four times per day and analyzed for nitrate, nitrite, and ammonia. The test was completed within 5 days (for a detailed outline of the test method, see Appendix G). Features to note are:

- Initially, nitrite/nitrate response shows an increase to a maximum. During this phase, the mass of nitrifiers in the system is able to oxidize the influent ammonia.
- A maximum nitrite/nitrate concentration is followed by a semi-exponential decrease with time. During this phase, a significant amount of nitrifiers are being washed out of the system, and the remaining mass is not capable of oxidizing all of the influent ammonia.
- The nitrite/nitrate response is particularly suitable for parameter estimation.
- Test duration typically is 4 to 6 days.
- The test leads to an estimate of ($\mu_{AUT} - b_{AUT}$), not μ_{AUT} explicitly.

- The ($\mu_{AUT} - b_{AUT}$) value is estimated by fitting a model equation to the observed nitrite+nitrate response.

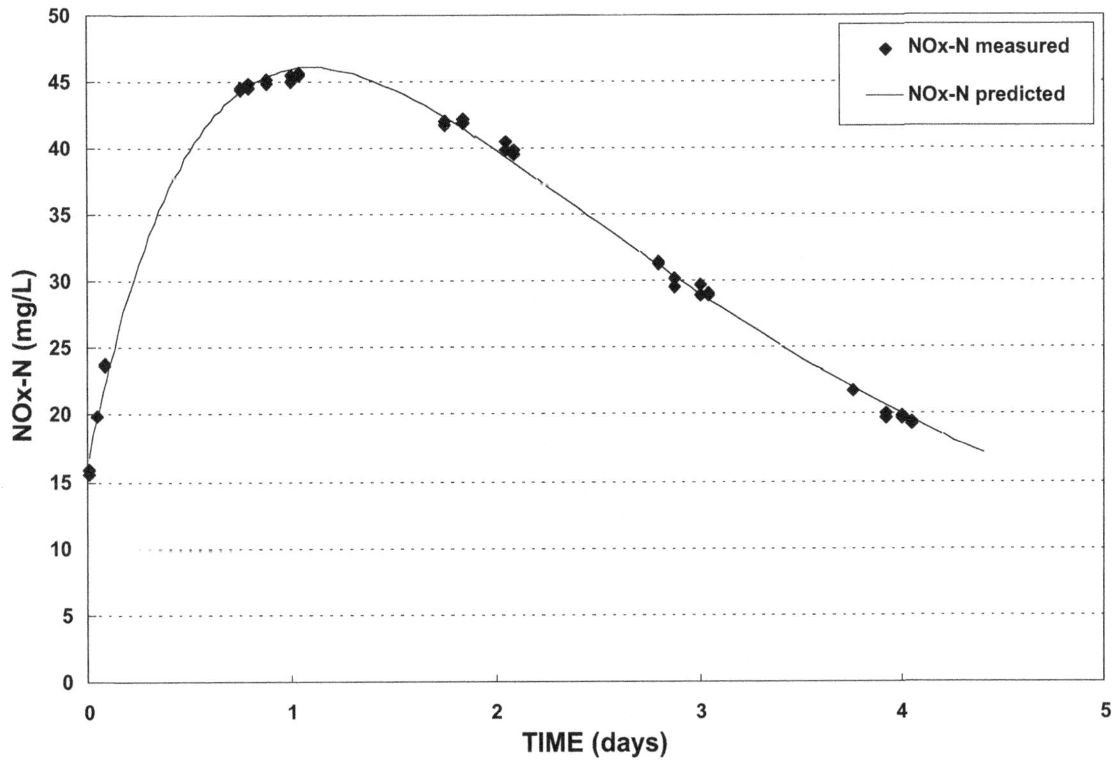

Figure 16-7. Nitrate Response in a Washout Test for Estimating μ_{AUT}

16.7.1 Brief Equation Formulation

Consider a completely-mixed flow-through reactor of volume V. Assume that a sample of nitrifying mixed liquor (from a nitrifying full-scale plant, for example) is added to the empty reactor, and that the reactor is then filled with influent wastewater from the treatment plant. The reactor now contains a nitrifier organism concentration, X_{AUT}, and a nitrite/nitrate concentration, N_{NOx}.

Now assume that:

- At time zero, the flow of influent to the reactor is set at a constant rate (Q). (The intention here is to select Q such that the hydraulic retention time (HRT = V/Q), which equals the SRT for the flow-through reactor, and will result in washout of nitrifiers.)
- The influent ammonia concentration (at least from soon after the experiment commences) is such that the concentration within the reactor is maintained relatively high compared to the nitrifier half-saturation constant, K_{NH}, so that nitrifiers grow at the maximum rate, μ_{AUT}.

Solving mass balances for nitrifier concentration and nitrite/nitrate concentration (see equation development later) yields the following expressions for the responses from time *zero*:

$$X_{AUT,t} = X_{AUT,0} \cdot e^{\left(\mu_{AUT} - b_{AUT} - \frac{Q}{V}\right)t} \qquad (16.7.1\text{-}1)$$

where:

$X_{AUT,t}$ = nitrifier concentration at time t

$X_{AUT,0}$ = nitrifier concentration at time zero

Q = flow rate through washout reactor

V = washout reactor volume

$$S_{NO,t} = \frac{\mu_{AUT} \cdot X_{AUT,0}}{Y_{AUT} \cdot \left(\mu_{AUT} - b_{AUT}\right)} \cdot \left[e^{\left(\mu_{AUT} - b_{AUT} - \frac{Q}{V}\right)t} - e^{-\left(\frac{Q}{V}\right)t} \right] + S_{NO,0} \cdot e^{-\left(\frac{Q}{V}\right)t} \qquad (16.7.1\text{-}2)$$

where:

$S_{NO,t}$ = oxidized nitrogen concentration at time t

$S_{NO,0}$ = oxidized nitrogen concentration at time zero

Three factors essentially control the nitrite/nitrate response in the system: the magnitude of μ_{AUT}; the initial nitrifier concentration, $X_{AUT,0}$; and the hydraulic retention time, HRT. Figure 16-8 illustrates the predicted nitrite/nitrate concentration response for various combinations of these parameters (and the same initial nitrite/nitrate concentration). For example:

- A lesser μ_{AUT} value results in a smaller concentration "hump" because nitrate is generated less rapidly and nitrifiers wash out more rapidly.
- A shorter HRT results in a smaller concentration "hump" because nitrifiers wash out more rapidly.
- A decrease in the initial nitrifier concentration, $X_{AUT,0}$, results in more rapid washout. In fact, if $X_{AUT,0}$ is very small then no "hump" in nitrate concentration is observed, and there is a semi-exponential decrease in nitrate concentration from the start of the test.

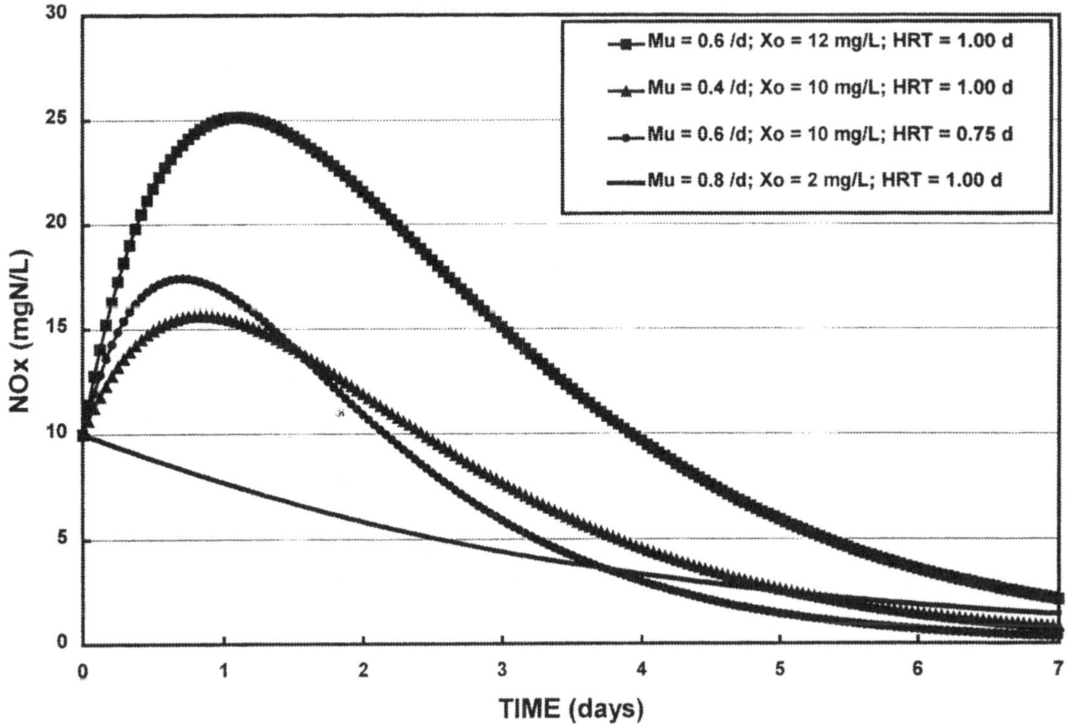

Figure 16-8. Influence of Initial Conditions on Washout Test Response

16.7.2 Washout Test Experimental Issues

A number of factors require close monitoring throughout the course of a washout experiment. These include:

- Influent flow rate: The influent flow rate should be set to a value that results in a reactor HRT (= SRT for a flow-through reactor) less than the minimum required to support nitrifier growth (i.e., the "washout" HRT). If the flow rate is too low (i.e., the HRT is too high), the nitrifiers will not wash out completely, and test results will be skewed.
- Ammonia concentration: At the outset of the test, the reactor should be supplemented with ammonia to ensure that initially the nitrifiers' growth is not substrate-limited.

Methods for addressing these experimental issues are discussed in detail in Appendix G.

16.8 Data Analysis for High F/M and Washout Methods

The data analysis method for estimating μ_{AUT} (actually $(\mu_{AUT} - b_{AUT})$) from nitrate response data gathered in high F/M and washout tests is discussed in detail in Appendix H. Two aspects should be highlighted:

- In addition to estimating the value for $(\mu_{AUT} - b_{AUT})$, the approach also provides confidence limits for the estimate.

- Residuals plot: A plot of residuals often can identify problems with the validity of the experimental data. In these cases, the residuals plot may exhibit a characteristic wave shape (e.g., a sinusoidal shape). An example is shown in Figure 16-9.

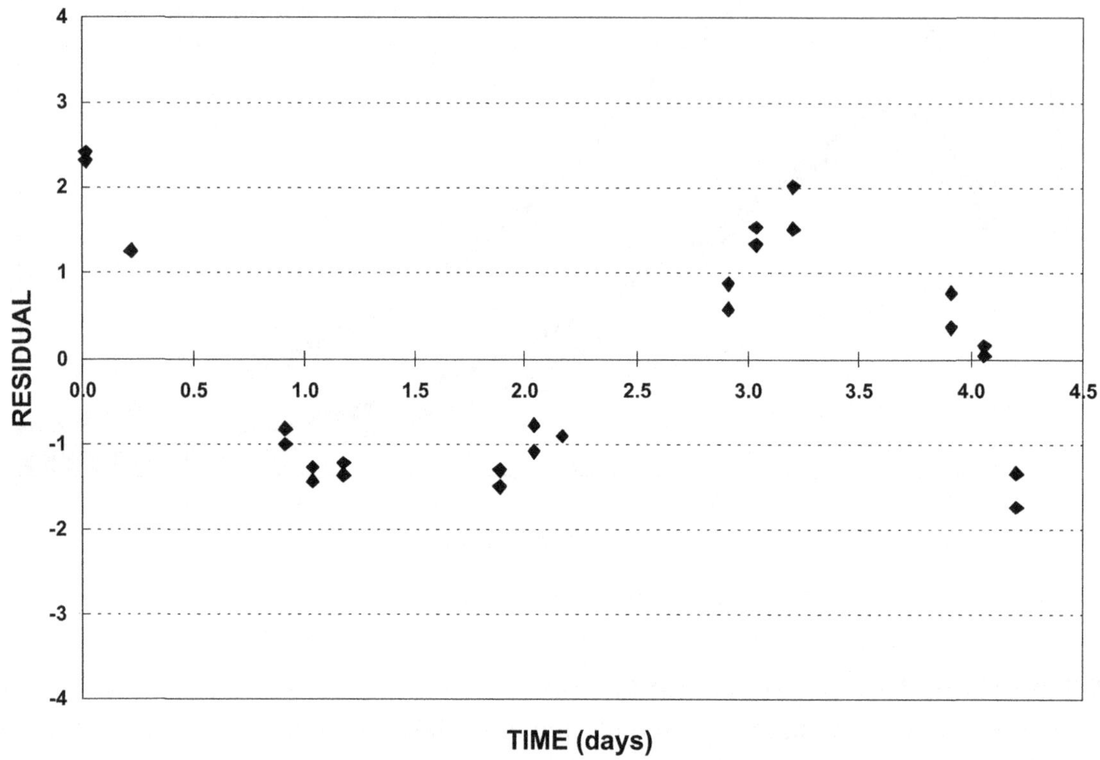

Figure 16-9. High F/M Residuals Plot Exhibiting a Characteristic Sinusoidal Wave Shape

16.9 Development of High F/M and Washout Equations

Consider a completely-mixed flow-through reactor of volume V containing a nitrifier organism concentration, X_{AUT}, and a nitrite/nitrate concentration, S_{NO}. Assume that:

- At time *zero*, the flow rate of influent to the reactor is set at a value Q.
- The influent does not contain nitrite/nitrate or nitrifiers.
- The ammonia concentration is such that the concentration within the reactor is relatively high compared to the nitrifier half-saturation constant, K_{NH}, so that nitrifiers grow at the maximum rate, μ_{AUT}.

A mass balance on nitrifier concentration, X_{AUT}, yields:

$$V \cdot \frac{dX_{AUT}}{dt} = 0 - Q \cdot X_{AUT} + \mu_{AUT} \cdot X_{AUT} \cdot V - b_{AUT} X_{AUT} \cdot V$$

$$\frac{dX_{AUT}}{dt} = \left(\mu_{AUT} - b_{AUT} - Q/V \right) X_{AUT} \qquad (16.9\text{-}1)$$

$$X_{AUT,t} = X_{AUT,0} \cdot e^{(\mu_{AUT} - b_{AUT} - Q/V)t}$$

This is the equation for predicting nitrifier concentration with time in a washout test (i.e., equation 16.7.1-1).

Note that for the high F/M batch test, the influent flow rate Q is zero. Therefore, for the high F/M batch test, equation 16.9-1 yields equation 16.6.1-3 for predicting nitrifier concentration with time.

A mass balance on nitrite/nitrate concentration yields:

$$V \cdot \frac{dS_{NO}}{dt} = 0 - Q \cdot S_{NO} + \frac{\mu_{AUT}}{Y_{AUT}} X_{AUT} \cdot V$$

$$\frac{dS_{NO}}{dt} = -\frac{Q}{V} \cdot S_{NO} + \frac{\mu_{AUT}}{Y_{AUT}} X_{AUT} \qquad (16.9\text{-}2)$$

Substituting for X_{AUT} from equation 16.9-1:

$$\frac{dS_{NO}}{dt} = -\frac{Q}{V} \cdot S_{NO} + \frac{\mu_{AUT}}{Y_{AUT}} X_{AUT,0} \, e^{(\mu_{AUT} - b_{AUT} - Q/V)t}$$

$$\frac{dS_{NO}}{dt} + \frac{Q}{V} \cdot S_{NO} = \frac{\mu_{AUT}}{Y_{AUT}} X_{AUT,0} \, e^{(\mu_{AUT} - b_{AUT} - Q/V)t} \qquad (16.9\text{-}3)$$

Grouping parameters, the equation can be written as:

$$\frac{dS_{NO}}{dt} + k_1 \cdot S_{NO} = k_2 \, e^{k_3 t} \qquad (16.9\text{-}4)$$

where:

$$k_1 = Q/V$$

$$k_2 = \frac{\mu_{AUT}}{Y_{AUT}} X_{AUT,0}$$

$$k_3 = \left(\mu_{AUT} - b_{AUT} - Q/V \right)$$

The integrating factor for the solution of this first-order linear differential equation is:

$$e^{\int k_1 dt} = e^{k_1 t} \qquad (16.9\text{-}5)$$

Multiplying both sides of the equation by the integrating factor:

$$e^{k_1 t} \cdot \left[\frac{dS_{NO}}{dt} + k_1 \cdot S_{NO} \right] = e^{k_1 t} \cdot k_2 \, e^{k_3 t} \tag{16.9-6}$$

$$e^{k_1 t} \cdot \frac{dS_{NO}}{dt} + k_1 \cdot e^{k_1 t} \cdot S_{NO} = k_2 \, e^{(k_1 + k_3) t}$$

We can re-arrange the left hand side of the equation by noting from the product rule that:

$$\begin{aligned} \frac{d}{dt}\left(e^{k_1 t} \cdot S_{NO}\right) &= e^{k_1 t} \cdot \frac{dS_{NO}}{dt} + S_{NO} \cdot \left[k_1 \cdot e^{k_1 t}\right] \\ &= e^{k_1 t} \cdot \frac{dS_{NO}}{dt} + k_1 \cdot e^{k_1 t} \cdot S_{NO} \end{aligned} \tag{16.9-7}$$

Hence:

$$\frac{d}{dt}\left(e^{k_1 t} \cdot S_{NO}\right) = k_2 \cdot e^{(k_1 + k_3) t} \tag{16.9-8}$$

Integrating and re-arranging:

$$\int d\left(e^{k_1 t} \cdot S_{NO}\right) = k_2 \cdot \int e^{(k_1 + k_3) t} \, dt$$

$$e^{k_1 t} \cdot S_{NO} = \frac{k_2}{k_1 + k_3} \cdot e^{(k_1 + k_3) t} + K \tag{16.9-9}$$

The constant of integration can be evaluated by noting that, when t=0 the nitrite/nitrate concentration is given by $S_{NO,0}$; that is:

$$1 \cdot S_{NO,0} = \frac{k_2}{k_1 + k_3} \cdot 1 + K$$

$$K = S_{NO,0} - \frac{k_2}{k_1 + k_3} \tag{16.9-10}$$

Hence:

$$e^{k_1 t} \cdot S_{NO,t} = \frac{k_2}{k_1 + k_3} \cdot e^{(k_1 + k_3) t} + S_{NO,0} - \frac{k_2}{k_1 + k_3}$$

$$e^{k_1 t} \cdot S_{NO,t} = \frac{k_2}{k_1 + k_3} \cdot \left[e^{(k_1 + k_3) t} - 1 \right] + S_{NO,0} \tag{16.9-11}$$

$$S_{NO,t} = \frac{k_2}{k_1 + k_3} \cdot \left[e^{k_3 t} - e^{-k_1 t} \right] + S_{NO,0} \cdot e^{-k_1 t}$$

Now:

$$\frac{k_2}{k_1+k_3}=\frac{\dfrac{\mu_{AUT}}{Y_{AUT}}X_{AUT,0}}{\dfrac{Q}{V}+\left[\mu_{AUT}-b_{AUT}-\dfrac{Q}{V}\right]}=\frac{\mu_{AUT}\cdot X_{AUT,0}}{Y_{AUT}\cdot\left(\mu_{AUT}-b_{AUT}\right)} \quad (16.9\text{-}12)$$

Hence:

$$S_{NO,t}=\frac{\mu_{AUT}\cdot X_{AUT,0}}{Y_{AUT}\cdot\left(\mu_{AUT}-b_{AUT}\right)}\left[e^{(\mu_{AUT}-b_{AUT}-Q/V)t}-e^{(-Q/V)t}\right]+S_{NO,0}\cdot e^{(-Q/V)t} \quad (16.9\text{-}13)$$

This is the equation for predicting nitrite/nitrate concentration with time in a washout test (i.e., equation 16.7.1-2).

Note that for the high F/M batch test, the influent flow rate Q is zero. Therefore, for the high F/M batch test equation 16.9-13 yields equation 16.6.1-4 for predicting nitrite/nitrate concentration with time.

16.10 Estimating μ_{AUT} by Calibration with a Simulator

A conceptual approach for estimating μ_{AUT} is through matching modeled activated sludge system dynamic response behavior (ammonia and nitrate) to experimental observations. Figure 16-10 shows an example of fitting a model to measured ammonia data monitored over 24 hours at a full-scale nitrification plant.

With this approach, difficulties often are encountered because:

- Many of the parameters impacting nitrification rate either are not known/measured or are not controlled. (See example in Chapter 18.0.)
- It is only possible to apply this approach if there is a reasonable level of diurnal ammonia response. In fully nitrifying plants operated at a high SRT, this may not be the case.
- Extensive data gathering is required for input to the simulation model.
- Usually full-scale treatment plants are operated at DO levels where the DO concentration switching functions have a significant impact on the simulated nitrification performance. Therefore, it becomes more difficult to separate out the effects of the switching function parameters and μ_{AUT}.

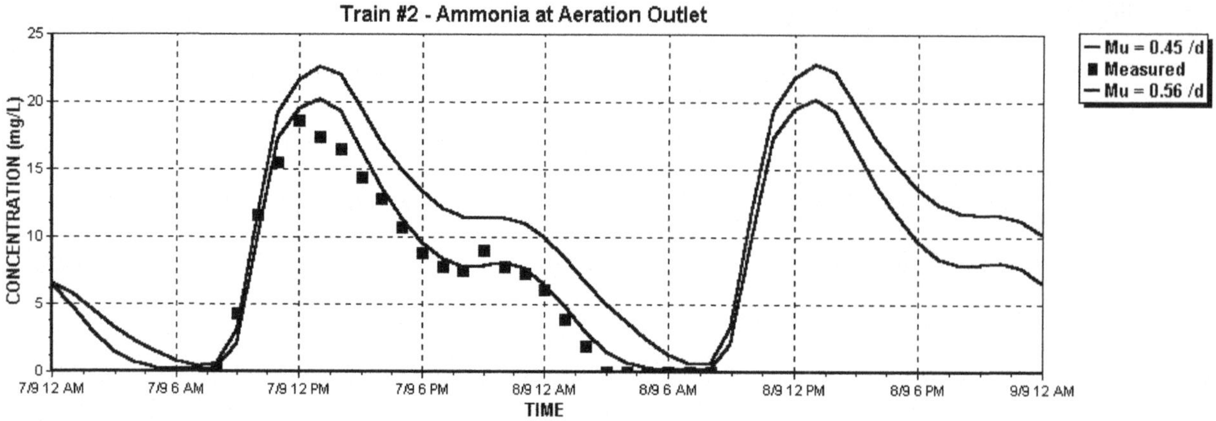

Figure 16-10. Example of Fitting Simulated Ammonia Response Data to Observed Data over 24 Hours in a Full-Scale Treatment Plant

16.11 Impact of Nitrifier Decay Rate

In all of the measurement methods, the estimate for μ_{AUT} is directly influenced by the assumed value for the nitrifier decay rate (b_{AUT}). The topic of nitrifier decay rate is discussed in detail in Chapter 17.0.

16.12 Growth Rate Temperature Dependency

Temperature impacts nitrification kinetics significantly. The EPA *Design Manual for Nitrogen Control* (U.S. EPA, 1993) suggests the following temperature dependency:

$$\mu_{AUT,T} = 0.47 \cdot e^{0.098(T-15)} = 0.47 \cdot 1.103^{(T-15)} \qquad (16.12\text{-}1)$$

On the basis of a reference temperature of 20°C, this equation becomes:

$$\mu_{AUT,T} = 0.77 \cdot e^{0.098(T-20)} = 0.77 \cdot 1.103^{(T-20)} \qquad (16.12\text{-}2)$$

There is considerable uncertainty in the literature regarding the impact of temperature on nitrifier growth rate. Researchers have suggested temperature dependency coefficients ($\theta = 1.103$ in above equations) in the range of 1.072 to 1.127. A higher temperature dependency factor implies a larger temperature dependency, and has a significant consequence on plant design and capacity rating at lower temperature systems because of the impact on the SRT required for nitrification at the winter temperature condition as shown earlier in Figure 16-1.

A sample of reported temperature sensitivity factors is recorded in Table 16.1. Note that the equations often are written in two different forms with reference temperatures of 15°C or 20°C:

$$\mu_{AUT,T} = \mu_{REF} \cdot e^{C(T-T_{REF})}$$
$$\mu_{AUT,T} = \mu_{REF} \cdot \theta^{(T-T_{REF})} \qquad (16.12\text{-}3)$$

where:

$C = \ln(\theta)$

or

$\theta = e^C$

Table 16-1. Examples of Reported μ_{AUT} Temperature Sensitivity Factors

Reference	Formulation	μ_{AUT} (d^{-1})		Temperature Dependency Factor	
		15°C	20°C	C	θ
Downing et al. (1964)	$\mu_{AUT} = 0.47 \cdot e^{0.098(T-15)}$	0.47	0.77	0.098	1.103
Downing & Hopwood (1964)	$\mu_{AUT} = 0.18 \cdot e^{0.116(T-15)}$	0.18	0.32	0.116	1.123
Hultman (1971)	$\mu_{AUT} = 0.50 \cdot 10^{0.033(T-20)}$ $= 0.50 \cdot 1.079^{(T-20)}$	0.34	0.50	0.076	1.079
Barnard (1975)	$\mu_{AUT} = 0.33 \cdot 1.127^{(T-20)}$	0.18	0.33	0.120	1.127
Painter & Loveless (1983)	$\mu_{AUT} = 0.18 \cdot e^{0.0729(T-15)}$	0.18	0.26	0.073	1.076
Water Research Commission (1984)	$\mu_{AUT} = \mu_{AUT,20} \cdot 1.123^{(T-20)}$	-	-	0.116	1.123
Barker & Dold (1997)	$\mu_{AUT} = \mu_{AUT,20} \cdot 1.096^{(T-20)}$	-	-	0.092	1.096

An interesting observation was derived from the analysis of nitrifier growth rate and decay rate data gathered. This aspect is discussed in detail in Chapter 17.0. Essentially the finding was that higher, more conservative μ_{AUT} temperature dependency (θ) factors reported in the literature may well be a mathematical artifact of calculating the temperature factor from either (1) changes in (μ_{AUT} - b_{AUT}) with temperature rather than changes in μ_{AUT}, and/or (2) incorporating the assumption that b_{AUT} is "negligible," i.e., zero.

During the project, limited parallel measurements of μ_{AUT} were conducted at 12°C and 22°C. The results indicated the following temperature dependency relationship; this is at the low end of the range previously reported:

$$\mu_{AUT,T} = \mu_{AUT,20} \cdot e^{0.070(T-20)} = \mu_{AUT,20} \cdot 1.072^{(T-20)} \qquad (16.12\text{-}4)$$

16.13 Comparison of the Methods

During this study, nitrification rate μ_{AUT} tests were performed at three locations (Hamilton, Ontario, Canada; Phoenix, Ariz.; Seattle, Wash.) using the three bioassay techniques: low F/M SBR, high F/M and washout.

Initially, estimates of μ_{AUT} were based on the assumption of a very low nitrifier decay rate ($b_{AUT} = 0.04\ d^{-1}$ at 20°C). Analysis of experimental results, all gathered at temperatures close to 20°C, indicated:

- Experimental methods involving systems at high solids retention times (SRT > 12 days) (i.e., low F/M) yielded low μ_{AUT} estimates; for example 0.45 d^{-1}.
- Experimental methods involving systems at low solids retention times (SRT < 3 days) (i.e., high F/M) yielded higher μ_{AUT} estimates; for example 0.75 d^{-1}.

An assessment of nitrifier decay rate was conducted after commencing the μ_{AUT} testing and method development. This involved investigating:

- The role of nitrifier decay rate, b_{AUT}, in estimation of nitrifier maximum specific growth rate, μ_{AUT}, in the High *versus* Low F/M tests.
- Methods for measuring the nitrifier decay rate, b_{AUT}.

The nitrifier decay rate issue is discussed in detail in Chapter 17.0. It suffices here to note that the nitrifier decay rate, b_{AUT}, is significantly higher than the value of 0.04 d^{-1} (at 20°C) assumed initially. A b_{AUT} value in the range, 0.15 to 0.17 d^{-1} (at 20°C), appears appropriate. This lead to an iterative re-analysis of all the data gathered in the study (several sets of data from the three locations, each using the three methods, with each set on systems receiving the same wastewater). Using a b_{AUT} value of 0.17 d^{-1} yielded μ_{AUT} estimates from the three different test methods that all corresponded closely. Examples of parallel μ_{AUT} estimates from high F/M tests, washout tests and the low F/M SBR test at one location are shown in Figure 16-11. The obvious outlier (low value) was observed in a washout test in which the nitrification rate initially was substrate limited.

Note: A principal objective of this study was to develop new, simpler methods for measuring nitrifier maximum specific growth rate, μ_{AUT}. It has been demonstrated that all three methods (low F/M SBR, high F/M, washout), when conducted in parallel, provide the same estimate for μ_{AUT}.

There was a second important observation evident from the assessment of nitrification rate data. As noted, μ_{AUT} measurements were conducted at the three locations using the three methods during this study. In addition, data from earlier low F/M SBR studies at six other locations were re-analyzed. In every instance, the μ_{AUT} estimates (at 20°C) all were very close, in the range, 0.90 to 0.95 d^{-1}. This analysis questions the widely-held perception that nitrifier growth rate varies significantly from plant to plant. Undoubtedly nitrifiers, being a specific, less diverse population than heterotrophs, are susceptible to inhibition and toxicity. However, the variability in the rate reported previously perhaps was an artifact of the incorrect assumption of a low decay rate.

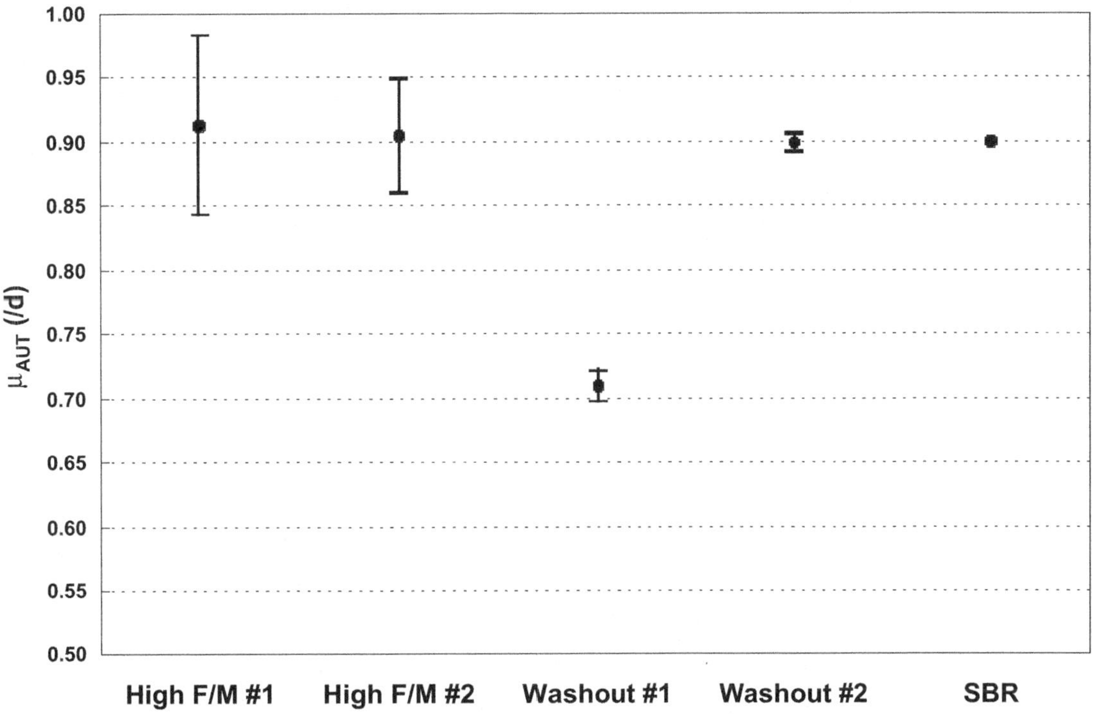

Figure 16-11. Examples of Parallel μ_{AUT} Estimates from High F/M, Washout and Low F/M Tests with Mixed Liquor from Processes All with the Same Influent Wastewater
(The obvious outlier (low value) was observed in a washout test in which the nitrification rate initially was substrate limited. Vertical lines on the plot represent 95 percent confidence intervals – no confidence intervals are calculated in the Low F/M SBR method)

Table 16-2 compares the three methods with advantages and disadvantages shown. Of the three test methods, the least preferred experimentally is the washout test, even though the response in the test perhaps is the best for parameter estimation. It involves a more complex set up and operation compared to the high F/M test. The high F/M test is relatively simple to perform and can be completed over a 4-7 day period with sampling three to four times per day. (It is strongly recommended that sampling be performed six times per day to increase confidence in the estimate.) Alkalinity adjustment for pH control is the main experimental issue, and requires careful attention. Also, with both the washout and the high F/M tests being conducted on a one-time wastewater samples, the tests should be conducted at least two times.

A major advantage of the low F/M SBR test is that it provides longer-term information about the wastewater characteristics (see Appendix D) and routinely assesses multiple samples for nitrification toxicity as a result of the daily batch feeding requirements. Thus nitrification kinetics testing can be coupled with wastewater characterization testing to determine other influent characteristics such as RBCOD and unbiodegradable COD fractions. Another advantage of the low F/M SBR method is that the bench-scale reactor can be operated at the same SRT expected for the full-scale plant design. When that is done, any possible uncertainty regarding b_{AUT} in calculating the μ_{AUT} does not significantly impact the plant performance predictions. Also, the reactor set up for the low F/M SBR method provides an opportunity to measure b_{AUT} in a batch phase at the end of the SBR study.

Table 16-2. Comparison of Advantages and Disadvantages of the Three Nitrification Rate Measurement Methods

Test Method	Advantages	Disadvantages
Low F/M SBR	• Simple operation • Provides other useful process data, such as RBCOD, non-biodegradable COD fractions. • Provides longer-term information on wastewater characteristics and possible nitrification inhibition. • Can be operated at the same SRT as full-scale plant to minimize the error from b_{AUT} assessment. • Can be used for b_{AUT} measurement, after being operated to steady state. • Simple pH control with aid of DO controller.	• Requires long-term (6 weeks) reactor operation. • Requires significant daily data collection to estimate amount of nitrifier biomass and to maintain SRT. • Requires intensive sampling for batch nitrification rate tests.
High F/M	• Simplest procedure. • Test is completed within 5-7 days. • No need for long-term wastewater and reactor sampling and analyses.	• More complex pH control • Requires careful selection of seed size (see "Simulated Batch Test Response" section of Appendix E). • Evaluates only one-time wastewater samples. Need to repeat at least two times. • Population limited by initial seed selection.
Washout	• Test is completed within 5-6 days. • Performed with mixed liquor from an actual full-scale system. • No need for long-term wastewater and reactor sampling and analysis. • Simple pH control. • Relatively simple to ensure non-ammonia-limited growth.	• More complex in terms of reactor operation and maintenance, feed preparation and HRT/SRT control. • Evaluates only one-time wastewater samples. Need to repeat at least two times. • More equipment - pumps and controller. • Difficult to estimate initial SRT/HRT correctly. • Population limited by initial seed selection.

The major disadvantage of the low F/M method is that the reactor operation is more demanding (monitoring, maintenance and SRT control), the system must be operated for approximately two months, and the nitrification rate tests are intensive. However, where longer-term information is desired on nitrification kinetics for a given location, the low F/M method provides a relatively simple means to meet such needs.

16.14 Concluding Remarks

The underlying approach in this study has been to model nitrification as a single-step process. In many of the experimental systems, nitrite accumulated to significant levels. As further information becomes available it is likely that there will be more reason to consider a two-population approach in modeling of nitrification, distinguishing between the behavior of ammonia- and nitrite-oxidizers.

The study has provided valuable information on nitrification kinetics and modeling, and two practical methods for measuring nitrifier growth rate have been developed. Based on the detailed experimental work in this study, a reasonable estimate of μ_{AUT} is 0.90 d^{-1}, with a temperature dependency coefficient (θ) of 1.072. The corresponding values for the nitrifier decay rate (b_{AUT}) and its temperature dependency coefficient are 0.17 d^{-1} and 1.029, respectively. The justification for this decay rate is presented in Chapter 17.0.

Nevertheless, there remain certain research questions:

- One issue to note is that the major portion of the experimental work involved conducting tests at temperatures close to the operating temperature of the plant from which mixed liquor was sampled. A question remains as to whether the nitrification rate data will be estimated correctly if mixed liquor withdrawn from a plant at, say, 12°C is tested at 20°C (a usual laboratory temperature). It is possible that population shifts with changing temperature may complicate the issue. Until this is resolved, it is suggested that testing of low-temperature systems should be conducted at the same low temperature.
- On a related issue, further information on temperature dependency should be gathered.

Note: The issue of uncertainty over temperature dependency has been noted. There may be cases where nitrification rate testing is being conducted at a plant that is operating at or close to the minimum annual temperature. In such a case, if temperature-controlled facilities are available, it is suggested that testing should be conducted at the same temperature condition of the full-scale plant (and the rate back-calculated to the reference temperature of say 20°C). This would remove the uncertainty over temperature dependence.

16.15 References

Barker, P.S., and P.L. Dold. 1997. General model for biological nutrient removal activated sludge systems: model presentation. *Water Environ. Res.* 69(5):969-984.

Barnard, J.L. 1975. Biological nutrient removal without the addition of chemicals. *Water Res.* 9:485-490.

Biesterfeld, S., L. Figueroa, M. Hernandez, and P. Russell. 2001. Quantification of nitrifying bacterial populations in a full-scale nitrifying trickling filter using fluorescent in situ hybridization. *Water Environ. Res.* 73:329-338.

Daims, H., J.L. Nielsen, P.H. Nielsen, K.-H. Schleifer, and M. Wagner. 2001. In situ characterization of *Nitrospira*-like nitrite-oxidizing bacteria active in wastewater treatment plants. *Appl. Environ. Microbiol.* 67:5273-5284.

Daims, H., P.H. Nielsen, J.L. Nielsen, S. Juretschko, and M. Wagner. 2000. Novel *Nitrospira*-like bacteria as dominant nitrite-oxidizers in biofilms from wastewater treatment plants: diversity and in situ physiology. *Water Sci. Tech.* 41(4-5):85-90.

Dionisi, H.M., A.C. Layton, G. Harms, I.R. Gregory, K.G. Robinson, and G.S. Sayler. 2002. Quantification of *Nitrosomonas* oligotropha-like ammonia-oxidizing bacteria and *Nitrospira spp.* from Full-scale Wastewater Treatment Plants by Competitive PCR. *Appl. Environ. Microbiol.* 68:245-253.

Downing, A.L., and A.P Hopwood. 1964. Some observations on the kinetics of nitrifying activated sludge plants. *Schweizerische Aeitschrift fur Hyddrologie* 26:271.

Downing, A.L., H.A. Painter, and G. Knowles. 1964. Nitrification in the activated sludge process. *J. Inst. Sew. Purif.* 64:130-158.

Hultman, B. 1971. Kinetics of biological nitrogen removal. Inst. Vattenforsorjmingsoch Avloppsteknik samt Vattenkemi, Stockholm: KTH Pub. 71(5).

Juretschko, S., G. Timmermann, M. Schmid, K.-H. Schleifer, A. Pommerening-Röser, H.-P. Koops, and M. Wagner. 1998. Combined molecular and conventional analyses of nitrifying bacterium diversity in activated sludge: nitrosococcus *mobilis* and *Nitrospira*-like bacteria as dominant populations. *Appl. Environ. Microbiol.* 64:3042-3051.

Painter, H.A., and J.E. Loveless. 1983. Effect of temperature and pH value on the growth-rate constants of nitrifying bacteria in the activated sludge process. *Water Res.* 17:238-248.

Scharmm, A., D. de Beer, M. Wagner, and R. Amann. 1998. Identification and activities in situ of *Nitrosospira* and *Nitrospira spp.* as dominant populations in a nitrifying fluidized bed reactor. *Appl. Environ. Microbiol.* 64:3480-3485.

U.S. EPA. 1993. *Manual-Nitrogen Control* (pp. 90-91). EPA/625/R-93/010. Washington DC: Office of Technology Transfer, U.S. Environmental Protection Agency.

Wagner, M., D.R. Noguera, S. Juretschko, G. Rath, H.-P. Koops, and K.H. Schleifer. 1998. Combining fluorescent in situ hybridization (FISH) with cultivation and mathematical modelling to study population structure and function of ammonia-oxidizing bacteria in activated sludge. *Water Sci. Tech.* 39(4-5):441-449.

Water Research Commission. 1984. *Theory, design and operation of nutrient removal activated sludge processes*. P. O. Box 824, Pretoria, South Africa: Water Research Commission.

CHAPTER 17.0

NITRIFIER DECAY RATE

17.1 Introduction

Assessment of nitrification behavior in activated sludge systems historically has focused on measuring the nitrifier maximum specific growth rate, μ_{AUT}. Very little attention has been directed at the importance of nitrifier organism decay rate, b_{AUT} (also referred to as k_D), and in many instances, the decay rate has been assumed negligible. However, incorrect assessment of decay rate leads to errors in the μ_{AUT} estimate; the magnitude of the error depends on the μ_{AUT} measurement method employed. This chapter illustrates why decay rate is important when measuring μ_{AUT}, and that the decay rate is significant. It also explains why measurement methods for nitrifier decay may have underestimated the decay rate. Analysis demonstrates that the magnitude of the μ_{AUT} temperature correction coefficient is impacted by the assumed decay rate. A low decay rate results in an overly severe temperature correction for μ_{AUT}; this has important implications for process analysis and plant capacity rating.

A number of factors appear to have contributed to the limited attention directed at the importance of nitrifier organism decay rate, b_{AUT}. For example:

- The early research on nitrification (Downing et al., 1964) reported that the decay rate "is so low that it may be neglected." Certain of the Downing group's studies were on essentially pure cultures of nitrifiers growing in river water. It could be argued that low decay rates would be anticipated because of minimal predation. (However, it should be noted that Copp (1993) and Copp and Murphy (1995) also studied pure cultures of nitrifiers, and reported b_{AUT} decay rates of the order of 0.15 d^{-1} at 20°C. That would contradict the predation argument.)
- Downing's studies on measuring nitrifier growth rate in activated sludge systems used the "washout" concept; namely, operating continuous flow systems at short SRT to identify the SRT below which nitrifier growth could not be sustained. (This washout approach was used widely as a means for estimating nitrifier maximum specific growth rate in continuous flow activated sludge systems, particularly during the 1970s. The washout method presented in Chapter 16.0 is based on this concept, but is implemented differently.) Necessarily, the impact of nitrifier decay in the low SRT systems would have been small.

- Several research studies during the 1970s and 1980s reported decay rates in the range zero to 0.1 d^{-1}; that is, low values compared to heterotrophs in activated sludge. For example, Rozich and Castens (1986) reported "a relatively insignificant value (0.0038 d^{-1}) for k_D."
- Certain of the growth rate measurement methods quantify the net growth rate ($\mu_{AUT} - b_{AUT}$), but this has been reported as μ_{AUT} (i.e., b_{AUT} is neglected).
- A number of textbooks and manuals have presented nitrification design equations and examples that exclude b_{AUT}, perhaps for the simple reason that it simplies the presentation. Although these cases were not necessarily stating that b_{AUT} may be ignored, this may have conveyed the idea that the impact of b_{AUT} is not significant.

During the 1990s, a number of studies reported significant b_{AUT} values. For example, 20°C b_{AUT} values for aerobic systems of 0.20 d^{-1} were reported by Nowak et al. (1994) and 0.21 ± 0.05 d^{-1} by Siegrist et al. (1999). A significant observation in both of these European studies was that the decay rate under anoxic conditions apparently is approximately half that under aerobic conditions. Copp and Murphy (1995) reported 20°C b_{AUT} values in the range 0.14 to 0.17 d^{-1} for studies with nitrifier enrichments. Lee and Oleszkiewicz (2002) measured an average value of 0.15 d^{-1} in bench scale aerobic batch systems. (ASM2d and ASM3 suggest a 20°C b_{AUT} aerobic value of 0.15 d^{-1} and an anoxic value of 0.05 d^{-1} (Henze et al., 2000).)

Setting aside the validity of the argument that the importance of nitrifier decay rate has been underemphasized, reported values for b_{AUT} have varied substantially. Higher values, as suggested by the more recent research, have a very significant impact on nitrification and nutrient removal system design and analysis. This highlights the importance of clarifying the issue(s) on nitrifier decay rate so that practioners can apply the models with confidence.

17.2 Initial Empirical Finding in This Study

Three methods (low F/M, high F/M, washout) for measuring and/or estimating nitrifier maximum specific growth rate μ_{AUT} in systems treating municipal wastewater were evaluated in this project. Details of the experimental protocols are provided in Chapter 16.0 and Appendices D, F and G. With each method, the μ_{AUT} value is based on an assumed value for the nitrifier decay rate (b_{AUT}). Initially in this project, estimates of μ_{AUT} were based on the assumption of a very low nitrifier decay rate (b_{AUT} = 0.04 d^{-1} at 20°C). Analysis of experimental results, all gathered at temperatures close to 20°C, indicated:

- Experimental methods involving systems at high solids retention times (SRT > 12 days) (i.e., low F/M) yielded low μ_{AUT} estimates; for example 0.45 d^{-1}.
- Experimental methods involving systems at low solids retention times (SRT < 3 days) (i.e., high F/M) yielded higher μ_{AUT} estimates; for example 0.75 d^{-1}.

Experimental work by Katehis et al. (2002) suggested that the maximum growth rate appears to change with SRT of the system in which μ_{AUT} is measured (but treating the same influent), and that μ_{AUT} decreases with increasing SRT. Katehis et al. (2000) also conducted tests to measure the nitrifier decay rate, b_{AUT}, and reported values of approximately 0.09 d^{-1} (at 20°C) – more than double the value initially assumed in this study.

These observations lead to an iterative re-analysis of all the data gathered in this study (several sets of data from three locations, each using the different experimental methods for measuring μ_{AUT}, with each set on systems receiving the same wastewater), but for a range of decay rate values. By trial, it was found that assuming a b_{AUT} value of 0.17 d^{-1} yielded μ_{AUT} estimates from the three different test methods that all corresponded closely. The μ_{AUT} estimates at 20°C all were in the region of 0.90 d^{-1}.

This empirical finding suggested the need to investigate:

- The role of nitrifier decay rate b_{AUT} in estimation of nitrifier maximum specific growth rate, μ_{AUT}, in the high versus low F/M tests.
- Methods for measuring the nitrifier decay rate, b_{AUT}.

17.3 Decay Rate in the Measurement of Maximum Specific Growth Rate

Bioassay methods for measuring nitrifier maximum specific growth rate μ_{AUT} conceptually involve two possible measurement regimes, both of which are based on combining nitrifier biomass with ammonia (at a concentration well in excess of the half-saturation coefficient, K_{NH}, so that the rate is at the maximum):

17.3.1 Low F/M Tests

The change in nitrifier mass during the test is small, and there is a *linear* response in the measured parameter (usually nitrite+nitrate concentration, S_{NO}). At the start of a low F/M test (such as the bench-scale SBR method – see Chapter 16.0 and Appendix D) the concentration of ammonia S_{NH} in the reactor is high enough (relative to K_{NH}) to ensure that the nitrification rate is at the maximum. That is, the specific growth rate ($\mu_{AUT} \cdot S_{NH}/(K_{NH} + S_{NH})$) essentially is equal to the maximum specific growth rate (μ_{AUT}). There is a linear increase in oxidized nitrogen (nitrite + nitrate, S_{NO}) concentration; this is termed the nitrate production rate (NPR):

$$NPR = \frac{dS_{NO}}{dt} = \mu_{AUT} \cdot \frac{X_{AUT}}{Y_{AUT}} \qquad (17.3.1\text{-}1)$$

where:

X_{AUT} = nitrifier concentration (mgCOD/L)

Y_{AUT} = nitrifier yield coefficient

Rearranging:

$$\mu_{AUT} = NPR \Big/ \frac{X_{AUT}}{Y_{AUT}} \qquad (17.3.1\text{-}2)$$

Details of the experimental method are not considered here. It suffices to note that the quotient (X_{AUT}/Y_{AUT}) is a function of the mass of nitrate generated, the operating SRT and the decay rate b_{AUT}:

$$\frac{X_{AUT}}{Y_{AUT}} = \frac{Q \cdot (S_{NO,EFF} - S_{NO,INF}) \cdot \theta_X}{(1 + b_{AUT} \cdot \theta_X) \cdot V_P} \quad \textit{[no denitrification]} \qquad (17.3.1\text{-}3)$$

where:

Q = influent flow (L/d)

V_P = system volume (L)

$S_{NO,INF}$ = influent nitrite+nitrate concentration (mgN/L) (usually zero)

$S_{NO,EFF}$ = effluent nitrite+nitrate concentration (mgN/L)

θ_X = SRT (days)

The concentration of nitrifiers in a given system (with fixed volume, load, SRT) and the quotient (X_{AUT}/Y_{AUT}) from equation 17.3.1-3 are determined by the decay rate coefficient, b_{AUT}. Obviously a larger b_{AUT} value implies a lower nitrifier concentration and a lower quotient (X_{AUT}/Y_{AUT}). Figure 17-1 shows a plot of the nitrifier mass generated per unit nitrate formed as a function of SRT. The effect of the assumed b_{AUT} value is very evident. At an SRT of approximately 12 days, a decay rate of 0.04 d^{-1} predicts double the nitrifier concentration compared to a decay rate of 0.17 d^{-1}. In the low F/M test, the μ_{AUT} value is obtained from equation 17.3.1-2, dividing the *observed* NPR by a *calculated* (X_{AUT}/Y_{AUT}).

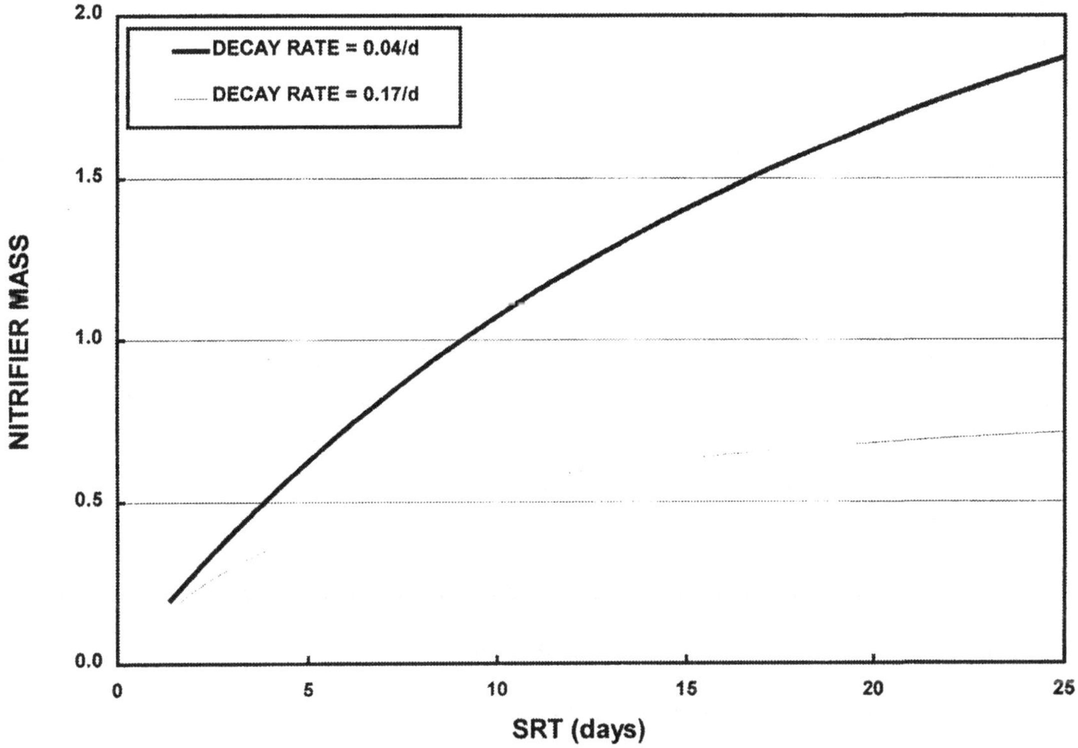

Figure 17-1. Nitrifier Mass as a Function of SRT for Different Nitrifier Decay Rates

Figure 17-2 was generated by (1) assuming a μ_{AUT} value of 0.95 d^{-1} and a b_{AUT} value of 0.17 d^{-1}, and (2) estimating the μ_{AUT} value as a function of SRT based on an incorrect b_{AUT} value of 0.04 d^{-1}. The effect obviously is substantial. The estimated μ_{AUT} value decreases with SRT; at a 12-day SRT (typical for the SBR measurement protocol), the estimated μ_{AUT} is half the actual μ_{AUT}. It is evident that the assumed decay rate b_{AUT} is very important in estimating the μ_{AUT} value correctly in low F/M measurement methods.

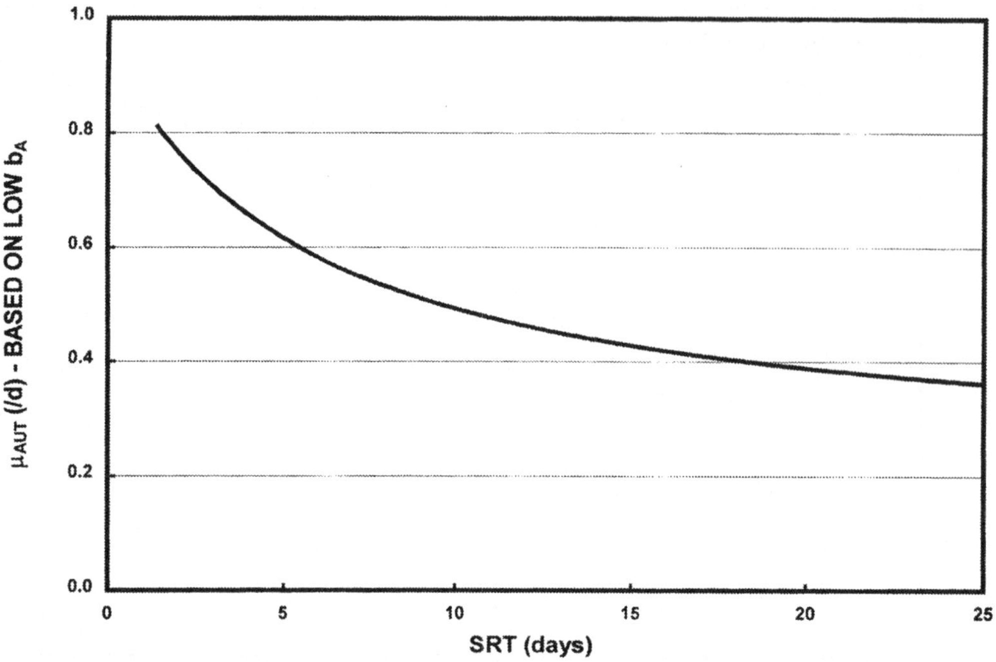

Figure 17-2. Estimating Maximum Specific Growth Rate Based on a Low Decay Rate (0.04 d^{-1})
(Assuming actual μ_{AUT} = 0.95 $d^{-1}$and b_{AUT} = 0.17 d^{-1})

17.3.2 High F/M Tests

The initial nitrifier mass is small relative to the mass produced during the test (i.e., exponential growth), and there is an *exponential-type* response in the measured parameter (usually nitrate concentration, or rather, nitrite plus nitrate). In the high F/M test (such as the batch test method – see Chapter 16.0 and Appendix F) the concentration of ammonia S_{NH} in the reactor again is high enough to ensure that the nitrification rate is at the maximum. The increase in oxidized nitrogen (nitrite + nitrate) concentration typically is monitored over a period of several days:

$$S_{NO,t} = S_{NO,0} + \frac{\mu_{AUT} \cdot X_{AUT,0}}{Y_{AUT} \cdot \left(\mu_{AUT} - b_{AUT}\right)} \cdot \left[e^{\left(\mu_{AUT} - b_{AUT}\right)t} - 1\right] \qquad (17.3.2\text{-}1)$$

where:

$S_{NO,0}$ = nitrite/nitrate concentration at time 0 (mgN/L)

$S_{NO,t}$ = nitrite/nitrate concentration at time *t* (mgN/L)

$X_{AUT,0}$ = nitrifier concentration at time 0 (mgCOD/L)

The test leads to an estimate of ($\mu_{AUT} - b_{AUT}$), not μ_{AUT} explicitly. For estimating ($\mu_{AUT} - b_{AUT}$), non-linear regression is used to fit equation 17.3.2-1 to observed oxidized nitrogen data. The effective SRT in the test is very low, so decay does not have a significant impact on nitrifier concentration per se. Nevertheless, any error in b_{AUT} will bias the estimate of μ_{AUT} by the amount of the error in b_{AUT}. For example, if the actual decay rate b_{AUT} value is

0.17 d^{-1}, but a b_{AUT} of 0.04 d^{-1} is assumed, the μ_{AUT} value will be underestimated by 0.13 d^{-1} (e.g., 0.77 d^{-1} versus 0.90 d^{-1} for a measured $\mu_{AUT} - b_{AUT}$ of 0.73 d^{-1}).

17.4 Experimental Methods for Determining Decay Rate

A number of methods for measuring nitrifier decay rate, b_{AUT}, have been proposed, but all use the same experimental concept. A sample of mixed liquor from a nitrifying system is placed in a batch reactor that is aerated continuously for a period of several days (e.g., 7 days). (Over the course of the experiment alkalinity is consumed, so pH must be controlled by adding alkalinity. Fillos et al. (2000) and Katehis et al. (2001) either added soda ash or injected CO_2 to control pH). The test involves monitoring the decline in nitrifier activity with time. At intervals, a sample of mixed liquor is withdrawn from the batch reactor, and placed in a *secondary* batch reactor. The sample is spiked with ammonia (and alkalinity) sufficient to ensure non-limited nitrifier growth. (In certain cases the ammonia spike is added directly to the primary batch reactor, but this may distort results because it results in nitrifier growth.) Different methods have been used for monitoring the decline in nitrifier activity.

Consider the response in the *secondary* batch reactor:

- Prior to adding the spike of ammonia, oxygen utilization rate (OUR) will be at a base level (OUR_{BASE}), essentially reflecting oxygen consumption for endogenous respiration of heterotrophs.
- When the spike of ammonia is added, the ammonia concentration in the secondary test is high compared to the nitrifier half-saturation constant, K_{NH}, so that nitrifiers grow at the maximum rate, μ_{AUT}. Over the course of a few hours (until ammonia is depleted) nitrate concentration will increase linearly. Also, OUR will increase to a maximum plateau value OUR_{MAX}). Once ammonia is depleted, OUR will decrease to the base level (OUR_{BASE}).
- Essentially the nitrifier activity at the time of the *secondary* test is reflected by either (1) the linear rate of increase in nitrate concentration with time (nitrate production rate – NPR), or (2) the step change in OUR ($OUR_{MAX} - OUR_{BASE}$).

Copp and Murphy (1995) and Siegrist et al. (1999) used the "delta OUR" approach. Nowak et al. (1994) applied a variation on that method. OUR_{MAX} was measured, and then nitrification inhibitor was added before measuring OUR_{BASE}. Fillos et al. (2000) and Katehis et al. (2001) measured NPR. Lesouef et al. (1992) applied a variation of the NPR method:

- Lesouef measured ammonia disappearance rate (ADR) rather than NPR; and
- Lesouef calculated b_{AUT} from the change in ADR between two data points (at time zero and after 1 day, say) whereas others have collected multiple data points over several days, and fit a line to the data.

The method used to determine nitrifier decay rate (b_{AUT}) from the measured data will be illustrated based on the NPR approach. NPR measurements are assumed directly proportional to the active nitrifier organism concentration, X_{AUT} (i.e., equation 17.3.1-1). This is reasonable provided the ammonia concentration in the NPR tests is relatively high compared to the nitrifier half-saturation constant, K_{NH}, so that nitrifiers grow at the maximum rate, μ_{AUT}. A straight line is fitted to a semi-logarithmic plot of NPR versus time, and the slope is taken as the nitrifier decay rate (b_{AUT}). That is, it is assumed:

$$X_{AUT,t} = X_{AUT,0} \cdot e^{-b_{AUT} \cdot t} \quad (17.4\text{-}1)$$

The basis of the method is that the nitrifier decay is a first order rate process, and the slope of the semi-logarithmic NPR plot reflects the rate. This would be appropriate if the only process in the test impacting nitrifier concentration is decay of nitrifiers introduced at the start. However, this method does not consider the growth of nitrifiers during the test. Decay of heterotrophic organisms during the test is accompanied by lysis of organic nitrogen. This, in turn, will be converted to ammonia, and then nitrified (with the associated yield of nitrifiers). That is, growth of nitrifiers during the test will offset decay of nitrifiers initially present. Hence:

- The estimation method will underestimate nitrifier decay rate, b_{AUT}. (This conclusion is not specific to the NPR approach. The same problem will be reflected in the decline in "delta OUR.")
- A semi-logarithmic plot of NPR versus time should not be linear.

An analysis was conducted of the expected response in the batch test to assess the extent to which overlooking heterotroph decay, and nitrification of lysed N, may impact the estimate of the decay rate coefficient b_{AUT}.

Consider a batch reactor of volume V. Assume that the reactor is filled with a sample of nitrifying mixed liquor (from a fully-nitrifying plant) at the start of the batch test. The reactor now contains a nitrifier organism concentration, $X_{AUT,0}$ mg COD/L, a heterotroph organism concentration, $X_{HET,0}$ mg COD/L, and a nitrate concentration, $S_{NO,0}$ mg N/L.

Solving the mass balance for nitrifier concentration (see Appendix to this chapter – Section 17.10), assuming that nitrogen lysed via heterotroph decay is nitrified, yields the following expression for the response from time *zero*:

$$\begin{aligned} X_{AUT,t} &= \frac{Y_{AUT} \cdot f_N \cdot (1-f) \cdot b_{HET} \cdot X_{HET,0}}{b_{AUT} - b_{HET}} \cdot \left[e^{-b_{HET} \cdot t} - e^{-b_{AUT} \cdot t}\right] + X_{AUT,0} \cdot e^{-b_{AUT} \cdot t} \\ &= \left(X_{AUT,0} - \frac{Y_{AUT} \cdot f_N \cdot (1-f) \cdot b_{HET} \cdot X_{HET,0}}{b_{AUT} - b_{HET}}\right) \cdot e^{-b_{AUT} \cdot t} \\ &\quad + \frac{Y_{AUT} \cdot f_N \cdot (1-f) \cdot b_{HET} \cdot X_{HET,0}}{b_{AUT} - b_{HET}} \cdot e^{-b_{HET} \cdot t} \end{aligned} \quad (17.10\text{-}17)$$

The corresponding nitrate production rate (NPR) in the *secondary* batch reactor will be:

$$NPR = \left(\frac{\mu_{AUT}}{Y_{AUT}}\right) \cdot X_{AUT,t} \quad (17.10\text{-}18)$$

From inspection of these equations, note that:

- A semi-logarithmic plot of nitrifier concentration or NPR versus time will NOT yield a straight line of slope ($-b_{AUT}$).
- The response is influenced, inter alia, by the ratio of heterotrophs to autotrophs at the start of the test. This, in turn, is influenced by the TKN/COD ratio of the influent to the system where the mixed liquor is generated and the SRT of the system.
- The response assumed by Katehis and Lesouef (equation 17.4-1) is applicable only for the special case where the initial heterotroph concentration, $X_{HET,0}$, is zero and/or where lysed organic nitrogen is not converted to nitrate.

Considering a specific example can assess the impact of heterotroph decay on the estimate of b_{AUT}. Consider a nitrifier decay rate batch test on mixed liquor from a nitrifying activated sludge system with the following concentrations of nitrifiers, heterotrophs and nitrate:

Initial values:	X_{AUT}	= 42 mgCOD/L
	X_{HET}	= 1,800 mgCOD/L
	S_{NO}	= 31 mgN/L

These concentrations were derived from a simulation of an aerobic activated sludge system operated at a 10-day SRT at 20°C. Characteristics of the influent to the system were for a typical raw municipal wastewater (COD = 500 mg/L; TKN = 40 mg/L).

Equations 17.10-17 and 17.10-18 can be applied directly to illustrate the predicted response of nitrifier concentration and NPR. In this illustration, the following kinetic and stoichiometric parameters were assumed:

Nitrifiers:	Y_{AUT}	= 0.15 mgCOD/mgN
	b_{AUT}	= 0.15 /d
Heterotrophs:	f	= 0.20
	f_N	= 0.068 mgN/mg organism COD
	b_H	= 0.24 /d

Note: The values of the parameters f and b_H for the heterotrophs correspond to the "endogenous respiration" parameters rather than those used in the death-regeneration modeling of heterotroph decay. This means that these parameters may not be used directly in a model that employs the death-regeneration approach (e.g., ASM 1). However, the parameters can be converted to values suitable for use in the death-regeneration approach (Dold et al. (1980); Grady et al. (1999)).

Equation 17.10-6 (see Appendix to this chapter – Section 17.10) quantifies the increase in nitrate concentration in the batch test assuming that the ammonia derived from heterotroph decay is converted to nitrate. This is plotted in Figure 17-3. It is clear that nitrate generation is significant, and that nitrifier decay will be offset by nitrifier growth. In this case, the 80 mg/L of nitrate generated over 7 days will result in approximately 12 mg/L of X_{AUT}; that is, almost 30 percent of the initial X_{AUT} concentration:

$$S_{NO,t} = S_{NO,0} + f_N \cdot (1-f) \cdot X_{HET,0} \cdot \left(1 - e^{-b_{HET} \cdot t}\right) \qquad (17.10\text{-}6)$$

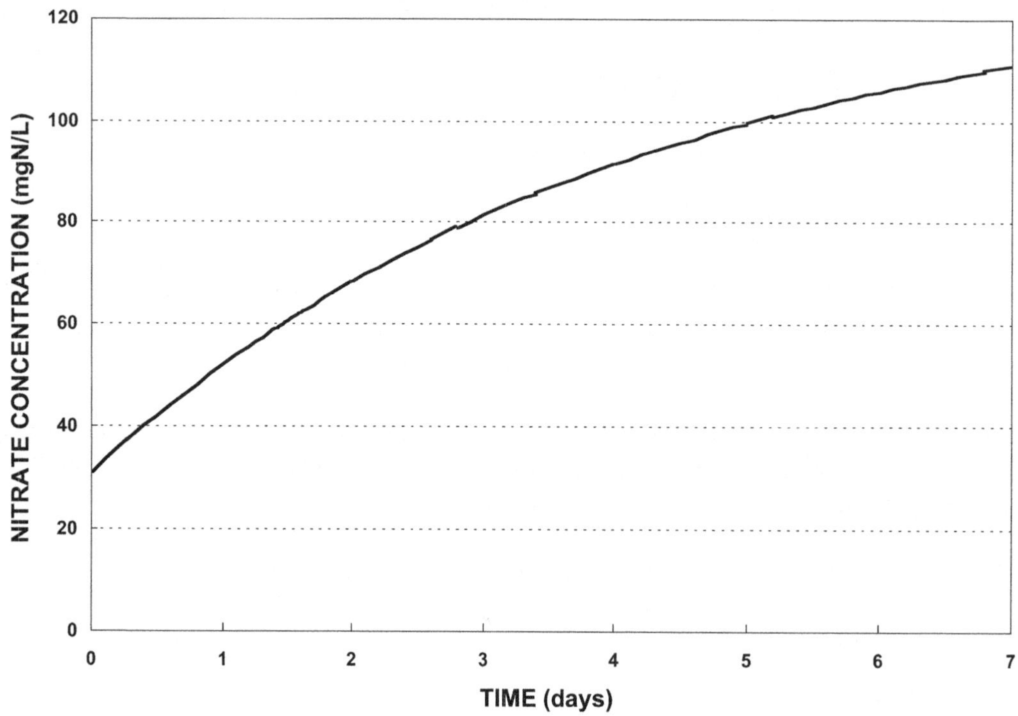

Figure 17-3. Nitrate Concentration Response in the Batch Reactor (equation 17.10-6)

Figure 17-4 shows a plot of the nitrifier concentration change in the batch test (equation 17.10-17). The dashed line corresponds to the case where there is no production of nitrate from lysis of organic nitrogen from heterotroph decay; that is, where the first term in equation 17.10-17 falls away and equation 17.4-1 applies. In that case, the nitrifier concentration obviously exhibits an exponential decrease – the basis of the methods for estimating decay rate.

The solid line in Figure 17-4 corresponds to the (actual) case where decay of nitrifiers is offset by growth of nitrifiers from lysed organic nitrogen. In this case, the decrease in nitrifier concentration essentially is linear (note that this is not a semi-logarithmic plot). This is merely an artifact of the ratio of heterotrophs to autotrophs at the start of this particular test. For example, were the $X_{HET,0} : X_{AUT,0}$ ratio higher (for a higher influent TKN/COD ratio), the X_{AUT} response would be concave-up initially.

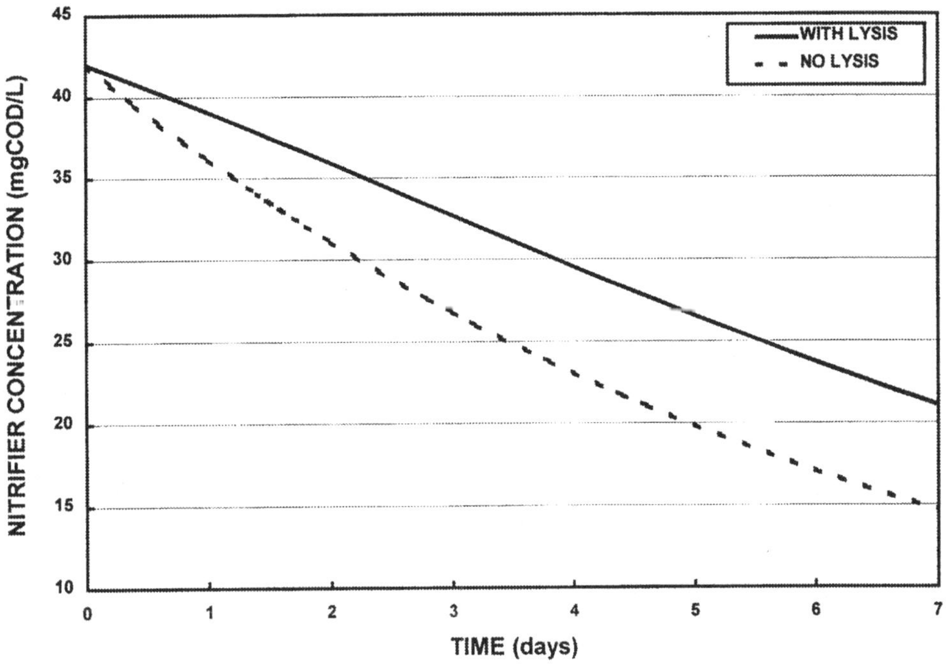

Figure 17-4. Nitrifier Concentration Response in the Batch Reactor (Linear Scale)

Figure 17-5 shows a semi-logarithmic plot of the nitrate production rate (NPR) that would be observed in secondary batch tests withdrawn from the batch reactor (equation 17.10-18). The dashed line (linear) again corresponds to the case of no "internal" growth of nitrifiers from lysis of organic nitrogen.

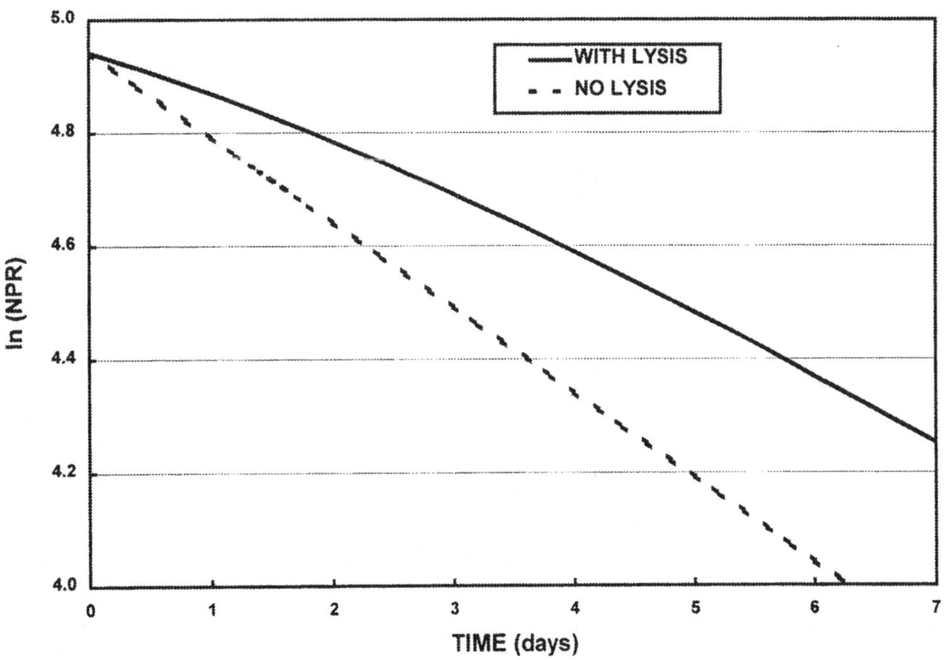

Figure 17-5. Semi-Logarithmic Plot of NPR Response

The method of Katehis/Lesouef estimates the nitrifier decay rate from the slope of the semi-logarithmic plot of NPR versus time:

$$b_{AUT} = \frac{\ln(NPR_0) - \ln(NPR_t)}{t}$$

The value calculated for the "no lysis" case would be:

$$b_{AUT} = \frac{4.94 - 4.01}{6.2} = 0.15 / day$$

That is, the value would correspond to the actual decay rate, b_{AUT}. However, the rate of decrease of NPR is not as marked, and the (incorrect) value calculated from the *actual* NPR response (i.e., accounting for heterotroph decay) would be:

$$\text{After 1.5 days}: \; b_{AUT} = \frac{4.94 - 4.83}{1.5} = 0.075 / day$$

$$\text{After 7 days}: \; b_{AUT} = \frac{4.94 - 4.25}{7} = 0.099 / day$$

It is important to note that:

- The actual ln (NPR) plot is not linear.
- The above estimates of b_{AUT} are substantially lower than the actual b_{AUT}.
- The (incorrect) estimate of b_{AUT} increases as the duration of the test increases. That is, the shorter the test duration (and the smaller the amount of nitrate generated), the larger the error in the b_{AUT} estimate.
- Katehis et al. (2001) calculated b_{AUT} values in the region of 0.09 d^{-1}.
- Nowak et al. (1994) and Siegrist et al. (1999) used the "delta OUR" approach, and did report "higher" estimates of 0.20 d^{-1} at 20°C. However, neither estimation procedure appears to have accounted for nitrifier decay being offset by nitrifier growth from lysed N.
- Copp and Murphy (1995) used the "delta OUR" approach, but with enrichment cultures of nitrifiers, and essentially no heterotrophs present. That yielded b_{AUT} estimates in the range 0.14 to 0.17 d^{-1} at 20°C.

17.5 Simulation of Batch Test Response

The test outlined above was simulated using a full activated sludge kinetic and stoichiometric model. The purpose was to confirm the response predicted by equations 17.10-17 and 17.10-6. The simulator configuration is shown in Figure 17-6. The bioreactor receives typical municipal wastewater influent, and is operated at a 10-day SRT. At time zero, a volume of waste mixed liquor is diverted to the batch reactor, and is aerated for a period of 7 days. Figures 17-7 and 17-8 show the simulated nitrate and nitrifier responses from time *zero*, respectively. Comparison of these figures with the solid lines in Figures 17-3 and 17-4 confirms that the mass balance analysis does in fact predict the anticipated response correctly. That is, the assumption that nitrogen lysed through heterotroph decay will be converted essentially immediately to nitrate.

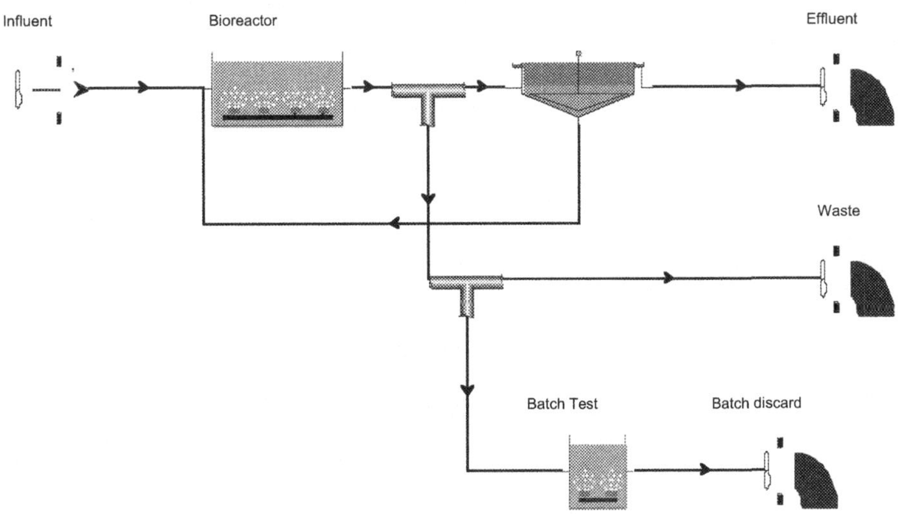

Figure 17-6. Configuration for Simulating Nitrifier Decay Rate Measurement Test

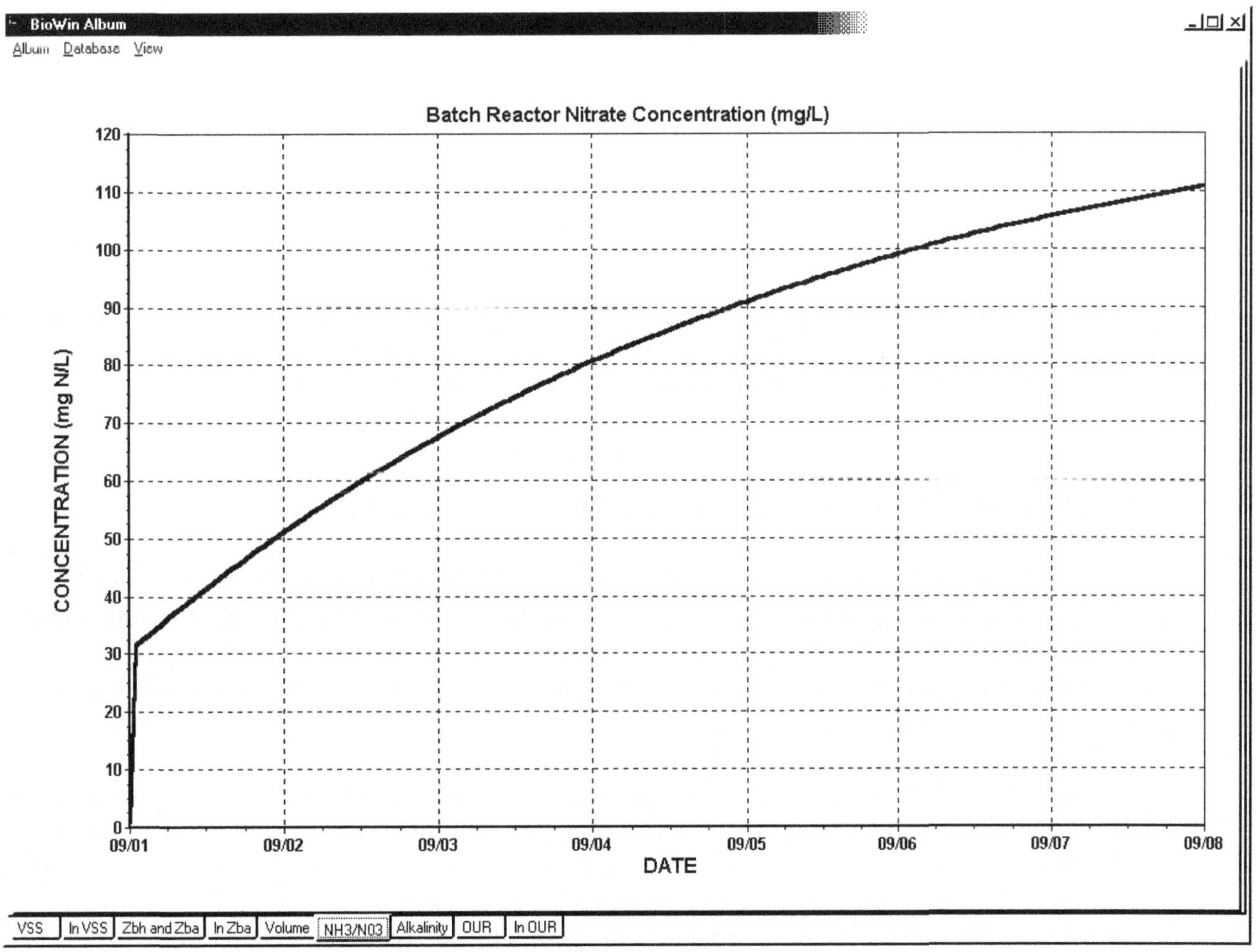

Figure 17-7. Simulated Nitrate Concentration Response in the Batch Reactor (c.f. Figure 17-3)

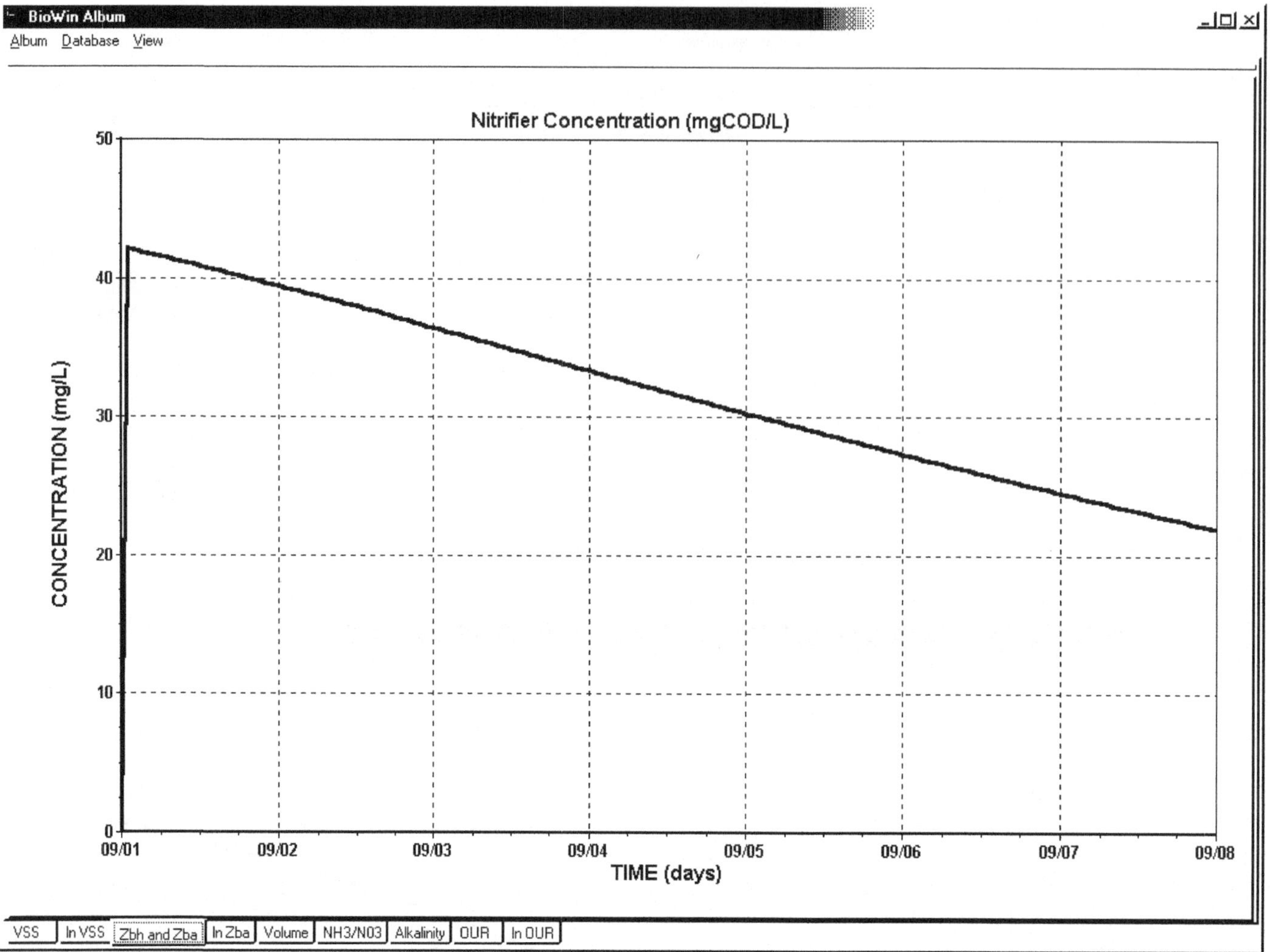

Figure 17-8. Simulated Nitrifier Concentration Response in the Batch Reactor (c.f. Figure 17-4)

17.6 Estimation of Nitrifier Decay Rate from Batch Test NPR Data

The nitrifier decay rate coefficient b_{AUT} value can be estimated using a non-linear least squares technique, fitting equation 17.10-17/17.10-18 *to the observed NPR data* from the batch test. In applying such a technique the method simultaneously estimates the initial nitrifier and heterotroph concentrations, $X_{AUT,0}$ and $X_{HET,0}$, respectively. Experience from evaluation of a number of data sets indicated:

- It is difficult to obtain a reliable estimate for b_{AUT} from experiments such as that shown in Figure 17-9, where there are limited data (e.g., five data points at intervals of 12 hours). The difficulty is that insufficient information is available to estimate all of the parameters simultaneously given the relatively small change in NPR.
- Any experimental scatter compounds the difficulties in parameter estimation.
- To obtain reliable estimates requires extensive data with NPR measurements at short intervals, and little scatter in the data; for example, Figure 17-10. (Collecting such a data

set is very labor intensive because each NPR measurement requires perhaps five or six nitrate analyses to assess NPR with reasonable accuracy.)

An alternate two-step non-linear least squares procedure was developed whereby nitrate response data is first used to estimate the initial heterotroph concentration, $X_{HET,0}$. Thereafter, the least squares technique of fitting equation A.17/18 is applied, but with $X_{HET,0}$ fixed. This approach utilizing *both the nitrate and NPR response data* appears to overcome some of the difficulties with the first method. Nevertheless, extensive NPR and nitrate data are required to obtain a reliable estimate for the nitrifier decay rate, b_{AUT}.

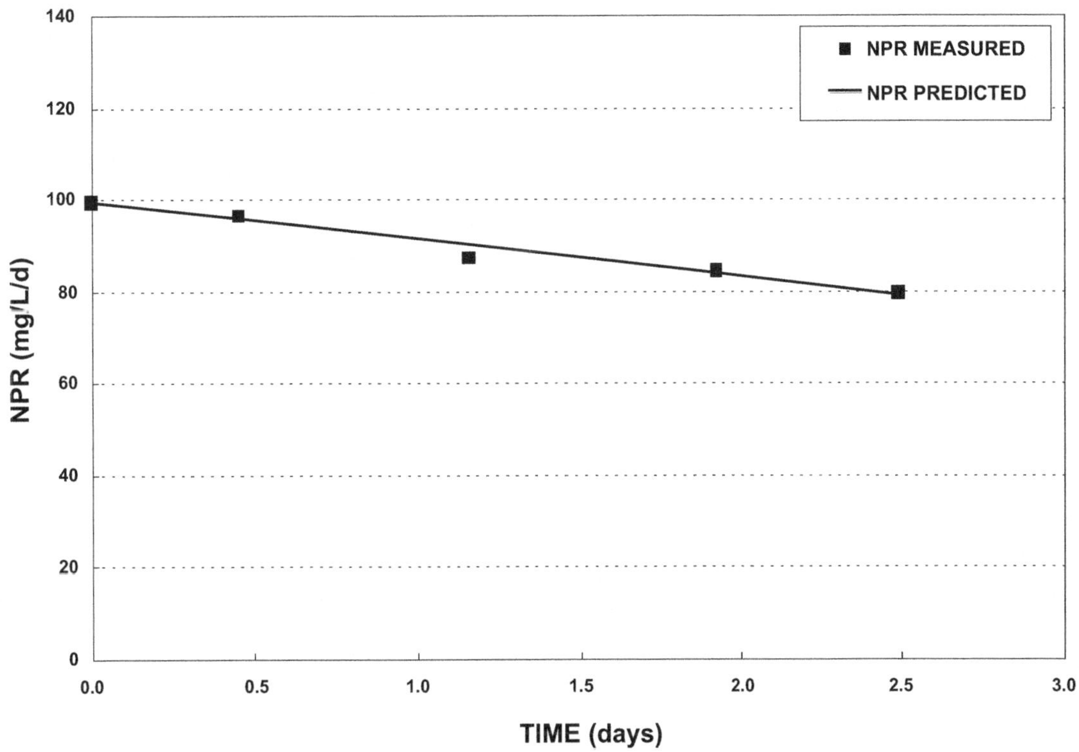

Figure 17-9. Least Squares Estimation of Nitrifier Decay Rate from Limited NPR Data

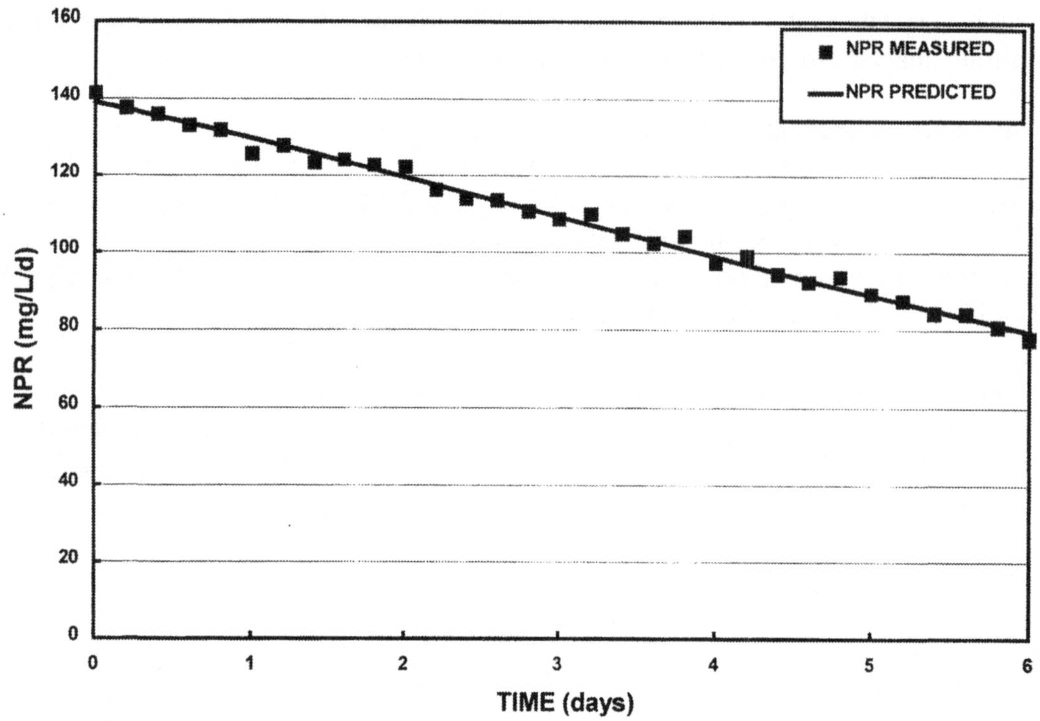

Figure 17-10. Least Squares Estimation of Nitrifier Decay Rate From Extensive NPR Data

A limited number of nitrifier decay rate tests were performed at 12°C and 22°C as part of this WERF study. The resultant b_{AUT} values were 0.14 and 0.19 d^{-1} at 12°C and 22°C, respectively. The specific endogenous decay temperature dependency (θ) coefficient was estimated at 1.029 and the 20°C b_{AUT} value as 0.17 d^{-1}:

$$b_{AUT,T} = 0.17 \cdot 1.029^{(T-20)}\ d^{-1} \qquad (17.6\text{-}1)$$

17.7 Decay Rate and Growth Rate Temperature Adjustment Interaction

Temperature sensitivity of nitrifier maximum specific growth rate was discussed in Chapter 16.0. An interesting observation was derived from the analysis of nitrifier growth rate and decay rate data gathered during this project. Higher, more conservative growth rate μ_{AUT} temperature dependency (θ) factors reported in the literature may well be a mathematical artifact of calculating the temperature factor from either (1) changes in (μ_{AUT} - b_{AUT}) with temperature rather than changes in μ_{AUT}, and/or (2) incorporating the assumption that b_{AUT} is "negligible," i.e., zero. Table 17-1 lists pairs of (μ_{AUT} - b_{AUT}) values gathered by Sözen et al. (1996) at temperatures of 10 and 20°C. The temperature dependency factors for each pair can be calculated based on differences in (μ_{AUT} - b_{AUT}); that is:

$$\theta = \exp\left[\frac{\ln\left(\mu_{AUT} - b_{AUT}\right)_{10} - \ln\left(\mu_{AUT} - b_{AUT}\right)_{20}}{10-20}\right] \qquad (17.7\text{-}1)$$

This yields an average θ value of 1.101. However, if the b_{AUT} value is incorporated in the calculation, and θ is calculated from equation 17.8 based on the μ_{AUT} values, then the average θ value decreases to 1.073. (In these calculations it is necessary to assume a temperature dependency factor for b_{AUT}. This was taken as 1.029. This value is from this project's decay rate tests conducted at 12°C and 22°C at the University of Washington. However, the value does not affect the conclusion concerning θ for μ_{AUT}.)

$$\theta = \exp\left[\frac{\ln(\mu_{AUT})_{10} - \ln(\mu_{AUT})_{20}}{10-20}\right] \qquad (17.7\text{-}2)$$

This effect is illustrated graphically in Figure 17-11 (noting that the y-axis is a logarithmic scale). This effect is demonstrated by noting that the slope of the line joining the μ_{AUT} values is less than the slope of the line joining the (μ_{AUT} - b_{AUT}) values.

Table 17-1. Temperature Dependency θ Factors Based on Change in (μ_{Aut} - B_{aut}) or μ_{Aut} with Temperature

Waste	μ_{AUT} - b_{AUT}		θ for	μ_{AUT} = (μ_{AUT} - b_{AUT}) + b_{AUT}		θ for
	10°C	20°C	μ_{AUT} - b_{AUT}	10°C	20°C	μ_{AUT}
Domestic	0.130	0.370	1.110	0.258	0.540	1.077
	0.100	0.360	1.137	0.228	0.530	1.088
	0.170	0.390	1.087	0.298	0.560	1.065
	0.150	0.390	1.100	0.278	0.560	1.073
	0.130	0.370	1.110	0.258	0.540	1.077
Synthetic	0.130	0.390	1.116	0.258	0.560	1.081
	0.330	0.560	1.054	0.458	0.730	1.048
Meat Processing	0.220	0.630	1.111	0.348	0.800	1.087
	0.200	0.430	1.080	0.328	0.600	1.062
		Average =	1.101		Average =	1.073

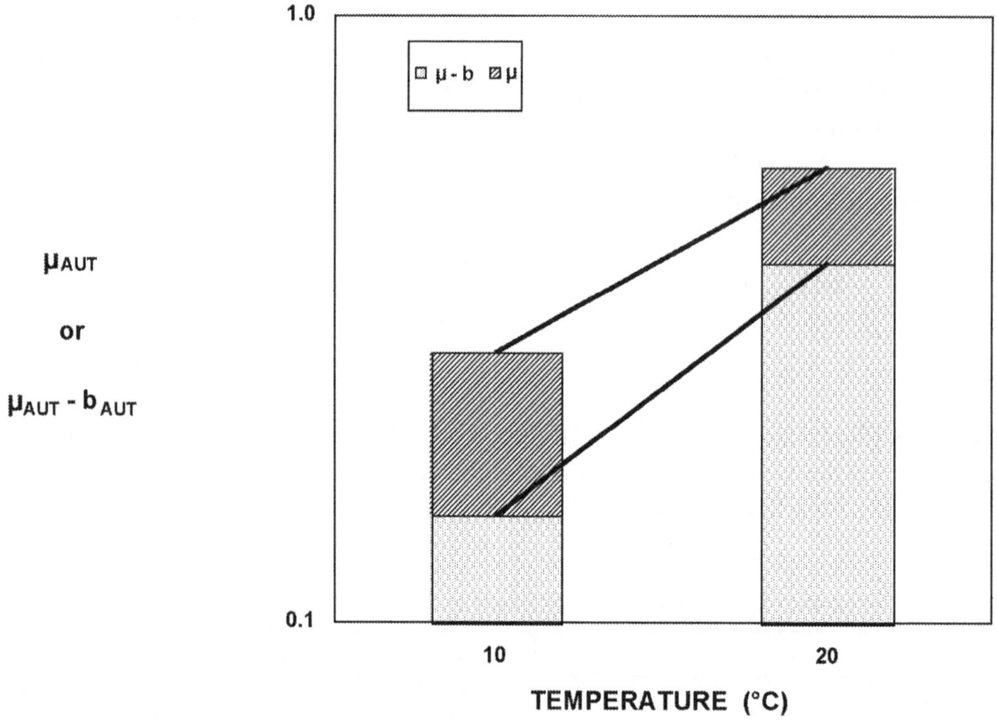

Figure 17-11. Impact of the Temperature Dependency Coefficient (θ) on Minimum SRT for Nitrification

17.8 Conclusions

Nitrifier decay rate, quantified by the first order decay rate parameter b_{AUT}, is important in assessing nitrification behavior for purposes of plant design and plant capacity rating. This research has identified and demonstrated a number of important aspects concerning decay rate, and also the interaction between decay rate and estimation of the nitrifier maximum specific growth rate, μ_{AUT}. The most important conclusions are:

- Experiments have demonstrated that the nitrifier decay rate is more significant than reported in many earlier studies.

- Data from this study indicate a value for the nitrifier decay rate b_{AUT} of 0.17 d^{-1} (at a reference temperature of 20°C) and an associated temperature dependency factor (θ) of 1.029:

  $$b_{AUT,T} = 0.17 \cdot 1.029^{(T-20)}\ d^{-1} \qquad (17.8\text{-}1)$$

- A correct estimate for b_{AUT} is important in estimating maximum specific growth rate, μ_{AUT}. In particular, in low F/M methods for measuring μ_{AUT} (such as the SBR method) applying too low a b_{AUT} value results in a significant underestimate of μ_{AUT}, perhaps by a factor of two. (Regarding past studies where μ_{AUT} perhaps has been underestimated because of assuming a low b_{AUT} value, this does not necessarily imply a problem. If the underestimated μ_{AUT} has been measured in a system of a certain SRT, and then applied to analysis of a system of similar SRT, then the underestimate is of no consequence. However, application of an underestimated μ_{AUT} to a lower SRT system would result in conservative predictions of nitrification performance.)

- The traditional experimental method for determining b_{AUT} from the slope of a semi-logarithmic NPR versus time plot underestimates b_{AUT} significantly. A new method for data analysis and estimating b_{AUT} has been presented.
- Higher, more conservative growth rate (μ_{AUT}) temperature dependency factors reported in the literature may well be a mathematical artifact of calculating the temperature factor from either (1) changes in (μ_{AUT} - b_{AUT}) with temperature rather than changes in μ_{AUT}, and/or (2) incorporating the assumption that b_{AUT} is "negligible."

Conclusions from this study should be viewed in the context of the assumption that it is sufficient to model nitrification as a single-step process. As further information becomes available it is likely that there will be more reason to consider a two-population approach, distinguishing between the behavior of ammonia- and nitrite-oxidizers.

It should be noted that all of the experimental work in this study considered aerobic-only systems. Further research should be undertaken on nitrifier decay under anoxic/anaerobic conditions in light of the findings of a reduced anoxic decay rate by Nowak et al. (1994), Siegrist et al. (1999) and Lee and Oleszkiewicz (2002).

17.9 References

Barker, P.S., and P.L. Dold. 1997. General model for biological nutrient removal activated sludge systems: model presentation. *Water Environ. Res.* 69:969-984.

Copp, J.B. 1993. *Nitrification parameter and biomass evaluation for the modeling of nitrifying activated sludge systems*. M. Eng. Thesis, McMaster University, Hamilton, Ontario, Canada.

Copp, J.B., and K.L. Murphy. 1995. Estimation of the active nitrifying biomass in activated sludge. *Water Res.* 29:1855-1862.

Dold, P.L., G.A. Ekama, and G.v.R. Marais. 1980. A general model for the activated sludge process. *Prog. Water Technol.* 12:47-77.

Downing, A.L., H.A. Painter, and G. Knowles. 1964. Nitrification in the activated sludge process. *J. Proc. Inst. Sew. Purif.* 64:130-158.

Fillos, J., D. Katehis, K. Ramalingam, L.A.Carrio, and K. Gopalakrishnan. 2000. Determination of nitrifier growth rates in NYC water pollution control plants. In *Proc. 73rd Annual Conference of the Water Environment Federation*, Anaheim CA, CD-ROM. Alexandria, VA: Water Environment Federation.

Grady, C.P.L, G.T. Daigger, and H.C. Lim. 1999. *Biological wastewater treatment,* 2nd Edition. New York, NY: Marcel Dekker, Inc. 1076 pp.

Henze, M., W. Gujer, T. Mino, and M.C.M. Loosdrecht. 2000. Activated Sludge Models ASM1, ASM2, ASM2d and ASM3. *IWA Scientific and Technical Report No. 9*, London, UK: International Water Association.

Katehis, D., J. Fillos, and L.A. Carrio. 2002. Comparison of bench scale testing methods for nitrifier growth rate measurement. *Water Sci. Tech.* 46(1-2):289-295.

Lesouef, A., M. Payraudeau, F. Rogalla, and B. Kleiber. 1992. Optimizing nitrogen removal reactor configuration by on-site calibration of the IAWPRC activated sludge model. *Water Sci. Tech.* 26(6):105-123.

Lee, Y., and J.A. Oleszkiewicz. 2002. Evaluation of maximum growth and decay rates of autotrophs under different physical and environmental conditions. In *Proc. 75th Annual Conference of the Water Environment Federation*, Chicago, IL, CD-ROM. Alexandria VA: Water Environment Federation.

Nowak, P., Schweighofer, and K. Svardal. 1994. Nitrification inhibition – a method for estimation of actual maximum autotrophic growth rates in activated sludge systems. *Water Sci. Tech.* 30(6):9-19.

Rozich, A.F., and D.J Castens. 1986. Inhibition kinetics of nitrification in continuous-flow reactors. *Journal WPCF* 58(3):220-226.

Siegrist, H., I. Brunner, G. Koch, P. Linh Con, and C. Van Chiew. 1999. Reduction of biomass decay rate under anoxic and anerobic conditions. *Water. Sci. Technol.* 39(1):129-137.

Sözen, S., D. Orhon, and H.A. San. 1996. A new approach for the evaluation of the maximum specific growth rate in nitrification. *Water Res.* 30(7):1661-1669.

Water Research Commission. 1984. *Theory, design and operation of nutrient removal activated sludge processes*. P. O. Box 824, Pretoria, South Africa: Water Research Commission.

17.10 Appendix – Derivation of Equations

Consider a batch reactor of volume V that is filled with a sample of nitrifying mixed liquor (from a nitrifying activated sludge plant). At the start of the test the reactor contains:

- A nitrifier organism concentration, $X_{AUT,0}$ mgCOD/L.
- A heterotroph organism concentration, $X_{HET,0}$ mgCOD/L.
- At time *zero*, the nitrate concentration is $S_{NO,0}$ mgN/L.

Assume nitrifier decay is a first order process:

$$\frac{dX_{AUT}}{dt} = -b_{AUT} \cdot X_{AUT} \tag{17.10-1}$$

If this were the only process impacting nitrifier concentration, then:

$$X_{AUT,t} = X_{AUT,0} \cdot e^{-b_{AUT} \cdot t} \tag{17.10-2}$$

Assume heterotroph decay is a first order process:

$$\begin{aligned} \frac{dX_{HET}}{dt} &= -b_{HET} \cdot X_{HET} \\ X_{HET,t} &= X_{HET,0} \cdot e^{-b_{HET} \cdot t} \end{aligned} \tag{17.10-3}$$

For each unit of X_{HET} disappearing in decay, a fraction (f) remains as endogenous residue, X_E. Assume that the organic nitrogen content of the endogenous residue is the same as in the active organisms, and that organic nitrogen is lysed from the decaying heterotrophs. If the nitrogen content of the active heterotrophs is f_N mgN/mg organism COD (say 0.068 mgN/mgCOD), then the rate of lysis of organic N from heterotroph decay is:

$$\begin{aligned} \frac{dN_{ORG}}{dt} &= -f_N \cdot (1-f) \cdot \frac{dX_{HET}}{dt} \\ &= f_N \cdot (1-f) \cdot b_{HET} \cdot X_{HET} \\ &= f_N \cdot (1-f) \cdot b_{HET} \cdot X_{HET,0} \cdot e^{-b_{HET} \cdot t} \end{aligned} \tag{17.10-4}$$

Assume that the organic nitrogen from lysis is converted to ammonia rapidly, and then nitrified rapidly. The rate of nitrate production will equal the rate of ammonia production from lysis:

$$\frac{dS_{NO}}{dt} = f_N \cdot (1-f) \cdot b_{HET} \cdot X_{HET,0} \cdot e^{-b_{HET} \cdot t} \tag{17.10-5}$$

Integrating from time zero to time t:

$$\begin{aligned} \int_{S_{NO,O}}^{S_{NO,t}} dS_{NO} &= f_N \cdot (1-f) \cdot b_{HET} \cdot X_{HET,0} \cdot \int_0^t e^{-b_{HET} \cdot t} dt \\ S_{NO,t} - S_{NO,0} &= f_N \cdot (1-f) \cdot b_{HET} \cdot X_{HET,0} \cdot \left(\frac{1}{-b_{HET}} \right) \left(e^{-b_{HET} \cdot t} - 1 \right) \\ S_{NO,t} &= S_{NO,0} + f_N \cdot (1-f) \cdot X_{HET,0} \cdot \left(1 - e^{-b_{HET} \cdot t} \right) \end{aligned} \tag{17.10-6}$$

The associated rate of production of nitrifiers from lysed nitrogen will be:

$$\frac{dX_{AUT}}{dt} = Y_{AUT} \cdot f_N \cdot (1-f) \cdot b_{HET} \cdot X_{HET,0} \cdot e^{-b_{HET} \cdot t} \tag{17.10-7}$$

Combining Eqs. 17.10-1 and 17.10-7, the overall rate of change of nitrifier concentration is:

$$\frac{dX_{AUT}}{dt} = -b_{AUT} \cdot X_{AUT} + Y_{AUT} \cdot f_N \cdot (1-f) \cdot b_{HET} \cdot X_{HET,0} \cdot e^{-b_{HET} t} \qquad (17.10\text{-}8)$$

Grouping parameters, this equation can be written as:

$$\frac{dX_{AUT}}{dt} + k_1 \cdot X_{AUT} = k_2\, e^{k_3 t} \qquad (17.10\text{-}9)$$

where:

$$k_1 = b_{AUT}$$
$$k_2 = Y_{AUT} \cdot f_N \cdot (1-f) \cdot b_{HET} \cdot X_{HET,0}$$
$$k_3 = -b_{HET}$$

The integrating factor for the solution of this first-order differential equation is:

$$e^{\int k_1 dt} = e^{k_1 t}$$

Multiplying both sides of the equation by the integrating factor:

$$e^{k_1 t} \cdot \left[\frac{dX_{AUT}}{dt} + k_1 \cdot X_{AUT}\right] = e^{k_1 t} \cdot k_2\, e^{k_3 t}$$
$$e^{k_1 t} \cdot \frac{dX_{AUT}}{dt} + k_1 \cdot e^{k_1 t} \cdot X_{AUT} = k_2\, e^{(k_1+k_3)t} \qquad (17.10\text{-}10)$$

The left hand side of the equation can be re-arranged by noting from the product rule that:

$$\frac{d}{dt}\left(e^{k_1 t} \cdot X_{AUT}\right) = e^{k_1 t} \cdot \frac{dX_{AUT}}{dt} + X_{AUT} \cdot \left[k_1 \cdot e^{k_1 t}\right]$$
$$= e^{k_1 t} \cdot \frac{dX_{AUT}}{dt} + k_1 \cdot e^{k_1 t} \cdot X_{AUT} \qquad (17.10\text{-}11)$$

Hence:

$$\frac{d}{dt}\left(e^{k_1 t} \cdot X_{AUT}\right) = k_2 \cdot e^{(k_1+k_3)t} \qquad (17.10\text{-}12)$$

Integrating and re-arranging:

$$\int d\left(e^{k_1 t} \cdot X_{AUT}\right) = k_2 \cdot \int e^{(k_1+k_3)t}\, dt$$
$$e^{k_1 t} \cdot X_{AUT} = \frac{k_2}{k_1 + k_3} \cdot e^{(k_1+k_3)t} + K \qquad (17.10\text{-}13)$$

The constant of integration can be evaluated by noting that, when t=0 the nitrifier concentration, $X_{AUT} = X_{AUT,0}$; that is:

$$1 \cdot X_{AUT,0} = \frac{k_2}{k_1 + k_3} \cdot 1 + K$$
$$K = X_{AUT,0} - \frac{k_2}{k_1 + k_3} \tag{17.10-14}$$

Hence:

$$e^{k_1 t} \cdot X_{AUT,t} = \frac{k_2}{k_1 + k_3} \cdot e^{(k_1+k_3)t} + X_{AUT,0} - \frac{k_2}{k_1 + k_3}$$
$$e^{k_1 t} \cdot X_{AUT,t} = \frac{k_2}{k_1 + k_3} \cdot \left[e^{(k_1+k_3)t} - 1 \right] + X_{AUT,0} \tag{17.10-15}$$

$$X_{AUT,t} = \frac{k_2}{k_1 + k_3} \cdot \left[e^{k_3 t} - e^{-k_1 t} \right] + X_{AUT,0} \cdot e^{-k_1 t}$$

Now:

$$\frac{k_2}{k_1 + k_3} = \frac{Y_{AUT} \cdot f_N \cdot (1-f) \cdot b_{HET} \cdot X_{HET,0}}{b_{AUT} - b_{HET}} \tag{17.10-16}$$

Hence:

$$X_{AUT,t} = \frac{Y_{AUT} \cdot f_N \cdot (1-f) \cdot b_{HET} \cdot X_{HET,0}}{b_A - b_H} \cdot \left[e^{-b_{HET} t} - e^{-b_{AUT} t} \right] + X_{AUT,0} \cdot e^{-b_{AUT} t}$$
$$= \left(X_{AUT,0} - \frac{Y_{AUT} \cdot f_N \cdot (1-f) \cdot b_{HET} \cdot X_{HET,0}}{b_{AUT} - b_{HET}} \right) \cdot e^{-b_{AUT} t}$$
$$+ \frac{Y_{AUT} \cdot f_N \cdot (1-f) \cdot b_{HET} \cdot X_{HET,0}}{b_{AUT} - b_{HET}} \cdot e^{-b_{HET} t} \tag{17.10-17}$$

The corresponding nitrate production rate (NPR) in the secondary batch reactor will be:

$$NPR = \left(\frac{\mu_{AUT}}{Y_{AUT}} \right) \cdot X_{AUT,t} \tag{17.10-18}$$

CHAPTER 18.0

DENITRIFICATION RATE

18.1 Background

Activated sludge processes that incorporate biological nitrogen removal are being used with increasing frequency at municipal wastewater treatment plants due to increased interest in water reuse, greater concerns about eutrophication and many other process benefits such as energy savings, improved sludge thickening and alkalinity production. Most activated sludge process designs used for biological nitrogen removal have a mixed, nonaerated anoxic zone before the aeration tank (anoxic-aerobic, A/O), where no oxygen is present and nitrate reduction occurs. Nitrate reduction can also occur in aeration basins at low dissolved oxygen (DO) concentrations due to anoxic conditions developed within the activated sludge floc or at lower rates limited by DO inhibition. Nitrate reduction is carried out by common facultative heterotrophic bacteria in the activated sludge process (Randall et al., 1992). In the absence of oxygen, these bacteria can use nitrate as an alternate electron acceptor at the end of their respiratory process.

Modeling of denitrification involves consideration of both the stoichiometry and the kinetics. This chapter deals with the kinetic aspects. Chapter 15.0 considers the stoichiometry.

Two principal factors control the rate of denitrification in activated sludge systems:

- The rate of utilization of readily biodegradable COD (RBCOD) derived from the influent wastewater (or added as a supplemental carbon source to enhance the extent of denitrification.)
- Once the RBCOD has been consumed, the rate of denitrification is controlled by the rate of hydrolysis of slowly biodegradable COD (SBCOD). This process controls the availability of substrate for denitrification.

Note: The extent of denitrification in systems receiving only municipal wastewater is strongly governed by the amount of readily biodegradable COD (RBCOD) in the influent. This endorses the importance of measuring RBCOD content for accurate modeling of nitrogen removal.

Note: The discussion of denitrification stoichiometry identified the phenomenon of a reduced yield coefficient under anoxic compared to aerobic conditions. It is important to consider this factor in the analysis of denitrification kinetics. The ASM1 and ASM2d models do not incorporate a reduced Y_{HET} for anoxic growth. Therefore, model parameters associated with modeling of denitrification rate are not necessarily equivalent between models.

18.2 Anoxic Growth - η_G

The rate of substrate utilization and endogenous decay activity has been observed to be lower when nitrate instead of oxygen is used as the electron acceptor. A significant result of the lower activity with nitrate present is that the rate of nitrate utilization, on an oxygen equivalent basis, is lower than the oxygen utilization rate that would occur under strictly aerobic conditions. A major factor causing the lower substrate utilization rate is that not all of the heterotrophic bacteria present in the activated sludge system mixed liquor are facultative organisms capable of nitrate reduction. Some portion of the mixed liquor contains strictly aerobic bacteria that cannot use nitrate or nitrite as an electron acceptor. Another factor to be considered is the synthesis yield factor, which based on relative free energy changes between oxygen and nitrate is lower when nitrate is the electron acceptor. In addition, it is not known how the maximum specific growth rate varies with the two electron acceptors. For present activated sludge design models based on the IWA structure, all of these factors are lumped into a single parameter, termed η_G, that accounts for a reduced effective substrate utilization rate or endogenous decay with nitrate as the electron acceptor instead of oxygen (Barker and Dold, 1997).

Note: The term η_g, is used to describe an effective specific growth rate for the system heterotrophic bacteria under anoxic versus aerobic conditions

The specific growth rate of heterotrophic biomass in the IWA-type activated sludge models are based on the soluble readily biodegradable substrate concentration according to Monod kinetics. This is shown in equation 18.2-1, ignoring the switching functions for the effect of DO and other limiting substrates at very low concentrations. The effective specific growth rate for the heterotrophic biomass will be lower when nitrate is used as the electron acceptor instead of oxygen and this is accounted for in equation 18.2-2 by the term η_G, which shows the rate of biomass growth under anoxic conditions.

$$r_{XHET} = \mu_{HET} \frac{S_S}{K_{SS} + S_S} X_{HET} \qquad (18.2\text{-}1)$$

$$r_{XHET} = \mu_{HET} \frac{S_S}{K_{SS} + S_S} X_{HET}(\eta_G) \qquad (18.2\text{-}2)$$

where:

r_{XHET} = heterotrophic organism growth rate

μ_{HET} = maximum specific heterotrophic growth rate

X_{HET} = heterotrophic biomass concentration

K_{SS} = readily biodegradable COD half-saturation coefficient

S_S = readily biodegradable COD concentration

η_G = ratio of growth rate with nitrate versus that for oxygen

---

Note: The value η_g may vary for different activated sludge systems depending on the design and system operation.

The value for η_G was found to vary from 0.30 to 0.85 for different activated sludge processes and was affected by the degree of anoxic substrate removal and process SRT (Stensel and Horne, 2000). A default value of 0.37 has been recommended in the IWA activated sludge model. A method to obtain site-specific η_G values would be useful for calibration of activated sludge models. The following method can be used to obtain a value for η_G using the site mixed liquor in parallel batch tests with nitrate and oxygen as electron acceptors. The relationship between the nitrate uptake rate and oxygen uptake rate to η_G is shown below.

The readily biodegradable substrate utilization rate is related to the biomass growth rate and oxygen uptake rate as shown in equation 18.2-3. By combining equation 18.2-3 with equation 18.2-1, the oxygen uptake rate is related to the Monod kinetics as shown in equation 18.2-4.

$$r_{SS} = \frac{r_{XHET}}{Y_{HET}} = \frac{OUR}{1-Y_{HET}} \qquad (18.2\text{-}3)$$

$$OUR = \mu_{HET} \frac{S_S}{K_{SS}+S_S} X_{HET} \frac{1\ \ Y_{HET}}{Y_{HET}} \qquad (18.2\text{-}4)$$

where:

r_{SS} = substrate utilization rate

Y_{HET} = synthesis yield of heterotrophic bacteria

OUR = oxygen uptake rate

A similar equation expresses the nitrate uptake for the same mixed liquor by including the factor η_G. as follows:

$$NUR = \mu_{HET} \frac{S_S}{K_{SS}+S_S} X_{HET} \frac{1-Y_{HET}}{Y_{HET}} \frac{\eta_G}{2.86} \qquad (18.2\text{-}5)$$

where:

NUR = nitrate-nitrogen uptake rate

2.86 = g oxygen equivalent per g nitrate nitrogen

By dividing equation 18.5 by equation 18.4 the value for η_G can be obtained from the batch tests with the same mixed liquor using nitrate and oxygen as electron acceptors:

$$\eta_G = \frac{2.86 \cdot NUR}{OUR} \qquad (18.2\text{-}6)$$

Note: Equation 18.2-6 is derived by dividing equation 18.2-5 (anoxic growth) by equation 18.2-4 (aerobic growth). The term Y_{HET} appears in both these equations. Therefore, if different values are used for aerobic and anoxic Y_{HET} in the model being applied, these terms will not cancel out in the analysis.

18.2.1 Approaches to Measuring η_G

Batch tests under anoxic and aerobic conditions are done using the same mixed liquor sample to determine NUR and OUR for calculating the η_G value. The batch tests are spiked with acetate so that the OUR and NUR in equations 18.2-4 and 18.2-5 are at their maximum rate and not affected by differences in readily biodegradable COD concentrations. Acetate was selected as the substrate as it represents a readily biodegradable substrate that could be present in wastewater influent or produced by fermentation in anoxic and anaerobic zones.

18.2.2 Aerobic and Anoxic Batch Tests

The same mixed liquor must be used within a similar time frame for each test. The mixed liquor sample can be collected and maintained under aeration for a short time before the tests. The oxygen uptake rate test is done as follows:

- An appropriate-sized reactor (500 mL to 2 L) containing a DO probe connected to a DO meter for continuous monitoring of the DO concentration is used. The reactor contains a magnetic stirring bar and is placed on a magnetic stir plate. An air diffuser stone and air pump or other air source is used to aerate the reactor mixed liquor.
- The mixed liquor sample is placed in the reactor and aerated and mixed with an initial DO concentration of 5 mg/L or greater. The temperature and pH are recorded.
- Acetate is added at a 30 mg/L dose and aeration is stopped, while mixing continues.
- The DO concentration with time is recorded. A linear decrease in DO concentration will be observed and this is used to calculate the OUR in mg/L-min.

A slightly modified reactor is used for the NUR test. The reactor is the same as described above for aeration and mixing. However, it has a cover with a small opening for gas venting and the inlet air line has a tee. On one side of the tee, air can be supplied and on the other, nitrogen. The test is done as follows:

- The mixed liquor is added and the reactor is aerated and mixed. NO_3-N is added to the reactor to reach a 10 to 15 mg/L concentration. The temperature and pH are measured.
- The feed gas is switched from air to nitrogen and the DO concentration decrease is observed.
- When the DO concentration is below 0.5 mg/L, acetate is added at 30 mg/L.
- Mixed liquor samples are obtained every 15 minutes or as appropriate and immediately filtered for NO_3-N analysis. Rapid filtration through a coarse media (such as a coffee filter) may be used first, followed by filtration with a glass-fiber filter paper.
- The linear decrease in NO_3 N concentration is determined from the data and is the NUR.

Equation 18.2-5 is used to calculate the η_G value.

18.3 Anoxic Hydrolysis – η_{HYD}

The anoxic utilization of RBCOD is slower than aerobic utilization as noted above. Nevertheless the anoxic utilization is still quite rapid. Therefore, once the RBCOD has been consumed, the rate of denitrification is controlled by the rate of hydrolysis of slowly biodegradable COD (SBCOD). This process controls the availability of substrate for denitrification.

In an analogous manner for reducing anoxic versus aerobic growth, the activated sludge models incorporate a η_{HYD} factor to reduce the anoxic hydrolysis rate for SBCOD relative to the aerobic rate. This is depicted in equation 18.3-1:

$$\begin{aligned} r_{DENITRIFICATION} &= r_{COD\ PRODUCTION} \cdot \frac{\Delta NO_3 - N}{\Delta COD} \\ &= r_{HYDROLYSIS} \cdot \frac{1}{\frac{2.86}{1-Y_H}} \\ &= \eta_{HYD} \cdot r_{AEROBIC\ HYDROLYSIS} \cdot \frac{1-Y_H}{2.86} \end{aligned} \qquad (18.3\text{-}1)$$

Information for determining the η_{HYD} factor in model calibration can be derived from anoxic batch tests combining mixed liquor from the plant with influent wastewater, and spiking the batch test with nitrate at the start. Nitrate concentration is monitored in the test. The VSS concentration in the batch test also should be measured. An example of an anoxic batch test is shown in Figure 18-1. In this test:

- Initially the nitrate profile is near linear, and nitrate concentration decreases rapidly. This corresponds to denitrification with RBCOD from the influent wastewater added to the test.
- Once the RBCOD has been consumed the nitrate response is again near linear, but the rate of denitrification is lower. In this phase, the rate is controlled by the rate of SBCOD hydrolysis.

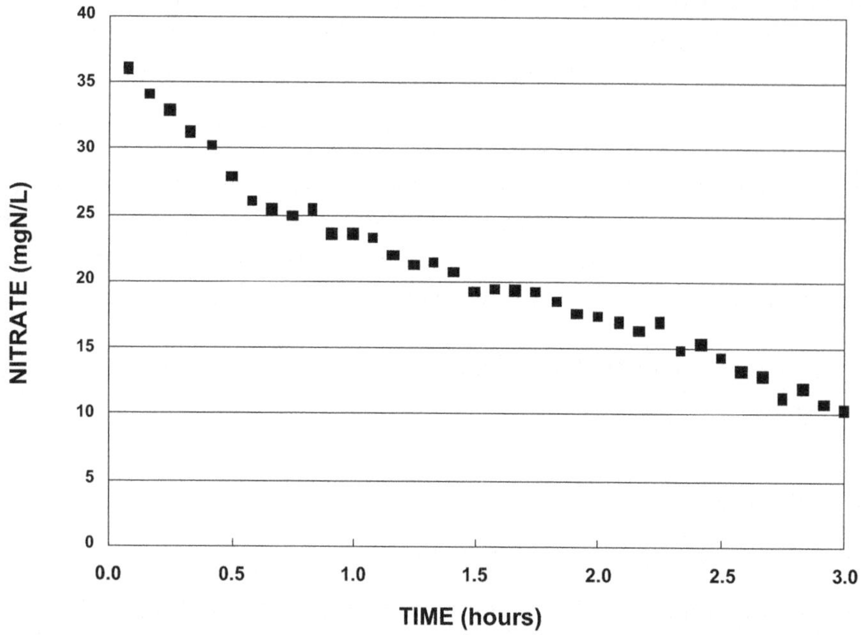

Figure 18-1. Nitrate Profile in an Anoxic Batch Test Combining Mixed Liquor and Municipal Wastewater

The η_{HYD} factor can be determined through calibrating the model to the nitrate response, adjusting the value of η_{HYD} to fit the data. This approach requires knowledge of the initial conditions in the test, in particular the concentration of active heterotroph biomass (and the RBCOD in the influent). Therefore, the composition of the mixed liquor in the plant from which mixed liquor is taken also must be known. This can only be determined if there is comprehensive information available on the loading to the plant as well as the operating conditions (SRT, etc.). Unfortunately, this information often is not known with confidence.

A convenient simulation approach is depicted in Figure 18-2 (assuming that the required data are known). The plant is included in the simulation configuration. Initially, a steady state simulation is executed to establish the composition of the mixed liquor in the plant. Then a volume of waste mixed liquor is diverted to the empty batch reactor, and the dynamic simulation of the batch test is performed. This procedure is repeated iteratively, adjusting the value of η_{HYD} to fit the data. Figure 18-3 shows the response to the batch test simulation.

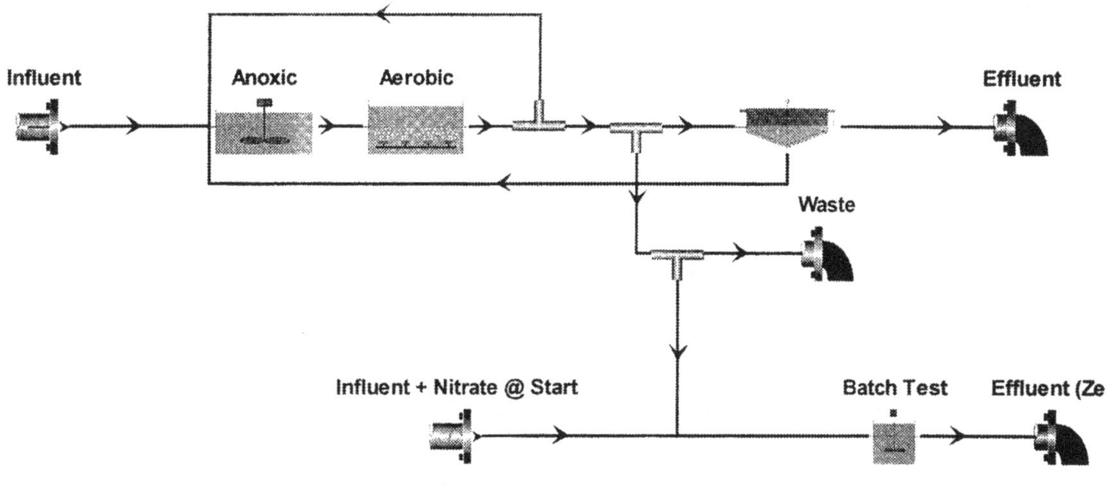

Figure 18-2. Configuration for Simulating Nitrifier Decay Rate Measurement Test

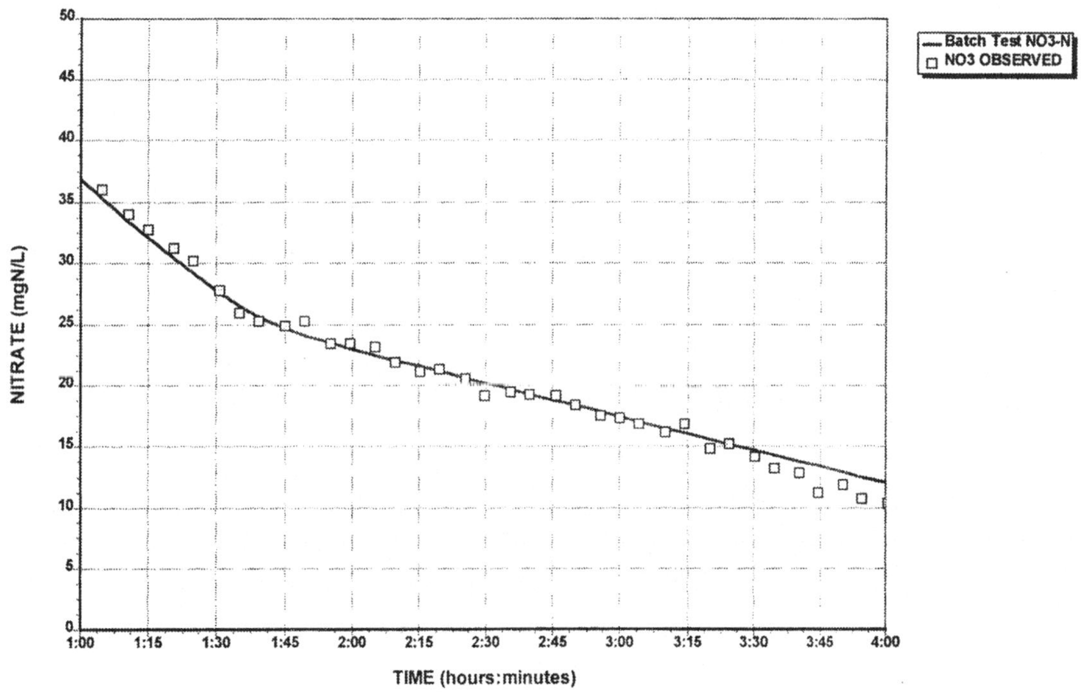

Figure 18-3. Simulation of Anoxic Batch Test Nitrate Response

18.4 Considerations

- Specific denitrification rate (SDNR), in terms of the rate of nitrate removal per unit VSS (mgN/mgVSS/hour), sometimes is quoted as a measure of denitrification rate. It should be recognized that the SDNR is not a model parameter as such. Nevertheless, it is useful to compare the model-predicted SDNR (calculated as a model output) to measured values.
- The phenomenon of a lower yield in anoxic growth compared to aerobic growth (see Chapter 15.0) holds important implications for modeling of systems with denitrification.
- Interpretation of denitrification rate data can be quite complex for anoxic-aerobic single sludge systems with substrate (such as methanol) addition to anoxic zones. For example, recent pilot-scale studies in New York City indicate that:

    1. Denitrification with methanol appears to be mediated by specific methanol-utilizing heterotrophs that do not grow efficiently in the aerated portions of the system. This introduces the concept of a minimum unaerated SRT to sustain growth.
    2. In cases of over-dosing with methanol, where methanol is not completely consumed under anoxic conditions and passes into aerated zones, the excess methanol is used for the growth of heterotrophs that are not able to denitrify.

The models perhaps should incorporate a further structuring of the heterotrophic biomass into denitrifying and non-denitrifying components.

18.5 References

Barker, P. S., and P. L. Dold. 1997. General model for biological nutrient removal activated sludge systems: model presentation. *Water Environ. Res.* 69(5):969-984.

Randall, C. W., J. L. Barnard, and H. D. Stensel. 1992. *Design and retrofit of wastewater treatment plants for biological nutrient removal.* Lancaster, PA: Technomic Publishing Company, Inc. 420 pp.

Stensel, H.D., and G. Horne. 2000. Evaluation of denitrification kinetics at wastewater treatment facilities. In *Proc. of 73rd Annual Conference of the Water Environment Federation,* Anaheim, CA, CD-ROM. Alexandria VA: Water Environment Federation.

CHAPTER 19.0

PROCESS MODELING USING A SIMULATOR

19.1 Background

Computer-based process simulation is a useful tool for a number of functions performed by wastewater treatment professionals. These include design, analysis, trouble-shooting, optimization, operator decision-making and operator training. A simulator allows the user to link various unit operations together, such as bioreactors and settlers, according to the flow scheme of the particular treatment plant, and then to mimic the performance of the plant for specified operational and influent loading conditions. Each unit in a plant configuration incorporates a mathematical model; of these, the bioreactor model usually is the most complex.

This chapter suggests five distinct stages to follow in a simulation exercise:

- Configuring the wastewater treatment process on the simulator;
- Gathering data to use for calibrating the simulator;
- Calibrating the simulator;
- Verifying calibrated simulator; and
- Applying the simulator for its intended purpose.

NOTE: Most simulators include suggested default values for stoichiometric and kinetic model parameters. Usually these are based on many experiments and model calibrations, and should allow reasonable estimation of process performance in most situations. That is, model parameters should not change significantly from case to case and therefore should not be changed without justification. If model predictions do not reflect observed behavior the first step should be to investigate other potential reasons (for example, poor information on sludge wastage, or whether the influent characteristics are unusual (e.g., a highly chlorinated influent that may have no impact on nitrification, but severely inhibit biological phosphorus removal)) rather than making adjustments to model parameters.

Care must be taken when using a simulator, particularly in the configuration and calibration of the WWTP under consideration, otherwise erroneous or misleading results can easily be obtained. This chapter identifies some of the important considerations when applying a wastewater treatment process simulator. The objective in this chapter is to provide an overview of the application of simulators. Chapter 21.0 provides more specific practical suggestions on model calibration.

19.2 Approach to Using a Simulator

Figure 19-1 highlights the requirements for setting up a simulation of a wastewater treatment process. Information is needed on the inputs to the process, the process configuration, and the process operating conditions.

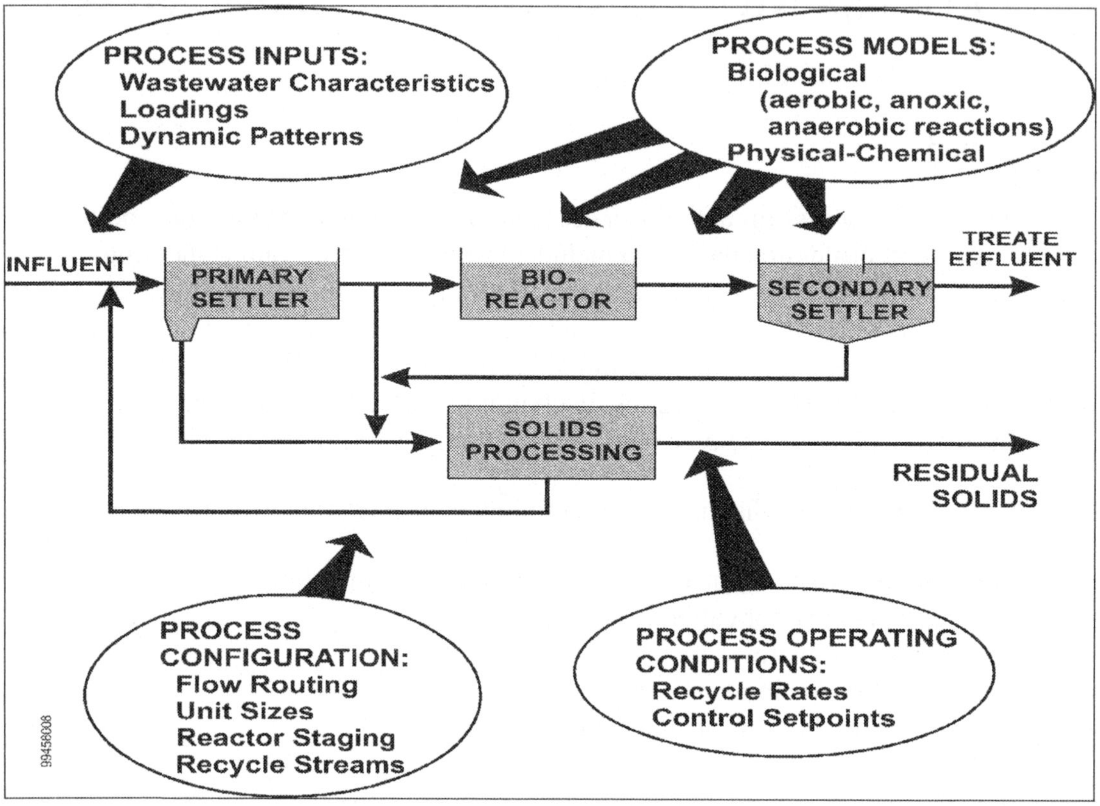

Figure 19-1. Essential Requirements for Wastewater Treatment Process Simulation

Within the context of the concept sketched in Figure 19-1, the specific approach to developing a task-by-task work program for the simulation exercise will depend on the objectives and purpose of the overall project. For example:

1. In the design of a *new treatment system*, given a process input specification, adjustments would be made to the process configuration and process operating conditions until a satisfactory treated effluent quality is predicted.
2. For *optimizing an existing process*, adjustments would be made to the process operating conditions, given a fixed process configuration and specified process inputs, to obtain an improved treated effluent quality or reduced operating costs.

3. To retrofit or *upgrade an existing facility*, modifications and/or additions to an existing process configuration would be considered, given certain process inputs and process operating conditions.
4. For *operational decision-making* or for *operator training*, the process configuration would be fixed and "what if ?" scenarios could be evaluated with different process inputs and process operating conditions.

Figure 19-2 illustrates a generalized methodology for setting up a simulation model for a particular plant. The steps are as follows:

1. The simulator is configured with the "physical design" characteristics of the treatment plant including unit sizes, flow routings and known flowrates, as well as the modeling framework (model selection for biological kinetics, settling, DO transfer, etc.)
2. Data are gathered on wastewater characteristics, plant performance and plant operating conditions for use in calibrating the simulator to the specific plant under consideration. These data may be assembled from historical plant operating records, and/or new measurements made to quantify certain unknowns, and/or assumptions made for certain parameters and variables for which either time or budget does not permit measurements to be made.
3. Calibration generally involves combining the "operational" or "controllable" aspects of the treatment plant with the input wastewater characteristics and making adjustments to selected parameters to fit a set of plant performance data. Sometimes after an initial calibration effort is made, it may happen that there is some inconsistency in the model predictions. Depending on the nature of the problem, it may be desirable either to re-assess the data or to gather new information to rectify the problem. Hence the upper dotted line on the left side of Figure 19-2.
4. Verification involves applying the calibrated simulator to a different set of operating data than that used for calibration. Usually at least one iteration between verification and calibration, as indicated by the lower dotted line on the left side of Figure 19-2, is required to obtain a good fit to the data. The types of model parameters that are adjusted during this portion of the calibration step are usually either those for which direct measurements cannot be made or those for which it would require a level of effort not suited to the task at hand.
5. Once the simulator has been adequately verified, it then can be used for the intended application.

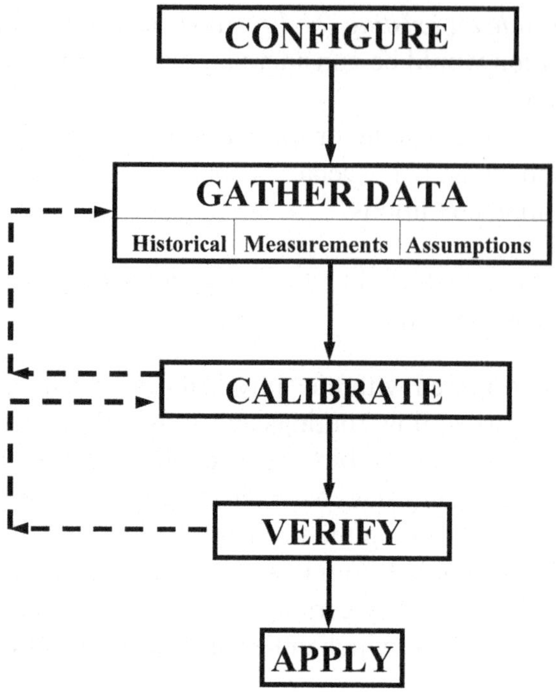

Figure 19-2. General Methodology for Using a Simulator

Steady state simulation analysis is appropriate for rough sizing of process units and for assessing solids production and sludge wasting quantities over the long term. On the other hand, dynamic simulation is useful for determining minimum and maximum aeration requirements, ammonia break-through under peak loading conditions, and evaluating the impact of stormflow or other transient loading conditions. In this regard, it is important to recognize that many plants will experience different diurnal patterns in flow and carbonaceous and nitrogenous loadings. Simulator projections using only the diurnal flow pattern with a constant COD and TKN concentration will often result in an over-prediction of the peak aeration requirements because the peak diurnal flowrate does not necessarily coincide with either the peak COD or the peak TKN loading times. If separate diurnal patterns are not used, this may lead to needless oversizing of the aeration system and blowers.

19.3 Configuration of the Plant in the Simulator

In configuring a plant in a simulator, the process arrangement is entered along with unit sizes and the introduction points of input and recycle streams. Special consideration must be given to configuring the bioreactor, which is modeled as one or more completely mixed vessels in series. Important considerations for bioreactors include compartmentalization into aerated and unaerated zones, the length:width:depth ratios of each zone, and the number and size of completely mixed vessels used to model the mixing regimes. These factors are of particular significance if the simulator will be used to predict spatial variations in a bioreactor. An example would be the determination of tapered aeration requirements along the length of a bioreactor as part of a design exercise for an aeration system. Another example would be the prediction of ammonia-nitrogen removal in a bioreactor.

A particularly useful feature when configuring and using a simulator is that the influent stream may be divided into various components, with each component having different characteristics. For example, Figure 19-3 shows a simulator configuration for a plant where the influent stream has been divided into four components: a municipal contribution, an industrial contribution (this industry discharged effluent over a single shift on weekdays only), an inflow and infiltration contribution (I & I was a significant problem at this plant), and a potential pre-fermenter addition component. By dividing the influent in this way, one can examine independently the impact of each component. For example, one could assess the need for providing equalization for the industrial discharge, and/or the benefits to be realized by mounting an intensive I & I reduction program, and/or the necessity of adding a supplemental source of fatty acids (VFAs) to improve EBPR performance.

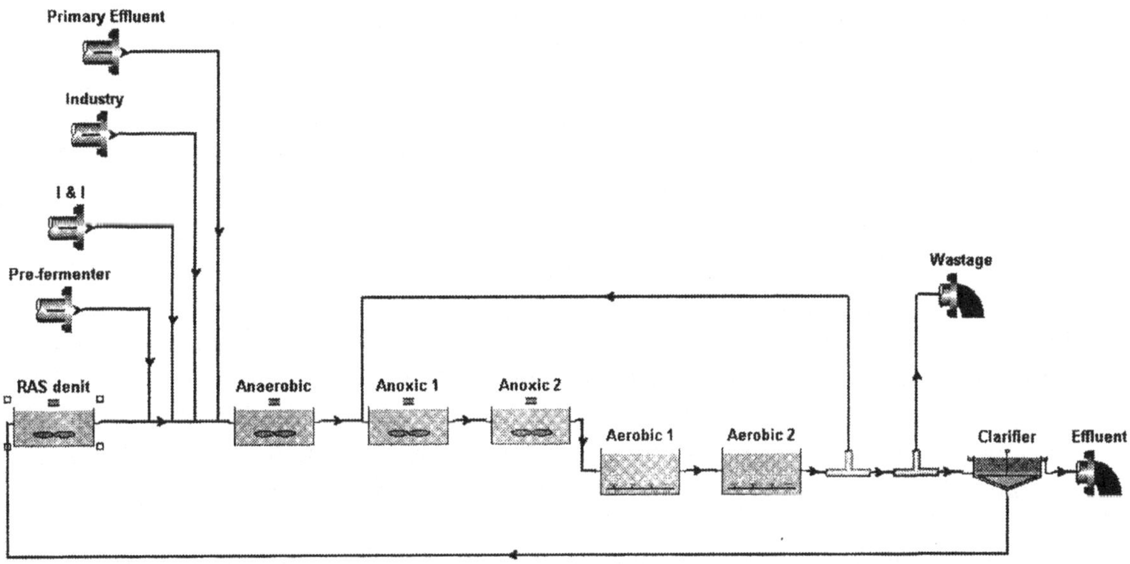

Figure 19-3. Example of Multi-Component Influent Stream in a Simulator Configuration

19.4 Data Gathering

If a new sewage collection system and treatment plant is to be designed, there will be no information available on wastewater quantity and quality. Neither would there be an opportunity to collect such information. In this case, one must make assumptions on wastewater quantities, characteristics and loading patterns to use in a simulation exercise and essentially no effort would be required for calibration and verification of the simulator. Furthermore, the design should include a comfortable margin of safety because of the considerable degree of ignorance about the true wastewater flow and loading patterns. However, if a treatment plant is to be designed and there already is a sewage collection system in place, then the opportunity exists to gather information on flow patterns and wastewater characteristics. In both of the preceding cases, there is no opportunity to gather stoichiometric and kinetic information to use for calibrating the models in the simulator and assumptions for these must be made. On the other hand, if the project involves upgrading or retrofitting an existing operating facility, or using a simulator as an operational decision-making tool or for operator training, there will be historical plant performance data. In addition, there will be the opportunity to gather additional data as desired for wastewater characterization and for the stoichiometric, kinetic and other parameters associated with the biological and physical-chemical process models used in the simulator.

NOTE: Chapter 21.0 provides more specific suggestions on data gathering needs for different types of activated sludge system.

Historical plant records generally provide useful information to make a crude calibration of a simulator. However, the data usually must be supplemented with special testing in order to make a refined calibration of a simulator that can be used with confidence to project process performance outside the operating range in which the calibration was made. Depending on the extent of the historical data available, in many instances it may be appropriate to design an intensive monitoring program that would:

- Gather 24-hour composite samples of untreated wastewater, primary effluent and secondary effluent for analysis of a range of parameters (e.g., COD, BOD, VSS, TSS, TKN, NH_3-N, NO_3-N, Total P and Soluble Reactive P, alkalinity). At the same time, also take measurements within the process of recycle rates and parameters such as MLSS and MLVSS. This data gathering exercise could extend as long as two weeks to provide sufficient data for a comprehensive calibration.
- During the composite sampling period described above, a 24- or 48-hour diurnal grab sampling program should be carried out on either an hourly or a bi-hourly basis. This information is extremely important in assessing the nature of the diurnal variations of the loading patterns and also the system's response in terms of nitrification and ammonia breakthrough, phosphorus removal in EBPR systems, and peak aeration requirements.

The duration of such a program, and how many parameters are measured in it, will be very site specific and depend on what plant data already are available. This may seem to be a costly and intensive undertaking; however, the quality of the data used to calibrate the simulator will have a direct impact on the reliability of the predictions made when using the calibrated simulator.

The data gathered in an exercise such as that described above provides information on the bulk composition and loading, for example, Total COD. Particularly for EBPR systems, it will be necessary to gather information on detailed wastewater characteristics for the carbon and nitrogen fractions as well as the specific growth rate of the autotrophic (nitrifying) organisms. For example, the readily biodegradable COD fraction (RBCOD fraction) is very important in determining nutrient removal performance, and therefore should be accounted for in the data gathering. Methodologies to characterize wastewaters are presented in Chapters 4.0 through 12.0.

In addition to the influent wastewater characteristics, there are a number of stoichiometric and kinetic parameters that must be defined in order to apply a model to predict system behavior. Sensitivity analysis can be used as a tool to determine which parameters have the largest effect on model predictions (see Chapter 20.0). Experience is that these stoichiometric and kinetic parameters do not change dramatically for different systems treating municipal wastewaters. In the past, it has been suggested that the maximum specific growth rate of the nitrifiers (μ_{AUT}) is an exception. This parameter was reported to show marked variations between systems treating different wastewaters; values for μ_{AUT} (at 20°C) ranging from 0.2 to 1.0 d^{-1} have been quoted. Undoubtedly nitrifiers are susceptible to toxicity and inhibition to a greater extent than heterotrophs, and will exhibit variability. For this reason variations in μ_{AUT} may correlate with the extent of the industrial component in the wastewater. It has been suggested that with increased industrial input there is an increased possibility of inhibition of nitrifying organisms. Another factor influencing the rate between wastewaters may be differences in pH. The nitrification rate has a major impact on nutrient removal system design and performance, and obviously on model predictions as well. Therefore, special attention should be paid to measurement or estimation of μ_{AUT}.

NOTE: Data presented in Chapter 17.0 indicate that some of the reported variation in μ_{AUT} may be the result of incorrect analysis based on an underestimate of the nitrifier decay rate.

In assessing the validity of historical or other data, it often is useful to examine the ratios of certain parameters rather than the absolute values of the parameters themselves. There usually is considerable day-to-day variation in concentration values; however, the ratio of TKN/COD, for example, should not show large fluctuations. Therefore, it is often helpful when identifying outliers or invalid data to look at the ratios. For example the COD/BOD_5 ratio for typical domestic wastewaters should be in the range of 2.0–2.2 and 1.9–2.1 for raw sewage and primary effluent, respectively. Likewise the expected COD/VSS ratio should be in the range of 1.42 to 1.48 mg COD per mg MLVSS. Significant deviations from these values would indicate that the data are suspect or that some influent wastewater characteristic is very different from the norm.

19.5 Calibration of the Simulator

Models are simplified descriptions of reality. Even if all of the input data to the simulator were perfect, the simulator would give an approximate prediction and not an exact match. This is compounded by uncertainties in the input data. This leads to differences between predicted and observed data either to a lesser or greater degree. In calibrating the simulator to take care of small discrepancies, it likely is necessary to make small adjustments to certain parameters in the simulation models until the predicted simulator outputs match the measured plant performance. The parameters to be adjusted should be those for which reliable data are not available from the data-gathering task described previously and have a large effect on the model predictions. Sensitivity analysis (Chapter 20.0) and engineering knowledge and experience can be used to determine if a certain parameter plays an important role in the performance of the model. Usually, very significant differences between predicted and observed are an indication of a gross error in the input information, very often in wastage rates (see example in Chapter 21.0).

If time and budget permit, it is preferred to estimate the parameters that must be determined by indirect techniques, such as f_{XI} and μ_{AUT}, using the bench-scale sequencing batch reactor (SBR) methodology developed specifically for this purpose (see Appendix D). Otherwise, these parameters can be estimated, using established parameter estimation techniques, in principle by adjusting them to fit long term plant historical data records on waste activated sludge production for the former, and diurnal treated effluent ammonia variations for the latter.

Calibration in a wide sense (including both measuring and estimating parameters) typically is the most time consuming part of a simulation exercise. For more details see Chapter 21.

19.6 Verification of the Calibrated Simulator

Once the simulator is calibrated, one or several separate sets of plant performance data collected under different loading conditions are used to confirm the calibration. It is up to a case-by-case evaluation by the modeling engineer if the results of the verification will be accepted. Iteration between verification and calibration may be necessary to fine-tune the simulator's calibration.

19.7 Application of the Simulator

A calibrated simulator can be used for a variety of purposes. For establishing the basic design sizing and operating conditions for treatment process components, steady state analysis can be used. Dynamic analysis can be used to size the aeration system for the desired maximum loading conditions. For process analysis, trouble-shooting and operational decisions, both steady state and dynamic analysis can provide insights to answer "what if?" questions. A simulator often is useful for assessing the capacity of an existing facility to meet a specified treated effluent quality limit. Finally, for operator training, both steady state and dynamic analysis can be used to demonstrate the impact of various operator decisions and to give an insight into the temporal and spatial variations of carbon, nitrogen, phosphorus and solids species throughout the treatment plant.

19.8 Recommended References

Barker, P.S., and P.L. Dold. 1997. General model for biological nutrient removal activated sludge systems: model presentation. *Water Environ. Res.* 69(5):969-984.

Ekama, G.A., P.L. Dold, and G.v.R. Marais. 1986. Procedures for determining influent COD fractions and the maximum specific growth rate of heterotrophs in activated sludge systems. *Water Sci. Tech.* 18(6):91-114.

Henze, M., W. Gujer, T. Mino, T. Matsuo, M. Wentzel, and G.v.R. Marais. 1995. Activated Sludge Model No. 2. *IAWQ Scientific and Technical Report No. 3*. London, UK: International Water Association.

Water Research Commission. 1984. *Theory, design and operation of nutrient removal activated sludge processes*. P.O. Box 824, Pretoria, South Africa: Water Research Commission.

Wilson, A.W., T. Marstaller, P.L. Dold, M.J. Tetreault, and D.W. DeFore. 1995. Use of a wastewater treatment process simulator as a tool for plant upgrading. In *Proc. 68th Annual Conference of the Water Environment Federation*, Miami, FL. Alexandria, VA: Water Environment Federation.

Wilson, A.W., and P.L. Dold. 1998. General methodology for applying process simulators to wastewater treatment plants. In *Proc. 71st Annual Conference of the Water Environment Federation*, Orlando, FL. Alexandria, VA: Water Environment Federation.

CHAPTER 20.0

SENSITIVITY ANALYSIS IN PRACTICE

20.1 Background

Sensitivity analysis can be used by the modeling practitioner to identify those model parameters that have the largest effect on the model variables. The term "parameter" is used in a general sense, referring to all parameters that are input in the model (i.e., not only model kinetics, but also influent characteristics and operational variables). Likewise, "model variables" are state variables in the model along with all other calculated outputs. The three main practical uses of sensitivity analysis are:

- Help in the selection of model parameters that can be estimated with the most accuracy, given a set of available measurements,
- Help in developing a sampling program to gather additional data for the particular process objective so as to target the most sensitive elements of the model.
- Identify those parameters that have negligible effect on model variables – or conversely, identify those variables that are not sensitive to any model parameters. These parameters usually can be left at default values and the variables do not warrant inclusion in a detailed sampling program.

Sensitivity analysis can be classified into non-linear and linear methods. Since biological and settling models applied in wastewater treatment modeling usually are fairly non-linear, non-linear analysis should provide a more information-rich snapshot of the model variable responses to changes in parameters. Non-linear sensitivity analysis is based on the calculation of the probability distribution of model variables from the probability distributions of the parameters. In theory, the derivation could be performed analytically, but this is impractical due to the large number of varying sets of equations that are employed by the typical models to describe different wastewater treatment plant configurations. Monte Carlo analysis will approximate the non-linear sensitivity functions numerically, and is used sometimes in model development exercises, but it is still impractical on a daily basis due to the large computational effort required. Consequently, non-linear sensitivity analysis will not be discussed any further in the context of this study.

Linear sensitivity analysis consists of developing (generally numerically approximating) sensitivity functions around a certain operating point (typical state of the model). There are different ways to express the results. In this study the relative change in state y_j for a 100% change in parameter θ_i is used, i.e.,:

$$\delta_{i,j} = \frac{\theta_i}{y_j}\frac{\partial y_j}{\partial \theta_i} \qquad (20.1\text{-}1)$$

where i=1…n (n number of parameters) and j=1…m (m number of states).

The advantage of this expression over other forms (containing absolute values) is that different parameters become easily comparable. Parameters for which $\delta_{i,j}$> 100 percent are generally considered very sensitive, though for a complete evaluation relative sensitivities between the parameters should be taken into account. An illustrative example is provided later in this chapter.

The analysis is performed by perturbing each individual parameter one at a time by an appropriately selected perturbation value $\Delta\theta$, i.e., approximate equation 20.1 by:

$$\delta_{i,j} = \frac{\theta_i}{y_j(\theta_i)}\frac{y_j(\theta_i) - y_j(\theta_i + \Delta\theta_i)}{\Delta\theta_i} \qquad (20.1\text{-}2)$$

The simplest practical approach is to develop the sensitivity functions in steady state, around a certain operating point. Since the wastewater treatment plant is governed by non-linear processes, and the models reflect this reality, the sensitivity functions will largely depend not only on the model structure, but on the operating point itself. A sensitivity function, developed around a low SRT, non-nitrifying process, will not be representative for the same plant when it does nitrify.

The result of a steady-state sensitivity analysis is a two-dimensional matrix, one dimension containing the parameters and the other the model states (n x m). Processing of such a matrix can provide valuable insight into the model structure and importance of parameters.

It is often important to observe the sensitivity functions under dynamic conditions. There are processes that are never in steady state, such as SBRs. For these, a simple steady-state analysis is not possible although it is still possible to generate an averaged sensitivity table. Also, the sensitivity functions will be quite sensitive themselves, not only to the long-term average operating point, but also to short term system disturbances, diurnal variation, storms and upsets. Dynamic sensitivity analysis will provide more detailed information. For example, a parameter would affect a variable differently before, during and after a high load event.

The reasoning in the previous paragraphs can be rigorously extended by incorporating one more dimension (t, time) into the sensitivity matrix. Then the same methodology is used to perform n dynamic simulations. The computational effort increases, however (unless the software used does not have a steady-state solver to start with, in which case steady-state analysis already takes a significant time). Also, multi-dimensional matrices rapidly lose their illustrative value. The proposed approach from the practical standpoint is to select the important parameters based on the steady-state sensitivity functions and develop dynamic sensitivity functions for this subset only.

Proper consideration must be given to the selection of the dynamic event, more precisely the forcing functions (influent and dynamically varying operational settings). Depending on the calibration task and data available, representative dynamic datasets should be chosen for the analysis. For example, if the task is to establish the effect of normal diurnal variation, then a typical diurnal flow and concentration profile, along with usual daily operational changes such as periodic de-sludging is required for a meaningful analysis.

20.2 Parameters and States in the Sensitivity Analysis

20.2.1 Model Parameters

Model parameters, in a general sense, are those constants and variables that are *input* into the calculations and have an effect on the calculated results of the model. Table 20-1 lists all types of parameters for a typical activated sludge plant model, and whether or not they are normally included in sensitivity analyses.

Table 20-1. Parameter Categories in Sensitivity Analysis

Parameter category	Parameter Sets	Examples	Used in Sensitivity Analysis	Note
Physical	Reactor configuration, environmental, physical constants	Volume, surface area, temperature, Henry constant	No	Exception: number of CSTRs in series sometimes
Driving Functions	Influent flow and concentrations	Influent COD, TKN, TP	Sometimes	
	Influent composition/fractions	Biodegradable fraction, soluble inert organic fraction	Yes	
	Operational variables	Recycle and wastage flows, airflow, chemical dosage, time schedules	Sometimes	
Model	Stoichiometric	N, P content, yields	Yes	
	Kinetic	Max. spec growth rates, half saturation constants	Yes	
	Settling	Vesilind and double-exponential constants	Yes	
	Initial conditions	Initial concentrations of heterotrophs, autotrophs	No	In special circumstances (e.g., simulation of short term batch tests)

Two parameter categories are typically *excluded* ab ovo from the analysis. One set is those that are supposed to be known to a high degree of accuracy – for example, physical parameters. One notable exception is the number of tanks in series in the representation of a plug-flow tank. Initial conditions is a special category that is typically not considered in the sensitivity analysis either. Steady-state initialization will remove the effect of these variables from a typical activated sludge plant model. In special cases (e.g., simulation of a batch test or respirogram), the initial conditions are meaningful, and can be included in the analysis. Even in this case though, they need to be properly developed and cannot be randomly selected.

The two most important parameter categories that are usually *included* in sensitivity analysis are the actual mathematical model parameters (stoichiometric and kinetic parameters) and influent composition (wastewater fractions). The actual parameters are model dependent. For example, ASM1 has 19 model parameters that can be incorporated in the analysis (Henze et al., 1987). Sensitivity analysis (or a priori knowledge of the model) can be used to select the most sensitive subset for more detailed analysis, as shown in the example in this chapter.

20.2.2 Calculated Variables

There are several categories of calculated variables in any model that are useful to include in the sensitivity analysis. The main categories and examples are:

- Model states in reactors (concentration of dissolved oxygen and heterotrophs).
- Combined variables in reactors (VSS, MLSS).
- Combined variables in settlers (effluent solids, RAS, sludge blanket).
- Rates, saturation and inhibition functions, other internal model expressions (OUR, NUR).

Inclusion or omission of any of these will not affect the results of the analysis. Consequently, the ones that will be displayed are selected based on the importance they play for a particular application. An important set is the model state variables. Most of the other categories can be calculated based on this set.

20.3 Sensitivity Analysis Example

20.3.1 The Wastewater Treatment Plant

To illustrate the practical application of sensitivity analysis, a simple nitrifying activated sludge plant is used, consisting of one process train, as summarized in Table 20-2.

Table 20-2. Activated Sludge Plant Specifications

	Item	Value	Units
Reactor	# of reactors	1	-
	Reactor volume	20,000	m^3
	HRT	6	hr
	DO setpoint	2.0	gO_2/ m^3
	Wastage (3.3 d SRT)	6,000	m^3/d
	Wastage (15 d SRT)	1,333	m^3/d
Influent	Flow (DWF)	80,000	m^3/d
	COD	500.0	$gCOD/ m^3$
	TKN	40.0	gN/ m^3
	Alkalinity	6.0	Mmole/L
Settler	Surface area	2,500	m^2
	Recycle flow	40,000	m^3/d

This plant was modeled as a single CSTR bioreactor, using the ASM1 model; and an ideal settler model with 99.8 percent solids capture and mass balance based RAS calculation. Wastage was withdrawn from the aeration tank to allow hydraulic SRT control. The plant was run to steady-state under two different operating conditions: 3.3 d SRT, partially nitrifying and 15 d SRT, complete nitrification. After the steady-state analysis results, a selected dynamic example is discussed.

20.3.2 Selected Parameters

The analysis was performed by providing a 1 percent positive perturbation to three different sets of parameters in the model:

Note: The nomenclature used in this chapter essentially is that used in ASM1, and may not agree with terms listed in Appendix A.

---

1. **Stoichiometric and kinetic parameters**

   The 19 stoichiometric and kinetic parameters in the ASM1 biokinetic model were analyzed. Default values were used in the base case as shown in Table 20-3. In an actual analysis, the default values might provide an initial starting point. Then, based on the results and the calibration values obtained, the sensitivity analysis can be repeated for the final calibrated model and parameter values.

Table 20-3. ASM1 Default Parameters Used in the Sensitivity Example

Parameters	Default value	Units
Stoichiometric		
N content of active biomass (i_{Xbn})	0.086	gN/gCOD
N content of endogenous/inert mass (i_{Xun})	0.06	gN/gCOD
Fraction of biomass leading to particulate products (f_u)	0.08	gCOD/gCOD
Heterotrophic yield (Y_H)	0.666	gCOD/gCOD
Autotrophic yield (Y_A)	0.24	gCOD/gN
Kinetic		
Heterotrophic maximum specific growth rate (μ_H)	6.0	1/d
Readily biodegradable substrate half saturation coefficient (K_{SH})	20	gCOD/m^3
Oxygen half saturation coefficient (K_{OH})	0.2	gO2/m^3
Nitrate half saturation coefficient (K_{NO})	0.5	gN/m^3
Anoxic growth factor (η_g)	0.8	-
Heterotrophic decay rate (b_H)	0.62	1/d
Autotrophic maximum specific growth rate (μ_A)	0.8	1/d
Ammonia half saturation coefficient for autotrophs growth (K_{NH})	1.0	gN/m^3
Autotrophic decay rate (b_A)	0.04	1/d
Oxygen half saturation coefficient for autotrophic growth (K_{OA})	0.4	gO2/m^3
Maximum specific hydrolysis rate (K_h)	3.0	1/d
Hydrolysis half saturation (K_X)	0.03	gCOD/gCOD
Anoxic hydrolysis factor (η_h)	0.4	-
Ammonification rate (k_a)	0.08	m^3/gCOD/d

2. **Influent (driving function) parameters**

   The effect of influent flow, organic and nitrogen loading as measured by total COD and TKN concentration, readily biodegradable fraction ($S_S/(S_S+X_S)$) and soluble inert fraction ($S_I/(S_I+X_I)$) was analyzed in this example. Table 20-4 provides default values. Perturbation was +1 percent.

Table 20-4. Default Influent (Driving Function) Values

Parameters	Default value	Units
Influent flow, Qinf	80,000	m^3/d
Influent COD, CODinf	500	gCOD/m^3
Influent TKN, TKNinf	40	gN/m^3
Readily degradable fraction, $S_S/(S_S+X_S)$	24.4	%
Soluble inert fraction, $S_I/(S_I+X_I)$	27.8	%

3. **Operational parameters**

   DO concentration setpoint and waste flow rate (defining a certain SRT) were selected as the two most important operational parameters on this simple plant. Perturbation was +1 percent.

Table 20-5. Operational Parameters Analyzed

Parameters	Default value	Units
DO setpoint	2.0	$gO2/m^3$
Wastage flow (3.3 d SRT)	6,000	m^3/d
Wastage flow (15 d SRT)	1,333	m^3/d

20.3.3 Sensitivity Analysis – the Procedure

The plant as specified above was implemented in the simulator (based on the ASM1 model equations) and run to steady state with the default set of parameters. Results for state variables and a few other selected operational variables are listed in Table 20-6. Values are recorded to 4 decimal digits even though there is no process meaning attached to this accuracy. For an accurate determination of the sensitivity functions a small perturbation must be chosen, thus necessitating the use of the maximum *numerical* accuracy.

The following steps were performed for the preparation of the steady-state sensitivity tables (Tables 20-7 to 20-10) around the two operating points:

1. Steady state was established (one for each base case and one for each perturbed parameter). The results of the runs were saved into a spreadsheet (two sheets – one for each operating point). Steady state can be achieved using an iterative method if available in the software, or with a long dynamic run on constant influent and operational settings. Attention has to be paid to numerical accuracy.
2. The sensitivity functions according to equation 20.1-2 were calculated in a matrix, one for each operating point, expressed as percent change in calculated variables for 100 percent change in parameters. Since the models are non-linear, and both parameters and calculated variables have physical bounds, generally it would be meaningless to change parameters by 100 percent. Expressing the results of a 1 percent perturbation scaled up to 100 percent change is solely for the purpose of presentation and must be taken as the approximation of the derivative at the parameter default value.
3. The matrix at this point contains all the information, but is not easily readable. The (arbitrarily chosen) sum of absolute values of sensitivities for each row and column was calculated as an additional row and column for the matrices. The absolute value of the calculated variables was also added as this contains important information in addition to the relative sensitivities.
4. The matrices were sorted by the sum of absolute values, both for parameters and calculated variables. Sorting left to right and top to bottom in descending order results in the most sensitive cells being moved to the top left corner. The order of parameters and variables, their *rank*, becomes an important aid for the analysis.

To further enhance readability of the sensitivity tables, low sensitivity values ($\delta_{i,j}$< 10 percent) were discarded, and the highest sensitivity fields appear in bold font ($\delta_{i,j}$> 150 percent).

Table 20-6. Results of the Two Base Cases

	Value		Units
Operational variables			
SRT	3.3333	15.0000	days
VSS[1]	2196.3930	6315.0828	$gVSS/m^3$
State variables			
Readily biodegradable substrate (S_S)	3.5166	2.4647	$gCOD/m^3$
Slowly biodegradable substrate (X_S)	40.1013	47.7716	$gCOD/m^3$
Soluble inert organics (S_I)	25.0000	25.0000	$gCOD/m^3$
Inert particulate organics (X_I)	834.9390	3,310.6961	$gCOD/m^3$
Heterotrophs (X_{BH})	2,030.9831	3,666.4450	$gCOD/m^3$
Autotrophs (X_{BA})	50.0365	207.8064	$gCOD/m^3$
Endogenous decay products (X_P)	294.6018	2,113.6034	$gCOD/m^3$
DO (S_O)	2.0000	2.0000	$gO2/\ m^3$
NH_3-N (S_{NH})	1.1238	0.2174	gN/m^3
NO_3-N (S_{NO})	12.7435	18.1798	gN/m^3
Particulate biodegradable organic N (X_{ND})	1.8672	2.7823	gN/m^3
Soluble biodegradable organic N (S_{ND})	0.6899	0.6410	gN/m^3
Alkalinity (S_{ALK})	3.0272	2.5741	$mole/m^3$

[1]VSS is calculated from particulate COD with 1.48 mgCOD/mgVSS conversion

20.3.4 Steady-State Sensitivity Analysis – Results

The sensitivity tables for both the 3.3 d SRT and the 15 d SRT operating points were divided into two tables for convenience. One type of table (Tables 20-7 and 20-9) contains the results of the analysis performed on model parameters, and the other (Tables 20-8 and 20-10) for influent and operational parameters.

As an example, the number -208 in the ammonia (S_{NH}) column and maximum specific autotrophic growth rate (μ_A) row in Table 20-7 represents a strong sensitivity of ammonium to the growth rate in a negative direction. For 1 percent increase in the growth rate, the ammonium variable will drop 2.08 percent. (-208 percent sensitivity for 100 percent change).

The conclusions that can be drawn for each table and for the whole of the example analysis are listed below.

20.3.4.1 Sensitivity to model parameters at 3.3 d SRT

1. Variables at the right side of Table 20-7 (X_I, S_I, S_O, SRT, ranked 12-15) are not sensitive at all to model parameters and cannot be modified by changes to the model parameter set. There are different reasons for this. Inert organics (X_I, S_I) are not undergoing any transformations in the model and their concentration will depend only on mass balance. DO and SRT, on the other hand, are controlled variables.
2. Variables with high ranks (X_{ND}, S_{NO}, X_{BA}, S_{NH}, X_S) can be readily affected by varying the model parameters. Some (ammonia, S_{NH}) are strongly dependent on one or two parameters (in this case μ_A, K_{NH}) only, while others are influenced by a whole range of parameters (e.g., X_{ND}).
3. The model parameters themselves are ranked – parameters on the top, for example, heterotrophic yield (Y_H), tend to have a larger effect on several variables, as opposed to the parameters ranked towards the bottom rows of the table. The parameters on the top obviously are the more important ones for the model calibration exercise, while the less important ones can be left at their default values.
4. The absolute values of the variables (listed under the "variables" row) provide an additional guideline when evaluating the impact of sensitivity. For a 1 percent increase in heterotrophic yield (Y_H) about the same relative change (1.5 percent) will occur to the endogenous products (X_P) and the heterotrophs (X_{BH}), but for X_P this means about 4 mg COD/L increase, while for X_{BH}, around 30 mg COD/L.
5. A few other practical points that can be obtained from Table 20-7 are listed below. It must be kept in mind that these are specific to the system and its operating point (one mixed tank, fully aerobic, 3.3 d SRT).

- Nitrate concentration is strongly dependent on heterotrophic yield – through N assimilation (since only the residual ammonia can be nitrified)
- If the predicted alkalinity is higher than the measured, there are three parameters available to reduce the predicted value in the model (Y_H , i_{Xbn} and μ_A). μ_A achieves this through more nitrification (strong effect on ammonia). If effluent ammonia was predicted correctly though, this points to a measurement error on influent or effluent ammonia or alkalinity.
- VSS (MLVSS) is mostly dependent on Y_H as expected. Based on practical experience VSS is even more strongly affected by influent loading and composition, as well as wastage. This will be discussed in association with Table 20-8.
- For the calibration of effluent nitrate and ammonia, the most important parameters are the μ_A, governing nitrification, and Y_H, i_{Xbn} which have a large effect on assimilation.

Table 20-7. Model Parameter Sensitivity Table (3.3 d SRT)

% Sensitivity	Rank	1	2	3	4	5	6	7	8	9	10	11	12	13	14	15
	Variables	X_{ND}	S_{NO}	X_{BA}	S_{NH}	X_S	X_P	S_{ND}	S_S	X_{BH}	S_{ALK}	VSS	X_I	S_I	S_O	SRT
Rank	Value	1.9	12.7	50.0	1.1	40.1	294.6	0.7	3.5	2031.0	3.0	2196	834.9	25	2	3.3
1	Y_H	83	**-241**	-123		17	**153**	-26		**154**	72	108				
2	b_H	95	23	23		72	57	87	76	-42		-20				
3	i_{Xbn}	95	-111	-78				86			33					
4	K_h	**-157**				**-162**										
5	μ_A		18	13	**-208**						-11					
6	K_X	96				99										
7	f_u	-11					94									
8	μ_H								-116							
9	K_{NH}				100											
10	k_a							-97								
11	K_{SH}								100							
12	Y_A			98												
13	η_g		-40													
14	K_{OA}				36											
15	b_A			-13	24											
16	K_{OH}															
17	i_{Xun}		-10													
18	η_h															
19	K_{NO}															

20.3.4.2 Sensitivity to influent and operational parameters at 3.3 d SRT

These parameters are presented in a separate table as "sensitivity" has a different meaning in relationship to influent and operational parameters. The model cannot be "calibrated" by changing influent COD, for example. Some general pointers to evaluate the information content in Table 20-8 are provided below.

1. In general, these parameters affect a large number of variables in the model. 41 percent of the fields have an entry, as compared to 14 percent in Table 20-7.
2. Ranking of the variables is different from Table 20-7, but can be used the same way to establish important and not as important parameters and variables.
3. Influent COD has the largest effect, even on nitrogen variables (except ammonia). The link between COD and N is through the assimilation process.
4. Ammonium is very sensitive (+182 percent) to wastage flow. This is an indication of the sensitivity of the nitrification process around the 3.3 d SRT operating point.
5. Wastewater soluble versus particulate fractions have a relatively small effect on the process.

Table 20-8. Analysis of Influent and Operational Parameters (3.3 d SRT)

% Sensitivity	Rank	1	2	3	4	5	6	7	8	9	10	11	12	13	14	15
	Variables	S_{NO}	X_{BA}	X_P	X_I	VSS	X_{BH}	S_{ALK}	X_{ND}	X_S	S_{NH}	S_I	S_O	SRT	S_{ND}	S_S
Rank	Value	12.7	50.0	294.6	834.9	2196	2031.0	3.0	1.9	40.1	1.1	25	2	3.3	0.7	3.5
1	CODinf	**-190**	-104	100	100	97	101	57	79	100		100			-32	
2	Qwas	-40	-120	**-152**	-95	-76	-57	17	-32		**182**			-99		40
3	Qinf		97	96	97	98	98		97	100						
4	TKNinf	**301**	**213**					**-162**	19						35	
5	DO setp	45						-15			-35		100			
6	$S_I/(S_I+X_I)$	13			-38							100				
7	$S_S/(S_S+X_S)$								-11	-29						

20.3.4.3 Sensitivity to model parameters at 15 d SRT

Several key conclusions can be drawn when comparing the results in Table 20-9, generated at 15 d SRT, with a system that is under higher loaded conditions (3.3 d SRT):

1. The list of insensitive (inert and controlled) variables did not change, and hence does not depend on the operating point.
2. There are changes in the rank of the parameters compared to the 3.3 d SRT operating point, but the general ranks did not change much: high-ranked, important parameters and less important ones are still in their respective positions.
3. Hydrolysis and decay parameters (K_h, K_X, f_u) gained a few positions in the ranking system under long SRT conditions, while the growth parameters (μ_A) decreased in importance.
4. When considering variables, the nitrification process is a lot less sensitive at 15 d SRT, based on the lower ranks of both ammonia and autotrophs.

20.3.4.4 Sensitivity to influent and operational parameters at 15 d SRT

1. The analyzed parameters generally retained their order of importance between the 3.3 d and 15 d SRT system.
2. The most important change in the rank of the variables concerns the two primary substrates, S_S and S_{NH}. Both become very insensitive at this high SRT.

Table 20-9. Model Parameter Sensitivity Functions (15 d SRT)

% Sensitivity	Rank	1	2	3	4	5	6	7	8	9	10	11	12	13	14	15
	Variables	X_{ND}	X_S	S_{ND}	X_P	S_S	S_{NO}	X_{BH}	S_{NH}	X_{BA}	VSS	S_{ALK}	X_I	S_I	S_O	SRT
Rank	Value	2.8	47.8	0.6	2113.6	2.5	18.2	3666.4	0.2	207.8	6315	2.6	3310.7	25	2.0	15
1	Y_H	**152**	88	-22	**201**		**-176**	**202**		-67	124	89				
2	b_H	62	55	95	26	99	12	-74		13	-23					
3	K_h	-138	-141													
4	i_{Xbn}	99		94			-39			-28		19				
5	K_X	97	100													
6	f_u	-15	-12		90			-10			16					
7	μ_A								-120							
8	μ_H					-111										
9	Y_A									98						
10	k_a			-98												
11	K_{NH}								101							
12	K_{SH}					100										
13	b_A								41	-34						
14	η_g						-38					19				
15	i_{Xun}						-13									
16	K_{OH}															
17	K_{OA}								23							
18	η_h															
19	K_{NO}															

Table 20-10. Analysis of Influent and Operational Parameters (15 d SRT)

% Sensitivity	Rank	1	2	3	4	5	6	7	8	9	10	11	12	13	14	15
	Variables	S_{NO}	X_{BA}	X_I	X_P	S_{ALK}	VSS	X_S	X_{BH}	X_{ND}	S_I	S_O	SRT	S_{NH}	S_{ND}	S_S
Rank	Value	18.2	207.8	3310.7	2113.6	2.6	6315	47.8	3666.4	2.8	25	2.0	15	0.2	0.6	2.5
1	CODinf	-110	-50	100	99	56	96	100	100	86	100				-20	
2	Qinf		92	90	87		92	99	97	96						
3	Qwas	-14	-69	-84	-107		-65	-10	-23	-18			-99	69		12
4	TKNinf	**213**	**153**			**-191**				12					22	
5	DO setp	39				-19						100		-18		
6	$S_I/(S_I+X_I)$			-38			-13				100					
7	$S_S/(S_S+X_S)$							-18								

20.3.5 Dynamic Sensitivity Analysis

Sensitivity functions in a dynamic model are sensitive not only to the general operating point of the system, but to the effect of time-varying forcing functions. A steady-state value of ammonia (1.1 mg N/L at 3.3 d SRT) can vary as much as 1-5 mg N/L under normal diurnal loading conditions. It is expected that the sensitivity of this variable will be different during the morning peak and the late night low loaded situation.

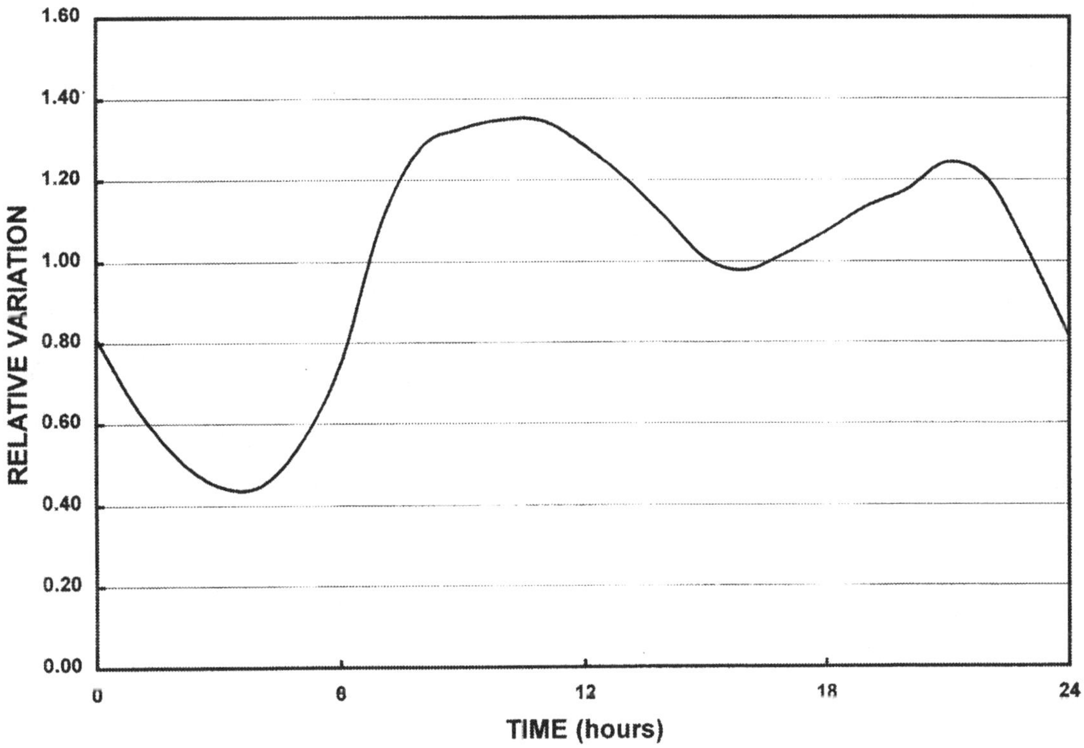

Figure 20-1. Normalized Diurnal Pattern Used for Dynamic Analysis

The steady-state sensitivity analysis performed above does not provide information on model sensitivity with respect to the effect of time-varying load. To illustrate a dynamic sensitivity analysis, the two most sensitive parameters, μ_A and K_{NH} as well as ammonia as output variable were selected. The steady-state average value of the influent was maintained, but diurnalized using the function in Figure 20-1. This diurnal profile was applied to influent flow, COD and TKN. 10 days of diurnal runs were performed with the same diurnal pattern, starting from the steady-state condition in the Table 20-7 (3.3 d SRT), and ammonia was recorded in the aeration tank during the last day. Following the base case ($\mu_A = 0.8\ d^{-1}$ and $K_{NH} = 1.0$ g N/m^3), two more runs were performed applying +1 percent perturbation to each parameter, individually. The resulting diurnal ammonia concentration, along with the perturbation cases, is plotted in Figure 20-2.

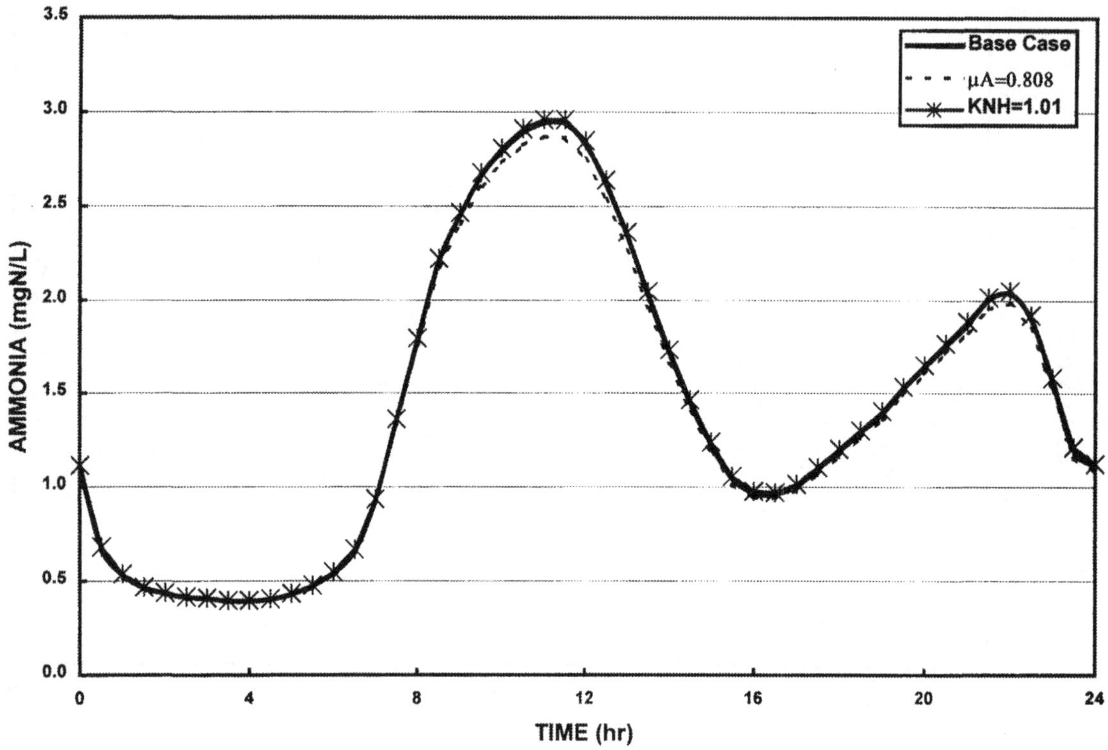

Figure 20-2. Diurnal Ammonia Profile

The sensitivity functions now can be approximated using equation 20.1-2 for these parameters. This is plotted in Figure 20-3.

The following conclusions can be drawn from Figure 20-3.

1. The average of the dynamic parameter sensitivity does not correspond exactly to the steady-state value (-222 percent instead of -208 percent, and 96 percent instead of 100 percent, for μ_A and K_{NH}, respectively), though there is a strong correlation. This is a well-known phenomenon in dynamic simulation of diurnal versus steady-state systems – steady state does not equal the average of a diurnal time series.
2. The variation in the sensitivity of μ_A is much larger than that of K_{NH}. If K_{NH} is increased in the model, it will result in an almost uniform increase of ammonia concentration throughout the day. If μ_A is increased, it will result in a slight decrease during low loaded periods (when most of the ammonia is oxidized) and a larger decrease and faster recovery of concentration peaks.
3. Ammonia is least sensitive to growth rate (consider the negative scale) during low loaded periods, more sensitive during high loaded periods, and reaches the highest sensitivity during recovery from a high loaded situation, when the rate of change of the concentration is the greatest in the negative direction.

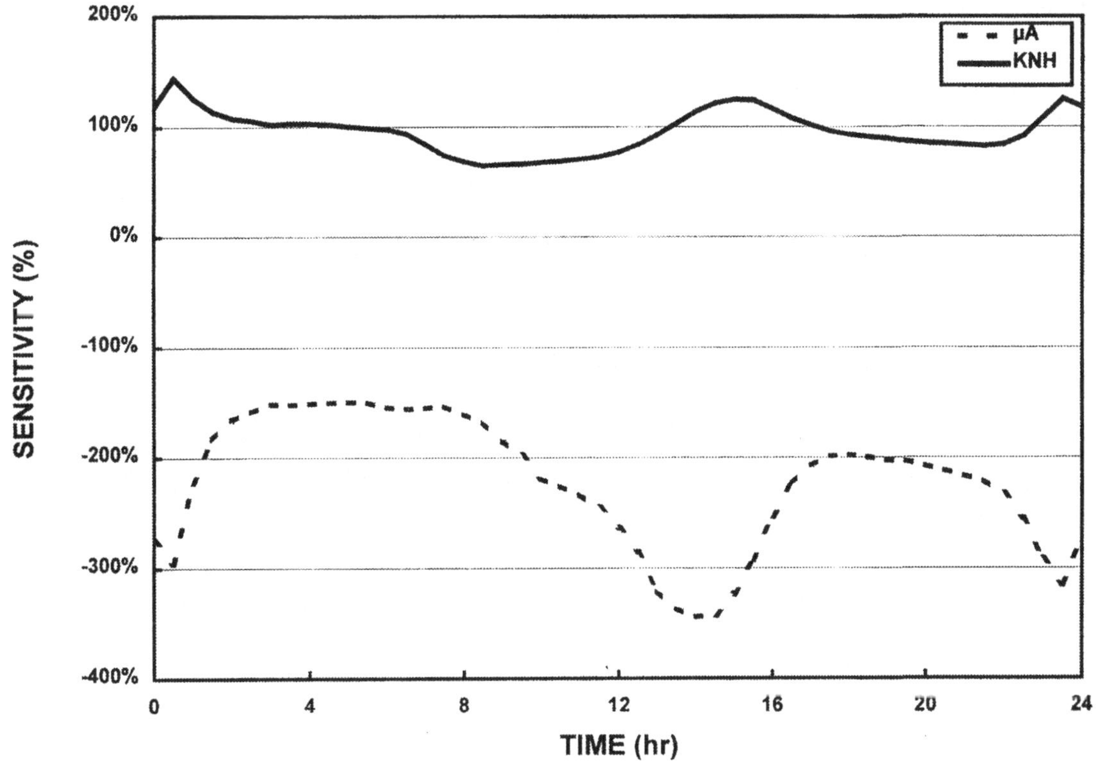

Figure 20-3. Diurnal Sensitivity Functions

20.4 Conclusions

This chapter describes the methodology of sensitivity analysis in modeling with a practical perspective, and provides examples for a simple aerobic system as simulated by the ASM1 model. Guidance is given on various forms of sensitivity analysis and how to deduce useful model and calibration information from the results of the analysis. The results as presented here are provided as examples to illustrate the methodology only – it is likely that parameter ranking and sensitivity would be significantly different for more complex systems describing nitrification, denitrification and excess phosphorus removal. In those cases (and even for simple aerobic systems under different loading conditions or simulated using other models than in the example in this chapter), the sensitivity analyses have to be performed with the actual model setup to generate valid sensitivity functions.

20.5 References

Dochain, D., and P.A. Vanrolleghem. 2001. *Dynamical modelling and estimation in wastewater treatment processes*. London UK: International Water Association.

Henze, M., C.P.L. Grady, W. Gujer, G.v.R. Marais, and T. Matsuo. 1987. Activated Sludge Model No. 1, *IAWPRC Scientific and Technical Report No. 1*. London, UK: International Water Association.

Jeppson, U. 1996. *Modelling aspects of wastewater treatment processes*. Lund, Sweden: Lund Institute of Technology.

Weijers, S.R., and P.A. Vanrolleghem. 1997. A procedure for selecting best identifiable parameters in calibrating activated sludge model no. 1 to full-scale plant data. *Water Sci. Tech.* 36(5):69-70.

CHAPTER 21.0

GUIDANCE FOR MODEL CALIBRATION

21.1 Introduction

Mathematical models, however complex, are necessarily simplified representations of the real system they approximate. The unavoidable simplifications lead to the need to adjust some of the model parameters on a case-by-case basis to match the performance of the process as represented by available data on a particular plant. This, in a broad sense (as opposed to the numerical technique of parameter estimation) is called *calibration* within the context of this guidance manual. Since this process is closely linked with data availability, various sources of data and data gathering procedures, those aspects also will be discussed as appropriate. There is more information on data collection in Chapter 19.0.

There is substantial literature available discussing the theoretical background for calibration, structural and practical identifiability of models, and optimal experimental design to obtain the most sensitive data from experiments. An excellent review of the fundamentals and specific examples is presented in Dochain and Vanrolleghem (2001). The purpose of this chapter is to provide *practical* advice to process engineers, designers and plant operators involved with process modeling of a wastewater treatment plant.

The IWA Scientific and Technical Report on Activated Sludge Model No. 2 (Henze et al., 1995) provides some general direction on model calibration. The report provides a number of key principles on which model calibration should be based; these principles are summarized here:

- Most of the parameters in the model do not change significantly from case to case and therefore should not be changed without justification.
- It is not advisable to adjust a parameter to fit data unless the model predictions of that experimental data are sensitive to changes in that parameter.
- Only one parameter should be changed at a time. If a parameter pair interacts (such as a growth rate and a decay rate), only the parameter with the largest relative influence should be changed.
- Experimental data for calibration of a parameter should be such that the effect of the parameter of interest dominates the response.

- In general, reasonable calibration of a model only is possible with a detailed understanding of the principles of the model. Without this, resulting parameter adjustments may distort the model far from reality.

In calibrating a simulator to an existing plant, there are several possible levels of complexity that can be adopted. This chapter outlines a *tiered approach* for practitioners to follow, depending on plant complexity, objective of the modeling exercise, available data and budget for additional data-gathering and analysis. Before addressing the tiered approach, it is appropriate to identify the categories of data that are required in a calibration exercise.

21.2 Distinguishing Between Categories of Data

Simulation of wastewater treatment plant performance requires information that can be categorized in four groups:

- **Physical plant data:** This relates to the size of units in the treatment system. For example, aeration tank dimensions (volume, length, width and depth), clarifier area and depths and so on. If the objectives of simulation include assessing aeration system performance, then information such as diffuser dimensions and densities also is required.
- **Plant operational data:** This includes data on operational aspects such as RAS flow rates, wastage rates, mixed liquor recycle flows and DO concentrations in aerated zones.
- **Influent loading data:** The influent flow and concentrations obviously are essential data. Diurnal variability of flow and composition are both important, as well as unusual events such as storm flows. An important part of the influent loading data is the wastewater characterization; that is, the division of bulk concentration measures such as total COD into the various characteristic sub-portions. Return flows from side stream treatment also may contribute significantly to activated sludge plant loading.
- **Kinetic and stoichiometric model parameters:** The activated sludge models incorporate a large number of kinetic and stoichiometric parameters.

21.3 Tiered Approach to Data Collection and Calibration

21.3.1 Objective of Calibration

Calibration typically is regarded as the process in which model parameters are adjusted until model predictions match selected sets of data linked to the performance of the actual plant. The *objective is to minimize the error* between the datasets and model predictions. It is important to remember that the objective is not to achieve a perfect fit, since the model is a simplified representation of the plant and ignores some of the inputs and processes occurring in the real world (deemed to be less important by the creators of the model). A perfect fit can be achieved with a simple set of polynomials of N degree where N is the number of data points to be matched by the model. However, the predictive power of such a model is quite limited. Overfitting, while it might reduce the total error for one particular dataset, invariably will reduce the model's predictive power and increase model error for other datasets.

When evaluating the match of the model against data, it is crucial to observe all the important variables. It is preferable to fit to most of the measured variables reasonably, rather than fit perfectly to one selected (however important) component concentration and poorly to others.

In addition to minimizing model error, there are two other objectives of the calibration procedure that often are ignored:

- The calibration should be used to *establish the field of validity* (i.e., the "design space") of the model. The authors of the model frequently include general guidelines regarding the circumstances under which the model can be used. Depending on the range of data it was calibrated to and verified on, the model might not be reliable for a particular application that is too far from the calibration conditions. For example, it is risky to develop process engineering conclusions from the behavior of a model during storm flow conditions if calibration was performed only on average or steady state data.
- The fit achieved during calibration (and verification) can indicate the *expected accuracy* of the model under the specific circumstances. Different model variables will have different error associated with them, but the most important performance parameters must be within a certain accuracy if the model is to be trusted for process work. In steady state situations (assuming good quality data), these variables should be matched within 5 percent to 20 percent overall, while during dynamic runs 10 percent to 40 percent temporary deviations are not unusual, even from well-established models. Secondary variables can vary even more in typical cases, without compromising the model's ability to predict general process performance truthfully.

21.3.2 Plant Type

The type of wastewater plant to be modeled has a large effect on the type of data collected and what is necessary for calibration. Small plants typically collect a smaller dataset than large plants. Unfortunately small plants normally are less stable, operating under higher diurnal and seasonal variation than large plants. The most important aspect of the process is the level of treatment the plant provides, e.g.:

- BOD removal only.
- Nitrification (fully aerobic plants).
- BNR plants which remove nitrogen (denitrifying plants with anoxic zones).
- BNR plants with EBPR (with anaerobic zones).
- Other side stream processes (e.g., sludge treatment and prefermenters) also may have an effect on the main plant process and should be taken into account.

21.3.3 A Generalized Procedure of Calibration

The conceptual procedure for calibration involves the following steps:

- Running a simulation for a scenario for which measured data are available: Care must be taken that the model setup reflects reality, so that differences can truly be attributed to the parameter values and not to modeling errors. Specifically, two important points must be mentioned:

- Steady state as opposed to long-term averages: Averaging several months' worth of data usually does not provide a dataset that can be used for steady state runs. Steady state is best approximated as the average of dry weather, normal operation and the dataset must be prepared according to guidelines in the *data conditioning* section.
- Initialization of the model: In dynamic simulation, the starting point (initial conditions) has a special importance. For example, if the initial autotrophic concentration is overestimated, it may result in significant underestimation of the growth rate. A dynamic run should be initialized properly to accurately estimate initial conditions for all variables. This can be achieved with a combination of steady state and dynamic runs, basically establishing the history of plant operation just before the simulation period that will be used for calibration. The importance of initial conditions decreases for longer dynamic runs.

♦ Comparing the simulation results with the observed data, and making some sort of error estimation. The error estimation often is simply a visual observation of time series plots or averages predicted by the model against the appropriate data. More rigorous error calculation methods are absolute or relative differences, sum of squares or the Maximum Likelihood criteria.

♦ Adjusting parameters, and returning to the first step.

During this process, the modeler frequently discovers that some of the data are "suspect," do not fall within normal ranges or do not satisfy a mass balance. This "automated consistency check" is a useful benefit of modeling and should be used to advantage. The errors discovered usually lead to a closer examination and reassessment of the dataset. However, more benefit can be derived from this if standard error checking, data reconciliation and data conditioning steps are performed *before* the modeling exercise. In other words, the modeler should use a conditioned, good quality dataset as opposed to the raw data. The data conditioning procedures are described later.

The sources of data that can be used in calibration can be divided into three categories:

♦ Historical data collected on the plant (laboratory logs, SCADA);
♦ Additional data collected on the plant, to supplement the historical data; and
♦ Direct parameter measurements from dedicated tests.

It is not possible and may not be necessary to use all potential sources of data every time. The project might not have an appropriate budget for full wastewater characterization and measurement of even the most important kinetic parameters. The following sections describe these situations in detail.

21.3.4 Calibration Level 1: Defaults and Assumptions Only

When modeling a green-field site, no process data are available. Modeling will still provide useful information that is not possible to obtain in any other way. When minimal data are available and the modeler has to work with an "uncalibrated" simulator, three points need to be kept in mind:

- Typical wastewater composition and default stoichiometric and kinetic data are available from the literature, or from experience with modeling plants in the same geographical area. While such default parameters, particularly for wastewater composition, have to be treated with caution and engineering judgment, they are (if the model is an internationally accepted one; see Chapter 3.0) usually based on thousands of experiments and model calibrations, and allow reasonable estimation of process performance far beyond what would be possible by using textbook design guidelines.
- Absolute performance numbers predicted by such a model are less reliable than relative trends. For example, two BNR configurations can be compared against each other using the same default parameter dataset. If one provides an estimate of 2 mg/L lower total effluent nitrogen than the other, it probably is a better configuration in this respect, even if the absolute 6 mg N/L versus 8 mg N/L effluent predictions need to be treated with caution without calibration of nitrification and denitrification kinetics.
- Confidence in the model predictions can be increased by running worst case scenarios. Selecting kinetic parameters on the low side of the default range will result in pessimistic process performance predictions – which can be used with more confidence for scenario analysis. In traditional design procedures, the approach is to use peaking and other safety factors. This procedure is somewhat different using a dynamic model. Safety factors should not be applied on top of worst case scenarios (such as overload) run on pessimistic model parameters, as this would result in significant overdesign of the plant.

21.3.5 Calibration Level 2: Historical Data Only

Most operating wastewater treatment plants have a sampling program and historical records in laboratory logbooks or spreadsheets. These data, while almost certainly incomplete for a full modeling exercise, can be used to obtain a rough calibration for the model.

The following types of data normally will be available:

- Flow measurements:
  - Influent daily average and typical diurnal flow. The weekday/weekend diurnal flow pattern can be established from detailed (e.g., hourly) flow data for an extended period (e.g., months).
  - RAS and mixed liquor recycle rates.
  - Primary and secondary wastage flows, and schedules in the case of intermittent wasting.
  - Return streams containing high solids, nitrogen and/or phosphorus loadings.
- Analytical measurements:
  - 24-hour composites (sometimes only grabs) of influent total BOD or COD; influent and primary effluent TSS, and perhaps VSS; process MLSS and perhaps MLVSS; DOs in aerated zones; effluent TSS and BOD (and ammonia, nitrate and phosphorus in BNR plants).

Useful information for the models that typically is missing includes flow splits if the plant is operating several trains, filtered CODs, OUR, N and P profiles within the process, diurnal and other dynamic sampling, aeration parameters (air flow, alpha), sludge blanket levels and so on.

Many plants monitor influent BOD_5 rather than COD, but a limited number of parallel COD and BOD tests can establish a COD/BOD ratio.

Waste activated sludge flow is the most common suspect variable due to its relatively small magnitude compared to other flows in the plant and inadequate instrumentation. This may add significant uncertainty because of the large effect it has on the process.

21.3.5.1 Data conditioning

In almost all cases, the raw data collected from the plant require some processing, filtering, cleaning and corrections. These adjustments are referred to as *data conditioning.*

Sampling procedures should be reviewed and verified. It is important to check the location where the samples are taken, if the sample is representative of the process (well-mixed sampling location), if all other standard procedures were followed (sample preservation, processing according to standard methods) and that analyses are repeatable.

A simple averaging of several months' worth of daily composite values is not satisfactory for process modeling. Basic statistical analysis should be performed to compute standard deviations and minimum/maximum values. This should be conducted with the objective of identifying obvious outliers and data entry errors. For dynamic analysis, trending and time-series analysis may detect instances of unreasonably fast change, completely "flat" (unchanging) data (sometimes due to disconnected sensors), too-noisy data or increasing bias due to sensor fouling.

The filtered dataset should be submitted to a simple *data reconciliation* step if possible. The principle of data reconciliation is to apply one or several independent checks to verify the consistency of the data. Independent checks that can be performed on typical wastewater process data in practice can be sorted in two categories.

- Fundamental checks (based directly on a law of nature):
  - Examples include performing flow balances or mass balances on inert components if possible. Solids can be considered inert around clarifiers. When performing an inert mass balance (such as non-volatile inorganic solids) around the whole process or plant, all inputs and outputs should be considered, including chemical dosages. Processes that generate inert non-volatile solids (e.g., bio-P removal, spontaneous precipitation) must be taken into account. If data are available for all inputs and outputs, it is also possible to use phosphorus or magnesium as tracers.
  - If a mass balance error is detected, it should be rectified before submitting the data to the model. Without going into the details of statistics and data reconciliation theory, it is clear that the error normally is the result of several incorrect measurements, which should all be adjusted in proportion to their reliability and noisiness (standard deviation). In practice it is frequently one flow variable (such as recycle or wastage flow) that will be modified to maintain the mass balance.
  - Consistency of kinetic parameters in single sludge systems also can be considered as strong checks – a single sludge system is expected to perform similarly under similar environmental conditions. For example, if two separate nitrification rates have to be used for the same sludge to fit ammonium concentrations in two locations in the same tank, and all environmental conditions (DO, mixing, pH, etc.) are accounted for, the

model representation or the data itself is most likely suspect (e.g., short circuiting within the tank).

- Empirical checks (based on engineering knowledge):
  - Experience-based "empirical" knowledge of the activated sludge process should be used to check the reliability of the dataset. Component ratios (influent COD/BOD, TKN/COD, TP/COD, VSS/TSS, particulate COD/VSS and mixed liquor VSS/TSS) and observed sludge production figures can be calculated and checked against typical values before trying to force a fit to data which might be outside the usual range.
  - Obviously there are situations when the collected data are correct but they are still outside normal ranges (e.g., significant industrial contribution).

For the purposes of steady state modeling, a filtered average value must be established for each monitored parameter. Data should be excluded for periods during which the plant is not at a reasonable steady state; for example, during storm events or process upsets. When evaluating data for an extended period, it may be appropriate to extract sets of data for several periods of steady state operation.

Frequently, the dynamic events that are removed from the steady state analysis are prime candidates for use in running individual dynamic simulations. The process upset or storm generates a highly variable output, and if the data collection frequency is high enough to describe the dynamics in sufficient detail, it provides a sensitive signal for calibration (assuming all other relevant variables are monitored). For example, effluent ammonia breakthrough during a storm is not valuable alone, without storm flow, MLSS and several other measurements.

The dynamic performance of the model also can be evaluated on long-term dynamic runs (e.g., datasets representing periods of several months to one year). These simulations usually are time-consuming and frequently lack some inputs. Also, operator decisions during the period may have an effect on the results. However, these longer runs are information-rich if conditions for a proper simulation can be met.

21.3.5.2 Parameters that can be extracted from historical data

The objective of day-to-day data gathering at plants usually is for monitoring effluent compliance and aiding basic operational decisions. The limited information usually does not provide sufficient data for detailed kinetic modeling. For example, generally it will not be possible to derive a complete influent wastewater characterization from the available data. Nevertheless, it is usually possible to deduce certain of the information with reasonable accuracy:

- If data on filtered effluent COD are available, the influent unbiodegradable soluble COD fraction (f_{SI}) can be estimated.
- If data on both influent COD and $cBOD_5$ are available, they can be used as a means for estimating the unbiodegradable particulate COD fraction (f_{XI}). A high unbiodegradable particulate COD fraction results in a high COD/BOD ratio. Appendix B presents an empirical method for calculating $cBOD_5$ based on total COD and the division of COD into biodegradable and unbiodegradable portions (an example is shown in the appendix).

The two soluble fractions (f_{SI} and f_{SS}) are specified, and the value of the unbiodegradable particulate COD fraction (f_{XI}) is adjusted iteratively until the observed BOD_5 is matched.

- In plants with primary sedimentation tanks (PSTs), usually it is possible to deduce information on certain other influent wastewater characteristics (i.e., the fractionation into soluble/particulate, and biodegradable/unbiodegradable portions). Consider the situation where data are available on PST influent and effluent (overflow) COD, BOD_5, TSS and VSS. The influent stream is specified, starting with typical default characteristics, and perhaps an adjusted f_{SI} from above. PST operation is simulated based on the observed percent solids removal, and adjustments are made to the influent fractions iteratively until a reasonable match to the observed output concentrations (COD, BOD, VSS, TSS) is obtained. This method takes advantage of the fact that soluble components flow through the PST, and only particulate components settle, leading to changes in the concentration of components that include both soluble and particulate matter (e.g., unfiltered COD and BOD).
- If activated sludge influent ISS data are available (derived from TSS – VSS), and process MLVSS and MLSS are known, a check can be performed on the wastage information (flow and concentration, i.e., mass of waste activated sludge). Since ISS is inert, simulation should result in an accurate prediction of mixed liquor ISS concentration (MLSS – MLVSS). If there is not a match, then the mass wastage in the simulation does not match the mass wastage on the plant. The most frequent reason is incorrect wastage *flow rate* data. However, the first step should be to check the RAS flow and TSS concentration (if wastage is withdrawn from the RAS). An error in RAS flow rate results in incorrect solids concentration in the model waste stream. If simulated RAS TSS concentration matches the plant data, then the reported wastage flow data likely is questionable. An indication of the error is derived by adjusting wastage rate in the simulation until the mixed liquor ISS concentration (MLSS – MLVSS) is predicted correctly. This may provide the basis for an investigation of the validity of wasting data on the plant. (There are other reasons that may explain the ISS discrepancy: (a) chemical addition in the plant generating more inorganics in the MLSS; (b) biological P removal, with stored P increasing ISS; (c) inorganics generated from precipitation; (d) wasting at a high rate for a short period each day, and withdrawing dilute RAS. Therefore, these factors should be excluded before making an adjustment to wastage rate in the simulation.)
- The unbiodegradable particulate COD fraction (f_{XI}) can also be estimated if MLVSS is known (provided that the wastage information is correct). The plant performance is simulated, making iterative adjustments to the value of f_{XI} until the simulated MLVSS matches the observed. Increasing f_{XI} will increase the VSS. This is equivalent to matching the observed volatile solids production.
- Sludge production (mass VSS or TSS generated / mass COD or BOD influent load) should be checked against rules of thumb for similar plants at similar SRT. Large deviations often indicate poor information on activated sludge wastage.
- In fully nitrifying plants (with low effluent ammonia), a lower bound for the nitrifier maximum specific growth rate (μ_{AUT}) can be deduced. The μ_{AUT} value at least exceeds that which would result in washout at the operating SRT. It may be possible to refine the estimate for μ_{AUT} from analysis of trends in temperature and effluent ammonia data on seasonally nitrifying plants.

- Half-saturation parameters for substrates (BOD, ammonia) can be estimated by matching concentrations in the effluent under normal operating conditions.

The extent of the information that can be derived from historical plant records obviously depends on the extent and reliability of the data. Most model kinetic and stoichiometric parameters (and others such as settling parameters) cannot be estimated reliably from typical historical datasets.

21.3.6 Calibration Level 3: On-Site, Full-Scale Testing

The accuracy and reliability of the calibration based on historical data alone can only be improved through supplementing the data with additional measurements and observations. The type of data required obviously depends on what is available already, and also on the type of plant. For example, data requirements for a nutrient removal plant will be more extensive than for a plant only concerned with removing BOD. The specific data needs should be evident from the analysis methods described in other chapters of this manual.

The purpose of these specific sampling campaigns might be multiple – to verify the accuracy of existing analytical and flow measurements, to close mass balances and determine typical fractions that cannot be derived from the laboratory dataset, and to sample specific events that provide a more sensitive signal for calibration than normal operating conditions.

There are two main categories of on-site test that are different in scope and objective: (a) additional composite and grab sampling to establish more detailed information on influent wastewater characteristics and provide a check on certain model stoichiometric parameters and (b) sampling during dynamic events (beyond the normal diurnal variation) to get better estimates of kinetic model parameters.

21.3.6.1 Additional sampling

The extent of the sampling program obviously depends on how much historical data is available, the need to verify historical data and the specific objectives for which the calibrated model is to be used. Experience has shown that the requisite data usually can be gathered over a two-week monitoring campaign. This should perhaps be repeated for different operating conditions (e.g., winter and summer, low and high load) if the model is to be used for predictions over a range of conditions. The monitoring campaign should include a number of components:

- Collecting average daily flow data on all streams.
- Collecting daily composite samples of raw wastewater, primary effluent and secondary effluent, and analyzing these for a number of parameters.
- Collecting daily grab samples of mixed liquor.
- Conducting spot checks on DO concentration levels in aerated zones.
- Monitoring RAS recycle rate and concentration, and whether the rate is constant or flow-paced.
- Gathering information on activated sludge wastage, including volume and time of wasting.

- Collecting data on primary sedimentation tank underflow, and whether sludge withdrawal is intermittent.
- Sampling to characterize side streams.

During the period of daily monitoring at least two 24-hour diurnal grab sampling (hourly or bi-hourly) exercises should be performed. The objective is to gather information on the diurnal variation in flow and influent/effluent concentrations.

The specific data requirements will always be case-specific. Table 21-1 is presented as a suggested checklist for a comprehensive data-gathering program. These guidelines are suggested for municipal wastewater treatment plants, and a distinction is made between requirements for different types of plant.

In principle, this sampling program applies to one liquid line process train, or those that can be lumped together due to similarity in operation (loading, SRT). If the plant has process trains that are significantly different (e.g., nitrifying and non-nitrifying trains), a separate sampling program needs to be set up for each.

21.3.6.2 Stress tests

Sampling dry weather normal operating conditions does not provide sufficient information about the behavior of the plant under stressed conditions; this may be necessary for assessing plant capacity.

There are two principal ways to collect data under higher-loaded conditions – sample *high flow events* that occur naturally (storm flow or sustained high load periods) or perform a *stress test.*

In both cases, sampling frequency is important since sampling that is too infrequent can miss peaks and cause incorrect calibration results. The best method is to pre-run the anticipated event in the uncalibrated simulator and determine the required sampling frequency that will reflect dynamic changes properly. This can be on a variable time scale, for example 30 minute samples during the duration of the highest flows, and hourly or bihourly samples during the following day while the plant recovers. Storms are sometimes difficult to prepare for as they occur outside the control of the modeling engineer, but if sampled properly, they provide useful information for calibration.

Stress tests are easier to design but in some cases are difficult to justify and may require special approvals. The principle is to try to overload a smaller section of the plant temporarily, while maintaining overall performance so that the whole plant will not violate its effluent permit during the test.

Sometimes, disturbances caused by planned maintenance (e.g., taking a tank off-line or bringing it back on-line) or even unintended equipment failure can present an opportunity to collect sensitive stress test data.

21.3.6.3 Other sources of information for calibration

Several other model parameters and inputs can be extracted from various data sources if they are available:

- *Step-feeding a plug-flow tank* results in a decreasing MLSS concentration along the length of the process. It often is difficult to determine the exact flow distribution, particularly if there are more than two flow input locations along the tank. However, it is simple to collect MLSS samples along the train and then fit various input flow fractions to match the MLSS profile. This method is particularly sensitive if performed on a dynamic basis (fitting MLSS variations at different locations along the tank in time).
- *Intermittently aerated plants* and SBRs provide a very strong signal for calibrating nitrification/denitrification parameters. If a tank has separate mixing and aeration facilities, this signal (increasing ammonia and decreasing nitrate concentrations during non-aerated periods) can be generated in aerobic systems as well. The length of the non-aerated cycle depends on plant conditions — typically 1 to 2 hours every 3 to 5 hours will provide a good signal. If mixing is provided by the aeration system, phase separation will occur very soon after ceasing aeration, and the tank cannot be modeled as ideally mixed anymore.
- In long, narrow plug-flow tanks that have no dividing walls between cells, the *number of CSTRs in series* becomes a model parameter. In typical tanks, this number (which is a simple representation only, since it will depend on flow, mixing, aeration and other conditions) can be estimated from width/length ratios.

The Water Research Centre in the United Kingdom proposed an empirical relationship to estimate the number of equivalent tanks-in-series for a diffused air aeration system:

$$N = 7.4 \cdot \frac{L \cdot Q \cdot (1 + R_R)}{W \cdot H} \qquad (21.3.6.3\text{-}1)$$

where:

N	=	equivalent number of tanks-in-series
L	=	aeration tank length (m)
Q	=	wastewater flow (m^3/s)
R_R	=	RAS recycle ratio (dimensionless)
W	=	aeration tank width (m)
H	=	water depth (m)

A more rigorous test approach is to perform a tracer test. (Environment Canada, Ontario Ministry of Environment and Energy, 1995). A slug of inert tracer (such as lithium or fluorescein) is added to the head of the tank and the washout curve is measured at the outlet. The measured curve can be simulated using an inert soluble component in the model to establish the optimum number of tanks in series.

If it is possible to stop influent flow to an activated sludge tank by diverting it to other parts of the plant for a short time (~0.5 hours), the resulting RAS buildup in the head of the tank can be used to the same effect. During the short period without influent flow (but with the recycle flow) the solids concentration increases in the head of the tank, and when the influent flow is resumed, this higher concentration "plug" will be shifted and dispersed along the tank. MLSS samples should be taken frequently at two locations (midpoint and outlet) and can be matched by modeling this experiment with different numbers of CSTRs in series.

Most full-scale tanks even if considered well mixed, are better described as two CSTRs in series due to non-ideal mixing.

Table 21-1. General Data Gathering Guideline for Different Types of Municipal Activated Sludge Wastewater Treatment Plants

PARAMETER	PLANT TYPE				COMMENTS
	BOD Removal	Nitrification	Nit - Denit	EBPR	
RAW INFLUENT:					
Flow (average daily)	✓	✓	✓	✓	
TSS	✓	✓	✓	✓	
VSS	✓	✓	✓	✓	For calculation of ISS
COD (total)	✓	✓	✓	✓	
COD (GF filtrate)	✓	✓	✓	✓	For calculation of influent solids COD/VSS
ffCOD	✓	✓	✓	✓	For RBCOD
COD (0.45 μm filtrate)*	✓	✓	✓	✓	
BOD_5 (total)	✓	✓	✓	✓	COD/BOD for estimating inert particulate COD
BOD_5 (GF filtrate)*	✓	✓	✓	✓	
VFA				✓	
TKN (total)	✓	✓	✓	✓	
TKN (GF filtrate)	✓	✓	✓	✓	For calculation of influent solids N/VSS
Ammonia		✓	✓	✓	
Nitrate			✓	✓	Spot checks – usually zero
TP	✓	✓	✓	✓	
TP (GF filtrate)	✓	✓	✓	✓	For calculation of influent solids P/VSS
Ortho-P (PO_4-P)				✓	
Alkalinity	✓	✓	✓	✓	
pH	✓	✓	✓	✓	
Mg				✓	
Ca				✓	

Table 21-1. General Data Gathering Guideline for Different Types of Municipal Activated Sludge Wastewater Treatment Plants (cont.)

PARAMETER	PLANT TYPE				COMMENTS
	BOD Removal	Nitrification	Nit - Denit	EBPR	
Diurnal flow pattern(s)	✓	✓	✓	✓	Weekday / weekend differences
Diurnal concentration pattern(s) – COD, TKN, NH_3, TP	COD	COD, TKN, NH_3	COD, TKN, NH_3	COD, TKN, NH_3, TP	Hourly or bi-hourly grab samples on 2 or 3 typical dry weather flow days
PRIMARY EFFLUENT:					
Flow (average daily)	✓	✓	✓	✓	
TSS	✓	✓	✓	✓	
VSS	✓	✓	✓	✓	For calculation of ISS
COD (total)	✓	✓	✓	✓	
COD (GF filtrate)	✓	✓	✓	✓	For calculation of influent solids COD/VSS
ffCOD	✓	✓	✓	✓	For RBCOD
COD (0.45 μm filtrate)*	✓	✓	✓	✓	
BOD_5 (total)	✓	✓	✓	✓	COD/BOD for estimating inert particulate COD
BOD_5 (GF filtrate)*	✓	✓	✓	✓	
VFA				✓	
TKN (total)	✓	✓	✓	✓	
TKN (GF filtrate)	✓	✓	✓	✓	For calculation of influent solids N/VSS
Ammonia		✓	✓	✓	
Nitrate			✓	✓	Spot checks – usually zero
TP	✓	✓	✓	✓	
TP (GF filtrate)	✓	✓	✓	✓	For calculation of influent solids P/VSS
Alkalinity	✓	✓	✓	✓	
pH	✓	✓	✓	✓	
Ortho-P (PO_4-P)				✓	
PST UNDERFLOW:					
Flow	✓	✓	✓	✓	Details of intermittent sludge removal
TSS	✓	✓	✓	✓	
PROCESS:					
MLSS	✓	✓	✓	✓	Profile in step feed systems
MLVSS	✓	✓	✓	✓	
D.O.	✓	✓	✓	✓	Zone to zone, and variation over 24 hours

Table 21-1. General Data Gathering Guideline for Different Types of Municipal Activated Sludge Wastewater Treatment Plants (cont.)

PARAMETER	PLANT TYPE				COMMENTS
	BOD Removal	Nitrification	Nit - Denit	EBPR	
Mixed liquor COD	✓	✓	✓	✓	Spot checks – for calculation of mixed liquor COD/VSS
Mixed liquor TKN		✓	✓	✓	Spot checks – for calculation of N/VSS content
Mixed liquor TP				✓	Spot checks – for calculation of P/VSS content
RAS flow	✓	✓	✓	✓	Details of flow-pacing
RAS TSS	✓	✓	✓	✓	
WAS flow	✓	✓	✓	✓	Details of intermittent wasting
WAS TSS*	✓	✓	✓	✓	May duplicate RAS TSS
Mixed liquor recycle flow(s)			✓	✓	
Temperature	✓	✓	✓	✓	
Profiles (zone to zone) – NH_3, NO_3, PO_4		NH_3, NO_3	NH_3, NO_3	NH_3, NO_3, PO_4	Grab samples, filtered immediately (noting time of sampling)
Clarifier sludge blanket level	✓	✓	✓	✓	
Mass of solids in the secondary clarifier	✓	✓	✓	✓	See Appendix 9 (Estimation of SRT at Full-scale Plants) for estimation method
SECONDARY EFFLUENT:					
TSS	✓	✓	✓	✓	
COD (total)	✓	✓	✓	✓	
ffCOD	✓	✓	✓	✓	For RBCOD
COD (0.45 μm filtrate)*	✓	✓	✓	✓	
BOD_5 (0.45 μm filtrate)	✓	✓	✓	✓	
TKN (0.45 μm filtrate)		✓	✓	✓	
Ammonia		✓	✓	✓	
Nitrate		✓	✓	✓	
Nitrite		✓	✓	✓	Spot checks – should be zero
TP (0.45 μm filtrate)				✓	
Ortho-P (PO_4-P)				✓	
Alkalinity	✓	✓	✓	✓	
pH	✓	✓	✓	✓	
Diurnal concentration pattern(s) – COD, NH_3, PO_4-P	COD	COD, NH_3, NO_3	COD, NH_3, NO_3	COD, NH_3, NO_3, PO_4	
RETURN STREAMS:					
Flow	✓	✓	✓	✓	Details of intermittent return

Table 21-1. General Data Gathering Guideline for Different Types of Municipal Activated Sludge Wastewater Treatment Plants (cont.)

PARAMETER	PLANT TYPE				COMMENTS
	BOD Removal	Nitrification	Nit - Denit	EBPR	
TSS	✓	✓	✓	✓	Sampling needs depend on stream type
VSS	✓	✓	✓	✓	
COD (total, GF filtrate)	✓	✓	✓	✓	
ffCOD, VFA				✓	For fermenter streams
BOD_5 (total)	✓	✓	✓	✓	
TKN, NH_3, NO_3		✓	✓	✓	
TP, PO_4-P				✓	
Alkalinity	✓	✓	✓	✓	
pH	✓	✓	✓	✓	
SPECIFIC LABORATORY STUDIES:					
SBR for comprehensive wastewater characterization					Need is case-specific
μ_{AUT} testing		✓	✓	✓	

* - Non-essential tests that provide additional or confirmatory information

GF – glass-fiber

ffCOD – flocculated and filtered COD

21.3.7 Calibration Level 4: Direct Parameter Measurements

In many cases, it is unfeasible or too expensive to obtain a consistent dataset from the full-scale plant to provide a full wastewater characterization and kinetic evaluation. The reason may be that the system is large and complex, many disturbances are outside the control of the modeler and mass balances are difficult to close or require a large number of expensive samples. If these difficulties can be overcome, the advantage of tests performed on the full-scale plant is that they provide information directly from the system under study.

If a monitoring program either does not provide all the requisite data, or perhaps provides unusual data, then bench-scale tests should be considered. The advantage of these tests as compared to the full-scale testing is that all aspects of the bench tests are easily controllable, and the number of samples is greatly reduced. A potential criticism of these tests is that they may develop a unique microbial community that might not necessarily be the same as that of the full-scale plant. However, there are many cases where these methods have been applied successfully to model calibration.

The following are bench-scale testing procedures that will provide estimates of specific parameters and wastewater characteristics:

- SBR procedure (Appendix D) for influent wastewater characterization (i.e., the COD and TKN fractions): A comprehensive plant data-gathering exercise should provide all the requisite information on wastewater characteristics. In cases where the derived characteristics are very different from those for typical wastewaters, then the costs of the

SBR study likely would be justified to confirm or refute the unusual value. For example, a high value for the particulate unbiodegradable COD fraction may imply millions of dollars in construction costs. The SBR protocol includes measurement of nitrifier maximum specific growth rate (μ_{AUT}).

- Measurement of nitrifier maximum specific growth rate (μ_{AUT}): It is strongly recommended that one of the rapid testing protocols for measuring μ_{AUT} is conducted in the case of nitrifying plants (Appendices D, F and G). Knowledge of this kinetic parameter is essential in plant analysis and optimization.

21.3.8 Steps in Calibration – A Typical Procedure

This chapter has covered many aspects and procedures associated with calibration. As outlined above, due to the large variation regarding the processes applied on plants, data availability and quality, and different objectives of the modeling job, it is impossible to give an all-inclusive job flow or a rigorous block diagram for all calibration tasks. The engineer, based on guidance provided here and his/her experience, will have to make the final judgment when setting up the exact steps. The following is a list of steps for a "*typical*" plant, and "*typically available*" data to provide an initial blueprint for a calibration exercise.

Consider a hypothetical BNR plant (providing BOD removal, nitrification, denitrification and EBPR) that has a log of daily composite samples for the typical parameters. The following step-wise procedure suggests the sequence to follow in a calibration exercise. This example focuses on the activated sludge system only, and excludes calibration of systems including primary settlers. Also, it is presumed that the initial steps for estimating wastewater characteristics outlined earlier have been conducted: namely, estimating the unbiodegradable particulate COD fraction (f_{XI}) from the observed COD/BOD_5 ratio and estimating COD fractions from the balance around the PST (if the plant includes a PST).

Step 1:
Set up the plant configuration in the simulator. Assume typical influent wastewater characteristics, and assume typical or default kinetic and stoichiometric model parameters.

Step 2:
Specify data on temperature, input flows and known influent characteristics (COD, ISS, Alkalinity, etc.).

Step 3:
Calculate influent soluble inert fraction (f_{SI}) from effluent filtered COD data. Modify the influent characteristics accordingly.

Step 4:
Fit the secondary effluent solids concentration. Adjust percent solids removal or sludge settling parameters for the secondary clarifier until the predicted TSS matches the observed. (Steps 3 and 4 essentially account for the COD leaving with the effluent – this is the first step for the mass balances later).

Step 5:
If influent ISS data are available, and mixed liquor VSS and TSS are known, a check can be performed on the wastage information. Since ISS is inert, simulation should result in

an accurate prediction of mixed liquor ISS concentration (MLSS – MLVSS). If there is not match, possible reasons could include (a) incorrect waste flow or mass flow data; (b) chemical addition in the plant (which generates more inorganics in the MLSS – this has to be added to the model); (c) mismatch in Poly-P data and prediction – this can be rectified later when the model bio-P performance is calibrated; (d) inorganics generated from spontaneous precipitation. The most frequent reason is incorrect sludge wastage data. Another common reason for a mismatch in ISS mass balance is an error in RAS flow rate that results in incorrect solids concentration in the model waste stream. This usually can be checked if the aeration tank and RAS solids concentrations are known (see Step 10).

Step 6:
If the ISS mass balance around the plant requires adjustment due to incorrect sludge wastage data, then adjust wastage flow. (At the same time, an investigation on the plant is suggested to determine correct wastage information, and whether the model adjustment is justified).

Step 7:
Check whether BOD removal is achieved. If not, something likely is wrong with the model setup. Only the highest loaded plants require adjustment to aerobic growth parameters. Check to ensure that flows are specified correctly, as well as unit volumes and process DO concentrations.

Step 8:
Check nitrification performance. Adjust the autotrophic growth rate (μ_{AUT}) to achieve the desired effluent ammonia concentration.

Step 9:
Check model predictions for MLVSS (and OUR if available). If the model does not match (too high for one, too low for the other) adjust influent inert COD fraction (f_{XI}). A higher inert fraction in the same influent COD will produce more sludge and less oxygen demand and vice versa.

Step 10:
Check RAS solids concentration (and sludge blanket level if the settling behavior is being modeled). If applying a flux model with the Vesilind settling velocity function (or one of its derivatives), adjust the hindered settling parameter to match sludge blanket (adjust the equivalent parameter if an alternative settling model is used). If the settler is not critically loaded, the hindered settling parameter will not have any effect on the RAS concentration. In this case, RAS concentration is simply mass balance governed. Check the solids mass balance around the settler (MLSS in; RAS flow; RAS and effluent solids out). An error in RAS flow rate results in incorrect solids concentration in the RAS, and hence the model waste stream.

Step 11:
Check the ammonia profile in the plug flow tank to fine tune the maximum autotrophic growth rate. Large dynamic changes in effluent ammonia can be used to the same effect.

Step 12:
Check the nitrate profile to fine-tune denitrification predictions. The following parameters should be examined closely:

- Influent RBCOD fraction – this will have a large impact on the extent of denitrification and the denitrification rate at the front end of the anoxic zone.
- Check that the model applies separate aerobic and (lower) anoxic heterotrophic yields for N/DN systems.
- In models that have separate anoxic yield, reduce the value if significant amounts of methanol are added (anoxic yield on methanol may be as low as 0.4 mg biomass COD/mg methanol COD utilized).
- Adjust anoxic hydrolysis rate factor if the denitrification is limited by release of stored substrate for denitrification.

Step 13:
Check the orthophosphate profile along the tank length. If there is no P release/uptake, check anaërobic tank DO, VFA, and nitrate. The model parameters can be used to fine-tune the P profile and effluent quality.

Step 14:
Adjust P half-saturation or PAO maximum specific growth rate to match phosphate profile and effluent P.

Steps 1 to 7 and 9 and 10 must be performed for all plant types. Steps 8 and 11 should be conducted for all nitrifying plants. Step 12 is specific to BNR plant modeling. Finally, Steps 13 and 14 apply only to EBPR plant modeling.

At any given time during this procedure it may be necessary to return recursively to Step 4 as the processes are interconnected and changing parameters in later steps will have an effect on effluent solids, for example.

If a reasonable match between data and model cannot be achieved, ensure that:

- The plant data are of good quality (refer to data conditioning section).
- An attempt is not being made to fit a process that is not accounted for in the model. Examples may include whether a significant concentration of nitrite is generated in the process, chemical precipitation is practiced, GAO competition occurs, an inhibitory industrial component is present and so on. The models must be used within their validity field.

21.4 Common Pitfalls in Calibration

This section describes selected examples where deficiencies in sampling, data collection or analysis have led to difficulties in model calibration exercises.

21.4.1 Sampling and Analysis Problems

Samples gathered and analyzed for simulator calibration, as for any other purpose, must represent the true conditions in the treatment process. Three examples of problems in sampling and analytical data are related below.

Figure 21-1 illustrates a six-week plot of mixed liquor suspended solids (MLSS) concentration in a bioreactor as sampled by an automated device drawing from the bulk of the mixed liquor. The MLSS concentration covered the range from about 3,200 to 4,900 mg/L during the interval plotted; the variation was a consequence of changes in plant operation. During the period, two 48-hour intensive diurnal sampling exercises were conducted, on July 18-20 and August 3-5. By coincidence, the MLSS was close to 4,000 mg/L on each occasion. Figure 21-2 shows two plots of bi-hourly MLSS data gathered from the same bioreactor during the intensive sampling.

The bi-hourly data in Figure 21-2 indicate that the MLSS concentration was in the 2,500 to 3,000 mg/L range on those days. Perhaps one data point is an outlier; however, the remaining 49 values suggest that the data are consistent and accurate. An obvious discrepancy existed between these data and the auto-sampler 4,000 mg/L data. This required resolution before using any MLSS data for simulator calibration. On investigation, it was determined that the diurnal data were based on grab samples collected near a convenient corner of the bioreactor where inadequate mixing allowed partial separation of the mixed liquor to occur, thereby resulting in an erroneously low measurement of mixed liquor concentration during the intensive sampling program.

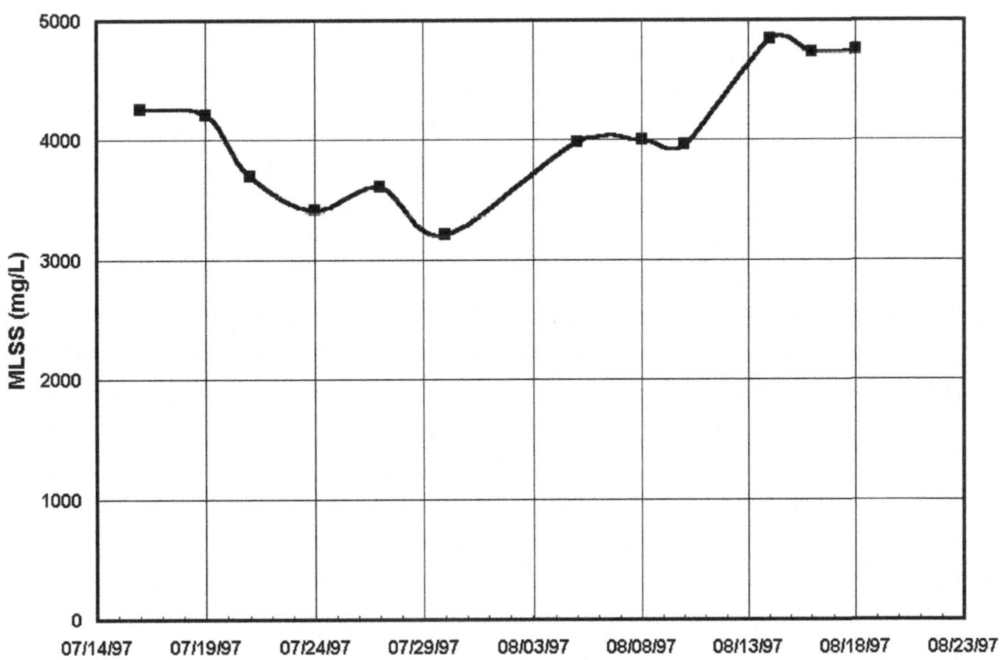

Figure 21-1. Example of Daily Composite MLSS Sampling

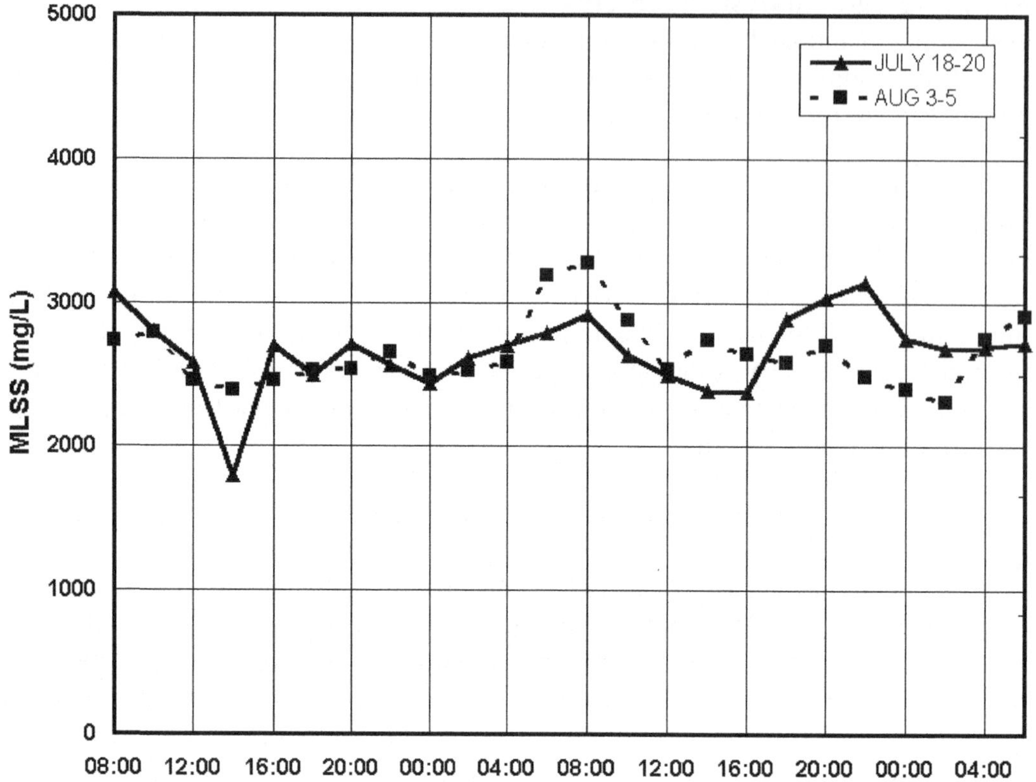

Figure 21-2. Example of Non-Representative Bi-hourly Grab MLSS Sampling

In another case of non-representative sampling, the historical primary effluent concentrations at a plant to be retrofitted with an EBPR process were estimated from the analysis of samples collected by an automated flow-proportional composite sampling device drawing sample aliquots from near the bottom of the primary effluent channel. Since the primary effluent channel was not well-mixed, a stratified flow regime existed and thus the samples contained a disproportionate amount of particulate matter. However, this was not the only problem. The automated sampling device relied on an over-600 mm Hg vacuum to withdraw and transport the sample aliquots from the primary effluent channel to the receiving sample vessel. Each sample aliquot was evacuated into the receiving vessel with considerable velocity and violent turbulence. This caused much of the more volatile fraction of the sample to "flash" as it entered the receiving vessel. The laboratory analysis of the samples gave $cBOD_5$ values that were more or less identical to the true values as measured later on a homogeneous primary effluent stream sampled by another technique. This was because the disproportionate amount of suspended matter in the sample was compensated for by the loss of volatile matter. Accordingly, the readily biodegradable fraction estimate for the primary effluent was erroneously low, while the particulate inert organic fraction estimate was erroneously high, resulting in a much higher MLVSS/MLSS ratio predicted by the simulator than actually occurred in the plant.

As an example of an analytical problem, an unusually high value of about 0.35 was estimated using a bench-scale SBR testing protocol to determine the particulate inert organic fraction of the secondary influent at a large U.S. mid-west city (Wilson et al., 1995). It was not immediately apparent why the value was so high. It was only after inconsistencies were noted in volatile fatty acid recoveries in the routine COD analysis conducted at the plant that the COD testing results became suspect. Further investigation revealed that a non-standard COD procedure was causing erroneously low estimates of the true COD values. A repeat of the bench-scale SBR protocol resulted in a much lower (and typical) inert particulate COD fraction estimate of 0.13.

21.4.2 Misleading Results from Deficiencies in Plant Operation

Figure 21-3 illustrates a diurnal data set for the ammonia-nitrogen concentration leaving a bioreactor. This information was to be used for adjusting the nitrification rate μ_{AUT} as part of the calibration step in a simulation exercise. An unrealistically low estimate of the nitrification rate would have resulted had not an examination of corresponding nitrate-nitrogen data, as plotted in Figure 21-4, been made. The nitrate-nitrogen data indicated a period of little or no nitrification from about 28 to 36 hours. Investigation at the plant revealed a limitation in air supply that caused an inhibition of nitrification during that period. Therefore, this data set could not be used to estimate μ_{AUT}.

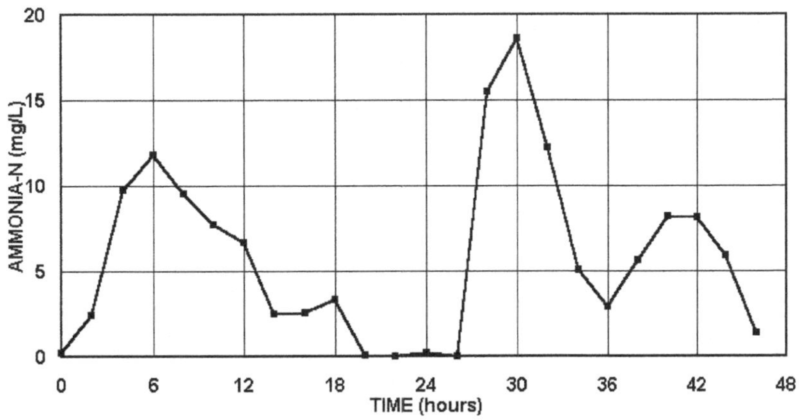

Figure 21-3. Example of Diurnal Treated Effluent Ammonia-N Variations

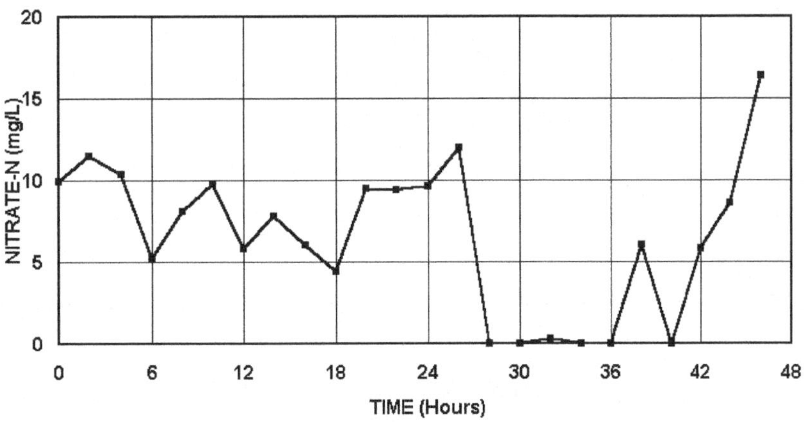

Figure 21-4. Example of Diurnal Treated Effluent Nitrate-N Variations

21.4.3 Waste Activated Sludge (WAS) Measurements and Sludge Age

Perhaps the most important step in simulator calibration is establishing the correct sludge age (SRT) in the activated sludge process. That is, the amount of excess biomass generated for various loading conditions and for various process options and operating conditions. In this case, it is useful to make a calibration check using long-term historical plant sludge production data. Quite often, however, such data are not reliable and this can be due to many reasons including faulty flow meter calibration and faulty suspended solids monitoring.

In one case for a treatment plant with a constant return activated sludge (RAS) pumping rate, WAS was withdrawn from the RAS line using a separate pump that operated for two hours once per day at a relatively high pumping rate. The high pumping rate caused a drawdown of the sludge blanket in the secondary settlers and a "rat-holing" effect through the sludge blanket. This resulted in an insufficient wasting of sludge mass from the system and the plant records were effectively reporting an overestimate of the actual solids production.

If available, ISS data can be used as a check on the solids balance because ISS will be a conservative substance in biological waste treatment systems. Cumulative ISS data over a several month period for the plant in question are plotted in Figure 21-5. Ideally, there should be a 1:1 relationship between input and output measured ISS values. However, in this case the plant data show a "production" of ISS in the system. Situations like this warrant further investigation to find a remedy before using the plant's historical solids production data for calibration purposes.

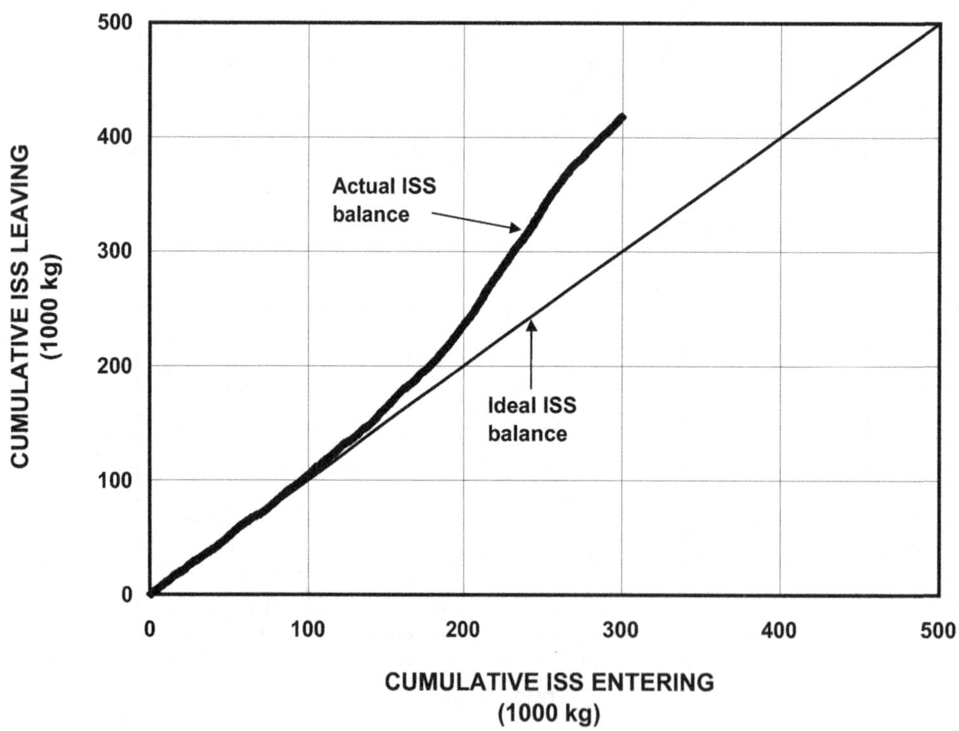

Figure 21-5. Example of Cumulative ISS Balance

21.5 References

Dochain, D., and P.A Vanrolleghem. 2001. *Dynamical modelling and estimation in wastewater treatment processes.* London, UK: International Water Association.

Environment Canada, Ontario Ministry of Environment and Energy. 1995. *Guidance manual for sewage treatment plant liquid train process audits.* Prepared by CH2MHill and Hydromantis, Inc. Toronto ON, Canada: Ontario Ministry of Environment and Energy.

Henze, M., W. Gujer, T. Mino, T. Matsuo, M. Wentzel, and G.v.R. Marais. 1995. Activated Sludge Model No. 2. *IAWQ Scientific and Technical Report No. 3*. London, UK: International Water Association.

Wilson, A.W., T. Marstaller, P.L. Dold, M.J. Tetreault, and D.W. DeFore. 1995. Use of a wastewater treatment process simulator as a tool for plant upgrading. In *Proc. 68th Annual Conference of the Water Environment Federation*, Miami FL. Alexandria VA: Water Environment Federation.

Wilson, A.W., and P.L. Dold. 1998. General methodology for applying process simulators to wastewater treatment plants. In *Proc. 71st Annual Conference of the Water Environment Federation*, Orlando FL. Alexandria VA: Water Environment Federation.

CHAPTER 22.0

Illustration of Model Calibration to Bench and Pilot Scale Systems

22.1 Background

This chapter shows model calibration using two different experimental systems. The first example is a bench scale activated sludge process with one complete-mixed, aerated bioreactor. The second example is a pilot scale process of five equal-sized bioreactors in series, with the first bioreactor operated as an unaerated anoxic zone. These are relatively simple cases that primarily show the use of average, steady state data for the determination of influent wastewater characteristics. In both cases, there were insufficient data to allow accurate estimation of kinetic parameters such as the nitrifier maximum specific growth rate.

22.2 Example 1: Bench Scale Aerobic System

This section illustrates the calibration of an activated sludge model to a simple aerobic bench scale system operated for an extended period of time on a municipal wastewater. The system was operated at McMaster University, Hamilton, Canada by Gibson (1991).

22.2.1 System Description and Operation

The laboratory activated sludge system in this example consisted of a 10 L glass carboy equipped with fittings for feed/recycle streams, and air supply. Feed was stored in a 30 L mixed and refrigerated tank, and pumped continuously to the reactor using a peristaltic pump. The reactor was mixed with a magnetic stirrer, and aerated through a fine pore diffuser. The liquid volume in the reactor was 6.35 L. Mixed liquor overflowed to a 2 L conical glass settling tank. Settled solids were recycled to the reactor using a second head on the feed pump.

Reactor feed was collected from the influent of a municipal wastewater treatment plant. A volume of wastewater sufficient for 7 to 10 days of operation was collected each time and collected at the same time of day to ensure consistent wastewater characteristics. The wastewater was diluted to a target COD concentration before feeding to the reactor.

The solids retention time (SRT) was maintained by wasting 0.6 L of mixed liquor once per day. This resulted in a target SRT of 10.6 days.

System design and operating parameters are summarized in Table 22-1. Table 22-2 summarizes the system parameters that were measured on a daily basis.

Table 22-1. Design and Operating Parameters for the Bench Scale Activated Sludge System

Parameter	Value
Reactor liquid volume	6.35 L
Daily influent volume	15.0 L
Hydraulic retention time	10.2 hrs
Return activated sludge (RAS) recycle ratio	1:1
Operating Temperature	20°C
Target solids retention time (SRT)	10.6 days

Table 22-2. Bench Scale Activated Sludge System Daily Monitoring Parameters

Measurement Location	Parameter
Influent	COD (unfiltered)
	TKN (unfiltered)
Effluent	COD (unfiltered)
	COD (filtered)
	TKN (filtered)
Reactor	Oxygen Utilization Rate (OUR)
	MLVSS

22.2.2 Operation Results

Table 22-3 summarizes the long-term averages for the monitoring parameters. The values for the influent total COD and the influent TKN were entered directly into the model. Next, the soluble unbiodegradable COD concentration (S_I) and the fraction of the total influent COD which is soluble unbiodegradable (f_{SI}) were estimated as follows:

Table 22-3. Summary of Monitoring Results

Parameter	Average Value
COD (mg/L)	
Influent (unfiltered)	322
Effluent (unfiltered)	71
Effluent (filtered)	41
TKN (mg/L)	
Influent (unfiltered)	36
Effluent (filtered)	2.2
Reactor VSS (mg/L)	1,784
OUR (mg/L/h)	24.1

$$S_I = Filtered\ Effluent\ COD$$

$$f_{SI} = \frac{Filtered\ Effluent\ COD}{Unfiltered\ Influent\ COD}$$

For this system:

$$S_I = 41\ mg/L$$

$$f_{SI} = \frac{41}{322} = 0.13$$

The fraction of the total influent COD which is soluble readily biodegradable (f_{SS}) was estimated in a separate aerobic batch bioassay, and the following value was found:

$$f_{SS} = 0.027$$

Steady state simulations of this particular system are quite insensitive to this value. This is due to both the overall calibration procedure used and the system configuration. In this calibration procedure, the slowly biodegradable fraction of the influent COD is calculated by difference. Therefore, adjustment of this fraction does not impact the total concentration of biodegradable material. For the completely-mixed system simulated here, it is the total amount of biodegradable material that is most important for system parameters such as the OUR. This would not be the case if other parameters were calculated by difference (i.e., X_I), if the system configuration was plugflow, or if the configuration included unaerated zones (i.e., in the next example).

22.2.3 Model Calibration

Model calibration proceeded by simulating the bench scale system and adjusting parameters until simulated and actual measurements were in reasonable agreement. The simulator configuration for this system is shown in Figure 22-1. Simulation of this system required a model for the secondary clarifier, which in this case used a simple model with constant solids removal efficiency. The initial values for influent wastewater fractions, clarifier solids removal and nitrifier kinetic parameters at 20°C are shown in Table 22-4. The values in Table 22-4 are a combination of initial estimates as derived above and model default values. The system was nitrifying fully in these experiments, and therefore there was insufficient information in the measurements to derive an accurate estimate of the nitrifier growth rate parameters. The results only allowed a minimum value for the nitrifier growth rate to be inferred.

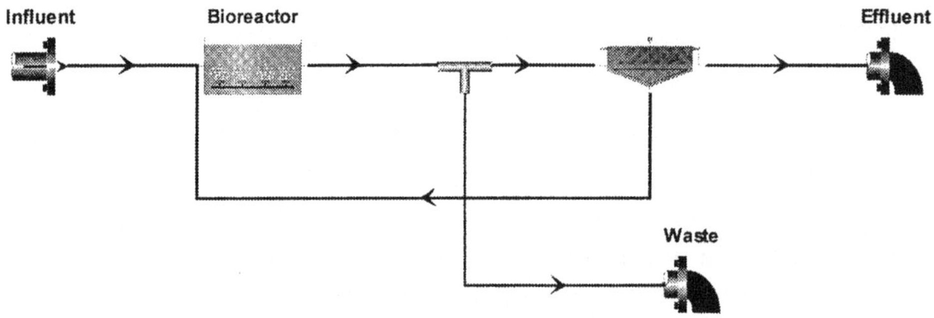

Figure 22-1. Simulator Configuration for the Bench-Scale Aerobic System

Table 22-4. Initial Estimates of Model Parameters

Parameter	Initial Value
Total influent COD	322 mg/L
Fraction which is soluble biodegradable (f_{SS}) *	0.027
Fraction which is soluble unbiodegradable (f_{SI}) *	0.13
Fraction which is particulate unbiodegradable (f_{XI}) **	0.13
Influent TKN	36 mg/L
Fraction which is soluble unbiodegradable (f_{SNI})**	0.0
Fraction which is ammonia (f_{SNH})**	0.75
Nitrifier kinetic parameters (@ 20°C)	
Maximum specific growth rate **	0.95 d^{-1}
Half saturation coefficient **	1.0 mg/L
Decay rate **	0.17 d^{-1}
Secondary clarifier	
Solids removal efficiency**	100 percent
* Estimated from monitoring results	
** Model default values	

Results from a steady state simulation conducted with the initial parameter estimates are shown in Table 22-5. In the first steps of the calibration, the soluble inert fractions, f_{SI} and f_{NSI}, were determined. The effluent soluble COD was slightly overestimated in the simulation. Therefore, f_{SI} was adjusted slightly downwards from the initial estimate. Although the system essentially was nitrifying completely, there was a slightly elevated concentration of residual effluent filtered TKN. Therefore, the fraction of the influent TKN, which is soluble unbiodegradable (f_{SNI}), was adjusted upwards.

Table 22-5. Comparison of Observed and Predicted Values Using Initial Parameters

Parameter	Observed Value	Predicted Value With Initial Parameters
VSS (mg/L)	1,784	1,920
OUR (mgO2/L/h)	24.1	28.4
Effluent NO3 (mgN/L)	25.2	26.5
Effluent soluble TKN (mgN/L)	2.2	1.8
Effluent total COD (mg/L)	71	44
Effluent soluble COD (mg/L)	41	44

Next, the clarifier solids removal efficiency was adjusted until there was agreement in the simulated and measured effluent total COD concentration. Note that with the suspended solids in the effluent included in the SRT calculation, the actual SRT value was about 8.3 days.

To estimate the fraction of the total influent COD which is particulate unbiodegradable (f_{XI}), the system was simulated and f_{XI} was adjusted until the simulated reactor VSS concentration was in agreement with the average measured value. Table 22-6 shows the results of a series of simulations conducted to estimate f_{XI}. It is apparent from these results that OUR can be used as an additional check on the f_{XI} estimate. There is also a slight impact on effluent nitrate concentration.

Table 22-6. Comparison of Observed and Predicted VSS, OUR and Effluent Nitrate for Different Estimates of Influent Particulate Unbiodegradable COD Fraction

Parameter	Observed Value	Predicted Value for Different f_{XI}			
		0.10	0.15	0.20	0.25
VSS (mg/L)	1784	1,545	1,671	1,797	1,923
OUR (mgO_2/L/h)	24.1	28.2	27.0	25.8	24.7
NO_3 (mgN/L)	25.2	25.6	25.4	25.2	25.0

22.2.4 Calibration Results

Table 22-7 shows the calibrated estimates of the model parameters. Also shown are those parameters that were adjusted in the calibration exercise, and which parameters were left at default values due to insufficient information. The results of a simulation using the calibrated parameters are compared to measured values in Table 22-8. This summary indicates that a good calibration was achieved for prediction of these observed values.

Table 22-7. Calibrated Estimates of Model Parameters

Parameter	Initial Value
Total influent COD	322 mg/L
Fraction which is soluble biodegradable (f_{SS}) *	0.027
Fraction which is soluble unbiodegradable (f_{SI}) *	0.12
Fraction which is particulate unbiodegradable (f_{XI}) *	0.20
Influent TKN	36 mg/L
Fraction which is soluble unbiodegradable (f_{SNI})*	0.012
Fraction which is ammonia (f_{SNH})**	0.75
Nitrifier kinetic parameters (@ 20°C)	
Maximum specific growth rate **	0.95 d^{-1}
Half saturation coefficient **	1.0 mg/L
Decay rate **	0.17 d^{-1}
Secondary clarifier	
Solids removal rate*	99.45 percent

* Calibrated

** Model default values

Table 22-8. Comparison of Observed and Predicted Values Using Calibrated Parameters

Parameter	Observed Value	Predicted Value
VSS (mg/L)	1,784	1,797
OUR (mgO2/L/h)	24.1	25.8
Effluent NO3 (mgN/L)	25.2	25.2
Effluent soluble TKN (mgN/L)	2.2	2.2
Effluent total COD (mg/L)	71	71
Effluent soluble COD (mg/L)	41	41

22.3 Example 2: Anoxic-Aerobic Pilot Scale System

This section describes calibration of an activated sludge model to a pilot scale system that was operated with one unaerated zone at the University of Cape Town, South Africa (Ekama and Marais, 1978).

22.3.1 System Description and Operation

The system used in this example was a pilot plant consisting of five bioreactors in series, each with a volume of 5 m^3. The first bioreactor was unaerated, resulting in an anoxic fraction of 20 percent. The process was fed with municipal wastewater at a constant flowrate. The SRT was maintained by wasting about 1390 L/d of mixed liquor from the last bioreactor, resulting in a target SRT of 18 days. Design and operating parameters are summarized in Table 22-9.

Table 22-9. Design and Operating Parameters for the Anoxic-Aerobic Pilot Scale System

Parameter	Value
Reactor liquid volume	5 m^3 per reactor x 5 reactors
Influent flowrate	40 m^3/d
Hydraulic retention time	15.0 hrs
Fraction unaerated	0.20
Return activated sludge (RAS) recycle ratio	3:1
Operating temperature	20°C
Target solids retention time (SRT)	18 days

22.3.2 Model Calibration

During steady state operation of the pilot plant, the following parameters were monitored in each bioreactor:

- VSS concentration
- OUR
- Soluble COD concentration
- Nitrate concentration
- Soluble TKN concentration
- Ammonia concentration

Figure 22-2 shows the simulator configuration of the pilot plant. Initial model calibration followed a similar procedure as the previous example:

- Average influent total COD and TKN concentrations were calculated.
- Simulated and observed soluble effluent COD and TKN were matched by adjusting the soluble unbiodegradable fractions of influent COD (f_{SI}) and influent TKN (f_{SNI}).
- Effluent solids concentration was matched by adjusting the secondary clarifier solids removal rate.
- Bioreactor MLVSS concentration was matched by adjusting f_{XI}.

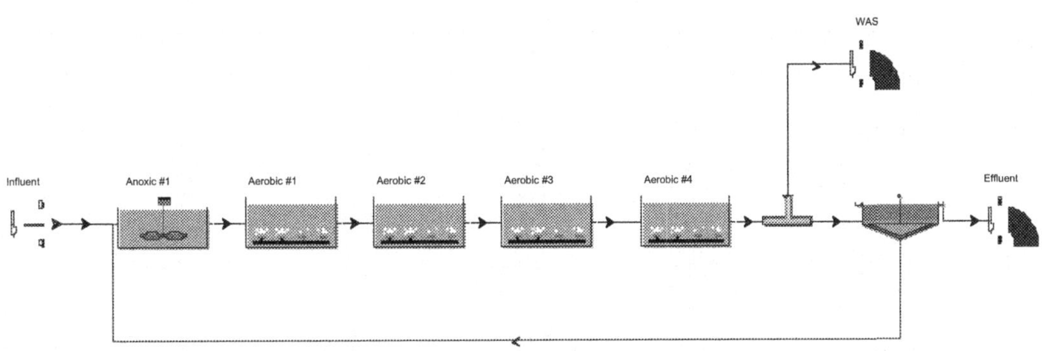

Figure 22-2. Simulator Configuration for the Pilot-Scale Anoxic-Aerobic System

An important aspect of the calibration procedure for this example was the determination of the fraction of the influent COD, which is soluble biodegradable (f_{SS}). Determination of this parameter follows procedures outlined in Chapter 7.0. Figure 22-3 provides an illustration of the sensitivity of the nitrate profile to f_{SS}, and therefore the importance of determining this parameter for a system with unaerated zones.

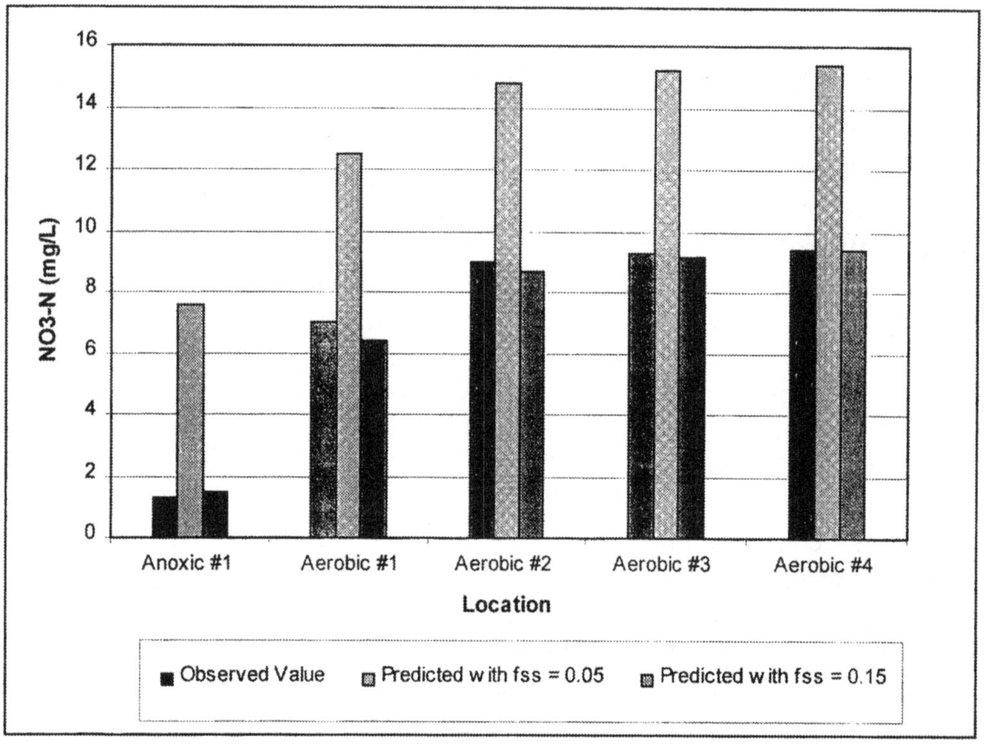

Figure 22-3. Sensitivity of the Bioreactor Nitrate Profile to the Value of f_{SS}

22.3.3 Calibration Results

The calibrated model parameters are summarized in Table 22-10. Similar to example 1, there was insufficient information in the results to estimate the nitrifier growth rate. Plots of the simulated versus observed bioreactor profiles are shown in Figures 22-4 to 22-8.

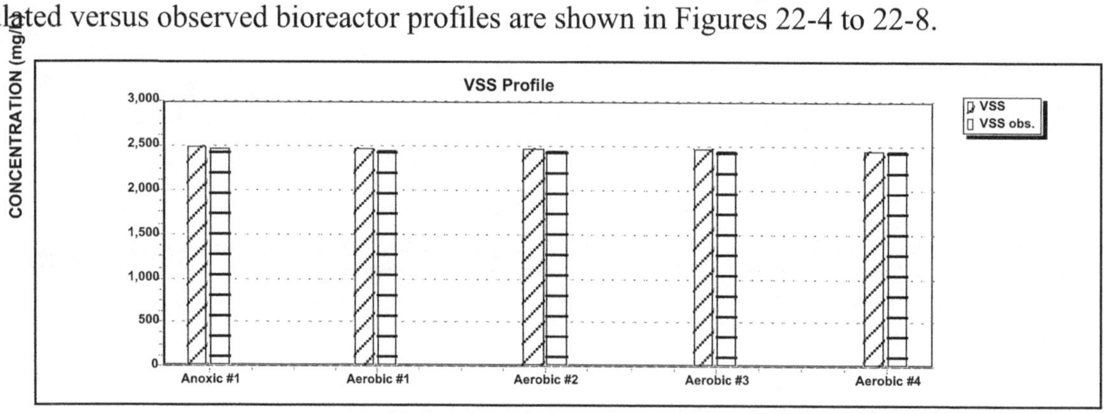

Figure 22-4. Simulated and Observed Reactor VSS Profile Using Calibrated Parameter Estimates

Table 22-10. Calibrated Estimates of Model Parameters for the Pilot Scale Anoxic-Aerobic System

Parameter	Initial Value
Total influent COD	477 mg/L
Fraction which is soluble biodegradable (f_{SS}) *	0.15
Fraction which is soluble unbiodegradable (f_{SI}) *	0.04
Fraction which is particulate unbiodegradable (f_{XI}) *	0.05
Influent TKN	45.1 mg/L
Fraction which is soluble unbiodegradable (f_{SNI})*	0.04
Fraction which is ammonia (f_{SNH})**	0.75
Nitrifier kinetic parameters (@ 20°C)	
Maximum specific growth rate **	0.95 d^{-1}
Half saturation coefficient **	1.0 mg/L
Decay rate **	0.17 d^{-1}
Secondary clarifier	
Solids removal efficiency*	99.97 percent
* Calibrated	
** Model default values	

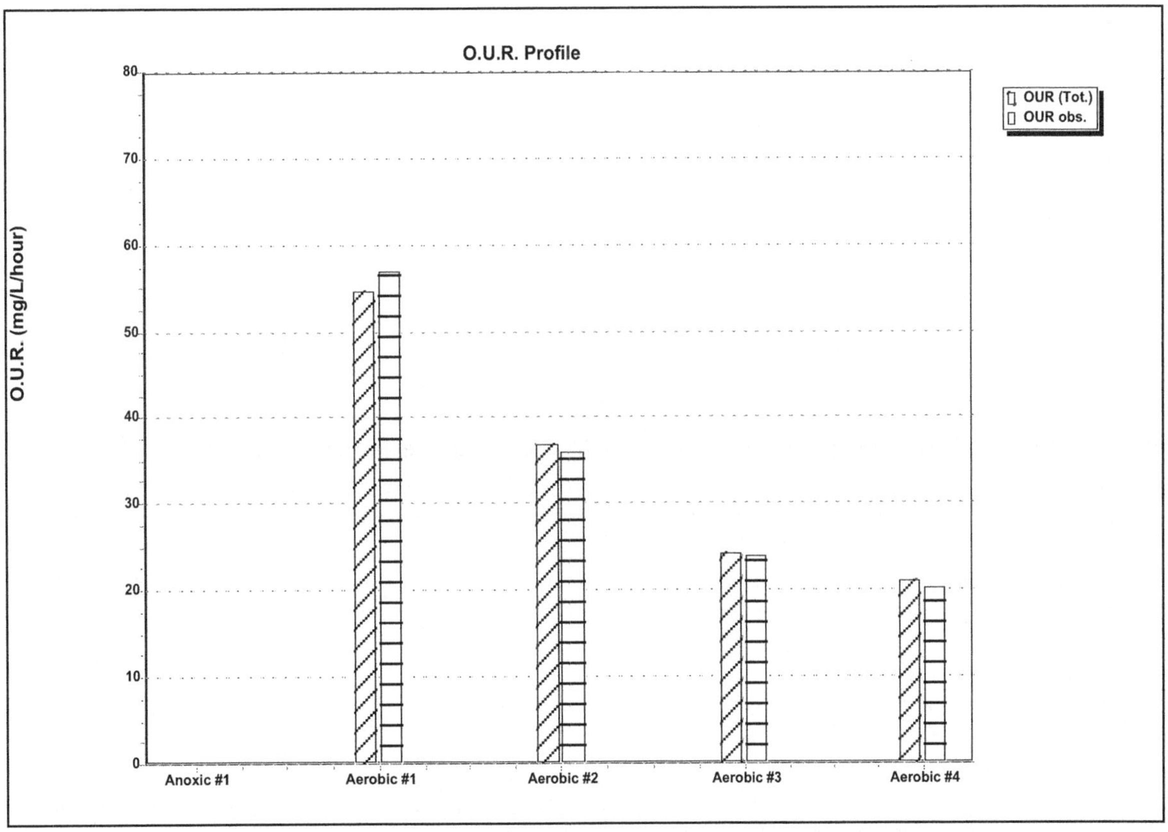

Figure 22-5. Simulated and Observed Reactor OUR Profile Using Calibrated Parameter Estimates

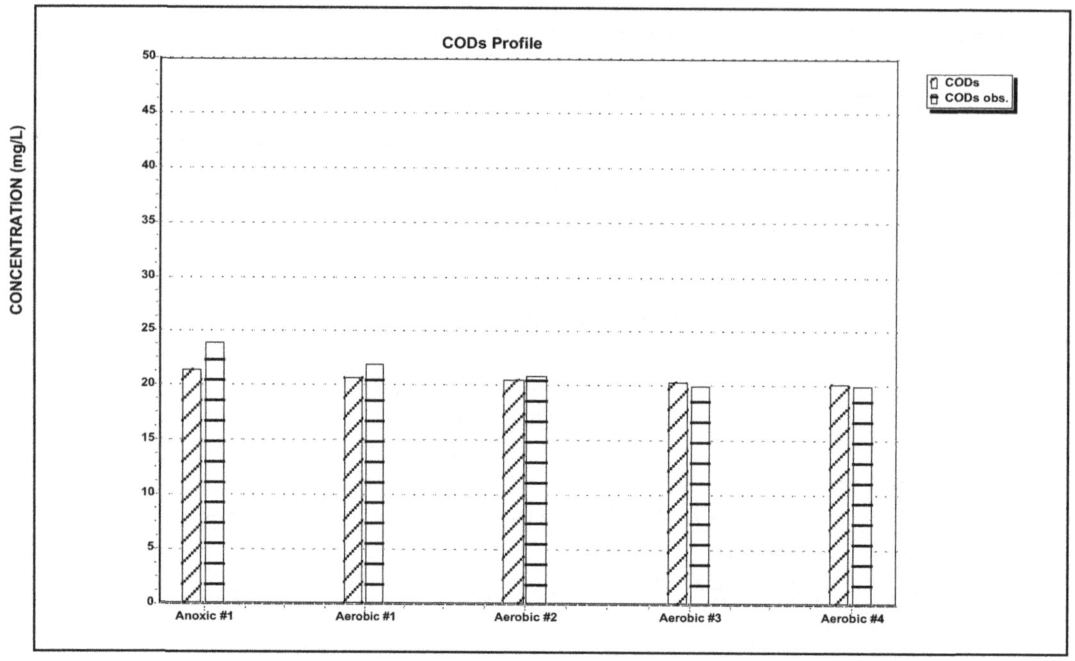

Figure 22-6. Simulated and Observed Reactor Soluble COD Profile Using Calibrated Parameter Estimates

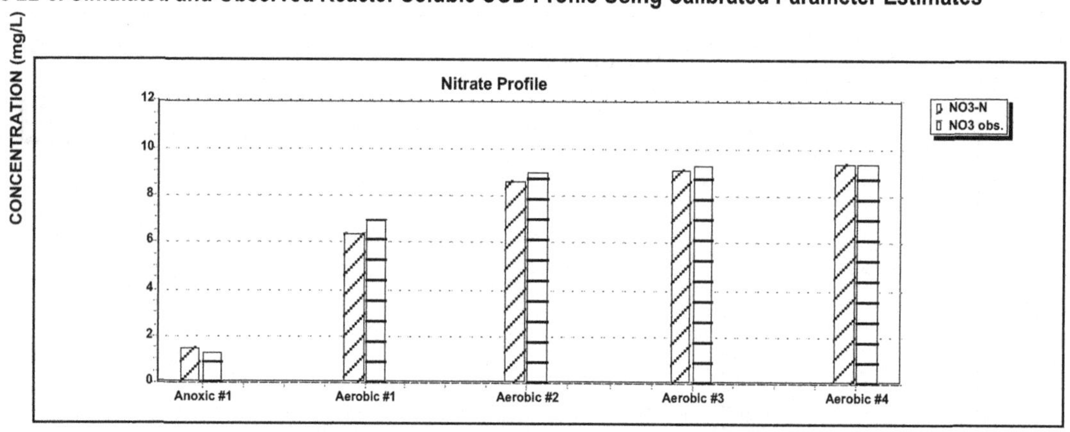

Figure 22-7. Simulated and Observed Reactor Nitrate-N Profile Using Calibrated Parameter Estimates

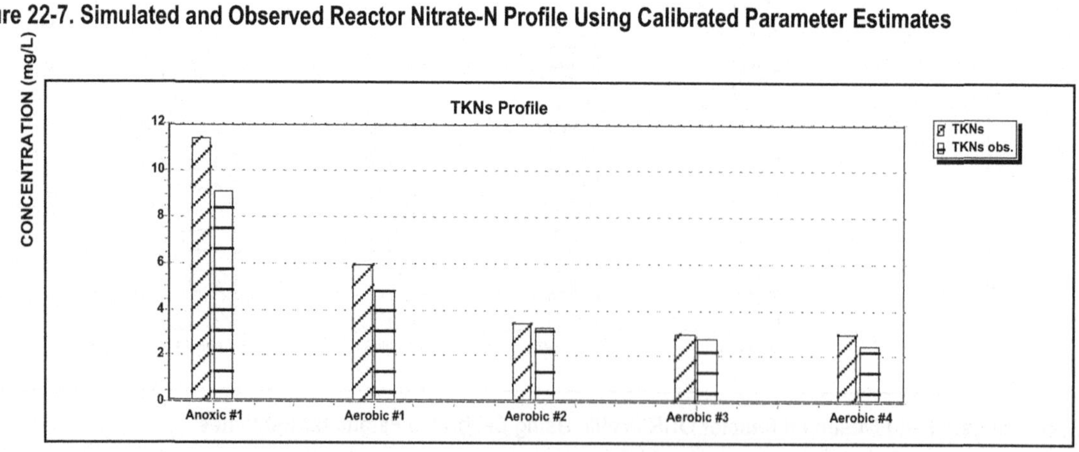

Figure 22-8. Simulated and Observed Reactor Soluble TKN Profile Using Calibrated Parameter Estimates

22.4 Summary

This chapter illustrates model calibration using data from two closely controlled experimental systems. The first system was a bench scale, complete-mix activated sludge process with one aerated bioreactor. This case study provided an illustration of a step-wise approach to calibration using long-term steady state data. The key steps involved were:

- Calculation and input of the long term average total COD and TKN concentrations.
- Adjustment of the soluble unbiodegradable fractions of COD (f_{SI}) and TKN (f_{SNI}) to match the effluent soluble COD and TKN concentrations.
- Adjustment of the clarifier solids removal rate to match effluent total COD (and suspended solids) concentrations.
- Estimation of the particulate unbiodegradable fraction of the COD (f_{XI}) by matching simulated and observed reactor MLVSS concentrations.

The second case study was a pilot scale process with five equal-sized bioreactors in series, and with an unaerated anoxic zone in the first tank. This case followed a similar calibration approach. While the first example was quite insensitive to the value of the influent soluble readily biodegradable COD fraction (f_{SS}), the second example illustrated the dramatic impact of f_{SS} on the predicted denitrification performance of the system. As a result, the predicted bioreactor nitrate profile could be used as a cross check on the estimated value of f_{SS}.

These case studies primarily are an illustration of the determination of influent wastewater characteristics using long-term operating data. Neither of these examples had the appropriate data to estimate kinetic parameters such as nitrifier growth rates, which would require separate testing as presented elsewhere in this manual.

22.5 References

Ekama, G.A., and G.v.R. Marais. 1978. *The dynamic behaviour of the activated sludge process*, University of Cape Town Research Report W 27. Rondebosch, South Africa: Department of Civil Engineering, University of Cape Town.

Gibson, J.D. 1991. *Activated sludge system modeling using the IAWPRC model.* M.Eng. Thesis, McMaster University, Hamilton, ON, Canada.

CHAPTER 23.0

EXAMPLES OF MODEL CALIBRATION IN MUNICIPAL PLANTS: AEROBIC SYSTEM

23.1 Background

This chapter provides an example of the calibration of a simulator-based activated sludge model for a conventional aerobic municipal activated sludge treatment plant. The plant is the Hastings Wastewater Treatment Plant (WWTP), which is operated by Minneapolis-St. Paul Metropolitan Council Environmental Services (MCES). The approach in this calibration exercise was to:

- Collect physical data on the full-scale plant such as tank sizes and flow configurations;
- Compile twelve months of influent, effluent and operational data from a typical year; and
- Calibrate an activated sludge model to measurements of primary and secondary effluent characteristics, mixed liquor suspended solids concentrations and thickened solids concentrations.

In this case study, only data from routine plant monitoring were available for model calibration. As a result, this case provides an example of calibration using typical routine monitoring data, the information that can be obtained from such an exercise and the difficulties that can be encountered. The objective of the simulation study, when it was originally conducted, was to provide an initial investigation of plant capacity during an area-wide sewage treatment capacity assessment.

23.1.1 Plant Description

The plant liquid train consists of two primary clarifiers in parallel, followed by four aeration tanks in parallel with two passes per aeration tank. Two circular secondary clarifiers follow the aeration tanks. Primary and waste activated sludge are combined before gravity thickening. Thickened sludge is anaerobically digested. Table 23-1 lists the tank volumes at the plant.

Table 23-1. Dimensions of Unit Tankage at the Aerobic System Case Study Plant

Process	Value
Primary Clarifiers (Circular)	
Diameter [ft]	60
Depth [ft]	11.4
Area, per unit [ft^2]	2,827
Number of units	2
Total primary clarifier area	5,654
Total volume [ft^3 (million gal)]	64,466 (0.482)
Aeration Tanks (2-pass)	
Volume, per pass [ft^3 (million gal)]	18,720 (0.140)
Volume, per tank [ft^3 (million gal)]	37,440 (0.280)
Number of 2-pass tanks	4
Total volume [ft^3 (million gal)]	149,760 (1.12)
Secondary Clarifiers (Circular)	
Number	2
Diameter [ft]	60
Depth [ft]	14
Surface area, each [ft^2]	2,827
Total surface area [ft^2]	5,655
Total volume [ft^3 (million gal)]	79,168 (0.592)
Gravity Thickener (Circular)	
Number	1
Diameter [ft]	45
Depth [ft]	12
Surface area [ft^2]	1,590
Volume [ft^3 (million gal)]	19,085 (0.143)

23.2 Calibration Approach

This section gives an overview of the approach used to calibrate a simulator for this plant.

23.2.1 Simulator Process Flow Configuration

Physical characteristics of the system, such as aeration tank volumes and settler dimensions, were defined in the simulator. Unless information is available on uneven flow splits between parallel unit processes, it is typical when simulating a plant to combine multiple parallel processes into one unit with a total volume and surface area equal to the individual units.

Figure 23-1 shows the simulator configuration of the plant. The simplifications summarized below explain the actual plant versus the simulated configuration:

- Two primary settling tanks were combined into one (2xPST in Figure 23-1).
- Plug-flow activated sludge tanks were modeled as four complete-mixed tanks in series.
- Three parallel aeration trains were modeled as one train with an equivalent volume of three trains (i.e., 3xA1, 3xA2, 3xB1, 3xB2 in Figure 23-1).
- Two final settling tanks were combined into one.
- The primary and secondary digesters were combined into one anaerobic digester (2 x Digester in Figure 23-1).

Although there are four parallel aeration tanks, only three were operating during the calibration period. Clarifier units in the simulations were modeled as ideal separators i.e., a solids removal and underflow rate is specified, and the simulator conducts mass balances on individual components entering the clarifier unit to determine effluent and underflow concentrations. The configuration also shows a digester element. Although the commercial simulator used has a digester model, there were no digester effluent data available so calibration of this model was not part of the exercise. This was not an important issue in this case because there were no streams (such as supernatant) returned to the process from the digester.

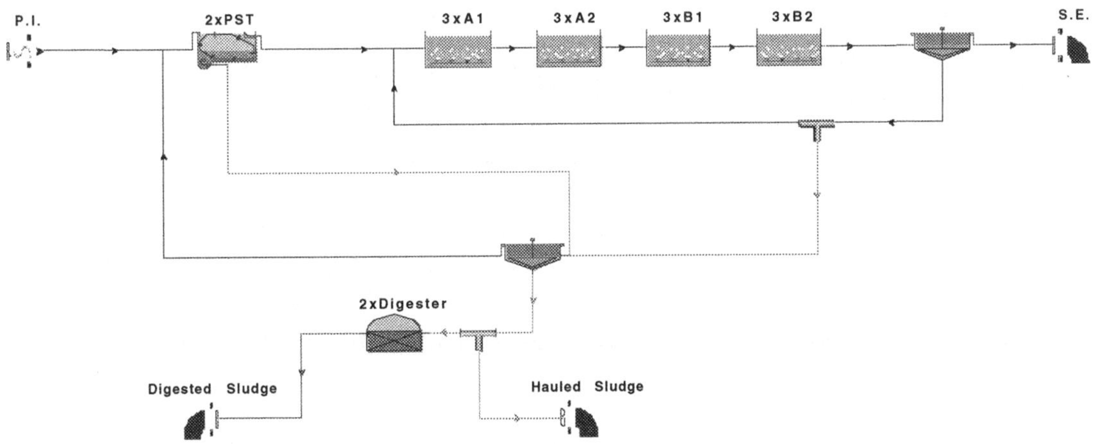

Figure 23-1. Simulator Configuration of the Aerobic System Case Study Plant

23.2.2 Data Collection

A recent year considered to represent typical operation was used as the basis for calibrating the simulator for this plant. The following information on plant operation and measured inputs was used to drive the simulation:

- Primary influent flow,
- Number of trains in service,
- Primary influent $cBOD_5$,
- Primary influent TKN, and
- Temperature.

Primary influent flow is plotted in Figure 23-2. There were 3 trains in service during the calibration period. Measured influent $cBOD_5$, TKN and ammonia-N concentrations are shown in Figures 23-3 and 23-4. The concentration measurements were all based on 24-hour composite samples. Process temperature is plotted in Figure 23-5.

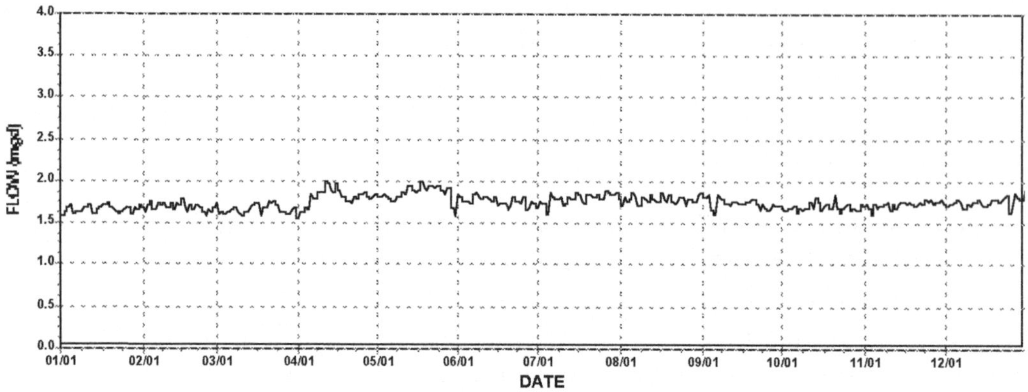

Figure 23-2. Calibration Period Measured Primary Influent Flow

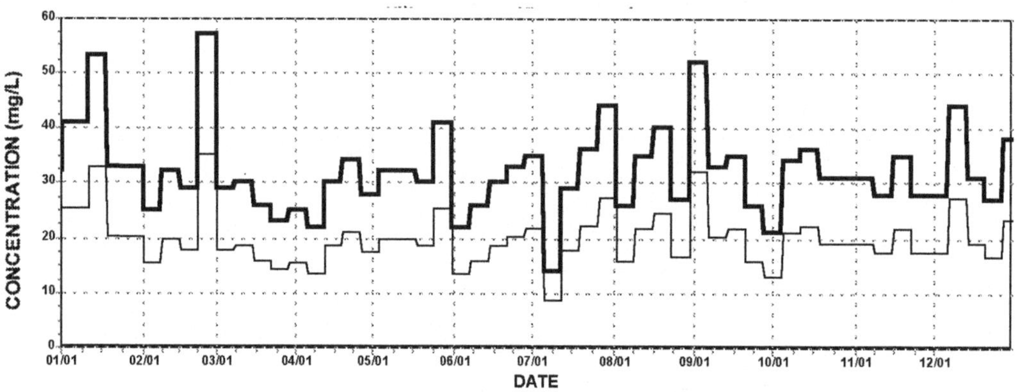

Figure 23-3. Calibration Period Simulated (Solid Line) and Measured (Data Points) Primary Influent $cBOD_5$

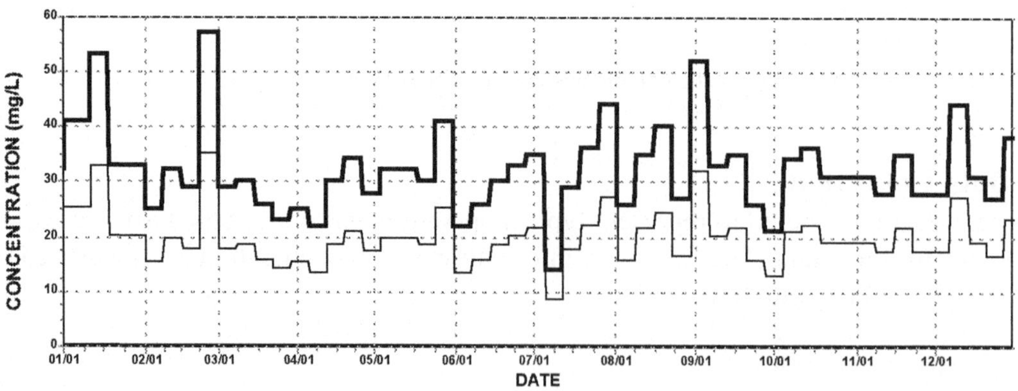

Figure 23-4. Calibration Period Measured Primary Influent TKN (Dark Solid Line) and Ammonia-N (Light Solid Line)

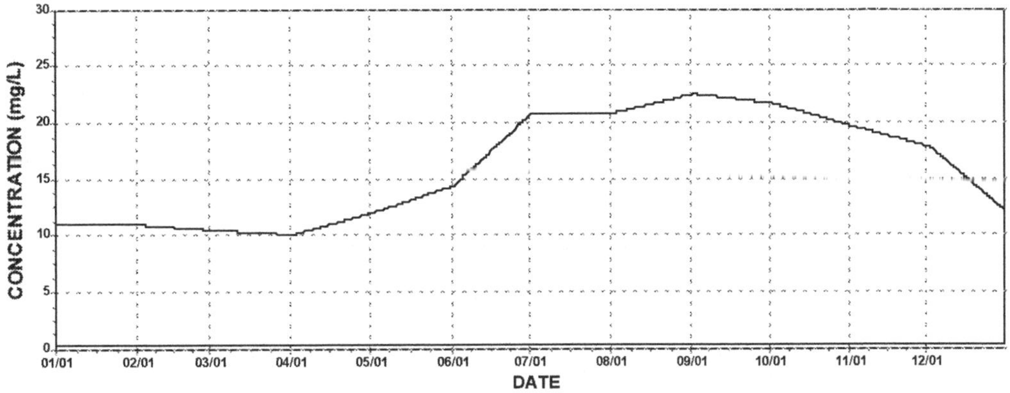

Figure 23-5. Calibration Period Measured Process Temperature

Note that $cBOD_5$ is not a direct input, but is calculated in the simulator based on the COD component concentrations. However, influent COD was not a routine measurement at this plant. To obtain wastewater characteristics for modeling, a limited number of parallel COD and $cBOD_5$ measurements were conducted. These measurements initially indicated a COD/ $cBOD_5$ ratio of 2.35. This is a relatively high value and could be indicative of a high fraction of unbiodegradable particulate COD (X_I) in the influent (see Chapter 6.0). However, a separate testing program by the plant operating authority had indicated that the addition of nitrification inhibitor to influent samples also inhibited the heterotrophs. The appropriate adjustment found in this study was to divide the measured influent $cBOD_5$ concentrations by 0.84. This adjustment resulted in a COD/$cBOD_5$ ratio of 1.97, which is more typical for a municipal primary influent. The COD component fractions could then be adjusted to give the measured COD/$cBOD_5$ ratio (see Appendix B). The resulting influent COD fractions are shown in Table 23-2. The measured volatile fraction of the gravity thickener underflow solids was used to indicate the concentration of ISS in the influent. The estimated ISS concentration, along with the particulate COD portions of the influent allowed an estimate of the primary influent TSS concentration (Figure 23-6). The calculated primary influent COD concentrations are shown in Figure 23-7.

Table 23-2. Summary of Influent Cod Fractions

COD Fraction	Symbol	Value
Readily biodegradable	f_{SS}	0.20
Soluble unbiodegradable	f_{SI}	0.08
Particulate unbiodegradable	f_{XI}	0.11
Fraction of slowly biodegradable COD that is particulate	f_{XS}	0.75

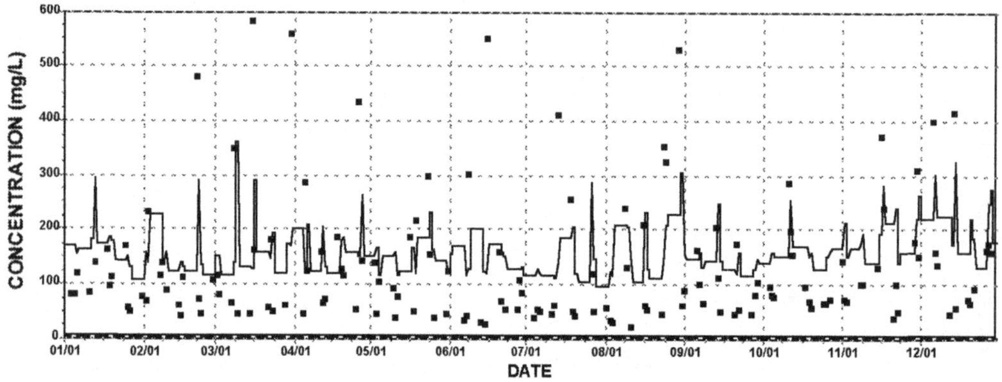

Figure 23-6. Calibration Period Measured (Data Points) and Simulated (Solid Line) Primary Influent TSS

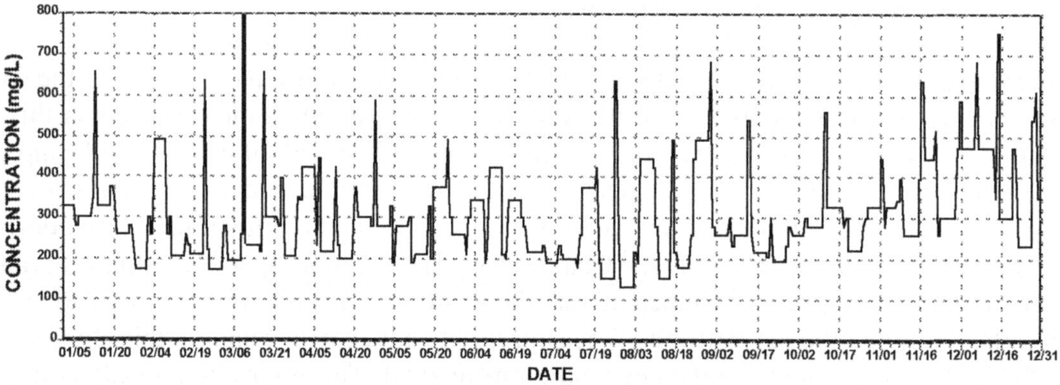

Figure 23-7. Calibration Period Calculated Primary Influent COD Concentration

After specifying the inputs, calibration proceeded by adjusting the primary clarifier TSS removal, WAS flow rate and gravity thickener underflow rate until simulated process and effluent variables were within reasonable agreement with measurements. There was not sufficient information available to make adjustments to kinetic parameters. For example, effluent ammonia concentrations based on 24-hour composite samples indicated year-round nitrification. Therefore, the value used for the nitrifier maximum specific growth rate was the default value in the simulator.

23.3 Calibration Results

The measurements available for calibration were:

- Primary effluent TSS (Figure 23-8);
- Primary effluent cBOD5 (Figure 23-9);
- MLSS (Figure 23-10);
- Effluent cBOD5 (Figure 23-11);
- Effluent TSS (Figure 23-12);

- Effluent TKN (monthly averages) (Figure 23-13);
- Effluent NH3-N (Figure 23-14);
- Effluent TP (Figure 23-15);
- Gravity thickener underflow TSS (monthly averages) (Figure 23-16); and
- Gravity thickener underflow VSS (monthly averages) (Figure 23-16).

All concentration measurements were based on 24-hour composite samples, except where indicated.

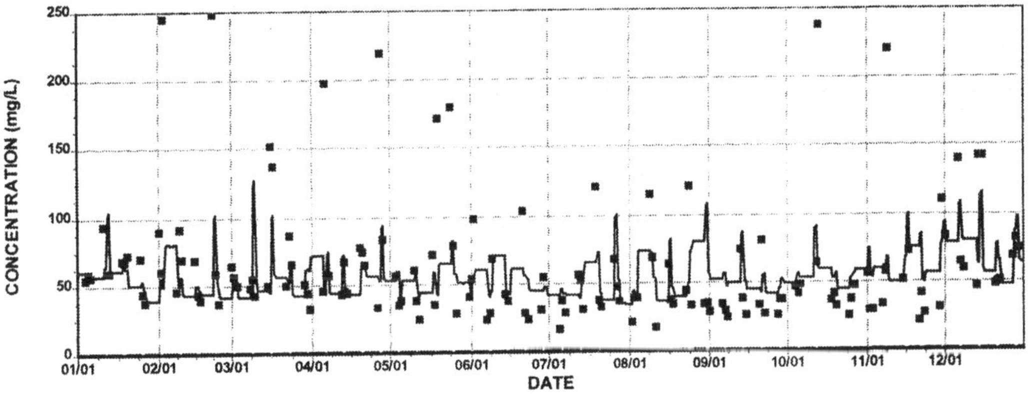

Figure 23-8. Measured (Data Points) and Simulated (Solid Line) Primary Effluent TSS

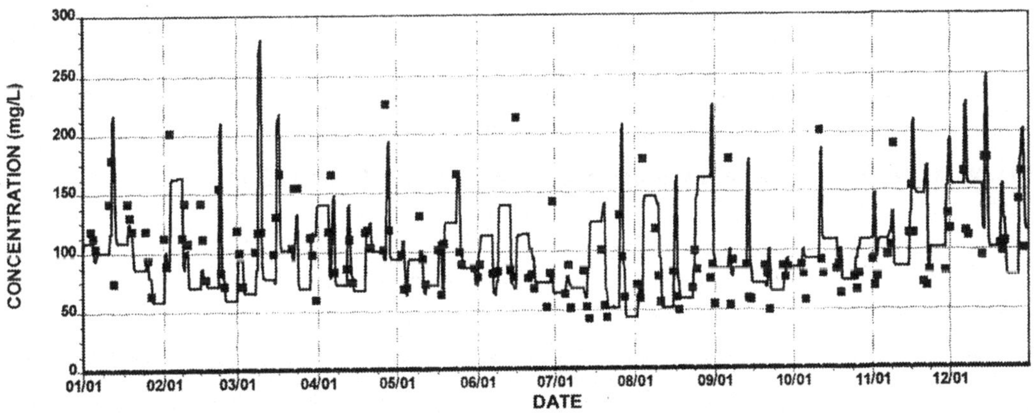

Figure 23-9. Measured (Data Points) and Simulated (Solid Line) Primary Effluent cBOD$_5$
(Note that the measured cBOD$_5$ has been corrected for the effect of nitrification inhibitor)

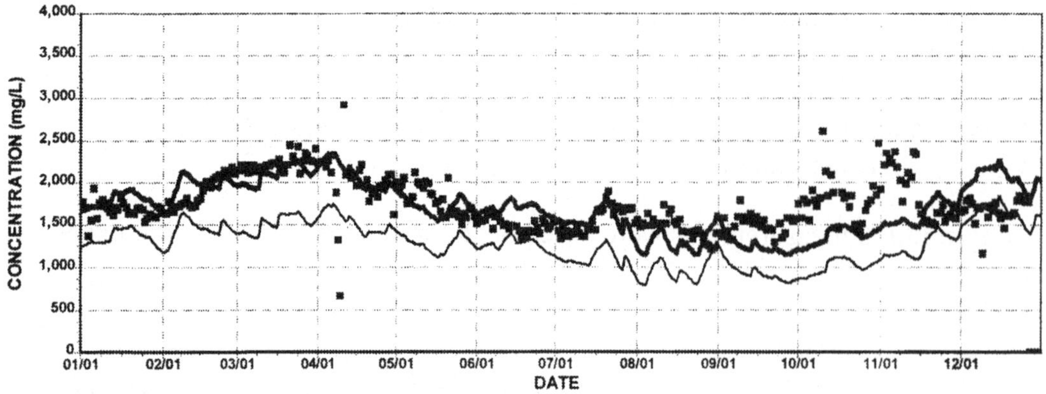

Figure 23-10. Measured (Data Points) and Simulated (Dark Solid Line) MLSS and Simulated MLVSS (Light Solid Line) as Monitored in the Last Aeration Tank

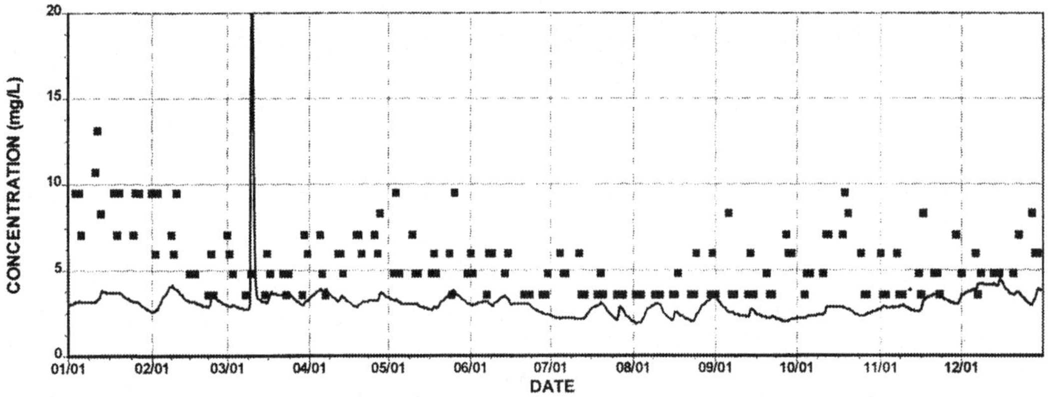

Figure 23-11. Measured (Data Points) and Simulated (Solid Line) Secondary Effluent $cBOD_5$

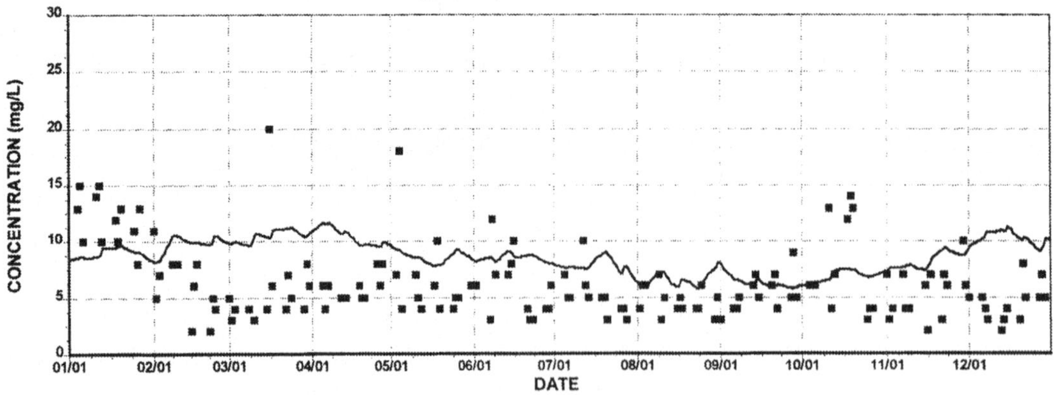

Figure 23-12. Measured (Data Points) and Simulated (Solid Line) Secondary Effluent TSS

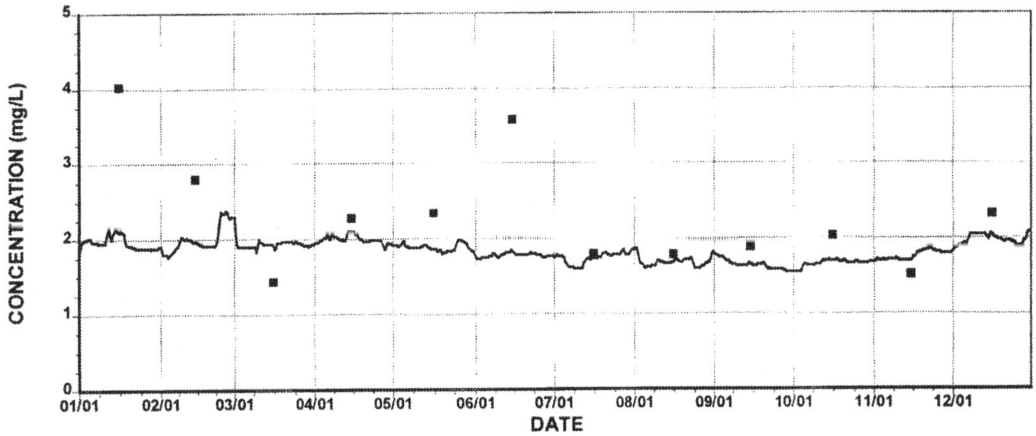

Figure 23-13. Measured (Data Points) and Simulated (Solid Line) Secondary Effluent TKN
(Note that the measured TKN values are monthly averages)

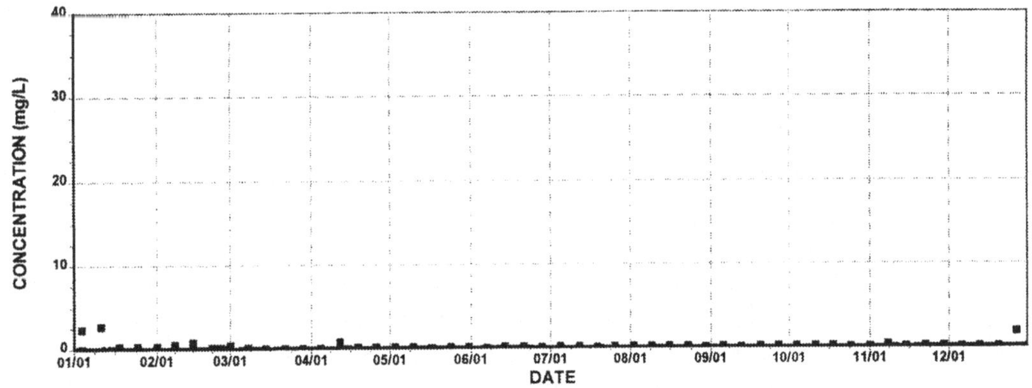

Figure 23-14. Measured (Data Points) and Simulated (Solid Line) Secondary Effluent Ammonia-N

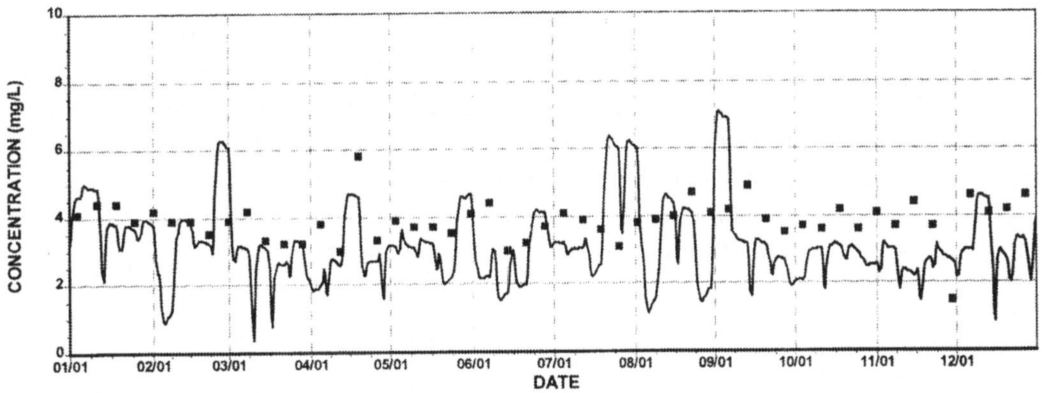

Figure 23-15. Measured (Data Points) and Simulated (Solid Line) Secondary Effluent Total Phosphorus

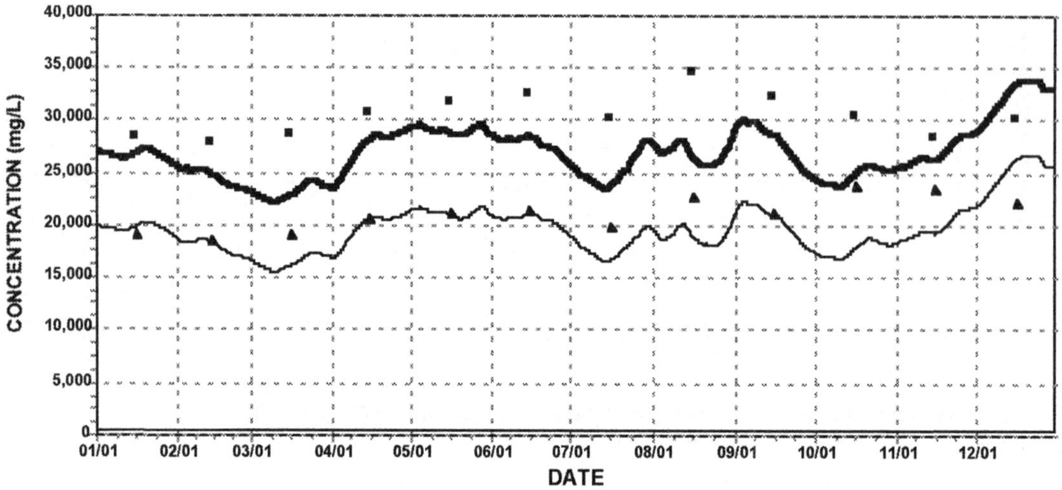

Figure 23-16. Gravity Thickener Underflow Measured (Square Data Points) and Simulated (Dark Solid Line) TSS, Measured (Triangle Data Points) and Simulated (Light Solid Line) VSS

23.4 Calibration Discussion

Although there was a limited data set for modeling, simulation was still a useful process analysis tool for this plant. Through the simulation results, a number of inconsistencies in the plant database were found. Before application of the model, assumptions were needed to overcome these data deficiencies. The simulation analysis was also a useful tool in evaluating these assumptions. These issues are discussed in this section.

23.4.1 Effluent $cBOD_5$

The simulated effluent $cBOD_5$ concentration displays a peak in the third month of the simulation (Figure 23-11). An examination of the influent data indicates that there was a very high influent $cBOD_5$ for one day at this point (Figure 23-3). Such a significant amount of soluble biodegradable material in the simulated effluent concentration (note that there was not a coincident increase in effluent TSS) often indicates that a "nutrient limitation" is being experienced in the model. As there was not a similar increase in the measured effluent $cBOD_5$, it is likely that the high influent $cBOD_5$ that is causing this response is a bad data point.

23.4.2 Plant Loading and Sludge Production

Initial simulations indicated inconsistencies in overall sludge production and plant loading. As a result, a "rule-of-thumb" analysis of expected BOD loading and overall sludge production was conducted based on averaged data for the year. This analysis is summarized in Table 23-3. Expected BOD loading rates for the population serviced (3,012 lb BOD/day) are 40 percent higher than the reported values (2,167 lb BOD/day). The reported sludge production rate (primary and waste sludge before digestion) was 3,994 lb TSS/day. This is almost twice the rate expected from the reported loading rate (2,167 lb TSS/day), based on an estimated production of 1 lb TSS/lb BOD.

Table 23-3. "Rule-Of-Thumb" Estimates of Loading and Sludge Production Rates

Loading		
Population served		17000
Estimated Domestic Loading		**0.17** lb BOD per capita per day
Industrial Load		240 lb COD/d
	COD/BOD	1.97
		121.8 lb BOD/d
Estimated Total Loading		**3012** lb BOD/d
Reported Loading		
	Average Conc (1999)	149 mg BOD/L
	Average Flow (1999)	1.744 mgd
	Reported Loading	**2167** lb BOD/d
Estimated flow		
	Assumed per capita	**100** gal/capita/d
	Industrial	0.093
	Est. Total	**1.793** mgd
	Actual flow (1999)	**1.744** mgd
Sludge Production		
Assumed Unit Sludge Production		**1.0** lb TSS/lb BOD
Estimated Total Sludge Production		
	Based on reported load	**2167** lb TSS/d
	Based on estimated load	**3012** lb TSS/d
Reported Sludge Production Calc.		
	1999 Thickener Underflow	0.0154 mgd
	Conc. (1999)	30492 mg/L
	Effluent Flow	1.73 mgd
	Effl. TSS	6.2 mg/L
	Total Reported Sludge	**3994** lb TSS/d

The influent flow data (approximately 1.75 mgd) appear reasonable based on the population estimate of 17,000 and a per capita contribution of 100 gal/day, plus the industrial contribution. Therefore it would appear that the influent BOD measurements might be underestimating the true concentrations. Possible reasons are a significant difference in the "0.84 BOD factor" for this wastewater or non-representative influent sampling.

23.4.3 RAS and WAS Flow Rates

Plant records reported RAS and WAS TSS concentrations double that of the reactor MLSS concentration, implying a 100 percent RAS recycle. However, plant records report the RAS rate as 50 percent. Furthermore, simulations (based on 100 percent RAS, i.e., low RAS TSS) strongly suggest that the WAS rate was approximately half of the reported value. (Simulations based on a 50 percent RAS rate and the reported WAS flows resulted in predictions of mixed liquor concentrations that were approximately one third of the observed MLSS). The reported waste flow measurements were adjusted by subtracting a constant value from each of the reported values. This resulted in assumed WAS rates approximately half the reported values. Applying this adjustment to the wastage rates resulted in a reasonable plant calibration. The need for this adjustment possibly is a consequence of the reported plant load being significantly underestimated. However, if the load is not underestimated the discrepancy could be caused by:

- A constant bias in the flowmeter calibration,
- Consistently wasting at a time in the day when the RAS TSS concentration is lower than average, and
- Wasting over a short period at a flow rate that is high enough to draw down the sludge blanket in the final settlers and cause a "rat-holing" effect through the sludge blanket.

The actual and adjusted WAS flows and the WAS TSS concentration are shown in Figure 23-17. The estimated average SRT using the adjusted WAS rates was 19 days (not including the final clarifiers).

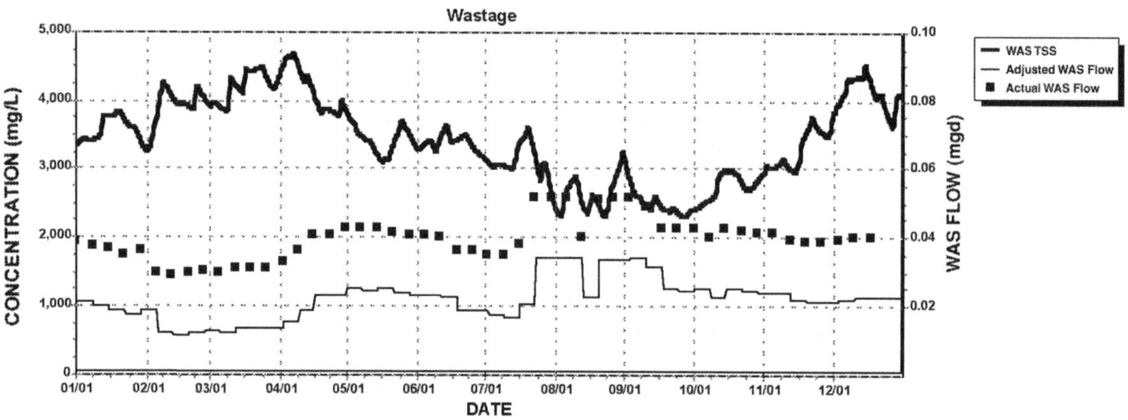

Figure 23-17. Actual and Adjusted WAS Flows and WAS TSS Concentrations

23.5 Conclusions

This case study provided an illustration of model calibration for a conventional activated sludge WWTP. The calibration data set was limited to typical routine monitoring data, and there was insufficient information in the plant response to adjust kinetic parameters such as the nitrifier growth rate. However, the calibration study was useful for finding inconsistencies in the plant data, and for evaluating assumptions for addressing the data problems. For example, sludge wastage rates were adjusted to better reflect the sludge production expected based on the measured process loading rates. After this adjustment, the model was found to give reasonable predictions of process behavior. However, reliable estimates of plant capacity would require additional testing such as measurement of the nitrifier maximum specific growth rate. Also, the discrepancies in plant loading and sludge production estimates would need to be investigated.

CHAPTER 24.0

Examples of Model Calibration in Municipal Plants: Anoxic System for Nitrogen Removal

24.1 Background

This chapter provides an example of the calibration of a simulator-based activated sludge model for biological nitrogen removal in a municipal activated sludge treatment plant. This investigation was conducted at the 91st Avenue Multi-Cities Wastewater Treatment Plant (WWTP). The City of Phoenix was an active partner in this WERF project; the data for this calibration example were collected solely for this WERF project. The approach in this activity was to:

- Gather data on the full-scale plant and the composition of individual waste streams;
- Calibrate an activated sludge model simulator to measured effluent characteristics; and
- Apply the calibrated model to estimate the maximum specific nitrifier growth rate.

24.1.1 Plant Description

The 91st Avenue Multi-Cities WWTP in Tolleson, Ariz. is owned by the Sub-Regional Operating Group cities of Phoenix, Glendale, Scottsdale, Mesa and Tempe, and operated by the City of Phoenix. Plant capacity was recently increased to 179 MGD. Biological nitrogen control is practiced year-round at the plant. The liquid stream facility is split into six sub-plants: 1A, 1B, 2A, 2B, 3A and 3B. All were converted to nitrogen control during the period 1992 to 1996. Each sub-plant is a two-stage nitrogen removal activated sludge facility with a mixed liquor return of approximately twice the primary effluent flow. The 91st Avenue WWTP is expected to receive an APP (aquifer protection permit) in 2002. The APP will have a nitrite-nitrogen limit of 1.0 mg/L, a nitrate-nitrogen limit of 10.0 mg/L and a total nitrogen limit of 10 mg/L, based upon a five month rolling average. The facility has operated to meet the APP limits since the last sub-plant was converted to nitrogen removal in 1996.

Primary effluent is directed to a four-pass aeration tank; the first one and one-third passes contain anoxic cells and the subsequent two and two-thirds passes are aerated. Mixed liquor is returned from the middle of the fourth pass to the head of the first pass as shown in Figure 24-1.

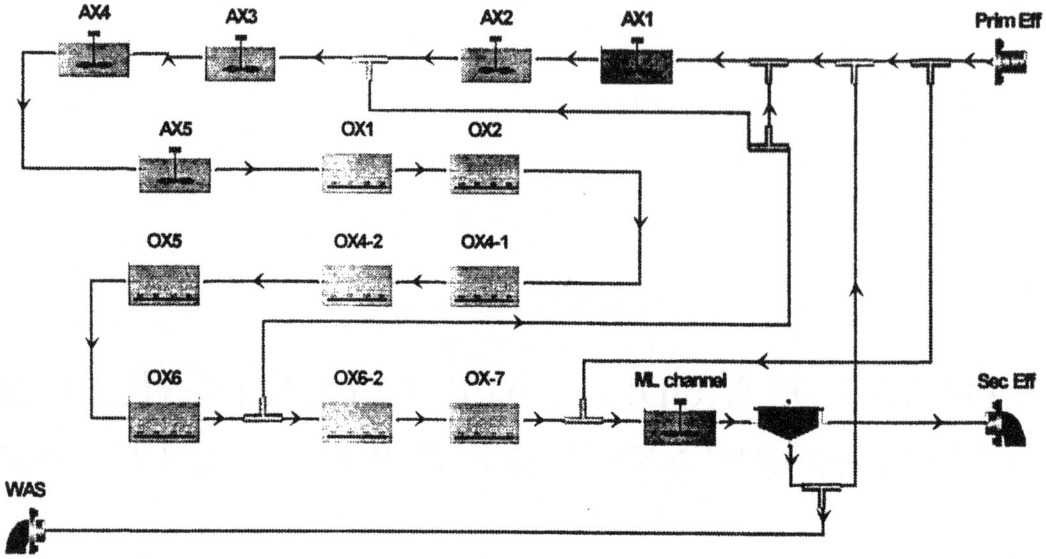

Figure 24-1. Simulator Configuration of the 91st Avenue WWTP

Advantage is taken of the volume that exists in the channels that convey mixed liquor to the secondary clarifiers to carry out a second stage of denitrification. Approximately 10 percent of the influent is fed to the end of the fourth pass to facilitate denitrification in the mixed liquor channel.

Table 24-1 lists the liquid volumes of the aeration tanks, and the dimensions of the secondary clarifiers. The historical database indicated that effluent ammonia concentrations ranged from about 1 mg/L in the summer to about 3 mg/L in the winter and effluent nitrate ranged from <1 mg/L in the summer to about 6 mg/L in the winter.

Table 24-1. Dimensions of Unit Tankage at the 91st Avenue WWTP

Process	Value
Aeration Tank (4-pass)	
Volume, per pass (m^3 (million gal))	25,630 (0.905)
Total volume (m^3 (million gal))	102,520 (3.62)
Clarifiers (rectangular)	
Number	8
Length (m (ft))	58 (190)
Width (m (ft))	12 (40)
Side water depth (m (ft))	3.9 (12.75)
Surface area, each (m^2 (ft^2))	706 (7,600)
Total surface area (m^2 (ft^2))	5,648 (60,800)

24.2 Calibration

This section gives an overview of the data that were collected to implement model calibration.

24.2.1 Process Flow Schematic

Physical characteristics of the system such as aeration tank volumes and settler dimensions were defined in the simulator, allowing the graphical interface to produce a process flow schematic of the aeration tank and clarifiers as shown in Figure 24-1. Flow rates for primary effluent, return activated sludge, internal mixed liquor recycles and sludge wastage were specified, as were dissolved oxygen concentrations in aerated zones.

24.2.2 Wastewater Characterization

Typically, plant performance data for a period of up to three years are reviewed for correctness, entry errors, extraordinary data or outliers, which are queried and screened out if appropriate. The data are then reduced to monthly average values, which are examined for trends and extremes in order to obtain a sense of plant performance. In particular, values of effluent quality characteristics are sought to exemplify typical summer and winter operating conditions, that is, operation under average dry weather conditions and low temperature conditions in order to establish the boundaries with which to form the basis of a steady-state model. These can include daily data on influent and effluent BOD, TSS, ammonia, nitrate and phosphorus, as well as operational data on solids balances, SRT, flow, temperature and SVI. They may be used to define initial conditions for the model. An approximate estimate of COD and TKN partitioning in the primary effluent stream may be made from average monthly data on wastewater characteristics. However, not all fractions of COD and TKN may be calculated from conventional plant data. Consequently, a comprehensive wastewater sampling program usually is undertaken in which COD, TKN and total phosphorus (TP) partitioning are examined for influent and effluent streams.

To address this need, a special set of data was collected and used in concert with the normally collected operational data. The special sampling work involved the collection of 13 consecutive days of 24-hour composite samples of secondary influent, anoxic effluent, mixed liquor and secondary effluent. In addition, one day of 2-hour grab samples were collected at the same locations defined above and at five additional locations along the length of the 4-pass aeration tank (sub-plant 3A) as shown in Figure 24-2. The samples were analyzed for the range of characteristics described above.

Analysis of all wastewater characteristics except readily biodegradable soluble COD (RBCOD) was according to Standard Methods (1995). RBCOD was measured by the procedure described by Mamais et al. (1993) as described in Chapter 7.0. To prevent actively respiring organisms from metabolizing ammonia and nitrate in the samples, the samples were filtered at the point of sampling for both composite and grab samples.

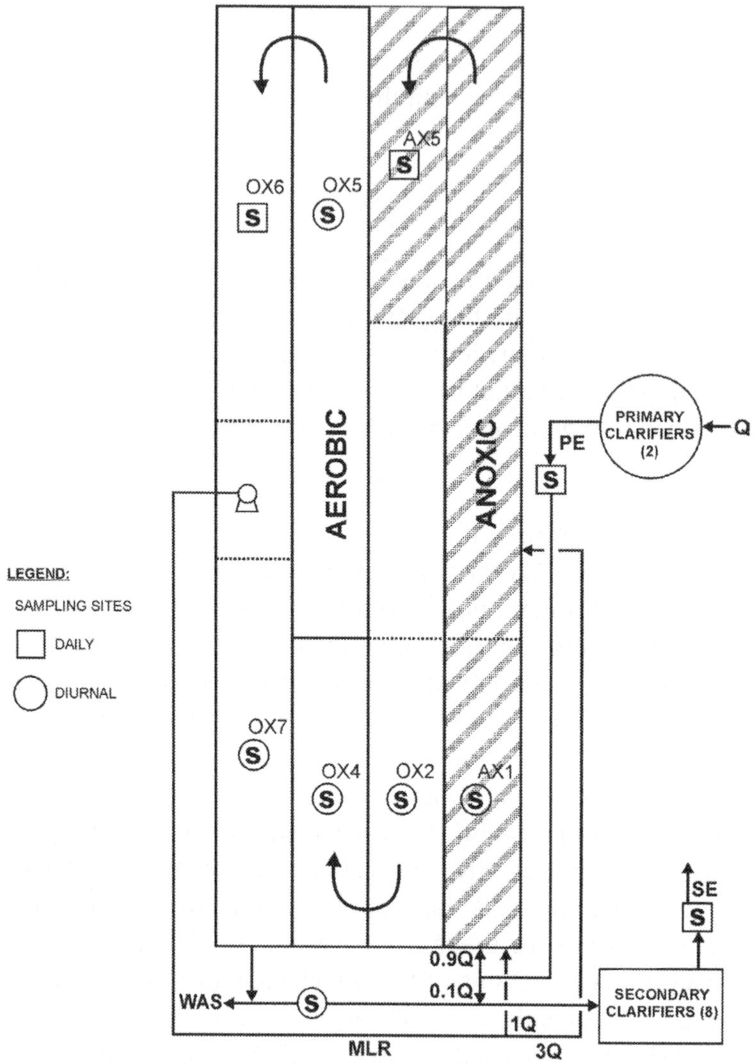

Figure 24-2. Sampling Locations in the 91st Avenue WWTP (24-hour Composite Samplers are Designated by the Letter "S" in a Square, Grab Sampling Locations by the Letter "S" in a Circle)

Table 24-2. Summary of RAS Characterization Data

Day	Flow (mgd)	TSS (mg/L)	VSS (mg/L)
5/10/2001	7.34	9,470	7,115
5/11/2001	7.48	9,640	7,220
5/12/2001	8.05	5,300	3,830
5/13/2001	7.57		
5/14/2001	7.52	10,600	7,880
5/15/2001	7.34	11,260	8,360
5/16/2001	7.20	9,280	6,940
5/17/2001	7.27	9,000	6,700
5/18/2001	7.21	9,080	6,660
5/19/2001	7.85	8,900	6,400
5/20/2001	7.64	9,200	5,980
5/21/2001	7.16	10,100	7,440
5/22/2001	7.27	10,360	7,680
Average	7.45	9,349	6,850

Daily composite wastewater characterization data for the liquid streams are summarized in Tables 24-2 and 24-3. RAS flow characterization data are included in Table 24-2.

Table 24-3. Summary of Wastewater Characterization Data for Primary Effluent and Anoxic Effluent (AX)

Primary Effluent

Day	Flow	TSS	VSS	COD	sCOD	ffCOD	BOD5	sBOD5	TKN	sTKN	NH3-N	NO2-N	NO3-N	TP	PO4-P	Alk
5/10/2001	15.0	120	90	322	137	94	123	51	33	32	26			5.6	3.7	272
5/11/2001	15.2	93	76	302	108	87	125		33		27			5.5	4.6	282
5/12/2001	16.0	94	72	324	130	88	129	52	33	25	27			7.0	3.7	284
5/13/2001	15.3	120	89	350	111	91	152	60	30	23	25			7.1	4.1	268
5/14/2001	15.4	120	94	346	154	84	165	58	36	30	26			6.8	3.8	270
5/15/2001	15.0	108	86	333	120	85			31.3	28	26			7.7	5.1	272
5/16/2001	14.7	98	82	326	99	89			32.5	28.5	26			6.7	5.2	260
5/17/2001	14.9	124	96	280	103	86			32.3	26.5	27			6.9	5.1	276
5/18/2001	14.8	100	82	279	94	86	136	62	27	22	24			6.8	5.4	262
5/19/2001	15.5	94	68	317	120	98	63	60	27	22	27					260
5/20/2001	15.1	122	88	322	106	78	136	59	30	23	23			7.1	4.1	264
5/21/2001	15.2	108	80	349	110	78	161	61	27	22	22			6.5	4.9	256
5/22/2001	14.7	96	84	341	106	80	140	51	28	20	22			6.6	5.7	264
Average	15.1	107	83.6	322	115	86.5	133	57.1	30.8	25.2	25.2			6.7	4.6	268

AX5 - Start of Aeration

Day	Flow	TSS	VSS	COD	sCOD	ffCOD	BOD5	sBOD5	TKN	sTKN	NH3-N	NO2-N	NO3-N	TP	PO4-P	Alk
5/10/2001					27				9.3		7.8	0.2	0.8		8.5	210
5/11/2001					49				7.9		7.6	0.2	0.4		9.3	216
5/12/2001					46				9.6		7.8	0.3	0.4		9.1	222
5/13/2001					41				7.9		7.4	0.2	0.5		10.5	206
5/14/2001					47				8.8		7.9	0.2	0.7		10.1	206
5/15/2001					55				9.2		7.9	0.3	1.1		9.7	204
5/16/2001					52				9.9		6.5	0.2	1.2		7.7	198
5/17/2001					48				7.9		6.9	0.4	1.6		6.8	195
5/18/2001					57				8.5		6.5	0.2	1.4		8.4	195
5/19/2001					45				10.9		8.7	0.2	1.2		8.3	210
5/20/2001					41				8.9		7.8	0.2	1.4		9.6	202
5/21/2001					47				8.2		7.4	0.2	1.6		9.9	200
5/22/2001					46				11.3		7.5	0.2	1.5		8.6	200
Average					46				9.1		7.5	0.2	1.1		9.0	205

Note: Flow in mgd units, all other wastewater characteristics in mg/L units. Alkalinity in mg/L as $CaCO_3$

Table 24-4. Summary of Wastewater Characterization Data for OX6 and Secondary Effluent

OX6 - Prior to mixed liquor return

Day	Flow	TSS	VSS	COD	sCOD	ffCOD	BOD5	sBOD5	TKN	sTKN	NH3-N	NO2-N	NO3-N	TP	PO4-P	Alk
5/10/2001		3,370	2,530		26						1.7	2.8	5.2		1.6	160
5/11/2001		3,640	2,720	3,950	32						1.0	2.4	6.0		2.2	170
5/12/2001		3,860	2,690	4,695	31						2.0	2.8	5.9		2.3	170
5/13/2001					31						1.8	1.5	5.5		2.4	164
5/14/2001		3,860	2,840	4,820	41						1.4	1.6	5.8		1.5	160
5/15/2001		3,610	2,620	4,955	55						1.1	1.9	6.2		2.2	160
5/16/2001		3,100	2,260	3,790	42						0.8	1.8	6.2		1.5	154
5/17/2001		3,500	2,600	3,760												
5/18/2001		2,490	1,840	2,715	50						1.1	1.8	6.5		2.9	155
5/19/2001		3,410	2,460	3,856	41						2.0	1.5	6.7		2.4	175
5/20/2001		3,340	2,460	4,245	37						2.6	1.0	5.7		3.8	168
5/21/2001		4,450	2,510	3,940	39						1.3	1.1	6.5		2.5	165
5/22/2001		3,480	2,560	4,200	42						2.3	1.4	6.5		1.8	155
Average		3,509	2,508	4,084	38.9						1.6	1.8	6.0		2.2	163

Secondary Effluent

Day	Flow	TSS	VSS	COD	sCOD	ffCOD	BOD5	sBOD5	TKN	NH3-N	sTKN	NO2-N	NO3-N	TP	PO4-P	Alk
5/10/2001		12	8	35	29	29	17		2.6	1.1		0.8	4.1	3.8	1.9	180
5/11/2001		10	8	43	26	27	25		4.4	1.1	4.3	0.6	2.8	4.1	1.8	160
5/12/2001		5	4	27	21	24	8		2.5	1.9	2.2	0.8	3.6	3.9	2.1	206
5/13/2001		6	4	35	29	29	6	<2	3	2.3	2.8	1.0	2.4	4.6	2.1	184
5/14/2001		5	3	29	22	14			4.9	2.2	5.3	0.8	2.4	3.4	2.0	180
5/15/2001		5	3	37	28	18			2.45	2.2	2.4	0.7	2.8	3.5	2.6	178
5/16/2001		4	3	30	26	25			3.1	1.9	2.2	0.5	2.8	4.0	1.7	178
5/17/2001		6	5	30	30	27			3.15	2.2	2.7	0.6	3.6	3.3	2.1	178
5/18/2001		5	4	34	34	34	5	2	3.2	1.7	3.0	0.5	3.5	3.7	2.2	176
5/19/2001		5	3	27	24	22	5	<2	3.3	1.9	3.3	0.6	3.8			176
5/20/2001		6	4	29	23	18	6	<2	3.4	2.4	3.1	0.8	3.0	4.0	2.3	175
5/21/2001		5	4	33	28	22			2.9	2.2	2.6	0.8	2.9	3.4	2.2	178
5/22/2001		6	5	30	22	17	4	<2	2.4	1.3	2.1	0.5	3.3	3.4	1.8	178
Average		6.2	4.5	32.2	26.3	23.5	9.5	2.0	3.2	1.9	3.0	0.7	3.1	3.7	2	179

Note: Flow in mgd units, all other wastewater characteristics in mg/L units. Alkalinity in mg/L as $CaCO_3$

Table 24-5. Summary of Diurnal Wastewater Characterization Data for Primary Effluent and AT AX5 (Anoxic Effluent)

Primary Effluent - Diurnal

Time	Flow	TSS	VSS	COD	sCOD	ffCOD	TKN	NH3-N	NO2-N	NO3-N	TP	PO4-P	Alk
9:00 AM	7.82	46	32	300	114	96	37	31.4			6.3	4.9	312
11:00 AM	13.32	52	42	268	76	65	46	34.7			6.1	4.5	324
1:00 PM	18.31	78	66	318	88	76	40	36.6			6.0	4.6	310
3:00 PM	18.26	156	88	328	70	61	40	34			6.4	3.9	305
5:00 PM	17.54	88	67	307	92	83	39	31.7			6.3	4.5	300
7:00 PM	16.66	76	58	333	106	94	35	29.3			6.1	4.7	298
9:00 PM	17.18	89	69	358	111	95	35.5	28.5			7.3	5.1	295
11:00 PM	18.05	98	76	367	115	102	35.5	27.9			5.8	5.1	302
1:00 AM	18.22	84	64	336	108	95	37	32.2			7.3	4.9	312
3:00 AM	15.54												
5:00 AM	11.12												
7:00 AM	7.88												
Average	15.0	85	62	324	98	85	38.3	31.8			6.4	4.7	306

AX5 - Start of Aeration - Diurnal

Time	Flow	TSS	VSS	COD	sCOD	ffCOD	TKN	NH3-N	NO2-N	NO3-N	TP	PO4-P	Alk
9:00 AM					27		6.2	4.8	0.18	0.86		5.5	192
11:00 AM					27		7.8	5.7	0.19	0.96		8.2	196
1:00 PM					40		10.8	10.2	0.33	1.16		6.2	208
3:00 PM					53		14.3	12.6	0.47	1.08		8.3	210
5:00 PM					79		15.5	12.9	0.52	1.81		9.0	210
7:00 PM					54		11.7	10.8	0.49	1.76		9.6	208
9:00 PM								7.8	0.43	2.03		10.1	198
11:00 PM							8.4	7.0	0.32	1.99		10.6	205
1:00 AM					40		6.6	6.6	0.22	1.47		11.0	208
3:00 AM					42		8.2	6.1	0.2	1.53		11.0	204
5:00 AM					35		5.9	5.0	0.16	1.52		9.8	200
7:00 AM					31		5.2	4.3	0.19	1.26		7.5	210
Average					43		9.1	7.8	0.3	1.5		9.0	204

Note: Wastewater characteristics in mg/L units. Alkalinity in mg/L as $CaCO_3$

Table 24-6. Summary of Diurnal Wastewater Characterization Data for OX6 and Secondary Effluent

OX6 - Prior to mixed liquor return - Diurnal

Time	Flow	TSS	VSS	COD	sCOD	ffCOD	TKN	NH3-N	NO2-N	NO3-N	TP	PO4-P	Alk
9:00 AM					39			0.07	0.1	5.6		0.89	156
11:00 AM								0.08	0.19	6.07		1.14	156
1:00 PM					58			1.47	1.58	7.19		2.58	153
3:00 PM					44			4.16	1.84	7.33		3.24	156
5:00 PM					47			4.65	2.02	7.81		3.42	162
7:00 PM					43			2.19	1.72	8.27		3.72	148
9:00 PM					46			0.38	0.72	9.22		3.43	150
11:00 PM					50			0.13	0.54	8.46		3.72	165
1:00 AM					54			0	0.74	7.95		3.97	162
3:00 AM					37			0	0.12	7.67		3.32	159
5:00 AM					29			0.07	0.13	6.89		2.83	170
7:00 AM					19			0.02	0.1	5.84		1.5	166
Average					42.4			1.1	0.8	7.4		2.8	159

Secondary Effluent - Diurnal

Time	Flow	TSS	VSS	COD	sCOD	ffCOD	TKN	NH3-N	NO2-N	NO3-N	TP	PO4-P	Alk
9:00 AM	2.7	2.7	30	26	24			1.06	0.42	2.11	3.99	2.8	184
11:00 AM	2	2	31	23	21			2.66	0.414	1.326	4.6	2.68	190
1:00 PM	2.3	2.3	27	22	21			2.2	0.28	2.39	3.96	1.68	184
3:00 PM	1.5	1.5	27	21	23			1.85	0.27	4.57	3.59	1.55	174
5:00 PM	1.3	1.3	29	24	20			1.74	0.31	5.69	3.87	1.65	167
7:00 PM	<1	<1	26	22	21			1.72	0.25	5.99	3.33	1.75	164
9:00 PM	<1	<1	27	22	20			1.88	0.25	5.67	3.9	1.77	164
11:00 PM	3.8	2.3	29	22	22			1.63	0.24	4.75	4.24	1.84	168
1:00 AM	3.8	3.8	32	25	22			1.5	0.25	4.03	4.39	2.1	170
3:00 AM	3.3	3.3	28	25	20			1.41	0.28	3.16	4.68	2.16	174
5:00 AM	1.3	1.3	28	23	20			1.51	0.25	3.66	5.3	2.33	176
7:00 AM			27	23	18			1.71	0.36	2.77	4.46	2.47	188
Average	2.4	2.3	28.4	23.2	21.0			1.7	0.3	3.8	4.2	2.1	175

Note: Wastewater characteristics in mg/L units. Alkalinity in mg/L as $CaCO_3$

24.2.3 Model Calibration

24.2.3.1 Steady State Modeling

Following the set up of initial conditions for the model that was based on historical data, the primary effluent composite samples were used to calibrate a steady state activated sludge simulator, by comparing composite measured values of specific wastewater characteristics to predicted values for the same characteristics at each location where samples were taken.

For a well-calibrated model, there should be close correspondence between simulated and observed behavior. When major discrepancies appear between measured and predicted values for effluent characteristics or major operating variables, such as mixed liquor volatile suspended solids (MLVSS) and sludge yield, investigation of plant data is initiated to determine the cause of such discrepancies.

An inorganic suspended solids (measured as the difference between TSS and VSS) mass balance across the activated sludge system showed a mass balance closure of 87 percent after allowing for excess biological phosphorus uptake. Small adjustments were made in the measured sludge wasting rate and fraction of influent COD that is particulate unbiodegradable (f_{XI}) in order to achieve agreement between measured and estimated MLVSS values. The results of steady state model calibration are shown in Table 24-7.

Table 24-7. Summary of Steady State Modeling

Parameter	Measured	Assumed	Predicted
Primary Effluent			
COD (mg/L)	322		
ISS (mg/L)	23.8		
f_{SS}	0.20	0.24	
f_{SI}	0.08		
f_{XI}	0.25	0.14	
WAS flow (mgd)	0.161	0.165	
OX6			
MLVSS (mg/L)	2,508		2,527
MLSS (mg/L)	3,509		3,604
Temperature (deg C)	29		
Secondary effluent			
COD (mg/L)	32.2		37.0
SCOD (mg/L)	26.3		30.1
BOD (mg/L)	9.5		3.1
SBOD (mg/L)	<2		1.4
TSS (mg/L)	6.2		6.1
VSS (mg/L)	4.5		4.2
NH_3-N (mg/L)	1.9		1.8
NO_3-N (mg/L)	3.9		4.0
TKN (mg/L)	3.2		3.2
STKN (mg/L)	3.0		2.7
TP (mg/L)	3.8		1.0
SP (mg/L)	2.0		0.6
Model parameters			
μ_{AUT} @ 20 deg C (d^{-1})		0.9	
θ (μ_{AUT})		1.045	
b_{AUT} @ 20 deg C (d^{-1})		0.17	

There was reasonably good agreement between measured and predicted values for most effluent characteristics in terms of steady state model analysis. The only apparent discrepancy was in effluent phosphorus concentration data.

The major difficulty encountered in model calibration stemmed from a lack of internal mixed liquor return flow monitoring. Other critical limitations in calibration included a lack of measurement of flow splitting of the primary effluent to the aeration tank and to the mixed liquor channel feeding the secondary clarifiers, and the internal mixed liquor return flow split between the first and third anoxic cells in the first pass of the aeration tank. Both flow-splitting operations are manual and estimates of the flow split were made by operations staff based on visual observation of the valve settings. Sensitivity analysis showed that the latter flow split impacted the denitrification process the most. Adjustments were made to flow splits to bring estimated ammonia and nitrate concentrations in line with measured values. The suitability of these assumptions was then tested in a dynamic model the development of which is described below.

With respect to the apparent discrepancy with effluent phosphorus, soluble phosphate data in Table 24-3 indicate that some phosphorus release was occurring in the first pass based on an average of 4.6 mg/L PO_4-P in the primary effluent and an average of 9.0 mg/L PO_4-P at AX5, the end of the anoxic cell. This was incidental to the process since there was no requirement for phosphorus removal nor was it the intent of the plant process design to remove phosphorus. The value of PO_4-P concentration at AX5 varied during the 13 days of composite sampling, suggesting some degree of instability in the biological phosphorus release that was occurring in the anoxic tank. A lack of mixing in the front end of the selector and visual evidence of stratification in that part of the selector (an effect that is difficult to model), gave credence to the instability of the BPR process and the apparent discrepancy between measured and modeled phosphorus concentrations in the context of a steady state analysis.

24.2.3.2 Dynamic Modeling

Grab samples were collected as described in the Wastewater Characterization section to determine the diurnal variability in the influent wastewater characteristics, which were used to formulate a dynamic model in the simulator. This was used to examine the plant response under varying influent conditions. Diurnal wastewater characterization data for the liquid streams are summarized in Tables 24-5 and 24-6.

To illustrate the dynamic modeling process, Figure 24-3 reproduces the nitrate concentration diurnal data from Tables 24-5 and 24-6 in concert with the grab-sampled diurnal data that were taken at locations AX1, OX2, OX4, OX5 and OX7 as shown in Figure 24-2, that is, it shows the variability in nitrate concentration profile at the sampling points along the activated sludge train during the day of diurnal sampling.

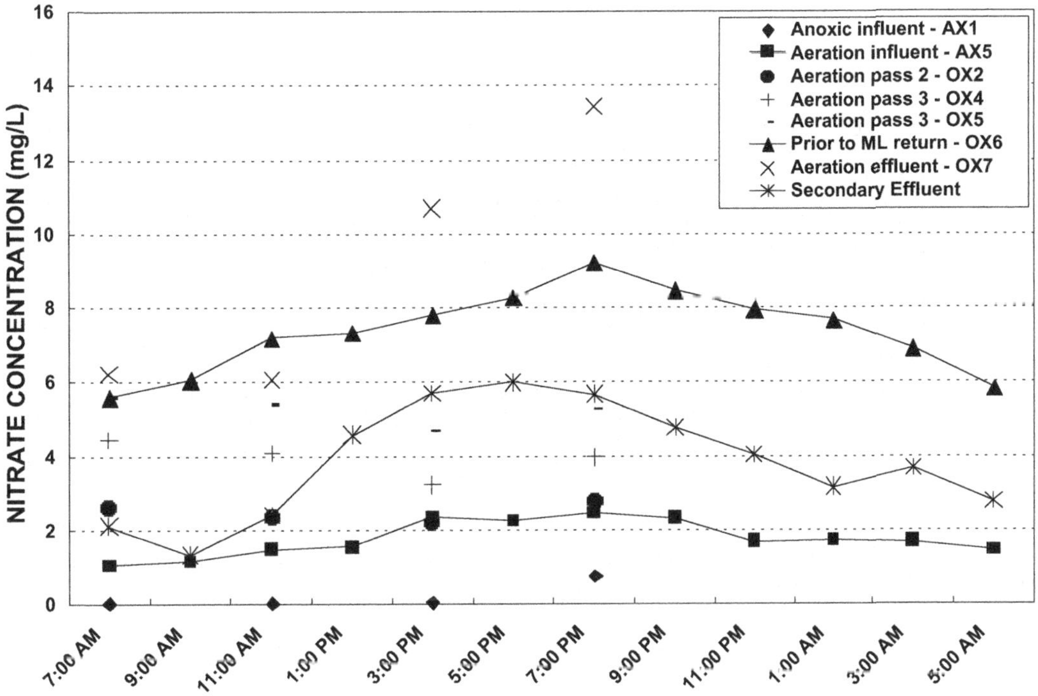

Figure 24-3. Diurnal Variability in Plant Nitrate Concentration

Only four two-hourly nitrate concentration values are shown for the grab samples for each of the locations, AX1, OX2, OX4, OX5 and OX7 because there were not enough automatic samplers available to sample all locations during the 24-hour period. The figure reflects the progressive oxidation of ammonia through the 4-pass aeration tank and the subsequent denitrification following passage through the anoxic portion of the aeration tank; further, the data demonstrate the attenuation of diurnal change in nitrate concentration through the aeration tank.

The diurnal primary effluent characteristics were fed to the simulator and predicted effluent concentrations compared with measured effluent concentrations for specific wastewater characteristics. An example of the result of this procedure is given in Figure 24-4, which shows the simulator output for ammonia and nitrate concentrations at the OX6 location, that is, the point at which mixed liquor is returned to the head of the aeration tank. Figure 24-4 includes measured nitrate concentration values (shown as squares) for each of the days when 24 hr composite samples were collected and for the 2-hourly diurnal day samples collected, and compares them to the modeled nitrate concentrations (shown as continuous lines). Shown also are ammonia concentrations at the same location for comparison.

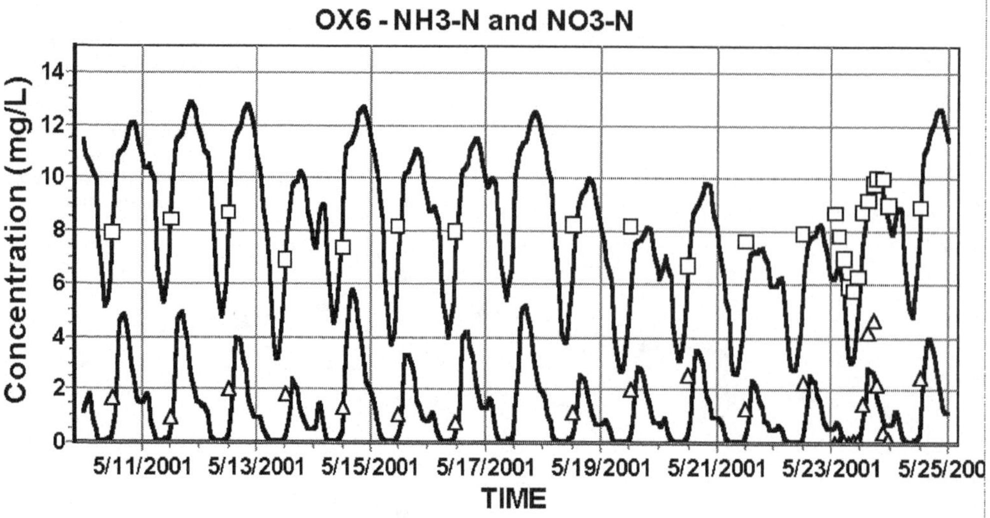

Figure 24-4. Simulation of Nitrate Concentration (Upper Curve) and Ammonia Concentration (Lower Curve) versus Observed Concentrations (as Squares and Triangles, Respectively)

The shape of the modeled nitrate concentration curve closely follows the measured values during the day of diurnal sampling. The goodness of fit demonstrates the success of model calibration.

24.3 Application

The simulator, once calibrated, may be used to simulate the effects of increasing plant capacity, or changes in plant loading or operational changes.

24.4 Acknowledgments

The WERF project team is indebted to the extensive help and support of the Process Laboratory staff at the 91st Avenue WWTP, in particular, Pat Wokulich, the plant operations staff and Jim Coughenour, Plant Superintendent, for carrying out the sampling investigation and sample analysis.

24.5 References

American Public Health Association. 1995. *Standard methods for the examination of water and wastewater (19th ed.).* Washington D.C.: American Public Health Association. 953 pp.

Mamais, D., D. Jenkins, and P. Pitt. 1993. A rapid physical-chemical method for the determination of readily biodegradable soluble COD in municipal wastewater. *Water Res.* 27(1):195-197.

CHAPTER 25.0

Case Study: Application of the Model to an N & P Removal Municipal Wastewater Treatment Plant

25.1 Introduction

This case study examines the application of the model to the City of Calgary's Bonnybrook Wastewater Treatment Plant. The objective of the case study is to assess the ability of the model to describe one full year of process performance at a typical, well-operated biological nutrient removal (BNR) plant, using the plant performance data routinely gathered by the plant operating and laboratory staff to monitor plant operation.

The methodology used in this case study employed a sequential approach as follows:

1. The plant's process performance database was examined to select a calendar year of relatively stable operation.
2. The simulator was configured in accordance with the design parameters of the full-scale plant.
3. The plant's database was critically assessed to prepare preliminary input variables and to develop a preliminary calibration of the model parameters.
4. The model was run to examine how well the predicted results match the plant performance in terms of solids production and treated effluent quality predictions.
5. The model inputs and model calibration were refined as appropriate to improve the fit of the model to the plant's output data.

25.2 The Bonnybrook Wastewater Treatment Plant

The Bonnybrook WWTP is one of two wastewater treatment plants owned and operated by the City of Calgary in the Province of Alberta, Canada. It has a nominal capacity of 500,000 m^3/d at maximum month conditions. The receiving stream for the treated effluent is the Bow River, which rises in the Rocky Mountains at Bow Lake, approximately 150 km west of Calgary, and generally flows eastward and northeastward, becoming a tributary of the Nelson River system and emptying into Hudson Bay. The treated effluent permit limits established by Alberta Environmental Protection are indicated in Table 25-1.

Table 25-1. Treated Effluent Permit Limits for the Bonnybrook WWTP

Variable	Concentration	Comments
$cBOD_5$	20 mg/L	Monthly arithmetic average
TSS	20 mg/L	Monthly arithmetic average
NH_3-N	5 mg/L Summer 10 mg/L Winter	Monthly arithmetic average
Total P	1.0 mg/L	Monthly arithmetic average
Total coliforms	1,000 CFU/100 mL	Monthly geometric mean
Fecal coliforms	200 CFU/100 mL	Monthly geometric mean

Note: All sampling is based on refrigerated 24-hour flow-proportional composite samples with the exception of the bacteriological samples, which are done on grab samples.

Figure 25-1 presents a simplified flow schematic for the Bonnybrook WWTP.

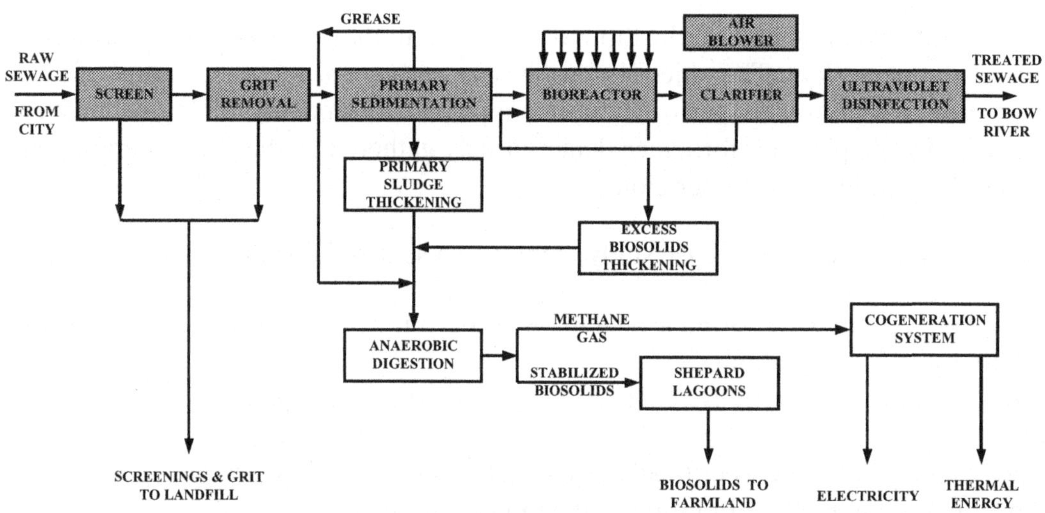

Figure 25-1. Flowsheet for Bonnybrook WWTP

The mainstream treatment process train at the Bonnybrook WWTP consists of the following steps:

- Pre-treatment using mechanically-raked bar screens and aerated grit removal. There are six bar screens and four aerated grit removal tanks.
- Primary treatment using gravity sedimentation. There are fourteen primary clarifiers.
- Secondary treatment using the activated sludge process modified for biological nutrient removal. There are ten bioreactors and thirty final clarifiers.
- Effluent disinfection using ultraviolet radiation. The UV system employs low pressure, low intensity lamps.

The sidestream sludge processing steps include:

- Fermentation and gravity thickening of about 20 percent of the primary sludge produced at the plant. There are two primary sludge fermentation trains, each consisting of a complete mix fermentation tank followed by a gravity thickener.
- Dissolved air flotation of waste activated sludge. There are four rectangular dissolved air flotation units.
- Mesophilic anaerobic digestion of blended primary and waste activated sludges. There are three clusters with four digesters in each cluster.
- Utilization of digester gas in a cogeneration system. There are four engine-gensets, each fitted with a heat recovery system on the engine and its exhaust.
- Gravity separation of digested sludge in open storage lagoons. The open lagoon system is located approximately 12 kilometers southeast of the plant site. Settled sludge is pumped out of the lagoons and hauled to farmland for subsurface injection.

There are three significant return streams to the mainstream treatment train from the sidestream sludge processing steps. They are:

- Dissolved air flotation underflow is directed to a primary effluent channel.
- Overflow from the gravity thickeners in the primary sludge fermentation system is pumped directly to the anaerobic zones of two bioreactors and also is added to the primary effluent channel leading to the other eight bioreactors to enhance biological phosphorus removal performance.
- Decant water from the open sludge storage lagoons used to thicken the digested sludge is directed to the headworks of thc plant via the City's sewer system.

The Bonnybrook plant has three distinct secondary treatment sections:

1. Secondary A was initially constructed in 1970 using a surface mechanical aeration system. It was first retrofitted to BNR technology in 1989, with subsequent process upgrades including a retrofit to fine bubble aeration in 1999. Secondary A consists of four bioreactors and twelve secondary clarifiers and is nominally rated at 170,000 m^3/d.
2. Secondary B was initially commissioned in 1984 using a jet aeration system. It was retrofitted to BNR and fine bubble aeration technology in 1997. Secondary B consists of four bioreactors and twelve secondary clarifiers and is nominally rated at 230,000 m^3/d.
3. Secondary C was commissioned in 1994 using BNR technology and fine bubble aeration from the outset. It consists of two bioreactors and six secondary clarifiers and is nominally rated at 100,000 m^3/d.

This case study focuses on one of the two bioreactor/secondary clarifier trains in Secondary C. Table 25-2 presents the design parameters for the tankage associated with Secondary C.

Table 25-2. Design Parameters for Secondary C at the Bonnybrook WWTP

Item	Description
Bioreactor:	
No. of Bioreactors	2
No. of unaerated zones in each bioreactor	4
Dimensions of each unaerated zone	15.2 m x 15.2 m x 6.0 m SWD
Volume of each unaerated zone	1390 m^3
No. of "swing" zones in each bioreactor	2
Dimensions of each "swing" zone	15.2 m x 15.2 m x 6.0 m SWD
Volume of each "swing" zone	1390 m^3
No. of aerated zones in each bioreactor	4
Dimensions of each aerated zone	23.0 m x 23.0 m x 6.0 m SWD
Volume of each aerated zone	3165 m^3
Features:	
Waste activated sludge	From mixed liquor channel
Internal recycle streams - NMLR	Nitrified mixed liquor recycle
- DNMLR	Denitrified mixed liquor recycle
Secondary Clarifiers:	
No. of secondary clarifiers	6
Dimensions of each secondary clarifier	35.0 m diameter x 5.0 m SWD
Features:	
Flocwell diameter	11.0 m
Sludge withdrawal	Two-arm tobrow mechanism
Surface skimming	Single-arm ducking skimmer
Primary Sludge Fermenters:	
No. of complete mix tanks	2
Dimensions of each tank	16.0 m diameter x 4.8 m SWD
Volume of each tank	992 m^3
Fermented Sludge Gravity Thickeners:	
No. of gravity thickeners	2
Dimensions of each gravity thickener	16.0 m diameter x 3.6 m SWD
Bottom slope	8.5 horizontal : 1.0 vertical
Volume of each gravity thickener	827 m^3
Mechanism	Rotating picket fence c/w scrapers

Note- "Swing" zone means one that can be operated in either aerated or unaerated conditions, depending on the preference of the operating staff.

25.3 Configuring Bonnybrook Secondary C in the Simulator

As indicated in Table 25-2, the Secondary C bioreactors include two internal mixed liquor recycle lines. The design includes flexibility to allow operation in several process modes, for example:

- Anaerobic/aerobic process,
- 3-Stage Bardenpho process,
- UCT process, and
- Modified UCT process.

The plant staff has found that operating the plant as a 3-Stage Bardenpho process offers the most stability and this is the configuration that has been employed for both halves of Secondary C since 1995.

Zone 5 and 6 are designated as "swing" zones because they are fitted with both diffusers and submerged mixers and can be operated in either aerated or unaerated modes. Usually zone 5 is operated in an unaerated mode while zone 6 is aerated.

Figure 25-2 illustrates the configuration of the Bonnybrook Bioreactor No. 10 as it appears on the graphical user interface of the process simulator.

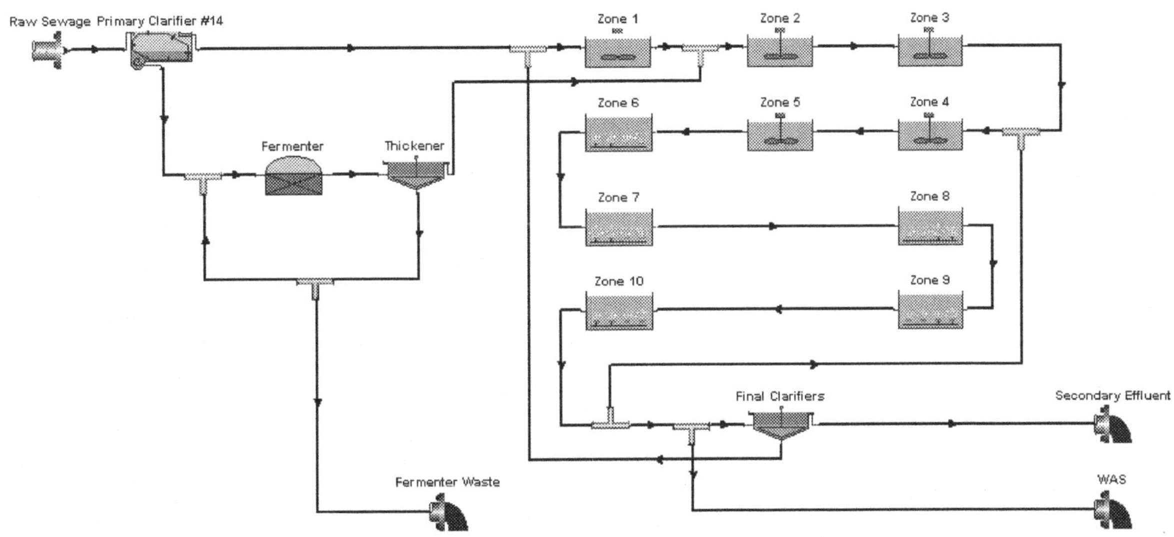

Figure 25-2. Bonnybrook Bioreactor No. 10 as Configured in the Simulator

25.4 Plant Process Performance Data

Table 25-3 lists the plant performance process data routinely gathered for Secondary C and its associated primary clarification and sludge fermentation systems. Individual daily measurements for $cBOD_5$, NH_3-N, NO_3-N and Total P on the secondary effluent for Secondaries A, B and C are not measured at the plant. Instead, an online monitoring system is used to measure secondary effluent transmittance as well as NH_3-N, NO_3-N and PO_4-P on each of the ten bioreactor and secondary clarifier trains. For regulatory compliance purposes, a 24-hour flow-proportional composite sample of the combined secondary effluents from all three secondary sections is collected; however, this sample is not useful for evaluating the performance of an individual bioreactor and secondary clarifier train.

Table 25-3. List of Data Routinely Collected for Bonnybrook Secondary C and Associated Primary Clarification and Primary Sludge Fermentation Systems

Item	Flow	TSS	VSS	$cBOD_5$	COD	solCOD	ffCOD	TKN	NH_3-N	NO_3-N	Total P	PO_4-P	VFA	Alk	pH
Raw sewage	7/wk	7/wk	-	7/wk	1/wk	-	-	1/wk	-	-	7/wk	-	-	-	-
Primary effluent	-	7/wk	1/wk	-	5/wk	3.5/wk	1/wk	1/wk	7/wk	-	7/wk	-	5/wk	1/wk	-
Primary sludge	7/wk	-	-	-	-	-	-	-	-	-	-	-	-	-	-
Fermenter complete mix tank	n/a	5/wk	-	-	-	-	-	-	-	-	-	-	-	-	-
Thickener decant	-	1/wk	-	-	-	-	-	-	-	-	-	-	5/wk	-	5/wk
Thickened sludge return	7/wk	-	-	-	-	-	-	-	-	-	-	-	-	-	-
Thickener waste sludge	7/wk	7/wk	-	-	-	-	-	-	-	-	-	-	-	-	-
Bioreactor – mixed miquor	n/a	7/wk	1/wk	-	-	-	-	-	-	-	-	-	-	-	-
Bioreactor – Nit. m.l. recycle	7/wk	-	-	-	-	-	-	-	-	-	-	-	-	-	-
Waste activated sludge	7/wk	-	-	-	-	-	-	-	-	-	-	-	-	-	-
Return activated sludge	7/wk	7/wk	-	-	-	-	-	-	-	-	-	-	-	-	-
Secondary effluent	-	7/wk	-	3.5/wk	-	3.5/wk	1/wk	-	O/L	O/L	-	O/L	-	-	-

Note: O/L indicates online analyzer used; however, data are not saved and entered into the plant's database. "ffCOD" indicates filtered and flocculated COD analysis by the method of Mamais et al. (1993). n/wk indicates that the measurement is recorded n times per week for the variable indicated.

Yearly plant performance data for Secondary C from 1995 to 2001 inclusive were examined as possible candidates to use in this case study. Plots were made of flowrates, concentrations and loadings for all measured variables over this 7-year period. The calendar year 1999 for Bioreactor No.10 and its associated Primary Clarifier No.14 and Fermenter System No.2 was chosen for this case study because all components of this system were in continuous operation throughout 1999 and the performance data indicated that relatively stable operation occurred. Plots of the 1999 raw sewage and primary effluent data recorded by the plant are illustrated in Figures 25-3 through 25-9. In preparing the plant data to make these plots, obvious outliers were deleted from the database. The input variables for the simulator taken from the database included:

- Wastewater temperature (from Figure 25-3);
- Primary effluent, nitrified mixed liquor recycle, return activated sludge and waste activated sludge flowrates (from Figure 25-4);
- Primary sludge and fermenter recycle and waste flowrates (from Figure 25-5); and
- Raw sewage COD, TKN, Total P and alkalinity concentrations (from Figures 25-6 through 25-9, respectively).

Figure 25-3. Wastewater Temperature for the Bonnybrook WWTP – 1999

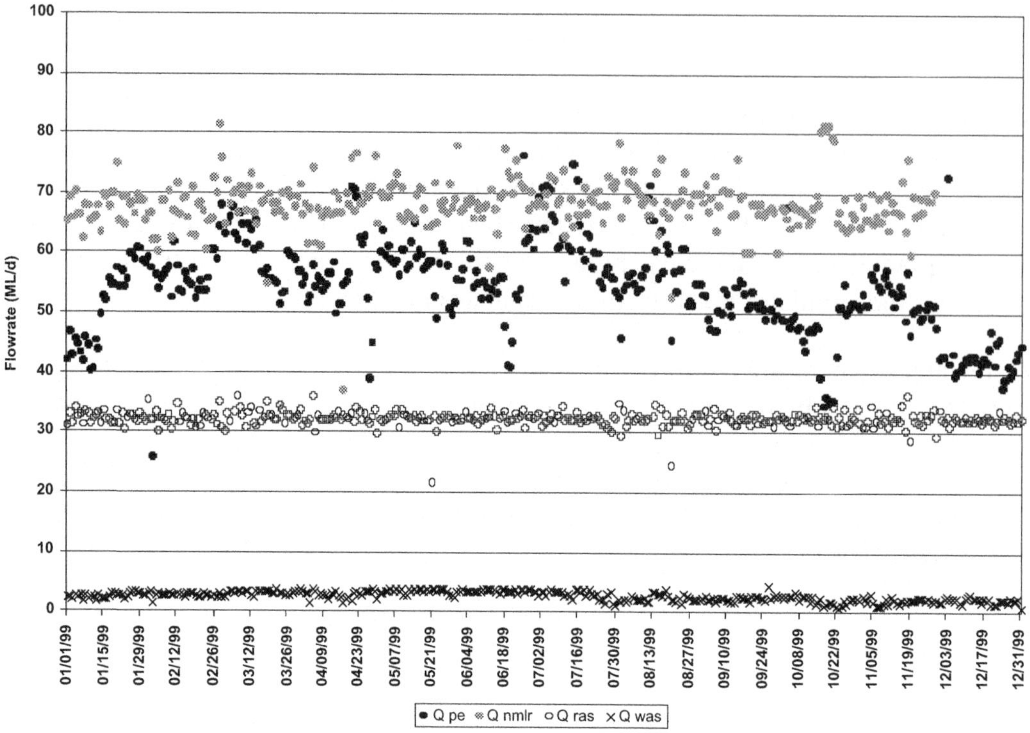

Figure 25-4. Primary Effluent, NMLR, RAS and WAS Flowrates in Bioreactor No.10 at the Bonnybrook WWTP – 1999

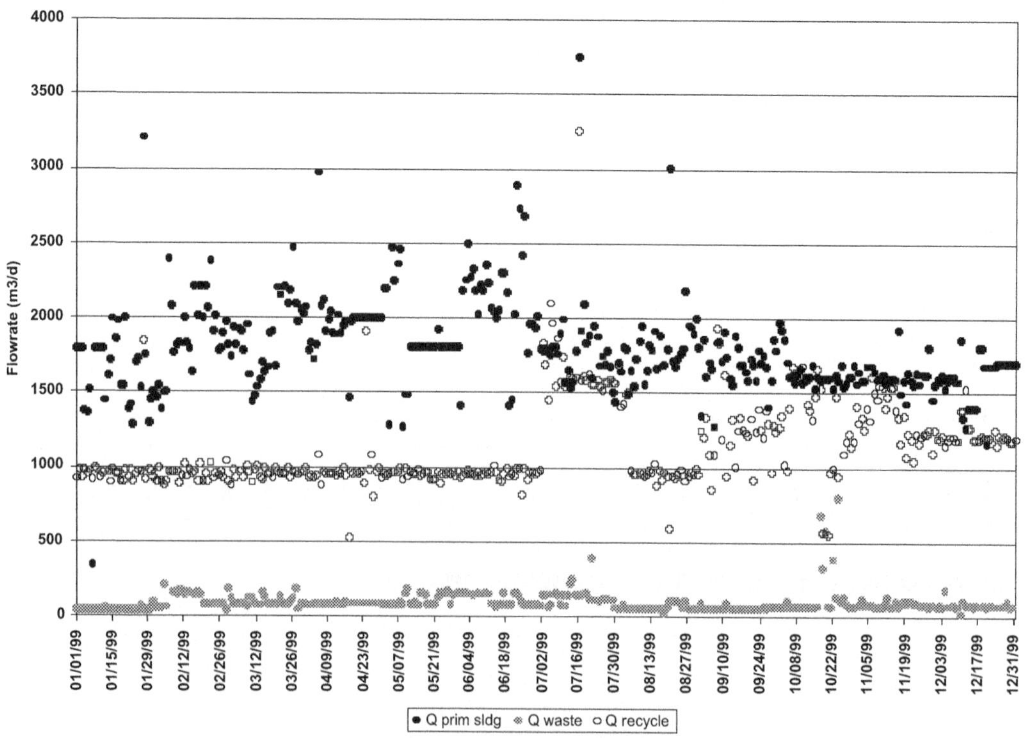

Figure 25-5. Primary Sludge Feed, Recycle and Waste Sludge Flows for Fermenter No. 2 at the Bonnybrook WWTP – 1999

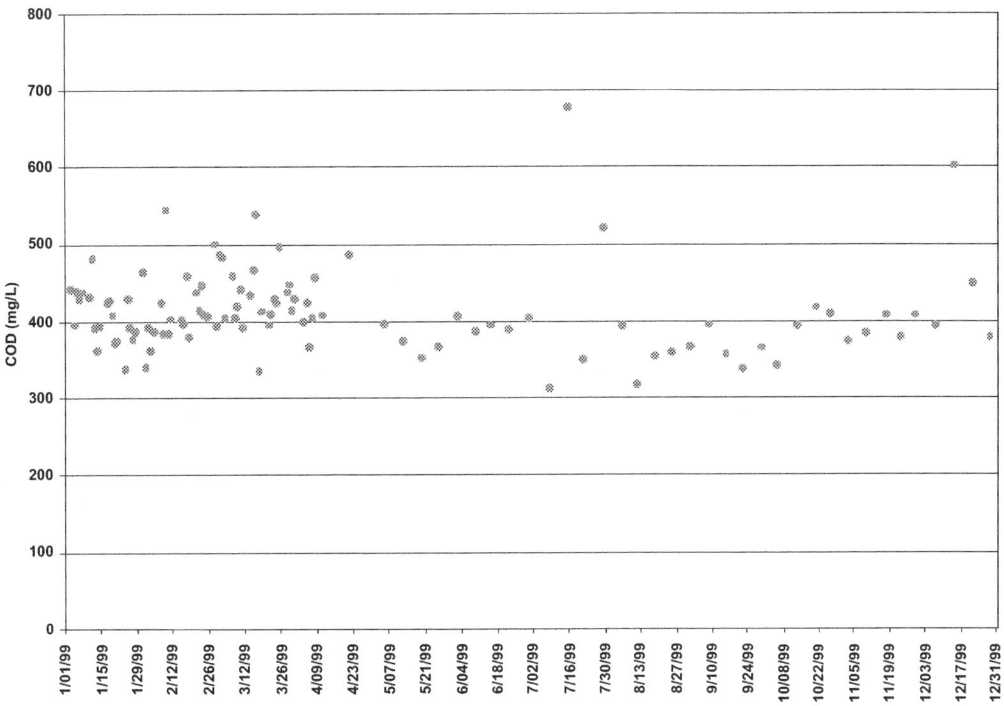

Figure 25-6. Raw Sewage COD Concentration in Bioreactor No.10 at the Bonnybrook WWTP – 1999

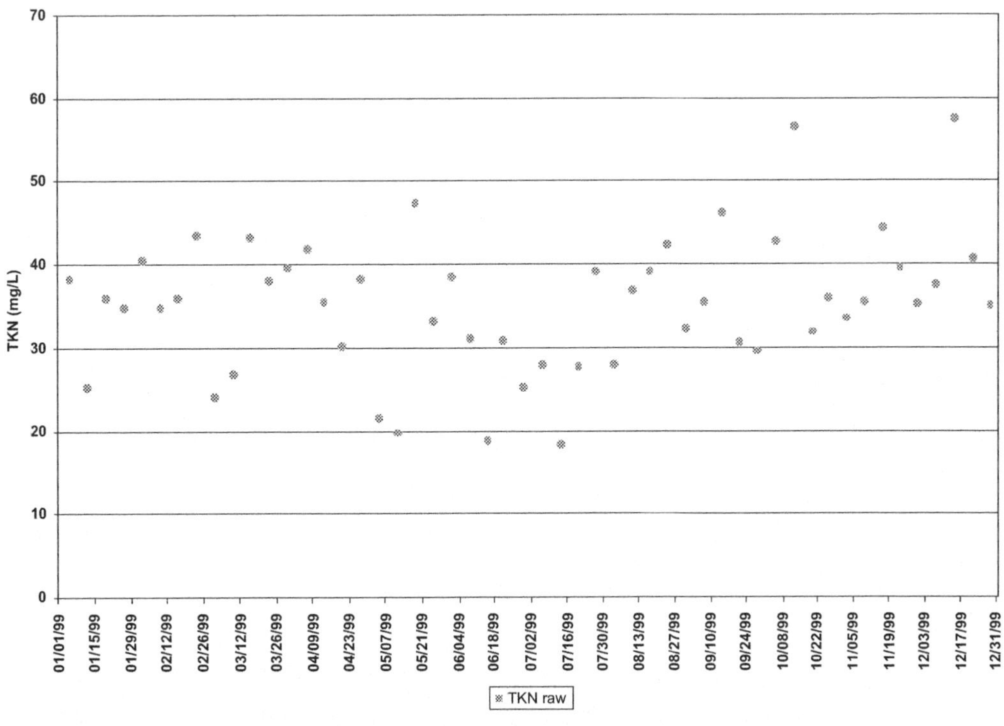

Figure 25-7. Raw Sewage TKN Concentrations in Bioreactor No.10 at the Bonnybrook WWTP – 1999

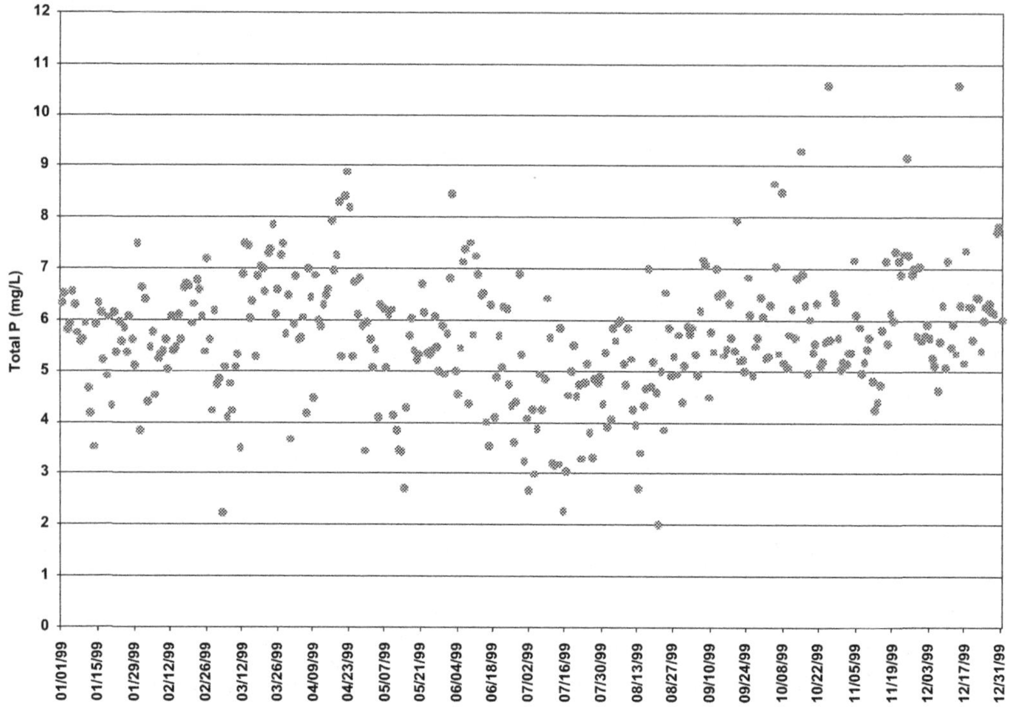

Figure 25-8. Raw Sewage Total Phosphorus Concentration for Bioreactor No.10 at the Bonnybrook WWTP – 1999

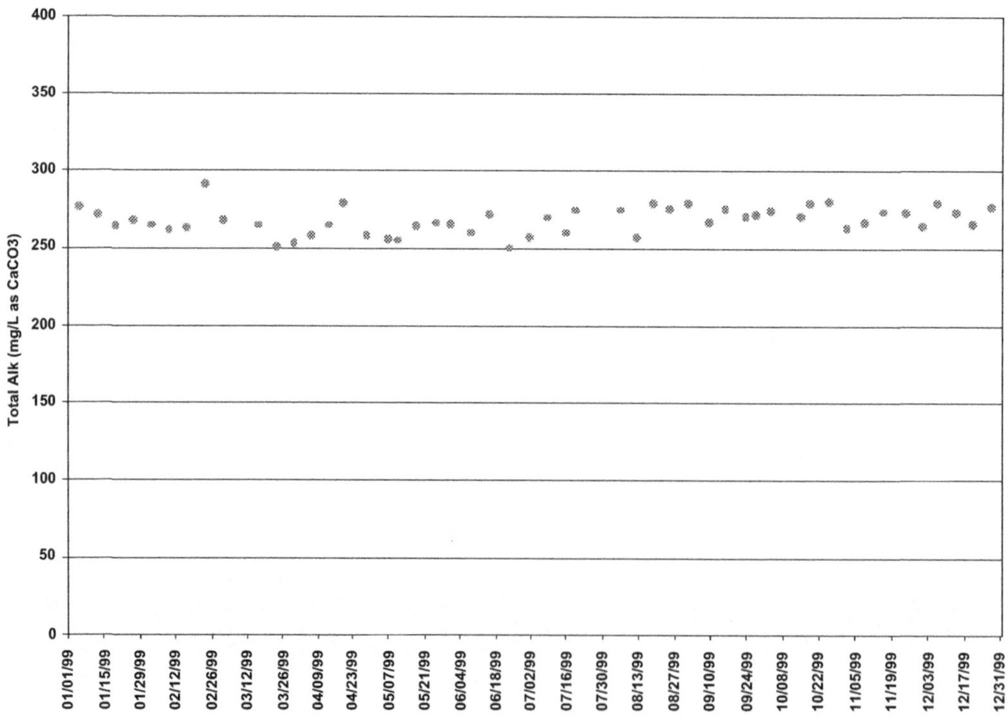

Figure 25-9. Primary Effluent Alkalinity to Bioreactor No.10 at the Bonnybrook WWTP – 1999

25.5 Initial Model Calibration

To make an initial calibration of the model for Secondary C, the plant process performance record was examined to determine the TSS removal efficiency of the gravity separation units. The estimates are indicated below and were entered into the simulator as part of the calibration exercise.

- Primary clarification TSS removal efficiency = 56.5 percent.
- Gravity thickener TSS removal efficiency = 87.5 percent.
- Secondary clarifier TSS removal efficiency = 99.8 percent.

Volatile Suspended Solids (VSS) are not measured directly but rather as the difference between the TSS and the ISS. The plant database has daily TSS values for both the raw sewage and the primary effluent streams; however, weekly ISS measurements are made only on the primary effluent. The VSS:TSS ratio in the primary effluent is 0.78. To estimate the raw sewage ISS concentration for input to the simulator, it was assumed that the raw sewage VSS:TSS ratio was 0.8 and that the raw sewage ISS could be approximated by multiplying the raw sewage TSS by a factor of 0.2 (Figure 25-10).

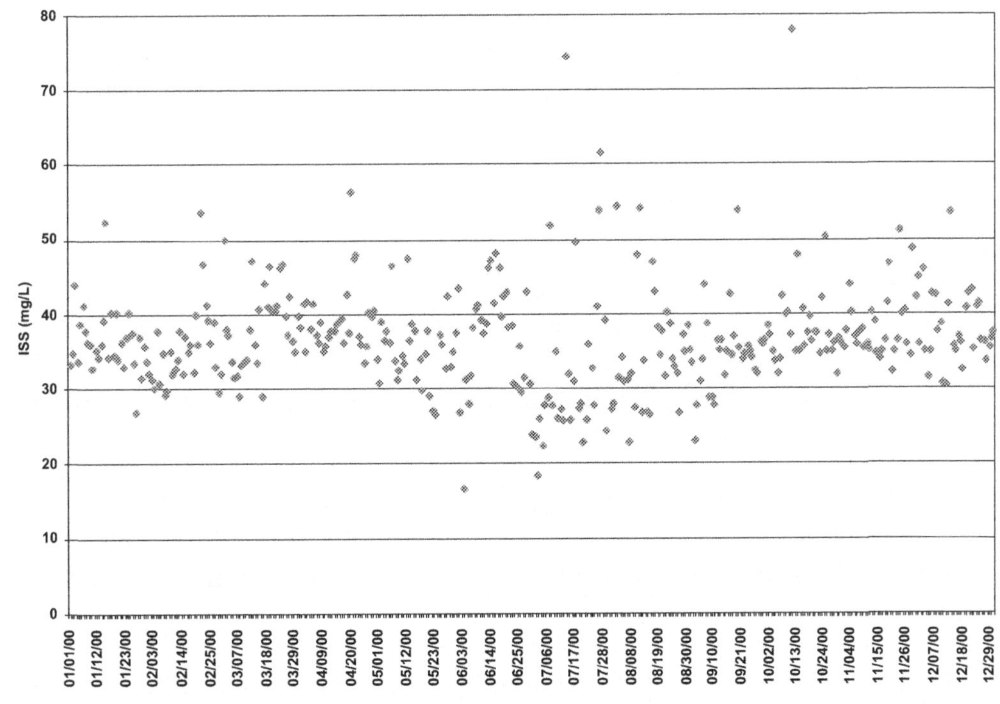

Figure 25-10. Raw Sewage ISS Concentration (Estimated) for Bioreactor No.10 at the Bonnybrook WWTP – 1999

Although not routinely collected by the plant staff, several flocculated and filtered COD (ffCOD) analyses (Mamais et al., 1993) were made by the plant on the primary and secondary effluent streams during 1999. These enabled estimation of certain COD fractions for the model. Table 25-4 lists the default values used in the simulator for the various COD, nitrogen and phosphorus fractions, together with those derived from the Bonnybrook database and used in the initial model calibration.

Table 25-4. Raw Sewage Carbon, Nitrogen and Phosphorus Fractions Used in the Initial Model Calibration (Prior to Fermenter Supernatant Addition)

Parameter	Simulator Default Value	Value Derived from Plant Data
f_{SS}: Fraction of total influent COD that is readily biodegradable.	0.200	Use Default Value
f_{SA}: Fraction of influent readily biodegradable COD that is VFA.	0.150	0.070
f_{XS}: Fraction of slowly biodegradable COD that is particulate.	0.750	0.670
f_{SI}: Fraction of total influent COD that is soluble unbiodegradable.	0.050	0.040
f_{XI}: Fraction of total influent COD that is particulate unbiodegradable.	0.130	Use Default Value
f_{SNH}: Fraction of influent TKN that is free and saline ammonia.	0.750	Use Default Value
f_{XNB}: Fraction of biodegradable organic TKN that is particulate.	0.500	Use Default Value
f_{SNI}: Fraction of influent TKN that is unbiodegradable and soluble	0.000	Use Default Value
f_{XNI}: N content of the particulate unbiodegradable influent COD.	0.068	Use Default Value
f_{SPO4}: Fraction of influent TP that is orthophosphate.	0.500	Use Default Value
f_{XPI}: P content of the particulate unbiodegradable influent COD.	0.021	Use Default Value

From a consideration of the ffCOD values reported in the plant database, an estimate for f_{SS} in the primary effluent was derived to be 0.26 as an average for 1999. No ffCOD values were reported for the raw sewage; therefore, no direct estimate of the f_{SS} in the raw sewage could be made. It is reasonable to expect that primary clarification would result in increasing the f_{SS} across the primary treatment system due to the removal of particulate matter. Therefore, the lower simulator default value for f_{SS} of 0.20 was used for the raw sewage in the initial calibration.

The f_{SA} value was estimated from VFA determinations as 0.07.

The f_{SI} value was estimated from the ffCOD measurements on the primary and secondary effluents as 0.04.

The f_{XS} value was estimated using the simulator by conducting a series of trial and error mass balances around the primary clarification system using the fraction values noted above and varying the f_{XS} value until the simulators estimate of raw sewage TSS matched that of the full-scale plant. The f_{XS} value estimated in this manner was 0.67.

No site-specific information is available on nitrification rates at the Bonnybrook WWTP. Therefore, the values for μ_{AUT} and b_{AUT} of 0.95 d^{-1} and 0.17 d^{-1}, respectively, as determined in the primary work of this overall project were used in the model.

25.6 Initial Modeling Results

The initial modeling runs showed that the measured MLSS was somewhat underpredicted by the model for approximately the first half of the year as indicated in Figure 25-11. On the other hand, the model's MLSS predictions for the latter half of the year looked reasonable when compared with the plant's measured data.

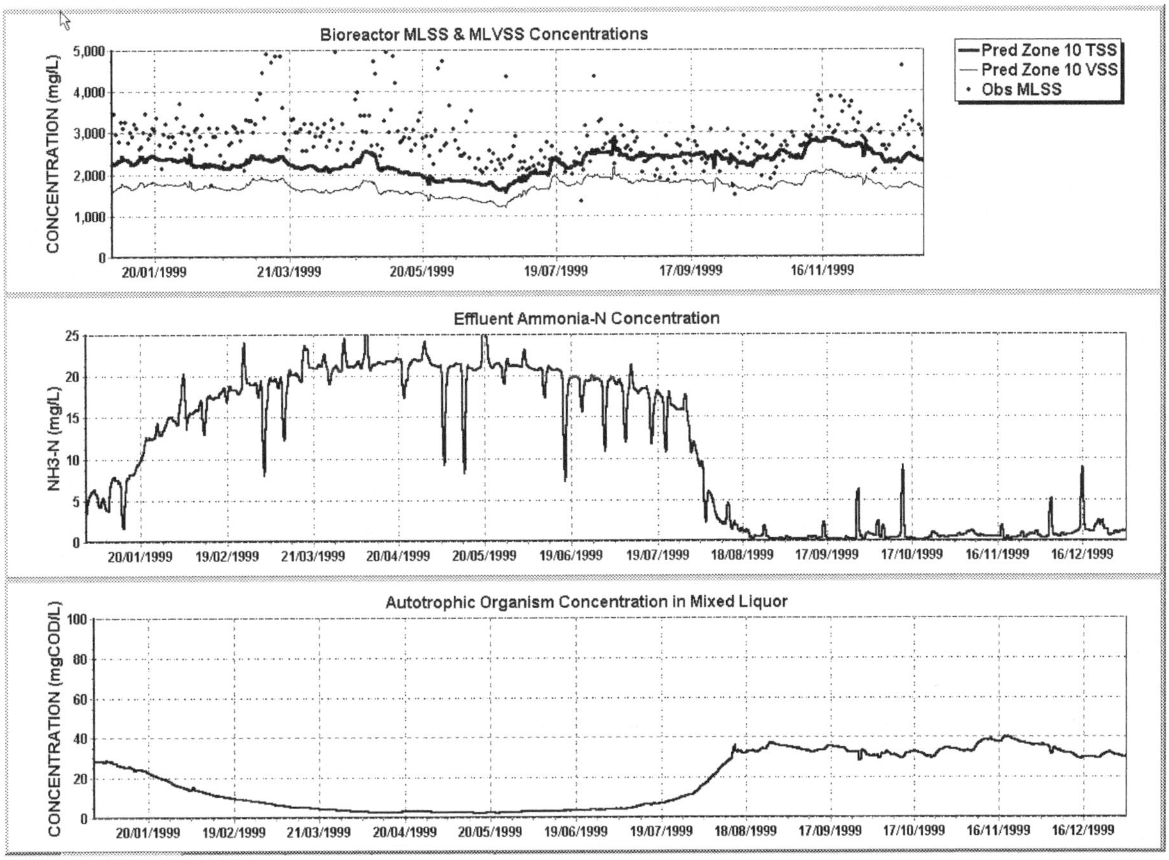

Figure 25-11. Initial Model Predictions for Mixed Liquor, Autotrophic Organism and Ammonia-Nitrogen Concentrations

The most noteworthy and surprising prediction of the model for the initial calibration runs was that the nitrifying organisms would be washed out during winter and nitrification would not be re-established until well into the summer months.

Such is not the case for the full-scale plant. As indicated previously, no permanent plant records are maintained for the secondary effluent NH_3-N concentration; however, monitoring by the online sensor units typically indicates a summer NH_3-N effluent concentration of <1.0 mg/L and winter NH_3-N concentrations in the range of 1.0 to 3.0 mg/L. The model was not accurately predicting the nitrification performance experienced at the plant. This is illustrated clearly in the two lower plots in Figure 25-11. Thus, further work to refine the model calibration was necessary. In the plots that follow, lines represent model predictions and discrete points represent plant data.

25.7 Model Refinements

25.7.1 Wasting Rate

Closer examination of the plant data revealed an anomaly in terms of the activated sludge wasting rate. The loading conditions on the system generally were the same throughout the year. Until about mid- to late July, the MLSS concentration averaged approximately 3,000 mg/L and the wasting rate for mixed liquor was about 3,000 m^3/d. However, after this time, the MLSS concentration averaged about 2,500 mg/L and the wasting rate for mixed liquor was about 2,000 m^3/d (see Figures 25-12 and 25-13). If the loading was approximately the same throughout the year, then the reported values are inconsistent with one another. That is, the higher wasting rate (and lower SRT) during the first part of the year resulted in a higher MLSS concentration than during the latter part of the year when the wasting rate was lower (with a correspondingly higher SRT).

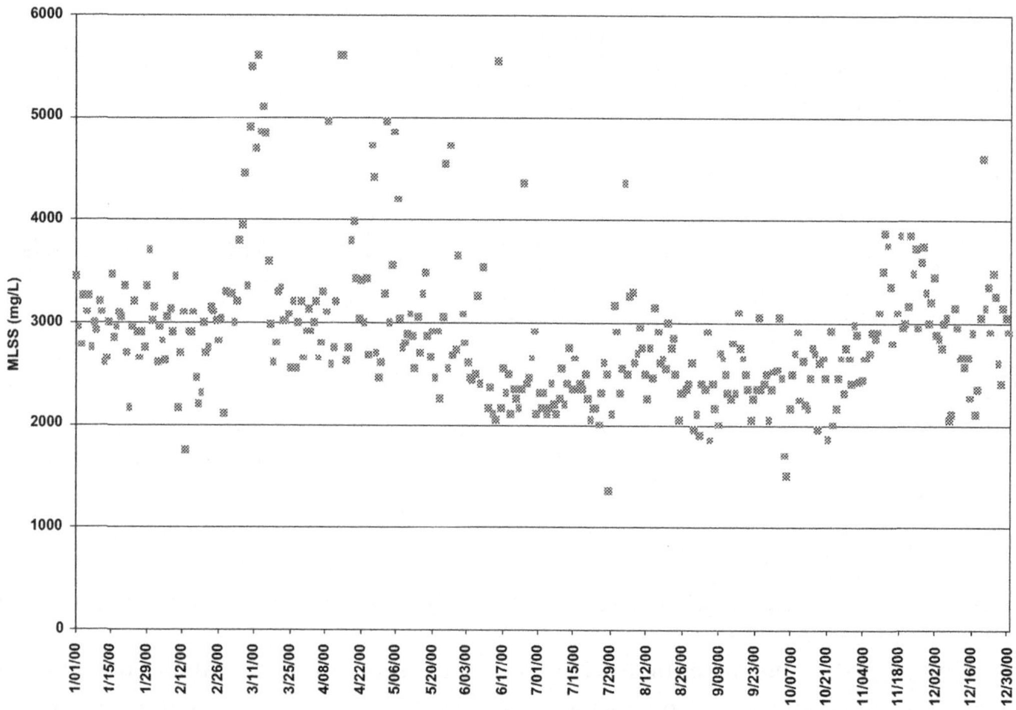

Figure 25-12. MLSS Concentration in Bioreactor No.10 at the Bonnybrook WWTP – 1999

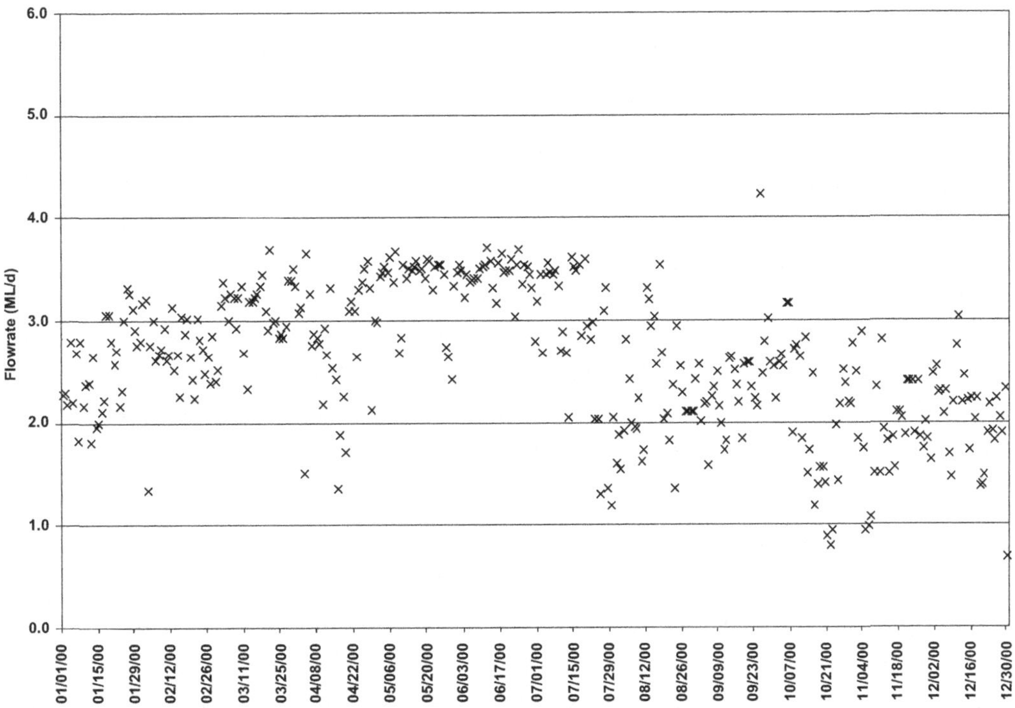

Figure 25-13. WAS Flowrate in Bioreactor No.10 at the Bonnybrook WWTP – 1999

Given a mixed liquor wasting rate of 3,000 m^3/d and a total bioreactor volume of 21,000 m^3, the estimated SRT would be 7.0 days. As the first five zones of the bioreactor are operated in an unaerated mode, the aerobic SRT is approximately two-thirds of the total SRT or about 4.7 days. Experience at other plants indicates that maintaining nitrification with wastewater temperatures in the order of 11°C over a long winter period operating with an aerobic SRT of 4.7 days is problematic. On the other hand, for a mixed liquor wasting rate of 2,000 m^3/d, the estimated total SRT is 10.5 days. For the same unaerated volume fraction in the bioreactor, the aerobic SRT is 7.0 days. Such a value is more suited to maintaining nitrification through the winter months.

Further evidence that the wasting rates reported in the plant database for the first part of the year are erroneously high is provided by an examination of VSS production in relation to COD load. During the first part of the year, the estimated VSS production for Bioreactor No.10 was about 6,000 kg/d while for the second part of the year, it was about 4,000 kg/d. Considering the COD load on the system, the latter figure gives a more realistic unit VSS production rate of approximately 0.28 kg VSS / kg COD.

Therefore, the wasting rates reported for the first part of the year are suspect. It is likely that the calibration of the WAS meter for that period was in error until on or about July 24, 1999. Thus, the simulator input values for WAS flowrate were adjusted by a factor of 0.666 for days prior to July 24, 1999 to provide in a more realistic estimate.

25.7.2 Raw Sewage TKN

The raw sewage TKN values are reported weekly and show considerable variability. Such a pattern results in unrealistic variability in the predicted treated effluent S_{NH} concentration.

The plant's database contains daily data on primary effluent NH_3-N concentrations. Therefore, in order to smooth the raw sewage TKN values, the f_{SNH} fraction for the primary effluent was estimated as 0.61 using weekly primary effluent TKN measurements. It was assumed that the raw sewage NH_3-N concentration was the same as the primary effluent NH_3-N concentration, implying that there was essentially no lysis occurring in the primary clarifier that would cause an increase in NH_3-N concentration. Therefore, to refine the TKN input to the simulator, the raw sewage TKN was estimated by dividing the primary effluent NH_3-N by the factor of 0.61. This resulted in a relatively smooth TKN pattern in the raw sewage with the annual average more or less equal to the annual average of the reported variable weekly TKN pattern.

25.7.3 Raw Sewage COD Data

After the first few weeks of 1999, raw sewage COD data were collected only once per week rather than five times per week. However, daily $cBOD_5$ data were reported for the entire year. To fill in the gaps in the COD data, a COD: $cBOD_5$ ratio was estimated from the available information and this ratio was used to generate raw sewage COD estimates for the days when no COD data were reported. This resulted in a more realistic COD loading pattern for the simulations.

25.7.4 COD Fractions

From a consideration of COD, $cBOD_5$ and ffCOD values reported in the database, adjustments were made to f_{SS} and f_{XI} to result in a better fit of the BOD exertion curve thereby making it consistent with the COD data. Accordingly the raw sewage carbon, nitrogen and phosphorus fractions used in the refined model calibration are listed in Table 25-5.

Table 25-5. Raw Sewage Carbon, Nitrogen and Phosphorus Fractions Used in the Refined Model Calibration

Parameter	Simulator Default Value	Value Derived from Plant Data
f_{SS}: Fraction of total influent COD that is readily biodegradable	0.200	0.187
f_{SA}: Fraction of influent readily biodegradable COD that is VFA	0.150	0.070
f_{XS}: Fraction of slowly biodegradable COD that is particulate	0.750	0.670
f_{SI}: Fraction of total influent COD that is soluble unbiodegradable	0.050	0.040
f_{XI}: Fraction of total influent COD that is particulate unbiodegradable	0.130	0.200
f_{SNH}: Fraction of influent TKN that is free and saline ammonia	0.750	0.610
f_{XNB}: Fraction of biodegradable organic TKN that is particulate	0.500	Use default value
f_{SNI}: Fraction of influent TKN that is unbiodegradable and soluble	0.000	Use default value
f_{XNI}: N content of the particulate unbiodegradable influent COD	0.068	Use default value
f_{SPO4}: Fraction of influent TP that is orthophosphate	0.500	Use default value
f_{XPI}: P content of the particulate unbiodegradable influent COD	0.021	Use default value

Selected plots of the simulation results using the refined calibration are presented in Figures 25-14 through 25-17. In these plots, the lines represent the model predictions and the discrete points represent actual plant data.

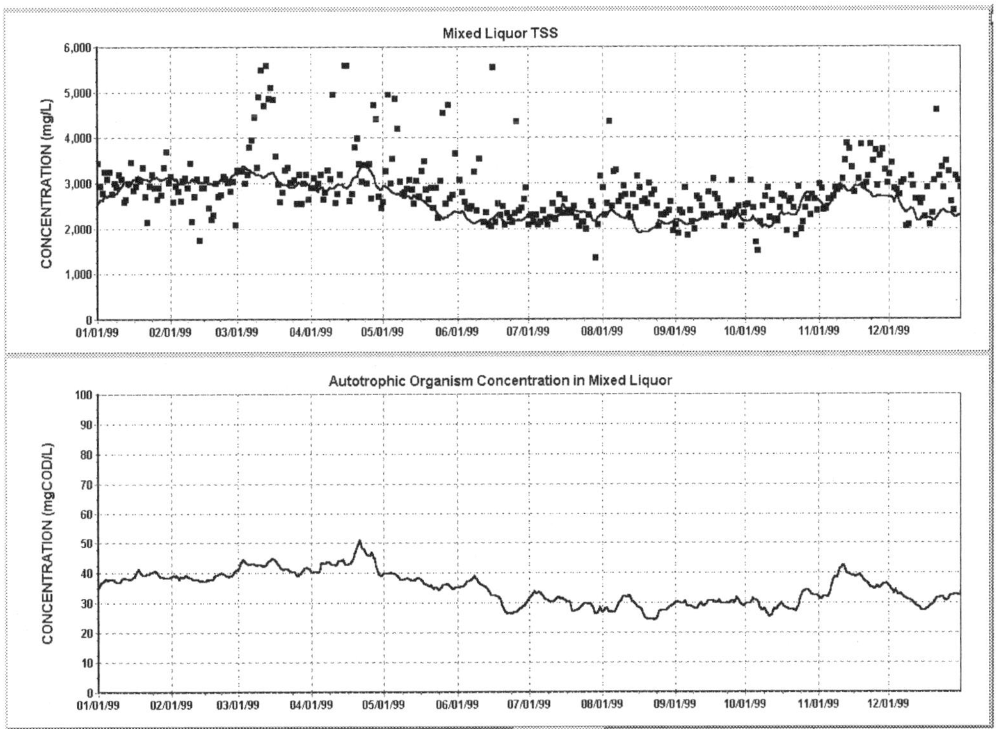

Figure 25-14. Predicted TSS and Autotrophic Organism Concentrations in the Mixed Liquor for the Refined Calibration

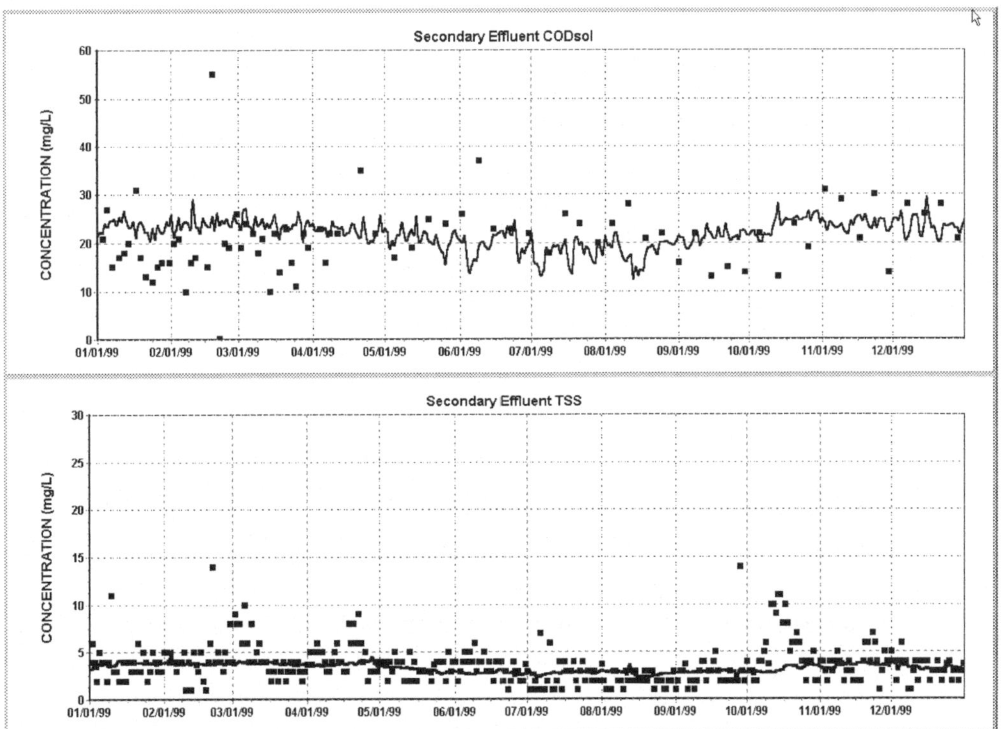

Figure 25-15. Predicted CODsol and TSS Concentrations in the Secondary Effluent for the Refined Calibration

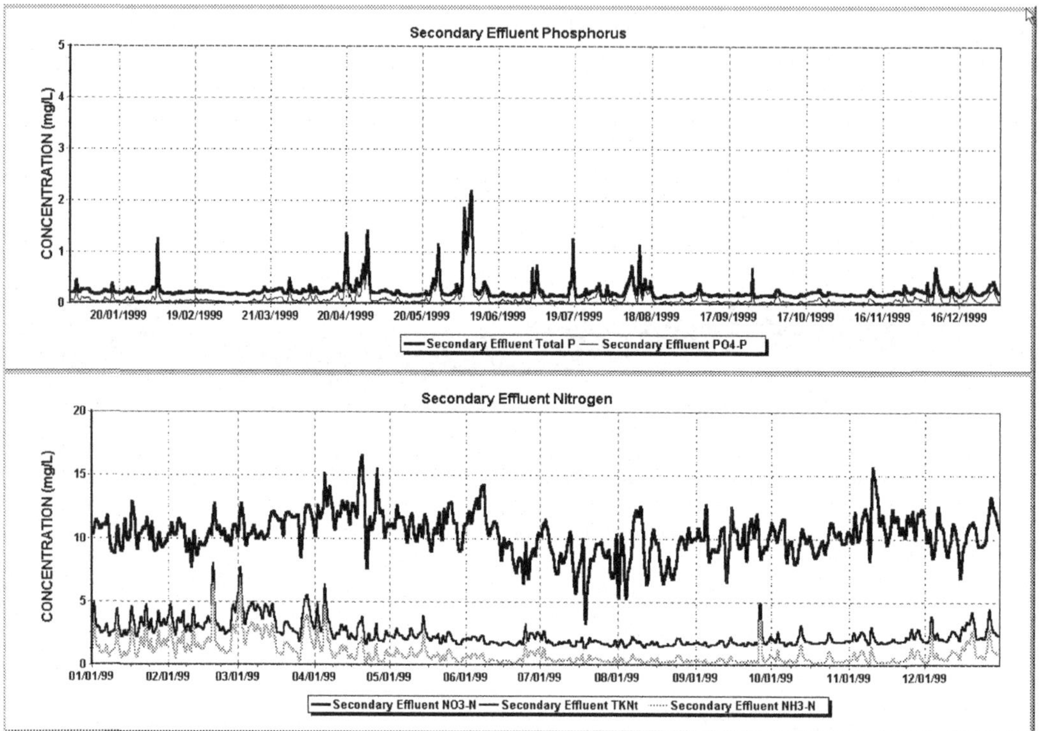

Figure 25-16. Predicted Ammonia, Nitrate and TKN Concentrations in the Secondary Effluent for the Refined Calibration

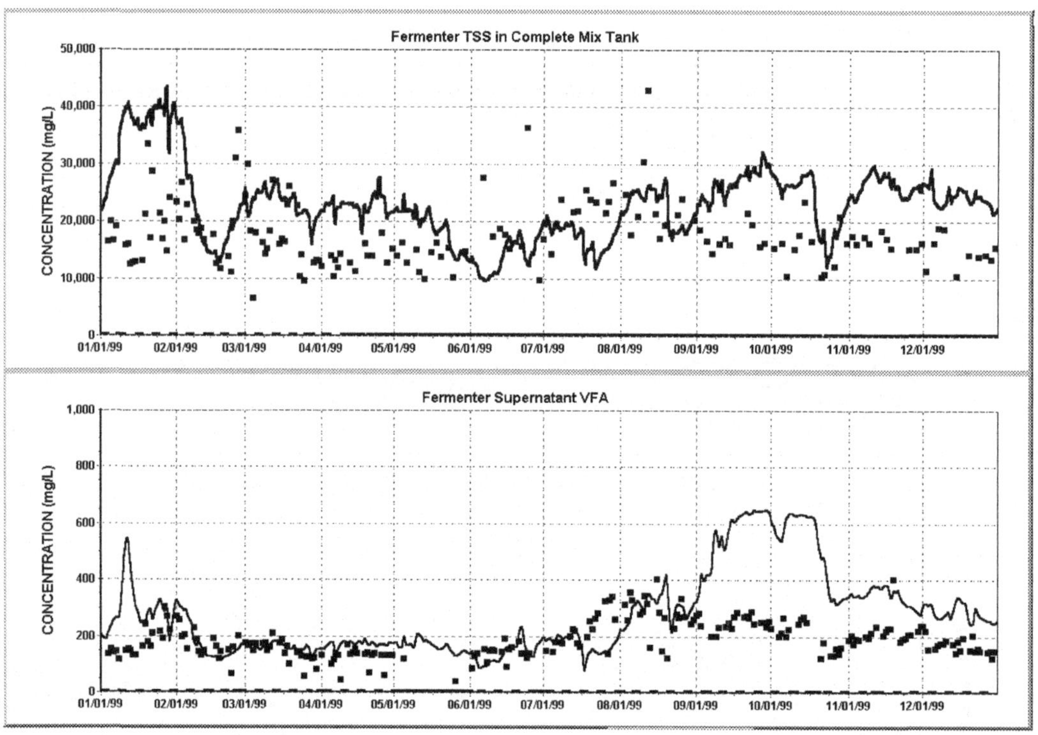

Figure 25-17. Predicted TSS Concentration in the Fermenter Complete Mix Tank and VFA Concentration in the Gravity Sludge Thickener

25.8 Concluding Remarks

The two-stage approach to model calibration consisting of making an initial fit of the model to the data to look for significant issues, followed by adjusting the model parameters and various input conditions as appropriate, resulted in a good fit of the model to the plant data for the Bonnybrook WWTP Secondary C plant data.

Critical and careful examination of raw plant data is essential to ensure that it is a realistic description of the true plant conditions. Adjustments can be made to the input database to get an improved fit of the model to the data as long as there is appropriate justification and/or a basis for doing so. In this case study, it was the calibration of the WAS pump that was suspect and caused the simulator to predict a lack of nitrification over the winter months.

The μ_{AUT} and b_{AUT} values of 0.95 d^{-1} and 0.17 d^{-1}, respectively, as determined in the other parts of this WERF research study, appeared to be appropriate to use at the Bonnybrook plant.

25.9 Acknowledgments

Special acknowledgement and thanks are given to the operating and laboratory staff at the City of Calgary's Bonnybrook WWTP for providing the data and for responding to the many questions that arose during the course of this work.

25.10 References

Mamais, D., D. Jenkins, and P. Pitt. 1993. A rapid physical-chemical method for the determination of readily biodegradable soluble cod in municipal wastewater. *Water Res.* 27(1):195-197.

CHAPTER 26.0

Industrial Wastewaters – General Discussion

26.1 Background

It is clear that an activated sludge mechanistic model applied to systems treating industrial wastewaters could provide many of the same benefits as in the municipal sector. It is likely that the municipal models such as ASM1 and ASM2 can be applied directly (with appropriate wastewater characterization) to certain industrial wastewater treatment systems such as food processing. This is possible if the same basic removal processes occur, if the wastewater can be characterized in the same general fashion as described in Section 2, and if oxygen and nitrate are the only significant electron acceptors in the wastewater. However, this is not the case for wastewater from all industrial sectors. In certain cases, the processes (biological and physicochemical) impacting system performance, and the kinetics and stoichiometry of the processes, may differ substantially from municipal treatment systems. For example, in the treatment of petroleum refinery wastes some of the considerations would be:

- Biomass yield coefficients for the unoxygenated hydrocarbons are lower than for organics in municipal wastewaters.
- Biological processes such as autotrophic oxidation of sulfides and other reduced sulfur compounds may impact oxygen demand significantly.
- Volatilization and stripping can be important "removal" mechanisms.

In such a case it may be necessary to develop an industry-specific model. However, the IWA-type models provide a useful starting point for model development. This is so because many of the basic processes used to model the treatment of municipal wastewater will apply in an activated sludge process used to treat the industrial wastewaters. For instance, biological growth and decay of heterotrophs occurs in activated sludge systems regardless of the nature of the wastewater. This chapter describes the framework for developing a model for an industrial system and highlights the differences and similarities to the development of the IWA municipal models.

In general, the objective of modeling is to be able to predict the response of a system during exposure to a particular set of conditions. For an activated sludge system these "conditions" refer to the operating mode (characteristics such as recycle rates, DO concentration levels in reactors and sludge age) and the influent loading (flow rate and composition). Similar to municipal systems, the concern for industrial systems is being able to predict the changes in concentration of certain components of interest within the bioreactor and in the effluent, and the associated impact on parameters such as oxygen utilization rate and the total mass of suspended solids.

The development of the model framework for an industrial treatment system follows the same approach as applied to the development of the IWA models. This approach comprises the following four steps:

- Selection of model structure and complexity,
- Gathering of experimental data,
- Mathematical model formulation, and
- Model calibration and verification.

The next sections describe each of these four steps.

26.2 Selection of Model Structure and Complexity

The required complexity of a model depends largely on the objectives for model use. Therefore, once the model application has been established, the complexity can be determined. However, for any level of complexity the aim should be to develop the simplest model structure that satisfies the objectives. In many cases, the objective will be to develop a mechanistic model with the capacity to predict the following system parameters:

1. Sludge production,
2. Oxygen demand,
3. Dynamic oxygen response, and
4. Effluent concentrations of selected parameters.

With the objectives established, the components and processes to include in the activated sludge model treating the industrial wastewater are selected. Establishing the desired model complexity also influences the parameters that must be measured in the experimental program in order to collect all the necessary data for model development.

Since often there are many different types of compounds and organisms in an industrial treatment system, the components of interest must be grouped together. Groupings of compounds should be based on those that undergo similar transformations. In some cases, specific compounds may be isolated as a component in the model if, for instance, there is a regulatory requirement to control that compound. As with the municipal treatment plant models, the use of the COD parameter to characterize the influent wastewater has many advantages. As a minimum, a COD fractionation similar to that used in municipal systems will be required. However, for certain industrial wastewaters, additional fractions will be needed. For instance, a significant removal mechanism for some of the organic compounds may be volatilization. Separate tracking of volatile compounds would be important in a system because of the concern regarding air emissions, and because organic compounds that are stripped from solution represent a COD loss that is not available for organism growth. Similarly, nitrogen and phosphorus bulk and fractionated wastewater characteristics are likely to be required. Additional groupings of inorganic compounds may also be important for some systems. As discussed below, one such grouping may be sulphur compounds.

Groupings of organisms will be based on those that mediate similar transformations of a certain group of compounds. In most systems, a minimum of two groups of organisms will be specified: (1) heterotrophic biomass resulting from growth on the influent organic material and (2) autotrophic biomass resulting from the oxidation of ammonia. In many industrial systems, the model will need to include additional groups of organisms to properly describe system behavior. For instance, the presence of significant concentrations of reduced sulphur compounds may result in the growth of sulphur oxidizing bacteria. The presence of these organisms and their substrates could result in significant additional oxygen requirements in the system.

26.3 Gathering of Experimental Data

Depending on the level of understanding of an industrial treatment system, the selection of an appropriate model structure and the gathering of experimental data may take several iterations. The initially selected structure will determine the appropriate data to collect, which could lead to a modification in the structure, more data gathering and so on. However, in many industrial situations, only minor modifications to existing municipal models will be required. In addition, the key processes to include in the industrial wastewater often will be evident by reviewing plant operating data and past experiences in characterizing activated sludge processes treating similar wastewaters.

In cases where insufficient prior information exists to formulate a model for the treatment system of interest, an experimental program will be required that consists of the operation of laboratory scale systems operated under closely controlled conditions. The experiments on the system will be designed to produce a database of performance information under closely controlled conditions and data on specific removal mechanisms of interest. For instance, batch tests may be designed to investigate the significance of stripping mechanisms in the removal of organic compounds in the wastewater.

In any case, one of the most critical aspects of an experimental program will be to appropriately characterize the wastewater to determine the concentrations of the components identified as important in the model. A specific example of such a characterization is discussed in the next chapter. However, in general, the wastewater characterization approach will be very similar to the fractionation of organic, nitrogen and phosphorus components described earlier for municipal systems.

26.4 Model Formulation

In this step, a system of equations is formulated to represent processes occurring in the activated sludge system. This involves determining stoichiometric expressions to identify the mass balance relationships and kinetic expressions to determine the rate-concentration dependence.

Models developed for municipal activated sludge systems will provide the framework of models for most industrial systems. As mentioned earlier, heterotrophic and autotrophic biomass growth should be included as key processes in an industrial model. Generally, Monod kinetics will be used to describe the mechanisms of biomass growth and substrate removal for the readily degradable organic compounds and for ammonia. Some specific biodegradable compounds may be inhibitory, and therefore an alternative expression such as the Haldane equation may be required to describe the removal of these components. Other mechanisms, which may need to be in the model, include stripping and adsorption.

26.5 Model Calibration

Unlike municipal systems, where most stoichiometric and kinetic parameters do not change appreciably from plant to plant, the parameters for industrial systems may be more site-specific. For example, the rate of aerobic growth of heterotrophs on readily biodegradable compounds specified in the wastewater may be very similar to the readily biodegradable compounds in municipal wastewaters. However, another group of biodegradable compounds may be inhibitory so that both the structure and parameters of the growth rate expression may be different for different wastewaters. Kinetic processes that were included to describe a specific case study will also require calibration for a specific plant or wastewater. Stripping of volatile organic compounds is one possible example.

Of the many stoichiometric parameters included in an industrial model, the yield is the most likely to require measurement for a particular type of wastewater. For instance, the heterotroph yield for petroleum refinery wastewaters has been found to be 10 percent to 15 percent lower than for municipal wastewaters. The yield determination can be conducted as described in earlier chapters. Model predictions are not as sensitive to other stoichiometric parameters such as the nitrogen and phosphorus content of the biomass and these can likely be set to the same values as used in municipal applications.

26.6 Summary

This chapter has provided a brief overview of the development and calibration of mechanistic models for the activated sludge process applied to industrial wastewaters. The subsequent chapters outline specific case studies in which models were developed for petroleum refinery and petrochemical wastewaters.

CHAPTER 27.0

Case Study: Model Development for Petroleum Refinery Wastewater Treatment

27.1 Background

This chapter builds upon Chapter 26.0 by providing a specific case where an activated sludge process mechanistic model was developed for petroleum refinery wastewaters. The development and calibration of the model followed the same general sequence as outlined in the previous chapter. Prior to the development of this particular model, a comprehensive mechanistic model of activated sludge processes applied to petroleum refinery wastewaters was not available. For this reason, the model development procedure included the operation of a laboratory scale activated sludge process in order to determine which mechanisms needed to be included in the model.

Note: The nomenclature used in this chapter necessarily is unique to the industrial sector, and may not agree with terms listed in Appendix A. However, care has been taken to define variables and terms as they occur.

27.2 Selection of Model Structure and Complexity

For the petroleum refinery model, the structure was determined through information obtained in the literature and from operation of bench scale continuous flow activated sludge reactors. The reactors were fed with wastewater collected at a refinery. Batch tests conducted on separate samples of influent wastewater and activated sludge supplemented the information collected on the continuous flow systems. The next section provides an overview of the experimental program conducted to develop the model.

27.2.1 Gathering of Experimental Data

The experimental program consisted of four components:

- Operation of the activated sludge system under constant input flow and COD loading

- Operation of the activated sludge system under daily cyclic square-wave loading
- Aerobic sludge digestion batch tests conducted over a period of several days to assess the decay rate of the activated sludge
- Batch tests to assess the potential for stripping of volatile organic compounds (VOCs) from the raw wastewater in the absence of sludge

27.2.1.1 Steady State Operation of Bench Scale Continuous Flow Systems

The constant loading experiments were designed to provide stoichiometric information on the process, such as biomass yield and oxygen requirements. Each day, a fixed influent volume of fixed COD concentration was pumped to the system at a constant rate over 24 hours. The system was operated at nominal sludge ages of 10 days and 20 days. In each case, operation continued for at least 3 to 4 sludge ages to attain steady state conditions.

27.2.1.2 Dynamic Operation of Bench Scale Continuous Flow Systems

After completion of steady state testing, the laboratory system was operated under dynamic loading conditions to assess the time-varying response and to provide data for development of a kinetic process model. The selected dynamic loading pattern followed a daily square-wave cycle. That is, each day the system received its fixed influent feed volume of fixed COD and TKN concentration at a constant rate, but over a 12-hour period; for the remaining 12 hours of the cycle there was no feeding. Information on the dynamic behavior of the system was obtained through monitoring the response of a number of selected parameters at short intervals over a 24 hour cycle. To assess a range of conditions, the system was operated at nominal sludge ages of 20 and 5 days.

27.2.1.3 Aerobic Sludge Digestion Batch Tests (Organism Decay Rate Tests)

Data on the decrease in organism activity with time were obtained by monitoring the change in oxygen utilization rate (OUR) in an aerated batch of activated sludge over a period of several days without external substrate addition.

27.2.1.4 Air Stripping Tests

Batch tests were conducted on a number of the collected raw wastewater samples to assess the potential for volatilization. These tests involved aerating a 3 L sample of raw wastewater in a 4 L cylindrical glass flask using the same aeration device as used in the reactor, and at approximately the same air flow rate as employed in the steady state tests. Samples were withdrawn from the batch at 15 minute intervals over two hours and analyzed for COD and TKN concentration.

27.3 Model Formulation

As mentioned earlier, the model was formulated based on information obtained through the literature and from the results of the operation of the bench scale processes. This section lists the components and processes in the model. For a more detailed description, and for a summary of the model stoichiometry, references are listed at the end of the chapter.

27.3.1 Model Components

Table 27-1 contains the list of components included in the petroleum refinery model. The partitioning of the organic, nitrogen and sulfur compounds into various fractions is illustrated in Figures 27-1 to 27-3.

Table 27-1. List of Components in the Petroleum Refinery Activated Sludge Model

Comp	Symbol	Definition	Units
1	Z_{BH}	Active heterotrophic biomass	mg COD/L
2	Z_{BN}	Active nitrifier biomass	mg COD/L
3	Z_{BS}	Active S oxidizing biomass	mg COD/L
4	Z_E	Endogenous mass	mg COD/L
5	X_S	Slowly biodegradable substrate	mg COD/L
6	X_{CH}	Adsorbed hydrocarbon substrate	mg COD/L
7	X_I	Particulate unbiodegradable matter	mg COD/L
8	S_{PH}	Phenolic biodegradable soluble compounds	mg COD/L
9	S_M	Mixed substrate biodegradable soluble component	mg COD/L
10	S_V	Volatile biodegradable soluble compounds	mg COD/L
11	S_{CH}	Soluble biodegradable hydrocarbon compounds	mg COD/L
12	S_{NH}	Ammonia nitrogen	mg N/L
13	S_{ON}	Soluble biodegradable organic nitrogen	mg N/L
14	S_{NO}	Nitrate nitrogen	mg N/L
15	S_{SR}	Reduced sulfur	mg S/L
16	S_{SO}	Sulfate	mg S/L
17	S_P	Soluble phosphate	mg P/L
18	S_I	Unbiodegradable soluble substrate	mg COD/L
19	S_O	Oxygen	mg (-COD)/L

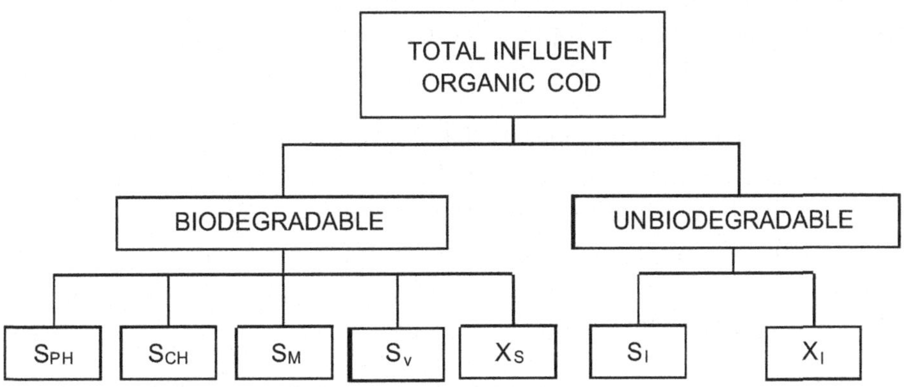

Figure 27-1. Division of Petroleum Refinery Wastewater COD into Constituent Fractions

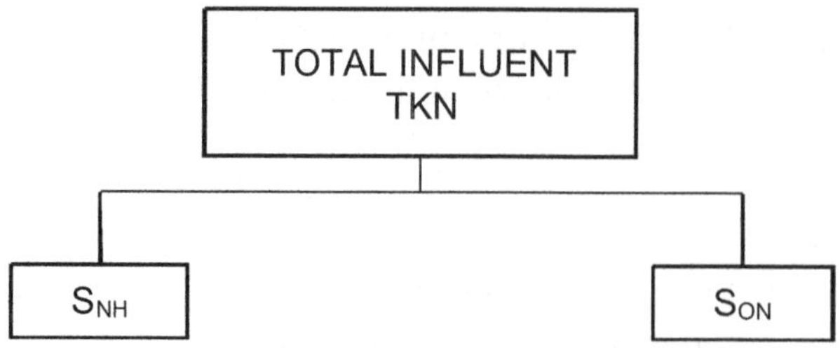

Figure 27-2. Division of Petroleum Refinery Wastewater TKN into Constituent Fractions

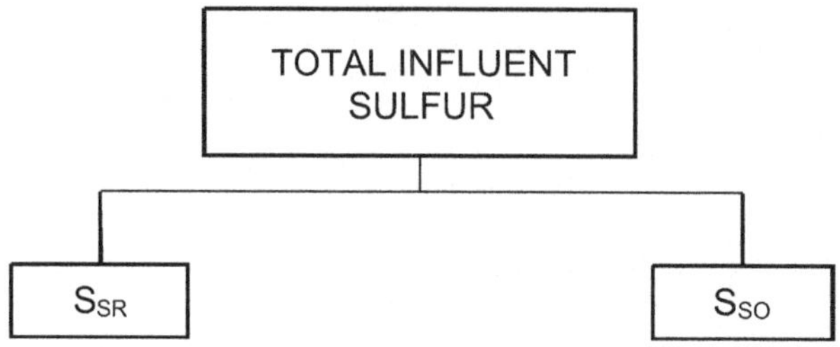

Figure 27-3. Division of Petroleum Refinery Wastewater Sulfur into Constituent Fractions

27.3.2 Model Processes

The following processes were considered to be key mechanisms in determining the behavior of an activated sludge system treating petroleum refinery wastewaters:

- Aerobic and anoxic growth of heterotrophs on mixed organic components,
- Aerobic and anoxic growth of heterotrophs on volatile organic components,
- Aerobic and anoxic growth of heterotrophs on phenolics,
- Aerobic and anoxic growth of heterotrophs on adsorbed hydrocarbons,
- Decay of heterotrophs,
- Hydrolysis of particulate COD,
- Adsorption of hydrocarbons onto sludge mass,
- Volatilization of volatile organic compounds,
- Ammonification of organic nitrogen to ammonia,
- Nitrification of ammonia (i.e., growth of nitrifiers),
- Oxidation of reduced sulfur compounds (i.e., growth of the sulfur oxidizers), and
- Decay of nitrifiers and sulfur oxidizers.

The experimental system used to generate data for model development was an aerated activated sludge process, similar to full-scale applications for refinery wastewater treatment. However, to account for the possibility of unaerated zones in refinery systems in future application of the model, the model structure included four anoxic growth processes to parallel each of the aerobic growth processes. It is stressed that this aspect of behavior in unaerated reactors was not investigated in this study.

Previous studies have considered many of the mechanisms listed above, and as a result, kinetic expressions have been proposed to represent these processes. In a number of cases, the kinetics were the same as those used in ASM1. The following sections identify the reactions incorporated in the petroleum refinery activated sludge model that are unique and describe the form of the rate equations used in the model to represent each process.

Note that a series of switching functions were incorporated into the kinetics to turn off rates if they are oxygen- or nutrient-limited. These switching functions are of the form used in ASM1. For example, in an aerobic system, a function to "switch off" the aerobic growth of heterotrophs as the dissolved oxygen concentration decreases to zero is expressed as follows:

$$AirYes = \frac{S_O}{K_O + S_O} \qquad (27.3.2\text{-}1)$$

where:

AirYes = switching function for dissolved oxygen (S_O)

K_O = switching constant usually of small magnitude (e.g.: 0.1 mg/L)

Similar switching functions were used to express ammonia and phosphorus limitations.

27.3.2.1 Aerobic Growth of Heterotrophs on Mixed and Volatile COD

For this model, it was assumed that heterotrophic organisms utilize soluble substrate for growth in the same fashion regardless of the nature of the organic compound, although the rate of growth may vary. In the petroleum refinery activated sludge system growth on mixed COD (S_M) and volatile organic COD (S_V) were modeled using the Monod expression for biological growth. For example, equation 27.3.2-2 shows the process rate equation for aerobic growth on S_M:

$$\rho_M = \mu_{hM} \frac{S_M}{K_{sM} + S_M} \cdot Z_{BH} \cdot AirYes \cdot NH_3Yes \cdot PO_4Yes \qquad (27.3.2\text{-}2)$$

where:

ρ_M = aerobic growth of heterotrophs on S_M, mg COD/L/d

μ_{hM} = maximum specific growth rate of heterotrophs on S_M, d^{-1}

K_{sM} = half-saturation coefficient for S_M, mg COD/L

AirYes = switching function for dissolved oxygen (S_O)

NH_3Yes = switching function for ammonia nitrogen (S_{NH})

PO_4Yes = switching function for phosphorus (S_P)

27.3.2.2 Aerobic Growth of Heterotrophs on Phenolic COD

Heterotrophic organisms are susceptible to inhibition effects when degrading phenolic compounds (S_{PH}). Current understanding is that the presence of phenolics slows the ability of the organism to degrade the organic and that this effect increases with increasing concentration. For low concentrations the specific growth rate increases as concentration of the inhibitory organic increases. However, a point is reached where the increased inhibition outweighs the increased growth rate, and there is a decline in specific growth rate for further concentration increases. To describe this behavior, an equation similar to the Haldane enzyme model was applied (Rozich et al., 1983, 1985). It is a modified version of the Monod equation and reduces to the Monod equation when the concentration of phenolics drops below inhibitory levels. As applied in this model, it is expressed as follows:

$$\rho_{PH} = \mu_{hPH} \frac{S_{PH}}{K_{sPH} + S_{PH} + \frac{S_{PH}^2}{K_I}} \cdot Z_{BH} \cdot AirYes \cdot NH_3Yes \cdot PO_4Yes \qquad (27.3.2.2\text{-}1)$$

where:

ρ_{PH} = aerobic growth of heterotrophs on S_{PH}, mg COD/L/d

μ_{hPH} = maximum specific growth rate of heterotrophs on S_{PH}, d^{-1}

K_{sPH} = half-saturation coefficient for S_{PH}, mg COD/L

K_I = inhibition coefficient for phenolics (S_{PH}), mg COD/L

27.3.2.3 Adsorption of Hydrocarbon COD

The fate of the hydrocarbon COD component (S_{CH}) was handled somewhat differently from other soluble substrate in the model. It was hypothesized that hydrocarbon COD is adsorbed onto the biomass and the adsorbed compound is utilized for growth. The soluble hydrocarbon (S_{CH}) then acts as a particulate form of COD and takes on the symbol X_{CH} in the model.

The rate of adsorption of hydrocarbon onto the biomass was assumed to follow first order kinetics with respect to the bulk concentration of hydrocarbon COD (S_{CH}) in the system and the concentration of heterotrophs (Z_{BH}). A driving force term was included in the model equation to account for the impact of the number of available adsorption sites. The rate of adsorption decreases as the number of available sites diminishes.

The rate of adsorption of S_{CH} is expressed as follows:

$$\rho_{AD} = K_A \cdot (f_{maxCH} - \frac{X_{CH}}{Z_{BH}}) \cdot S_{CH} \cdot Z_{BH} \qquad (27.3.2.3\text{-}1)$$

where:

ρ_{AD} = adsorption rate of S_{CH}, mg COD/L/d

K_A = adsorption rate constant, L/mg COD/d

f_{maxCH} = maximum adsorption sites factor

27.3.2.4 Aerobic Growth of Heterotrophs on Hydrocarbon COD

Growth on hydrocarbon COD was represented using a modified Monod expression. Instead of growth being a function of the bulk concentration of the soluble COD, the Monod equation was expressed as a function of the surface concentration of the hydrocarbon COD on the heterotrophic biomass. This concentration is estimated as X_{CH}/Z_{BH}, the concentration of adsorbed hydrocarbon COD per unit mass of heterotrophs. The rate is expressed as follows:

$$\rho_{CH} = \mu_{hCH} \frac{\frac{X_{CH}}{Z_{BH}}}{K_{sCH} + \frac{X_{CH}}{Z_{BH}}} \cdot Z_{BH} \cdot AirYes \cdot NH_3Yes \cdot PO_4Yes \qquad (27.3.2.4\text{-}1)$$

where:

ρ_{CH} = aerobic growth of heterotrophs on X_{CH}, mg COD/L/d

μ_{hCH} = maximum specific growth rate of heterotrophs on X_{CH}, d^{-1}

K_{sCH} = half-saturation coefficient for X_{CH}, mg COD/L

27.3.2.5 Volatilization of Volatile Organic COD

The volatilization of volatile organic compounds (S_V) from the wastewater was modeled using a first order equation with respect to S_V, as shown below:

$$\rho_V = K_V \cdot S_V \qquad (27.3.2.5\text{-}1)$$

where:

ρ_V = rate of volatilization of S_V, mg COD/L/d

K_V = volatilization rate constant, d^{-1}

In this case, the value of K_V was assumed constant. Although this is a significant simplification (K_V is a function of characteristics such as air flow rate, temperature and transfer efficiency of aeration equipment), this expression was used as an initial step to include volatilization as a removal mechanism. Experimentation confirmed that a first order representation adequately predicted the stripping of VOCs from an aerated batch of raw petroleum refinery wastewater.

27.3.2.6 Sulfur Oxidizer Growth

Sulfur oxidizers (Z_{BS}) serve to minimize the impact of reduced sulfur compounds on receiving waters by converting them to sulfate species. Sulfur oxidizer growth was expressed as the Monod equation for biological growth with respect to the bulk concentration of reduced sulfur species (S_{SR}) in the activated sludge system.

$$\rho_S = \mu_S \frac{S_{SR}}{K_{sS} + S_{SR}} \cdot Z_{BS} \cdot AirYes \cdot NH_3Yes \cdot PO_4Yes \qquad (27.3.2.6\text{-}1)$$

where:

ρ_S = growth of sulfur oxidizers, mg COD/L/d

μ_S = maximum specific growth rate of sulfur oxidizers, d^{-1}

K_{sS} = half-saturation coefficient for S_{SR}, mg COD/L

27.3.2.7 Sulfur Oxidizer Decay

Sulfur oxidizer decay was modeled in the same way as the decay of the heterotrophs and nitrifiers. These decay processes are modeled with the death-regeneration concept used in ASM1.

27.3.2.8 Additional Model Processes

The remaining processes were modeled using the same relationships as used in ASM1, and the details will not be repeated here. These processes are:

- Anoxic growth of heterotrophs,
- Decay of heterotrophs,
- Hydrolysis of particulate matter,
- Ammonification,
- Nitrifier growth, and
- Nitrifier decay.

27.3.3 Matrix Representation

The matrix format used to present the ASM models can also be used in this case to present all of the compounds, processes and reaction terms and the related stoichiometry in the petroleum refinery wastewater treatment plant model. This representation is shown in Baker (1994).

27.4 Model Calibration

Model calibration was conducted using a subset of experimental results collected for model development and calibration. This section provides a review of the calibration of stoichiometric and kinetic model parameters.

27.4.1 Stoichiometric Parameters

Yields were required for each of the 4 categories of substrates: (1) phenolics, (2) mixed organic compounds, (3) volatile organics and (4) enmeshed hydrocarbons. Yield values were determined by fitting the model predictions to the actual experimental results of both the volatile solids concentration and oxygen utilization rate. Both constant loading and dynamic loading results were used in this calibration. Because only one bulk parameter (VSS) was used to quantify all of the particulate matter accumulated in the system (i.e., active biomass, endogenous mass, unbiodegradable particulate mass), yields could not be determined individually. It was assumed during model calibration that the heterotrophic yield from each of the substrates was the same. A yield of 0.555 g cell COD per g COD utilized was measured, which is about 17 percent lower than the yield typically observed in municipal treatment systems.

Other important stoichiometric parameters that required measurement were determined as follows:

- The nitrogen component of the active organism mass was determined from the experimental systems by measuring the TKN/VSS ratio of the mixed liquor suspended solids.
- The COD/VSS ratio was determined by measuring it in the mixed liquor from the experimental systems. However, the ratio was found to increase with decreasing sludge age. It was hypothesized that this was due to less biodegradation of adsorbed hydrocarbons at lower sludge ages. For this reason, the COD/VSS ratio was set to a value that was comparable to those observed in municipal systems (1.42).

The values of other stoichiometric parameters in the model were found to not significantly affect the model predictions, so these parameters were set to similar values as observed in municipal systems.

27.4.2 Kinetic Parameters

The model contained growth rate parameters (i.e., maximum specific growth rate and half saturation constants) associated with each of the various substrates utilized by the organisms. In many cases, it was found that the model predictions were relatively insensitive to the values of these parameters. For utilization of mixed organics and phenolics for organism growth, the kinetic parameters were set to the same value used successfully in municipal models (i.e., $\mu_{het} = 3.2\ d^{-1}$ and $K_s = 5$ mg COD/L). A much lower value for the maximum specific growth rate was assumed for the utilization of the volatile compounds ($\mu_{het} = 1.5\ d^{-1}$), but again, the model was fairly insensitive to the value of this parameter.

Nitrification rate parameters were determined by fitting model predictions to effluent TKN and nitrate concentrations in the dynamic loading experiments. The nitrifier maximum specific growth rate could be estimated quite accurately because of the sensitivity of the model response to the value of this parameter. For example, Figure 27-4 shows the dynamic TKN response of a bench scale activated sludge system operated at a 5 day SRT and a daily cyclic square-wave loading pattern (the times that feed was started and stopped are indicated). The plot shows the model predictions for three different values of the nitrifier maximum specific growth rate (μ_{AUT}). A μ_{AUT} value of $0.36\ d^{-1}$ provided the best fit to the experimental data. Note that this model used a nitrifier decay rate of $0.04\ d^{-1}$.

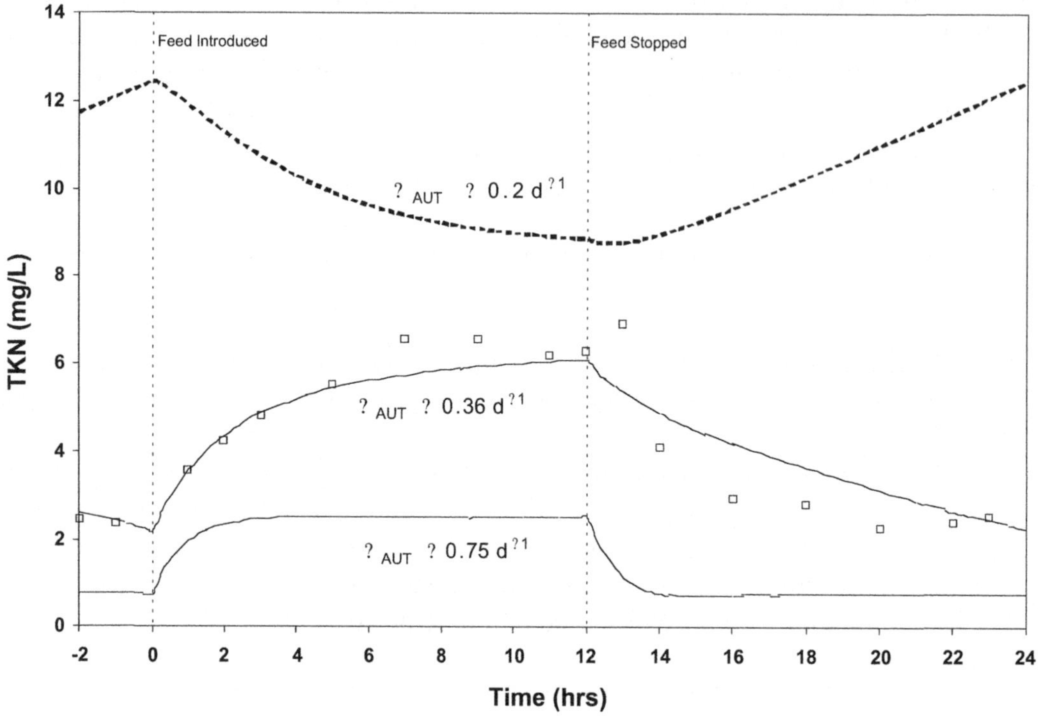

Figure 27-4. Example of Petroleum Refinery Model Calibration for Estimation of the Nitrifier Maximum Specific Growth Rate (μ_{AUT}) by Matching Measured TKN (o) to Model Predictions (Lines Shown for Various Values of (μ_{AUT})

Additional kinetic parameters that were set in the model and an indication of the method used to estimate them are listed below. For more details on the model calibration, the reader is referred to the references at the end of this chapter.

- The kinetic parameters for adsorbed hydrocarbon utilization were based on the kinetics of particulate matter utilization in municipal systems.
- Organism decay rates were determined from the results of the aerobic sludge digestion batch tests.
- Rate constants for hydrolysis of slowly biodegradable substrate were set to values that had been used to simulate municipal systems. However, due to the relatively small amount of this substrate fraction in the influent, the model performance was insensitive to the values of these parameters.
- Ammonification rate parameters were estimated by fitting the model predictions to the effluent TKN and nitrate concentrations in the constant and dynamic loading experiments.
- Nitrification rate parameters were determined by fitting model predictions to effluent TKN and nitrate concentrations in the dynamic loading experiments. The nitrifier maximum specific growth rate could be estimated quite accurately because of the sensitivity of the model response to the value of this parameter.
- There was limited information on the kinetics of sulfur oxidation because the importance of this mechanism only became apparent towards the end of the experimental program when the refinery changed feedstocks to a sour crude oil and the concentration of reduced

sulfur compounds in the influent increased substantially. The kinetic constants used in the model were initial estimates.

27.5 Model Verification

The petroleum refinery model was verified using a subset of the experimental data from the bench scale systems described earlier. This section provides two examples to illustrate the performance of the calibrated model.

For one of the constant loading periods, the bench scale activated sludge system was operated at a 20 day SRT. Table 27-2 shows steady state model predictions compared to measured values for a number of key process variables. All of the parameters were estimated to within 10 percent of the measured variables (the percentage deviation values are shown in parentheses with the predicted values). The largest relative deviation was for the effluent TKN concentration, but the percentage difference was large due to a small absolute value of this variable. The excellent predictions of the steady state process behavior provided a verification of stoichiometric parameters such as the heterotrophic yield of 0.555 g cell COD/g COD utilized.

Table 27-2. Comparison of Observed Values and Model Predictions for Steady State Operation of a Bench Scale Activated Sludge System Operated on Petroleum Refinery Wastewater at a 20-Day SRT

Parameter	Value		Units
	Observed	Predicted	
Mixed liquor volatile suspended solids	1,120	1,167.5 (+4.1%)	mg/L
Effluent COD (filtered)	43	43.4 (+0.9%)	mg/L
Effluent TKN (filtered)	1.2	1.1 (-9.1%)	mg/L
Effluent NO_3-N	10.0	10.9 (+8.3 %)	mg/L
OUR	24.0	25.8 (+7.0%)	mg/L/h

The 20-day SRT system also was operated under dynamic conditions using the cyclic square wave feed pattern. Figure 27-5 shows experimental data for the oxygen uptake rate (OUR) compared to model predictions for one 24 hour cycle. The OUR measured towards the end of the period, when the feed was off, provided an indication of the endogenous OUR. This level was slightly overpredicted. However, the rate of response and the maximum OUR during the feed period were extremely well predicted.

In general, the model developed for petroleum refinery activated sludge systems provided good predictions of process behavior. In the complete study described in Baker (1994), examples are shown of model predictions for a bench scale activated sludge system operated under steady state and dynamic conditions at 5, 10 and 20 day SRTs. A single set of kinetic and stoichiometric parameters were used to provide a reasonable fit to all of these conditions. The study indicated that a useful mechanistic model could be developed for petroleum refinery activated sludge systems using ASM1 as a basis.

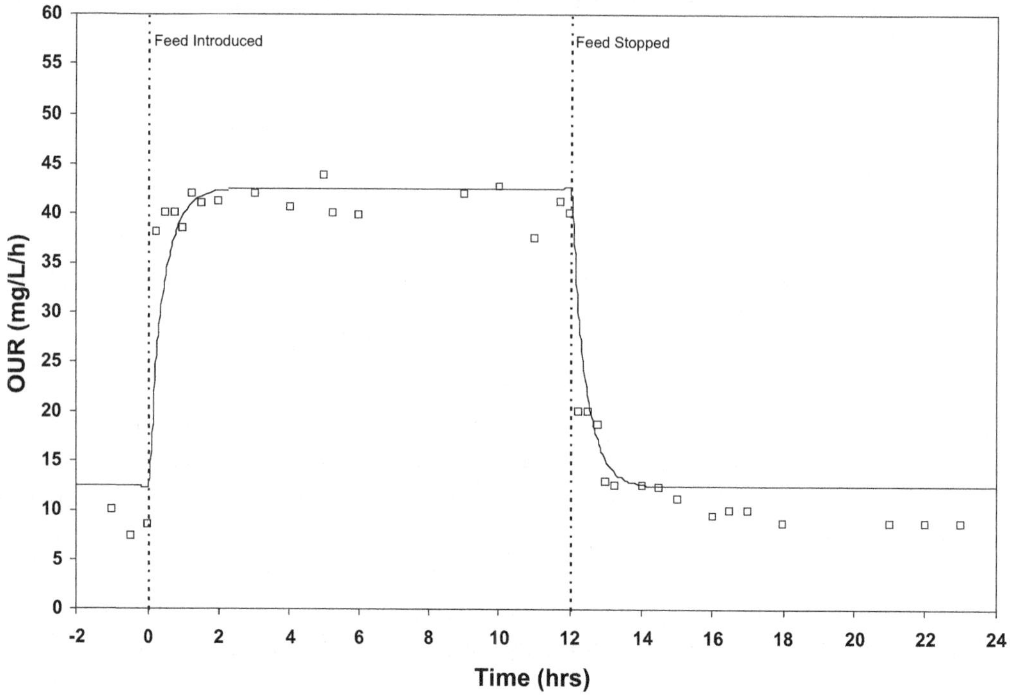

Figure 27-5. Comparison of Petroleum Refinery Predictions (Solid Line) and Measured Values (o Symbols) of OUR during a Cyclic Square Wave Feed Pattern on the 20-day SRT Bench-Scale System

27.6 References

The following references provide a more detailed description of the petroleum refinery and petrochemical plant models.

Baker, A.J. 1993. *Activated sludge treatment of petroleum refinery wastewater: experimental behavior.* Master's Thesis, McMaster University, Hamilton, ON, Canada.

Baker, A.J. 1994. *Modeling activated sludge treatment of petroleum and petrochemical wastes.* Ph.D. Thesis, McMaster University, Hamilton, ON, Canada.

Baker, A.J., and P.L Dold. 1992. Activated sludge treatment of petroleum refinery wastewater: part I – experimental behavior. *Water Sci. Tech.* 26(1-2):333-343.

Jones, R.M., P.L. Dold, A.J. Baker, and T. Briggs. 1995. Optimization of a biological wastewater treatment process at a petrochemical plant using process simulation. In *Proc. of the 68th Annual Conference and Exposition of the Water Environment Federation*, Miami Beach, FL. Alexandria, VA: Water Environment Federation.

The following is a list of references utilized in the development of the petroleum refinery and petrochemical plant models.

Dold, P.L. 1990. Current practice for treatment of petroleum refinery wastewater and toxics removal. *Water Poll. Res. J. Canada* 24(3):363.

Ford, D.L. 1971. Analytical parameters of petrochemical and refinery wastewaters. *J. Water Poll. Control Fed.* 43(8):1712-1723.

Gaudy, A.F., Jr. 1988. Practical methodology for predicting critical operating range of biological systems treating inhibitory substrates. *J. Water Poll. Control Fed.* 60(1): 77-85.

Godrej, A.N., and J.H. Sherrard. 1988. Kinetics and stoichiometry of activated sludge treatment of a toxic organic wastewater. *J. Water Poll. Control Fed.* 60(2):221-226.

Greenshields, J. 1987. *Environmental contaminants in petroleum refinery wastewater: an assessment of current information and a monitoring approach.* Study prepared for the Joint Industry/Government Technical Committee, Petroleum Refining Sector of the Municipal Industrial Strategy for Abatement Program. Toronto, ON, Canada: Ontario Ministry of the Environment.

Kincannon, D.F. 1983. Removal mechanisms for toxic priority pollutants. *J. Water Poll. Control Fed.* 55(2):157-163.

Mueller, J.A., K. Wu, and S. Kaczmarek. 1985. Nitrification in Refinery Wastewater Treatment. In *Proc 40th Annual Purdue University Industrial Wastewater Conference* (pp. 507-522). Ann Arbor, MI: Ann Arbor Science.

Rebhun, M., and N. Galil. 1988. Biotreatment inhibition by hazardous compounds in an integrated oil refinery. In *Proc 42nd Annual Purdue University Industrial Wastewater Conference* (pp. 163-174). Ann Arbor, MI: Ann Arbor Science.

Rebhun, M., and N. Galil. 1988. Inhibition by hazardous compounds in an integrated oil refinery. *J. Water Poll. Control Fed.* 60(1):1953-1959.

Rhee, C.H., P.C. Marty, and J.G. Kremer. 1987. Removal of oil and grease in the hydrocarbon processing industry. In *Proc 42nd Annual Purdue University Industrial Wastewater Conference* (pp. 143-150). Ann Arbor, MI: Ann Arbor Science.

Rozich, A.F. 1983. Predictive model for treatment of phenolic wastes by activated sludge. *Water Res.* 17(10):1453-1466.

Rozich, A.F., A.F. Gaudy, Jr., and P.D. D'Adamo. 1985. Selection of growth rate model for activated sludges treating phenol. *Water Res.* 19(4):481-490.

Rozich, A.F., and A.F. Gaudy, Jr. 1985. Response of phenol-acclimated activated sludge process to quantitative shock loading. *J. Water Poll. Control Fed.* 57(7):795- 804.

Stover, E.L., and D.F Kincannon. 1982. Biological treatability of specific organic compounds found in chemical industry wastewaters. In *Proc 36th Annual Purdue University Industrial Wastewater Conference* (pp. 1-16). Ann Arbor, MI: Ann Arbor Science.

Stover E.L., D.E. McCartney, F. Dehkordi, and D.F. Kincannon. 1983. Variability analysis during biological treatability of complex industrial wastewater for design. In *Proc. 37th Annual Purdue University Industrial Wastewater Conference* (pp. 773-783). Ann Arbor, MI: Ann Arbor Science.

CHAPTER 28.0

CASE STUDY: MODEL DEVELOPMENT AND CALIBRATION FOR A PETROCHEMICAL WASTEWATER TREATMENT PLANT (1)

28.1 Background

Many of the benefits of simulating municipal activated sludge processes using mechanistic computer models are applicable to industrial applications. However, while the municipal activated sludge models provide a suitable starting point for the formulation of an industrial model, often there are significant differences between the composition of the wastewaters, and a specific model formulation usually is required for each type of industrial wastewater. After formulation of a suitable model, wastewater characterization and parameter calibration similar to municipal applications is required.

This chapter provides an example of development and calibration of a simulation model for a petrochemical wastewater treatment plant in Canada. The approach in this exercise was to:

- Gather data on the full-scale plant and the composition of individual waste streams;
- Formulate a mechanistic model for the system that accounts for the fate of specific dominant components in the waste streams;
- Conduct a series of batch tests using sludge from the full-scale plant together with segregated wastewater streams to provide data for model calibration; and
- Apply the calibrated model to investigate changes in plant operating mode and loading.

The petrochemical treatment plant model was based on the petroleum refinery activated sludge model described in Chapter 27.0, with only minor modifications to account for the fate of different wastewater components in this system. The experimental program used in the development of this model involved a series of batch tests on various wastewater streams rather than operation of a continuous flow bench scale process as in the petroleum refinery treatment system model development.

Note: The nomenclature used in this chapter necessarily is unique to the industrial sector, and may not agree with terms listed in Appendix A. However, care has been taken to define variables and terms as they occur.

28.1.1 Plant Description

This case study is about an Ontario, Canada petrochemical that was concerned with the operation of the activated sludge system treating its wastewaters. The development of a model for the plant was part of an initiative to evaluate different operating strategies to optimize system performance. The activated sludge plant consists of four aeration basins and three clarifiers and receives a number of influent streams that feed into the system at different points. The typical configuration of the plant is shown in Figure 28-1. The tankage consists of an aerated equalization tank (PT) and four aeration tanks (AT-1, AT-2, AT-3 and AT-4). Although there are four secondary clarifiers, only three were operating. The fourth clarifier was used as a holding tank for spills and wastes trucked from various locations around the petrochemical complex (indicated as Stream 5 in the schematic). Table 28-1 lists the liquid volumes of the equalization and aeration tanks, and depth and surface areas of each of the secondary clarifiers. The historical database at the plant included information on a number of routine monitoring parameters, of which a summary is included in Table 28-2.

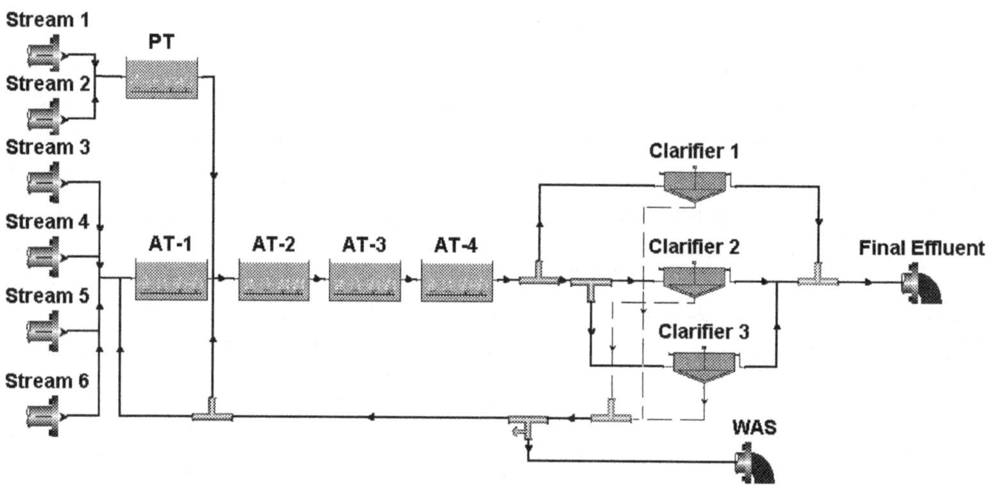

Figure 28-1. Typical Configuration of the Petrochemical Wastewater Treatment Plant

Table 28-1. Dimensions of Unit Tankage at the Petrochemical Wastewater Treatment Plant

Equalization Tank:	Liquid Volume (m³)	
T-1	3,206	
Aeration Tanks:	**Liquid Volume (m³)**	
T-2, T-3, T-4, T-5 (each)	6,000	
Secondary Clarifiers:	**Depth (m)**	**Surface Area (m²)**
(Each)	3.5	616

Table 28-2. Historical Averages for Routine Monitoring Parameters at the Petrochemical Wastewater Treatment Plant

Stream	Average Daily Flow (m^3/d)	DOC[1] (mg/L)	Phenolics (mg/L)	NH_3-N (mg/L)	TSS (mg/L)
1	4,240	31	0.07	3.9	20
2	12,600	47	0.4	1.6	119
3	775	440	0.07	3.5	99
4	2	81,000	-	-	-
5	500	884	8.7	3.1	-
6	60	2,000	0.3	3.9	1,230

[1]DOC: Dissolved organic carbon.

28.2 Model Formulation

The development of the petroleum refinery activated sludge model (as described in Chapter 27.0) significantly simplified the development of the petrochemical model. The key modifications involved the identification of components that are important in the petrochemical system, and then the identification of the processes acting on those components.

Development of the model for the activated sludge system treating the petrochemical wastewaters is described below.

28.2.2 Model Components

Wastewater characterization forms an integral part of model development. A structured model is based on differentiating between different components in the influent wastewater, and modeling the fate of each component within the system. The model for the petrochemical refinery wastewater treatment plant developed in this study differentiated between the wastewater components shown in Table 28-3. Also included in the table are the various biomass components expected to occur in the system. Figure 28-2 shows the division of influent COD, TKN and sulfur into the various components.

Table 28-3. List of Components in the Petrochemical Refinery Activated Sludge Model

Comp	Symbol	Definition	Units
1	Z_{BH}	Active heterotrophic biomass	mg COD/L
2	Z_{BN}	Active nitrifier biomass	mg COD/L
3	Z_E	Endogenous mass	mg COD/L
4	X_S	Slowly biodegradable substrate	mg COD/L
5	X_{CH}	Adsorbed hydrocarbon substrate	mg COD/L
6	X_I	Particulate unbiodegradable matter	mg COD/L
7	X_{ON}	Particulate organic nitrogen	mg COD/L
8	S_{TBA}	Tertiary butyl alcohol	mg COD/L
9	S_{AN}	Acetonitrile	mg COD/L
10	S_{PH}	Phenolic biodegradable soluble compounds	mg COD/L

Table 28-3. List of Components in the Petrochemical Refinery Activated Sludge Model (cont.)

Comp	Symbol	Definition	Units
11	S_V	Volatile biodegradable soluble compounds	mg COD/L
12	S_M	Mixed substrate biodegradable soluble compounds	mg COD/L
13	S_{CH}	Soluble hydrocarbon compounds	mg COD/L
14	S_{NH}	Ammonia nitrogen	mg N/L
15	S_{ON}	Soluble organic nitrogen	mg N/L
16	S_{NO}	Nitrate nitrogen	mg N/L
17	S_{SR}	Reduced sulfur	mg S/L
18	S_{SO}	Sulfate	mg S/L
19	S_P	Soluble phosphate	mg P/L
20	S_I	Unbiodegradable soluble substrate	mg COD/L
21	S_O	Oxygen	mg (-COD)/L

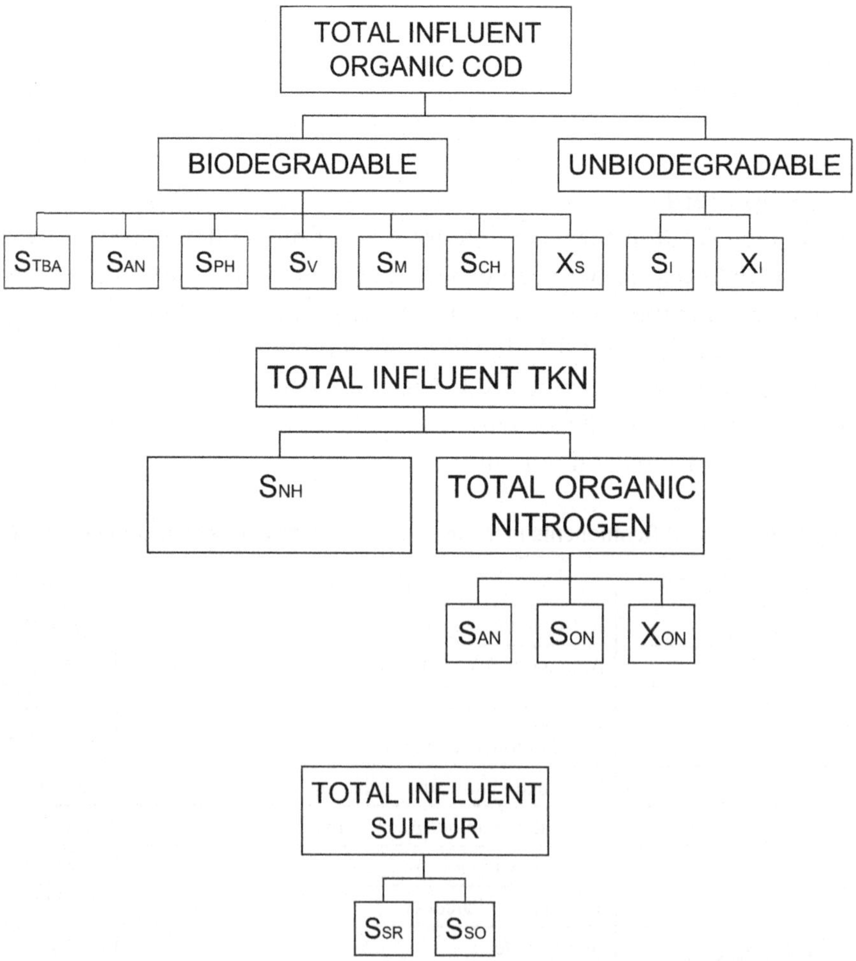

Figure 28-2. Petrochemical Plant Wastewater Characterization of the Total Influent COD, TKN and Sulfur

In order to determine the components to include in the model, a thorough analysis of the full-scale wastewater treatment system at the petrochemical site was conducted. This involved consideration of the six different streams entering the plant, each differing in composition and strength. Monitoring programs performed by the plant personnel provided extensive data on the operation of this facility and the types of compounds treated.

In general, wastewater input streams to the petrochemical activated sludge plant consist largely of organic carbon in the form of tertiary butyl alcohol, phenolics, volatile organics, acetonitrile and various other hydrocarbons and mixed organic compounds. Nitrogen input to the system is principally that associated with low concentrations of free ammonia and organic nitrogen. The major component of the organic nitrogen is the nitrogen content of the acetonitrile. Sulfur compounds were also identified as being important. Plant personnel indicated that sodium meta-bisulfite is added during one of the manufacturing processes as an oxygen scavenger, resulting in some residual material in the influent wastewater. Two types of sulfur species were therefore considered; these were reduced sulfur species and sulfate sulfur. Biomass in the system is composed of heterotrophs and nitrifiers.

28.2.3 Model Processes

The petrochemical plant activated sludge model is very similar to the petroleum refinery activated sludge model presented in Chapter 27.0. A number of modifications were required to account for the fate of the different wastewater components included. Listed below are the processes that are different in the petrochemical plant model. The details of the model are not presented here, but are available elsewhere (Baker, 1994).

28.2.3.1 Aerobic growth of heterotrophs

For this model, heterotrophic organisms utilize mixed soluble substrate (S_M) and volatile compounds (S_V) in the same manner as the petroleum refinery activated sludge model. However, the plant staff was interested in tracking tertiary butyl alcohol (S_{TBA}) as a separate soluble substrate. For this reason, a separate process was added for heterotrophic growth on this compound. The Monod expression for growth was used for this substrate, as shown in equation 28.2.3.1-1:

$$\rho_{TBA} = \mu_{hTBA} \frac{S_{TBA}}{K_{sTBA} + S_{TBA}} \cdot Z_{BH} \cdot AirYes \cdot NH_3Yes \cdot PO_4Yes \qquad (28.2.3.1\text{-}1)$$

where:

μ_{hTBA} = maximum specific growth rate on S_{TBA}, d^{-1}

K_{sTBA} = half-saturation coefficient for S_{TBA}, mg COD/L

AirYes = switching function for dissolved oxygen (S_O)

NH_3Yes= switching function for ammonia nitrogen (S_{NH})

PO_4Yes = switching function for phosphorus (S_P)

The switching functions were applied to represent growth limitations due to reduced concentrations of dissolved oxygen (S_O), ammonia nitrogen (S_{NH}), or phosphorus (S_P). The form of the switching functions is shown in Chapter 27.0.

28.2.3.2 Hydrolysis of acetonitrile

Acetonitrile was one of the key sources of nitrogen in the system, and so an appropriate representation of the behavior of this component was important. As an initial hypothesis, acetonitrile was assumed to undergo hydrolysis in the process, with the products being ammonia and mixed substrates (S_M). The hydrolysis was assumed to follow a first order reaction with respect to the concentration of acetonitrile (S_{AN}), and the concentration of heterotrophs (Z_{BH}), as shown in equation 28.2.3.2-1:

$$\rho_{AN} = K_R \cdot S_{AN} \cdot Z_{BH} \qquad (28.2.3.2\text{-}1)$$

where:

K_R = acetonitrile hydrolysis rate, L/(mg•d)

28.2.3.3 Hydrolysis of particulate organic nitrogen

The incorporation of particulate organic nitrogen as a nitrogen source was identified by analysis of the influent characteristics. This component was neglected in the development of the petroleum refinery model as it was found to be negligible in that situation. The hydrolysis process used the same kinetics as ASM1. That is, the hydrolysis rate of slowly biodegradable substrate (X_S) was multiplied by the ratio of particulate organic nitrogen (X_{ON}) to X_S.

28.2.3.4 Sulfur oxidation

All of the processes described above are biologically mediated. There are also physical-chemical processes that were considered important for describing the behavior of the treatment plant. These processes are, volatilization of volatile organic compounds (described in Chapter 27.0), and chemical oxidation of reduced sulfur.

The oxidation of reduced sulfur compounds in the petrochemical facility was modeled as a first order reaction with respect to the concentrations of reduced sulfur compounds (SSR). The rate equation for this process (ρ_{SO}) is shown in equation 28.2.3.4-1:

$$\rho_{SO} = K_{SR} \cdot S_{SR} \cdot AirYes \qquad (28.2.3.4\text{-}1)$$

where:

K_{SR} = sulphite oxidation rate, d^{-1}

28.2.3.5 Additional model processes

Additional model processes were as described in Chapter 27.0.

28.2.4 Matrix Representation

The matrix format used to present the ASM1 activated sludge models can also be used in this case to present all of the compounds, processes and reaction terms and the related stoichiometry in the petrochemical wastewater treatment plant model. This representation is shown in Baker (1994).

28.3 Calibration

This section gives an overview of the experimental procedures used to collect data for calibration of the petrochemical plant model, and the determination of wastewater components and model parameters from this data. The basis of this procedure was as follows:

- Sampling and analysis of each wastewater stream provided characteristics such as total and soluble COD, TKN, ammonia and sulfate, as well as concentrations of individual compounds such as tertiary butyl alcohol, phenolics, acetonitrile and the various volatile organic compounds.
- A series of batch experiments was conducted in which return activated sludge (RAS) was mixed with a measured volume of an influent stream, and the responses of soluble COD, soluble TKN and OUR were measured.

To illustrate the batch test calibration procedure, an example of the OUR response in a batch test conducted on Stream 2 is presented in Figure 28-3. Depending on what part of the batch test was being considered, different model coefficients were adjusted. This was because various organic compounds dominated different parts of the OUR response. Four main areas of interest are shown in Figure 28-3. In Area A, the readily biodegradable component (S_M) dominated the behavior of the batch test. The area under the OUR curve in this section of the response provided an estimate of the concentration of S_M. In Area B, more slowly biodegradable components dominated the behavior of the response. In this case, it was the volatile organic component (S_V). In other cases, tertiary butyl alcohol would control the response in this region of the curve. With a separate measurement of the concentration of these compounds, the OUR response can be used to estimate the values of the kinetic coefficients for the removal of these components. Area C of the response was indicative of the adsorption and utilization of the hydrocarbon component (S_{CH}), so adjustment of the relevant kinetic parameters calibrated the model to this part of the response. Finally, endogenous processes controlled Area D of the response. Separate batch tests on sludge only were conducted to estimate the endogenous decay rate. The volatilization rate parameter was estimated from a separate emissions study conducted on the plant, and confirmed in aerated batch tests.

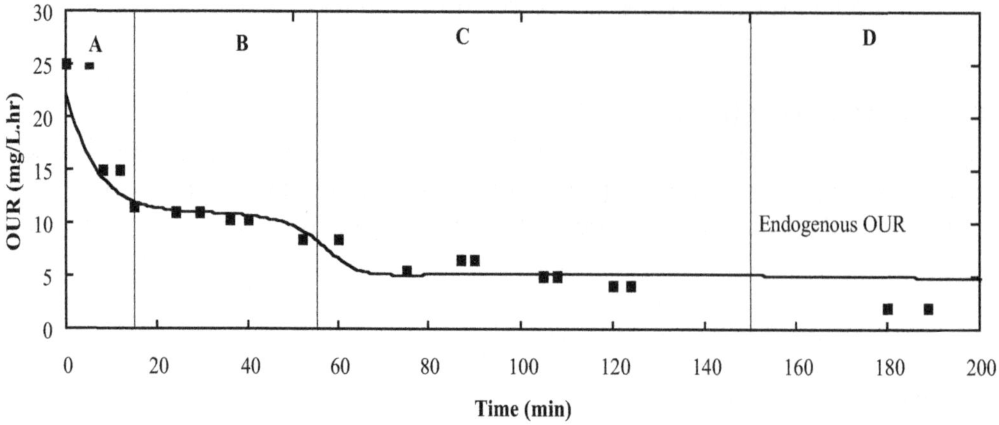

Figure 28-3. Simulation of a Batch Test for Stream 2, Illustrating the Different Regions of the OUR Response Curve

The Stream 4 batch test response is another useful case to examine. Stream 4 comprised the main input of phenolic compounds to the process so that batch tests on this influent could be used to calibrate the parameters for growth on phenolics. Initial parameter estimates were obtained from the literature to establish a range of reasonable values (Rozich et al., 1985). The parameter values were then adjusted to allow the simulated OUR response to match the actual. An OUR response from one such batch test is shown in Figures 28-4 and 28-5. The response is characteristic of a situation where the presence of an inhibitory component (S_{PH} in this case) initially suppresses the OUR, and the OUR gradually increases as the component is consumed by the organisms. The two simulation cases here indicate that the simulated response is quite sensitive to the values of K_I (Figure 28-4) and K_{SPH} (Figure 28-5). The lack of fit in these examples also indicates that an alternative form of kinetic equation may need to be investigated. One source of information on kinetic equations used to describe the effect of inhibitory substrates on growth rates is the text by Grady et al. (1999), where a number of other references on the topic are listed.

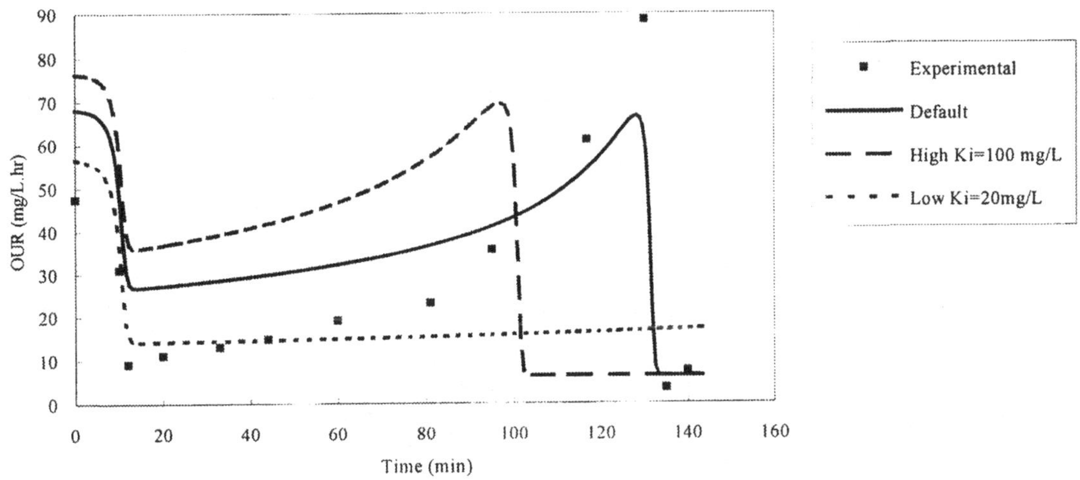

Figure 28-4. Impact of the Inhibition Constant, K_I, Associated with Growth of Organisms on Phenolic Compounds on the Simulated OUR Response of a Stream 4 Batch Test (Default K_I =60 mg/L)

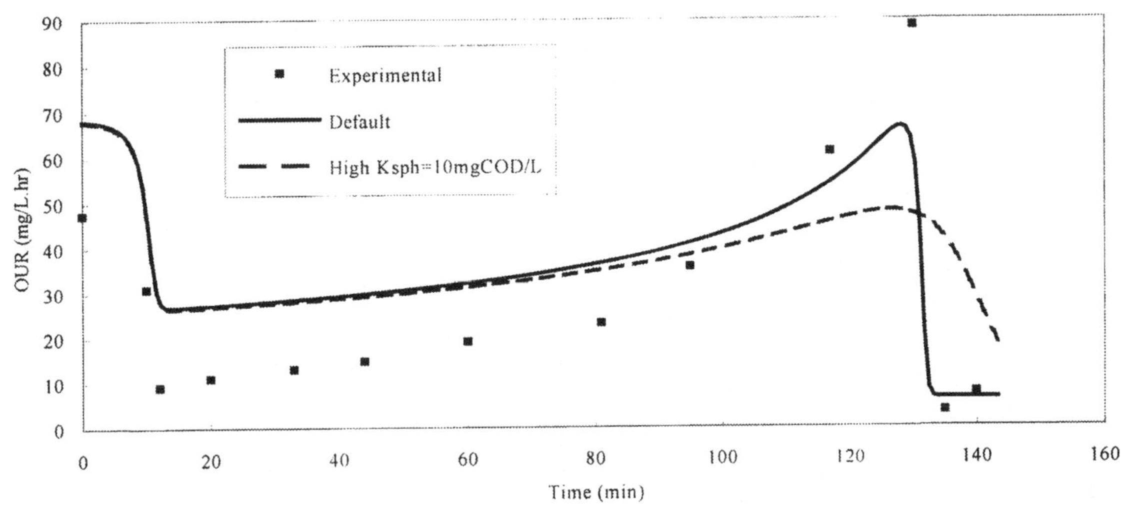

Figure 28-5. Impact of the Half-Saturation Coefficient, K_S, Associated with Growth of Organisms on Phenolic Compounds on the Simulated OUR Response of a Stream 4 Batch Test (Default K_{SPH} = 1 mg/L)

28.4 Application

This section provides two examples of how the calibrated model was used to evaluate operational concerns at this facility.

28.4.1 Case 1: Influence of Operating Configuration on Volatile Organic Compound (VOC) Emissions

Simulations of the plant in the current operating mode (Figure 28-1) indicated VOC emissions of 75 kg COD/d. An independent study conducted on the VOC emissions from this plant identified that the bulk of the emissions originated from the equalization tank (PT). In order to reduce these emissions it was proposed to convert PT into another aeration basin. A simulation was conducted where RAS was directed from the settling basins back into PT. This provided biomass to degrade the volatile compounds before being stripped from the surface of the tank. The impact of this change in configuration was to reduce the total plant VOC emissions to 20 kg/d.

28.4.2 Case 2: Spike Loading

Tertiary butyl alcohol contributes a significant portion of the organic load at the treatment plant. Most of the loading of this compound enters the facility in Stream 3. For the simulated spike load, the concentration of the alcohol in Stream 3 was increased to 20,000 mg COD/L for 2.4 hours from a baseline concentration of 1,380 mg COD/L. Plant operations staff had indicated that this was a realistic scenario for the process.

The rationale for the current configuration had been that the more concentrated, more slowly degradable or inhibitory streams should be fed to AT-1, where a longer hydraulic retention time could be provided. The simulated plant response shown in Figure 28-6 supports this choice of operating mode. Although a significant response in soluble COD was observed in AT-1, there was little or no response in the downstream aeration tanks. One concern that did arise during an examination of the results was the possibility of nutrient deficiencies during increases in organic load. The simulations in this case showed ammonia-N concentrations dropping below 0.02 mg/L in AT-2 through AT-4 during the spike loading.

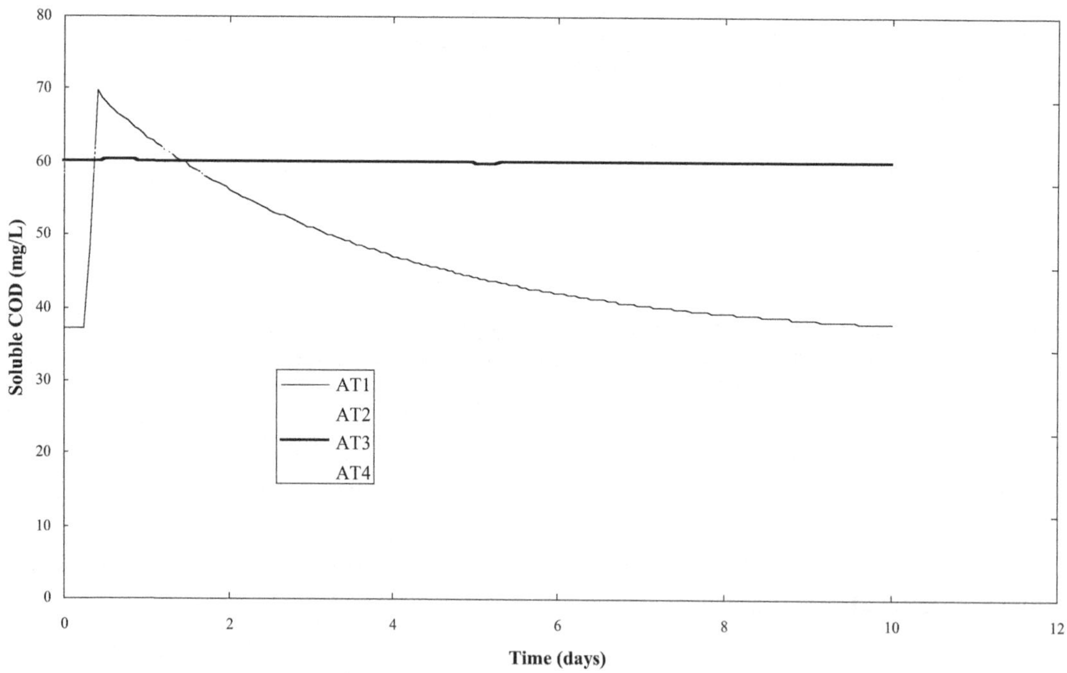

Figure 28-6. Impact on Bioreactor Soluble COD in Response to a 14-fold Increase in Tertiary Butyl Alcohol Concentration for a Period of 2.4 Hours

28.5 References

Baker, A.J. 1994. *Modeling activated sludge treatment of petroleum and petrochemical wastes.* Ph.D. Thesis, McMaster University, Hamilton, ON, Canada.

Grady, C.P.L, G.T. Daigger, and H.C. Lim. 1999. *Biological wastewater treatment* 2nd Edition. New York, NY: Marcel Dekker, Inc. 1076 pp.

Jones, R.M., P.L. Dold, A.J. Baker, and T. Briggs. 1995. Optimization of a biological wastewater treatment process at a petrochemical plant using process simulation. *Proc. of the 68th Annual Conference and Exposition of the Water Environment Federation*, Miami Beach, FL. Alexandria, VA: Water Environment Federation.

Jones R. M., P. L. Dold, A. Baker, and T. Briggs. 1994. Simulation for optimization of a biological wastewater treatment process at a petrochemical plant. In *Proc 49th Annual Purdue University Industrial Wastewater Conference* (pp. 737-744). Ann Arbor, MI: Ann Arbor Science.

Rozich, A.F., A.F. Gaudy, Jr., and P.D. D'Adamo. 1985. Selection of growth rate model for activated sludges treating phenol. *Water Res.* 19(4):481-490.

CHAPTER 29.0

CASE STUDY: MODEL DEVELOPMENT AND CALIBRATION FOR A PETROCHEMICAL WASTEWATER TREATMENT PLANT (2)

29.1 Background

This chapter provides another example of development, calibration and verification of a simulation model for a petrochemical wastewater activated sludge plant in Ohio. The approach in this exercise was to:

- Gather data on the full-scale plant and composition of the important waste streams
- Formulate a mechanistic model based on the ASM-2 logic (Henze et al., 2000) that accounts for the fate of specific dominant components in the wastewater
- Conduct a series of batch tests using full-scale plant biomass and segregated wastewater streams to provide data for model calibration
- Verify model calibration by evaluating the impact of shock loading on a pilot plant operated with full-scale plant biomass
- Apply the calibrated and verified model to investigate spill diversion / bleed back strategies to address shock-loading events in the full-scale plant

The steps in formulating this petrochemical treatment plant model were similar to those of previous cases but with several differences:

- There was no "slowly biodegradable COD" fraction in the influent; all degradable COD was readily biodegradable. Each unique wastewater fraction was assigned a "specific biomass" that was responsible for its degradation
- A "predator" fraction was added to account for biomass "death and regeneration" process in the existing ASM-2 Model

---

Note: The nomenclature used in this chapter is unique to this industrial sector, and may not agree with terms listed in Appendix A. However, care has been taken to define variables and terms as they occur.

29.2 Plant Description

This case study concerns a wastewater treatment facility serving a thermoplastic resin plant. The development of the model was part of the effort to provide the operators a better spill diversion and bleed back strategy to improve the performance of this activated sludge system. The plant consists of two trains of completely mixed activated sludge units. Two batch wastes were periodically released to the main wastewater stream. Referred to as V-353 and MeOH, these two waste streams are significantly different in organic concentrations and characteristics from the main wastewater stream. The typical dimensions and capacities of the activated sludge components are listed in Table 29-1.

Table 29-1. Dimensions and Capacities of Unit Processes at the Wastewater Treatment Plant for a Single Train

Process	Value	Units
Aeration Tank		
Volume	530,000	gallons
Air flow rate	368-917	SFCM
Secondary Clarifier		
Diameter	40	feet
Surface area	1,256	ft^2
SWD	16	feet
Return sludge flow rate	200-350	gpm
WAS flow rate	Sufficient to maintain SRT at 125 days	

The historical wastewater characteristics are summarized in Table 29-2.

Table 29-2. The Historical Average Values of the Wastewater Characteristics

Parameters	Values	Unit
Flow, average per train	200	Gpm
Flow, peak per train	300	Gpm
COD, total	261	mg/L
COD, soluble	223	mg/L
BOD_5, carbonaceous	150	mg/L
Oil and grease	4	mg/L
Ammonia-N	78	mg/L as N
TKN	80	mg/L as N
(Nitrite + Nitrate)-N	0.2	mg/L as N
pH	9.1	
Total dissolved solids	4,088	mg/L
TSS	38	mg/L
VSS	18.8	mg/L
Total alkalinity	222	mg/L as $CaCO_3$
Phosphate	Trace	Added as nutrient

29.3 Model Formulation

Using the ASM2 as a starting point, the key steps involved identification of the components that are important to the system, that is, the V-353 and MeOH organic content, and their associated removal mechanisms. The following sections describe model development.

29.3.1 Model Components

This structured model is based on modeling the fate of each component based on influent component concentrations, reasonable removal mechanisms and their associated rate constants. The influent and biomass components of this model are given in Table 29-3. This wastewater consists of largely simple organic acids, ketones, alcohols and other mixed organic solvents. High ammonia concentrations and some insoluble oil are present in the wastewater. The biomass in the system is assumed to be composed of four heterotrophic groups and one nitrifying group.

Table 29-3. List of Components in the Petroleum Wastewater Treatment Activated Sludge Model

Symbol	Definition
S_V	Soluble strippable or volatile COD
S_M	Soluble mixed COD
S_{MeOH}	MeOH spill COD
S_{chex}	V-363 Tank content COD
S_h	Hydrolysate COD - generated by hydrolysis of biomass
S_i	Inert COD
S_{NH}	Ammonia - N
S_P	Phosphate - P
S_{DO}	Dissolved oxygen – O_2
S_{NO}	(Nitrate + Nitrite) - N
Z_{bH}	Biomass - general heterotrophs, COD
Z_{bMeOH}	Biomass - specific for MeOH, COD
Z_{bChex}	Biomass - specific for V-363 content, COD
Z_{bHyd}	Biomass - specific for hydrolysate, COD
Z_{bN}	Biomass - nitrifiers, COD
X_S	Degradable volatile solids, COD
Z_E	Endogenous particulate residue, COD
FdAsh	Feed Ash content, TSS

Figure 29-1 shows the general relationship between the feed components and the corresponding biomass.

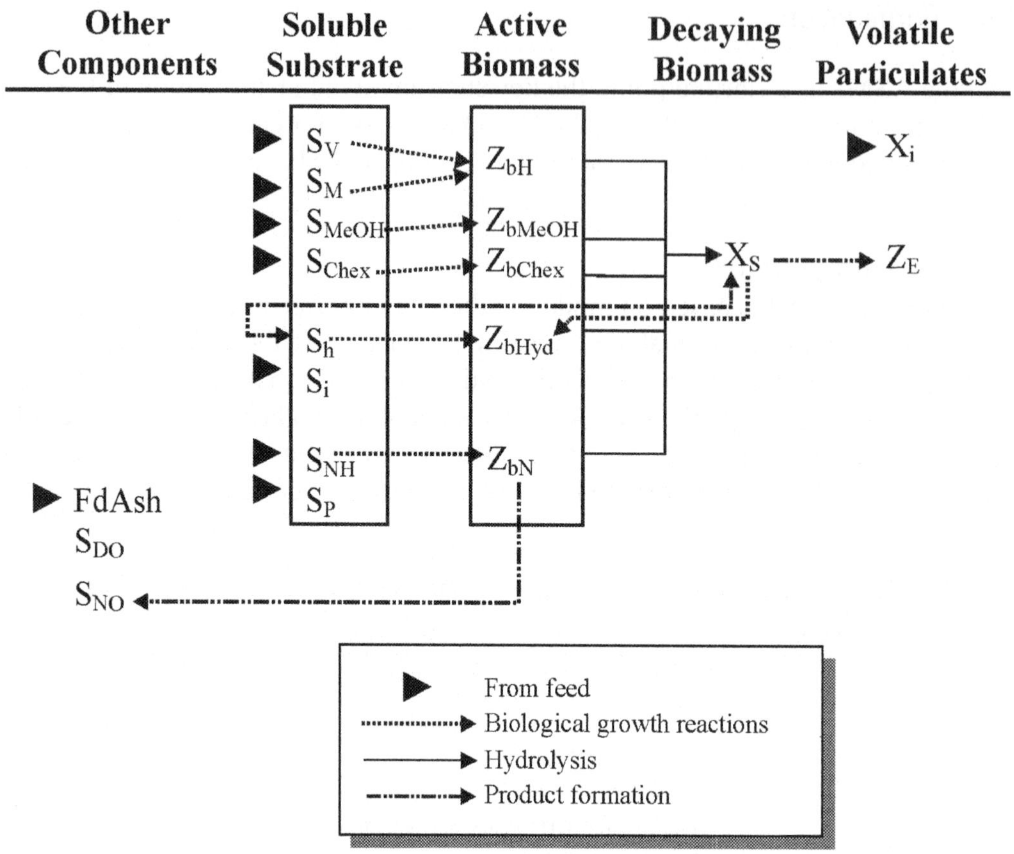

Figure 29-1. Graphical Representation of Model Components and Their Relationships

29.3.2 Model Processes

The basic biological kinetics equations used in the ASM2 were adapted in this study. Both biological and physico-chemical processes are described below.

For heterotrophs, growth was based on modified Monod kinetics and biomass decay, based on a modified decay-regeneration model with a "predator" biomass term introduced to account for the endogenous respiration of the whole biomass.

A Haldane expression was used to describe the nitrification kinetics because of the high concentration of ammonia-N in the feed stream.

All feedstream substrates were modeled as soluble components because of the low VSS and the absence of oil and grease components. Therefore, no substrate hydrolysis steps were incorporated, except for biomass particulate decay by the "predator" biomass component.

Stripping of volatile organic COD was also included in the matrix. This was estimated by assuming conditions based on equilibrium control where diffused air bubbles were saturated by the volatile components with respect to their liquid phase concentration. The interphase relationship was calculated based on an experimentally derived Henry's Law coefficient.

Dissolved oxygen concentration was modeled by using the field-derived first order oxygen transfer coefficient, k_La. An off-gas technique was used to estimate the mass transfer characteristics of the jet aeration system (Redmon et al., 1983).

29.3.3 Matrix Representation

The matrix format used in the ASM2 (Henze et al., 2000) can be used to express all the components, processes, reaction terms and the related stoichiometry of this model.

29.4 Calibration

This section gives an overview of the experimental procedures used to collect data for calibration of this petrochemical wastewater activated sludge model and the determination of wastewater components and model parameters from these data. The procedure was as follows:

- Sampling and analysis of the main wastewater stream provided characteristics such as total and soluble COD, and ammonia.
- A series of batch experiments with return activated sludge (RAS) from the full-scale plant were conducted to measure the response to background organics in the main wastewater stream and to those in the two batch streams. 3-liter reactor volumes were used and operated at a constant temperature typical of the full-scale plant following the procedures described by Ekama et al. (1986) and by Kappeler and Gujer (1992), respectively.

The general procedure is outlined graphically in Figure 29-2. Both model calibration and verification are shown in this figure; model verification is described in the next section.

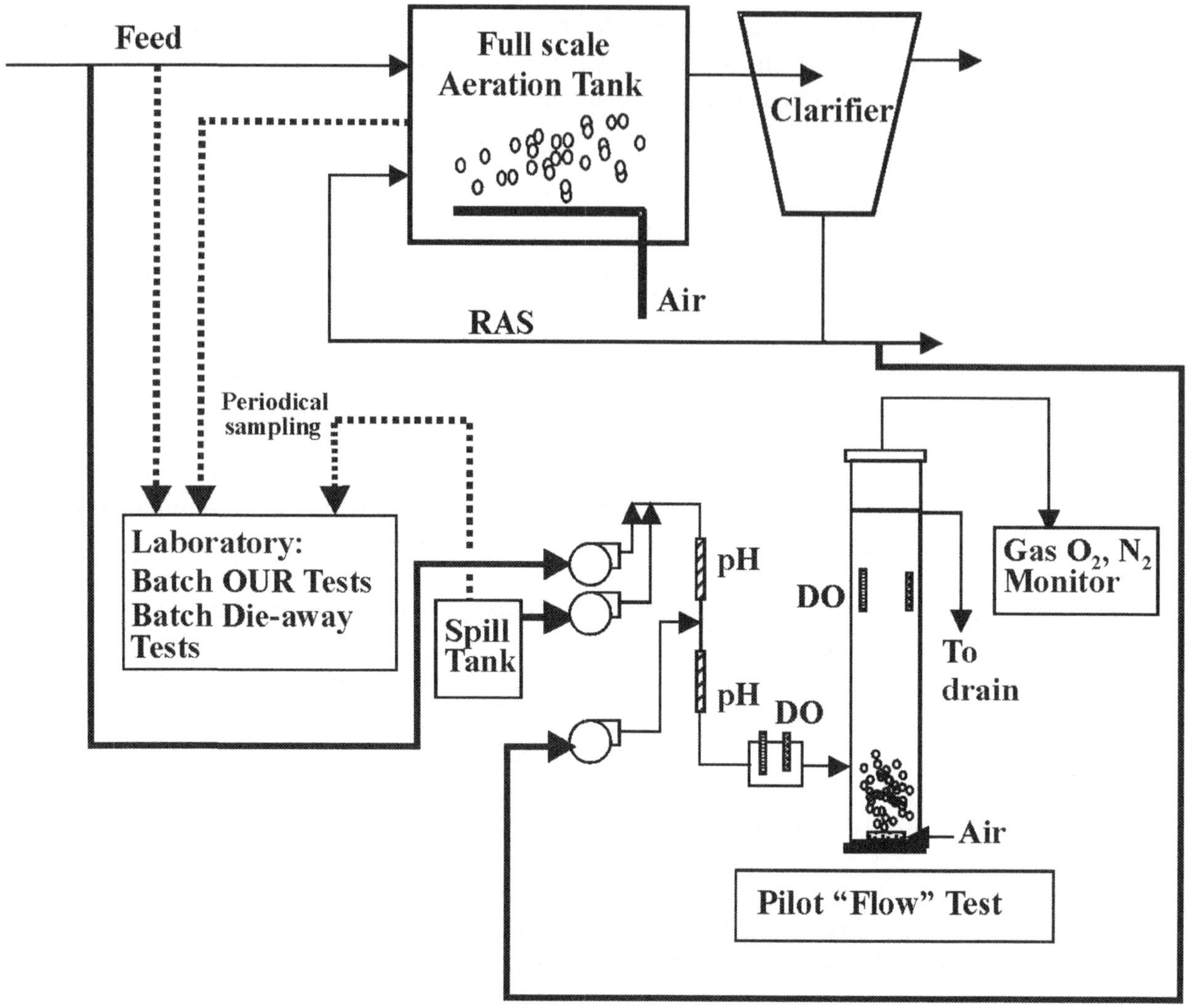

Figure 29-2. Schematic Diagram Demonstrates the Procedures Used in Model Calibration and Verification

The detailed steps for calibration are as follows:

1. Laboratory batch tests to measure oxygen uptake rate (OUR), soluble COD and ammonia concentrations were conducted using the protocol described in Figure 29-3.
2. Data from the batch tests were used to calibrate the model to full-scale performance data using the protocols described in Figure 29-4.

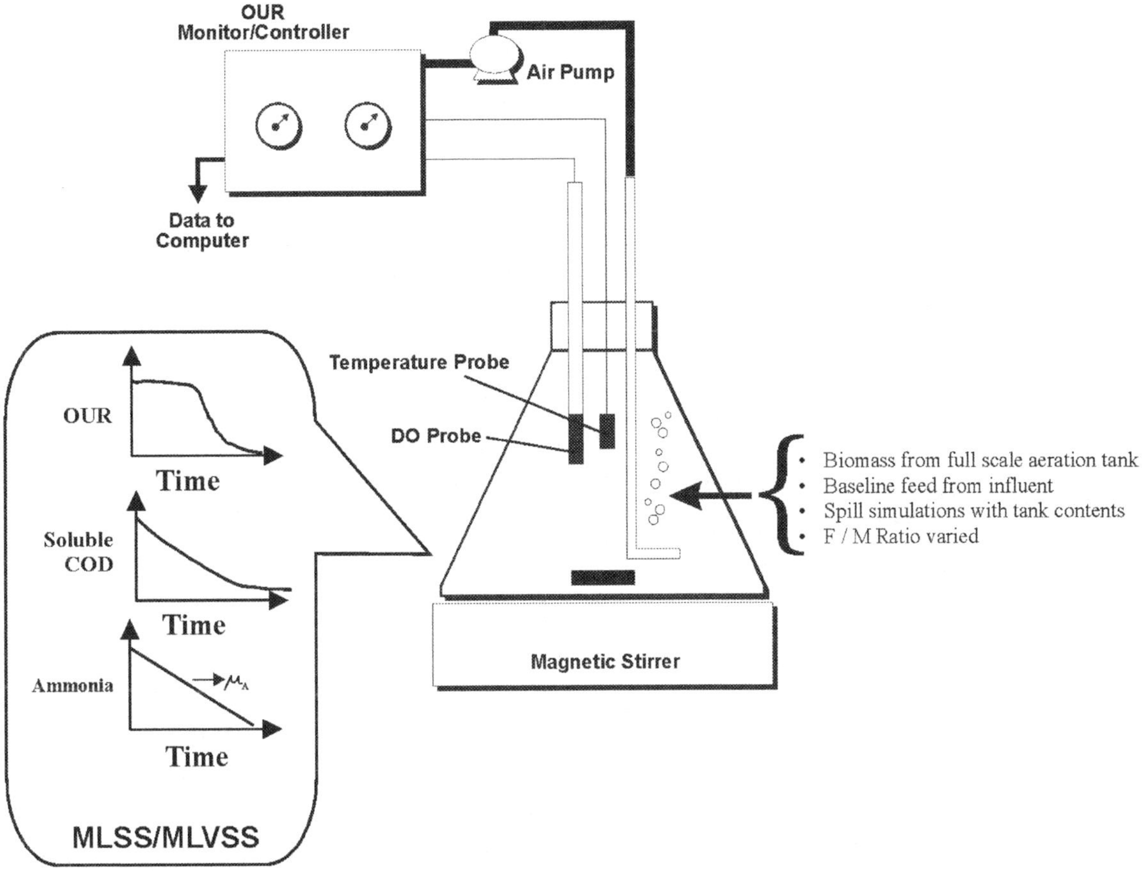

Figure 29-3. Laboratory Batch Experimental Procedure

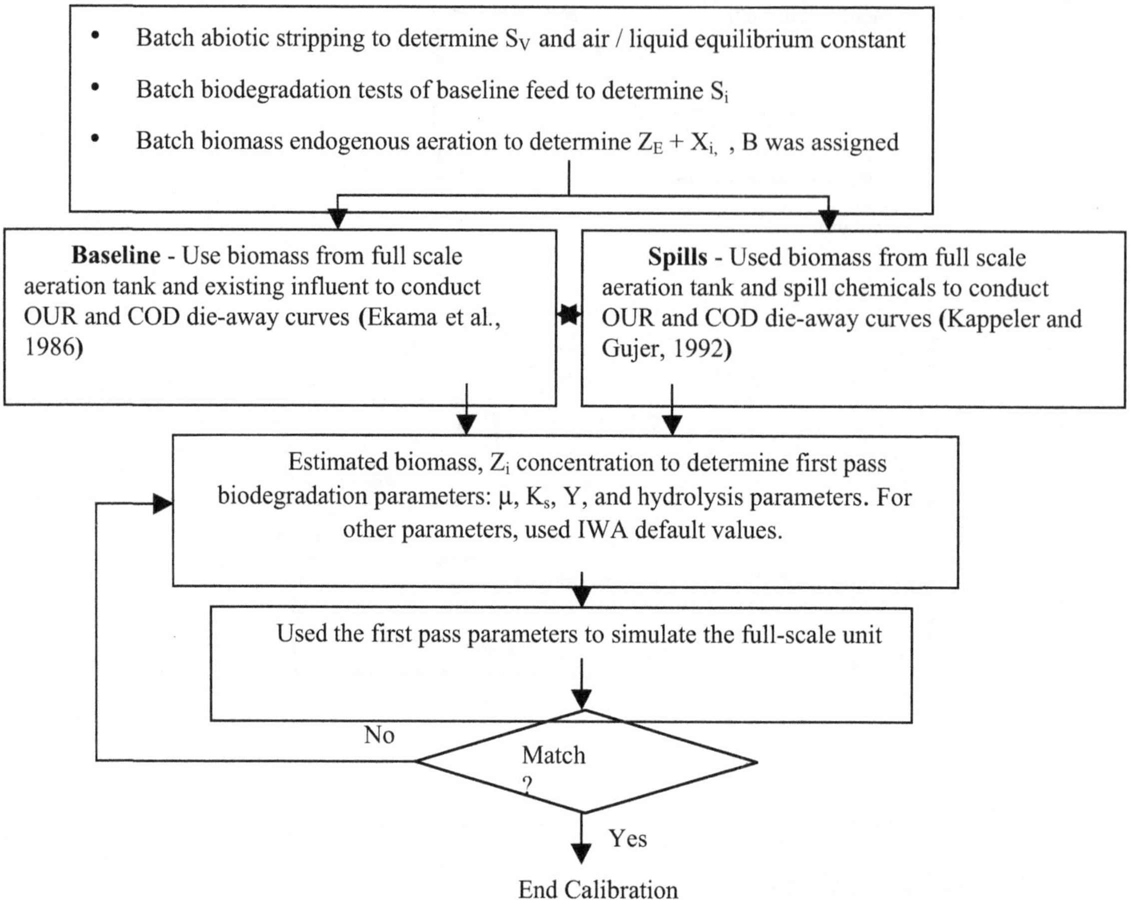

Figure 29-4. Model Parameters Estimation Procedure

29.4.1 Calibration Results

To illustrate the batch test calibration results, an example of the OUR and COD response in a batch test conducted on the V-353 waste is presented in Figure 29-5.

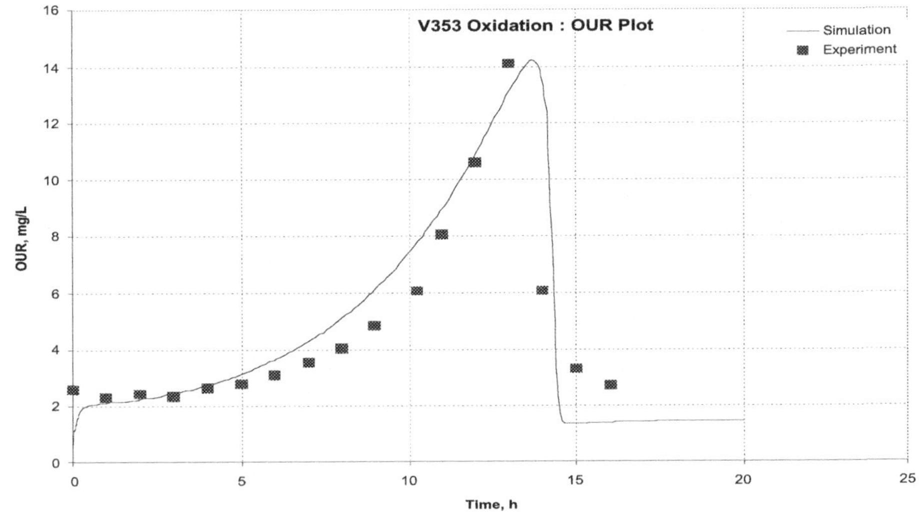

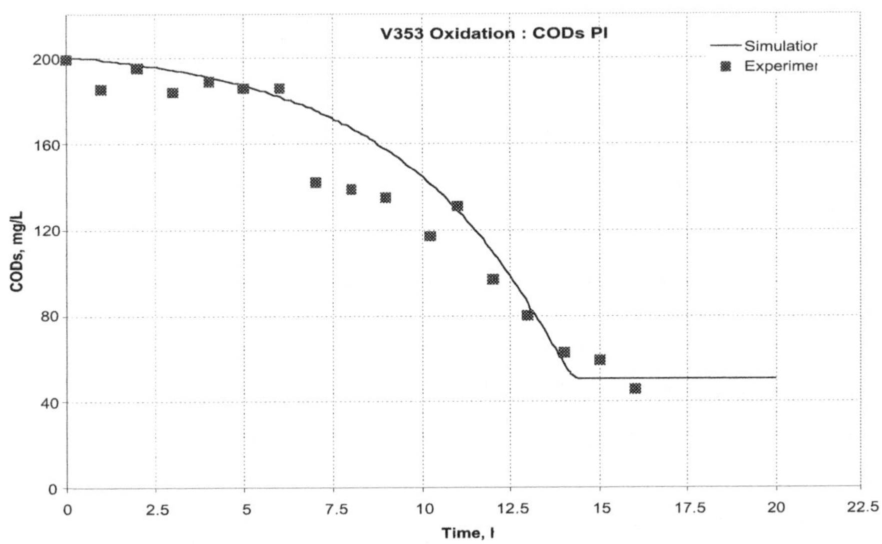

Figure 29-5. A Simulation of a Batch Test for the V-353 Waste, with OUR and COD Responses

The influent soluble COD was divided into five components:

$$S = S_V + S_M + S_{MeOH} + S_i + S_{Chex}$$

The procedures used to measure these components were are follows:

- S_V – Abiotic stripping tests
- S_{MeOH} – By GC analysis
- S_i – Residue of batch die-away tests
- S_{Chex} – Removal in feed stream during test period
- S_M – By difference

Table 29-4 provides the list of pertinent parameters estimated and used in the model.

Table 29-4. Estimated Parameters Used in the Final Model

Symbol	Description	Value	Units
Y_{hM}	Yield of heterotrophs for degradation of S_m	0.50	g cell COD yield per g COD utilized
Y_{hV}	Yield of heterotrophs for degradation of S_V	0.60	g cell COD yield per g COD utilized
Y_{hMeOH}	Yield of heterotrophs for degradation of S_{MeOH}	0.50	g cell COD yield per g COD utilized
Y_{hChex}	Yield of heterotrophs for degradation of S_{Chex}	0.55	g cell COD yield per g COD utilized
Y_{hHYD}	Yield of heterotrophs for degradation of S_h	0.55*	g cell COD yield per g COD utilized
f_{EpH}	Fraction of active mass remaining as endo residue	0.05*	g COD endo mass/gCOD active mass
Y_N	Nitrifier yield	0.17	g cell COD yield per g N utilized
f_{EpN}	Fraction of active nitrifier remaining as endo residue	0.08*	g COD endo mass/gCOD active mass
K_{OH}	DO switch constant (aerobic / anoxic growth)	0.2*	mg O / L
K_{NO}	NO_3 availability switch constant	1.0*	mg N / L
μ_{hM}	Max specific growth rate on mixed COD	4	per day
K_{SM}	Half saturation coefficient for mixed COD	1.0*	mg COD / L
μ_{hV}	Max specific growth rate on S_V COD	4*	per day
K_{SV}	Half saturation coefficient for S_v COD	1.0*	mg COD / L
b_H	Decay rate	0.4*	per day
K_H	Max specific hydrolysis rate	0.028	per day
K_X	Half saturation coefficient for hydrolysis	0.15*	mg COD per mg COD
$\mu_{n,T1}$	Max specific nitrifier growth rate @ temperature 32C	0.36	per day
K_{SNH}	Half saturation coefficient for nitrifier growth	0.25*	mg N / L
b_N	Nitrifier decay rate	0.062	per day
K_{Ni}	Inhibition coefficient for nitrifier growth	500	(mg N / L)^2
μ_{hMeOH}	Max specific S_{MeOH} growth rate	4.18	per day
K_{SMeOH}	Half saturation coefficient for S_{MeOH} growth	1.0*	mg COD / L
μ_{hChex}	Max specific S_{Chex} growth rate	6.2	per day
K_{SChex}	Half saturation coefficient for S_{Chex} growth	2.0	mg COD / L
μ_{hHYD}	Max specific S_h growth rate	6*	per day
K_{SHYD}	Half saturation coefficient for S_h growth	4.0	mg COD / L
b_{hHYD}	Decay rate for predator Z_{BHYD}	0.4*	per day
α	Alpha factor for oxygen transfer	0.44	fraction
β	Beta factor for oxygen transfer	0.95*	fraction
θ	Theta factor for oxygen transfer	1.024*	fraction
$S^*_{o,T}$	Saturation DO in clean water at 20 °C	8.88	mg O / L
SOTR	Standard Oxygen Transfer Rate	37.8	mg O / L /h
K_{eqV}	Equilibrium stripping constant for S_v	0.026	L_gas / L-aqueous

** Assigned values*

29.5 Model Verification

Once the model was calibrated reasonably well, it was verified by using a pilot-scale reactor located near the full-scale plant (See Figure 29-2). The reactor (with a volume of 227 L and air flow of 2,880 L/day) was fed with a sidestream of full-scale plant influent and a sidestream flow of the full-scale plant RAS. No internal solids separation was provided, that is, the system operated as a chemostat. The reactor was equipped with a DO and pH probes. Its off-gas stream was monitored with a gas analyzer where the oxygen content of the gas was continuously measured. OUR was estimated from the off gas oxygen content using the procedure outlined in Redmon et al. (1983). The mixed liquor biomass concentration was analyzed periodically. Also measured were the effluent filtered COD, filtered ammonia, and filtered nitrate and nitrite concentrations.

Verification was evaluated over four operating periods, each lasting approximately 100 hours. The pilot-scale reactor was not insulated, so that the reactor temperature deviated markedly from the full-scale aeration tank content. A conventional temperature correction formula was inserted into the model to compensate for this change. In addition, to simulate nitrification, an inhibition coefficient of 500 mg/l of ammonia-N was incorporated into the model and was derived by fitting data. Apart from these two deviations, all the parameters used in the model were derived from the laboratory batch studies as listed in Table 29-4.

The four experimental operating conditions are defined in Table 29-5.

Table 29-5. Pilot Reactor Experimental Conditions

Experiment	Influent flow, L/d	RAS flow, L/d	HRT, Days	MeOH Spike	NH4-N Spike	V-353 Spike
I	104	148	0.9	No	No	No
II	104	148	0.9	No	Yes	Yes
III	205	148	0.64	No	No	No
IV	205	148	0.64	Yes	No	No

The response of the pilot-scale reactor to steady state and shock loading conditions was monitored and compared with simulated results using the calibrated model. Figure 29-6 shows the pilot-scale reactor influent and effluent COD concentrations for the whole verification test period.

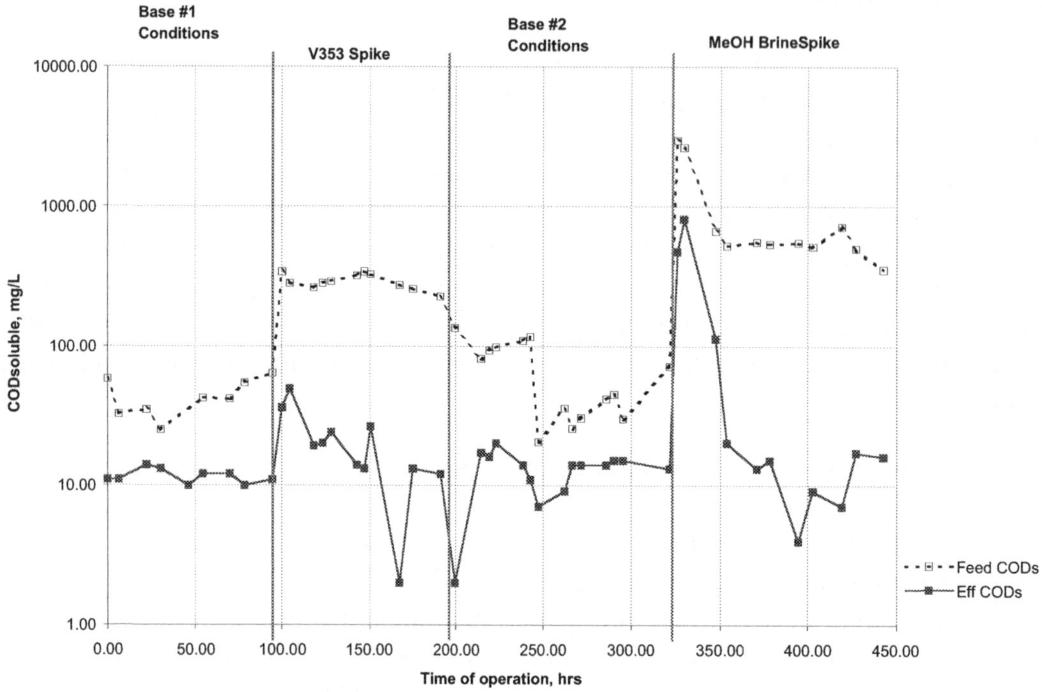

Figure 29-6. Shock Loading Conditions Throughout the Pilot Plant Test Period Showing Influent and Effluent Soluble COD Concentrations

Figure 29-7 compares the measured and simulated effluent soluble COD concentrations.

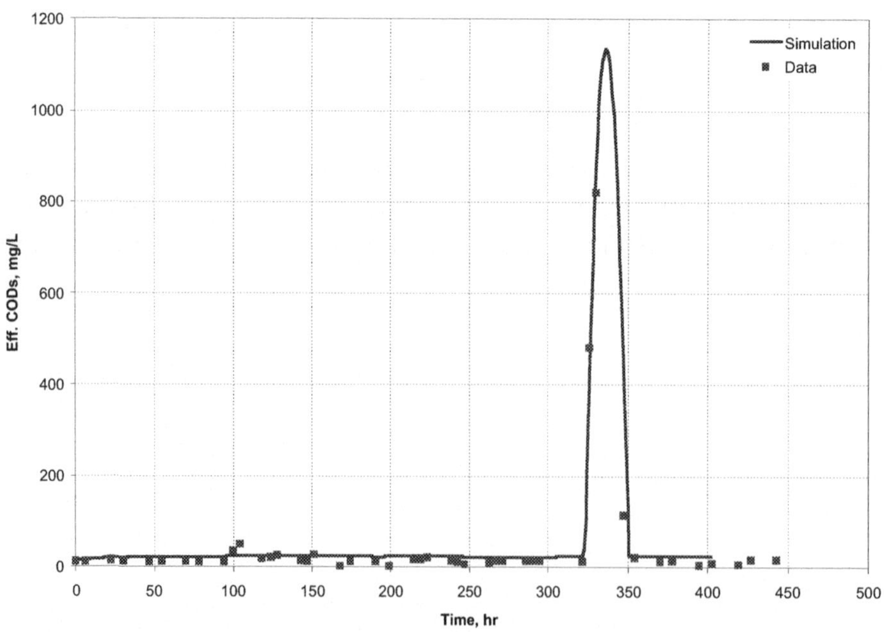

Figure 29-7. Simulated versus Measured Effluent Soluble COD Concentration

At Hour 320, soluble COD breakthrough occurred when the MeOH shock load exceeded the system capacity. The model predicted this event well. The DO concentration of the reactor was always maintained above 2 mg/L, so the COD leakage could be attributed to biokinetic limitations rather than a DO mass transfer limitation.

Figure 29-8 shows the measured and simulated influent and effluent ammonia concentrations during the verification period. The model successfully simulated system responses to both steady state and shock loading operating conditions.

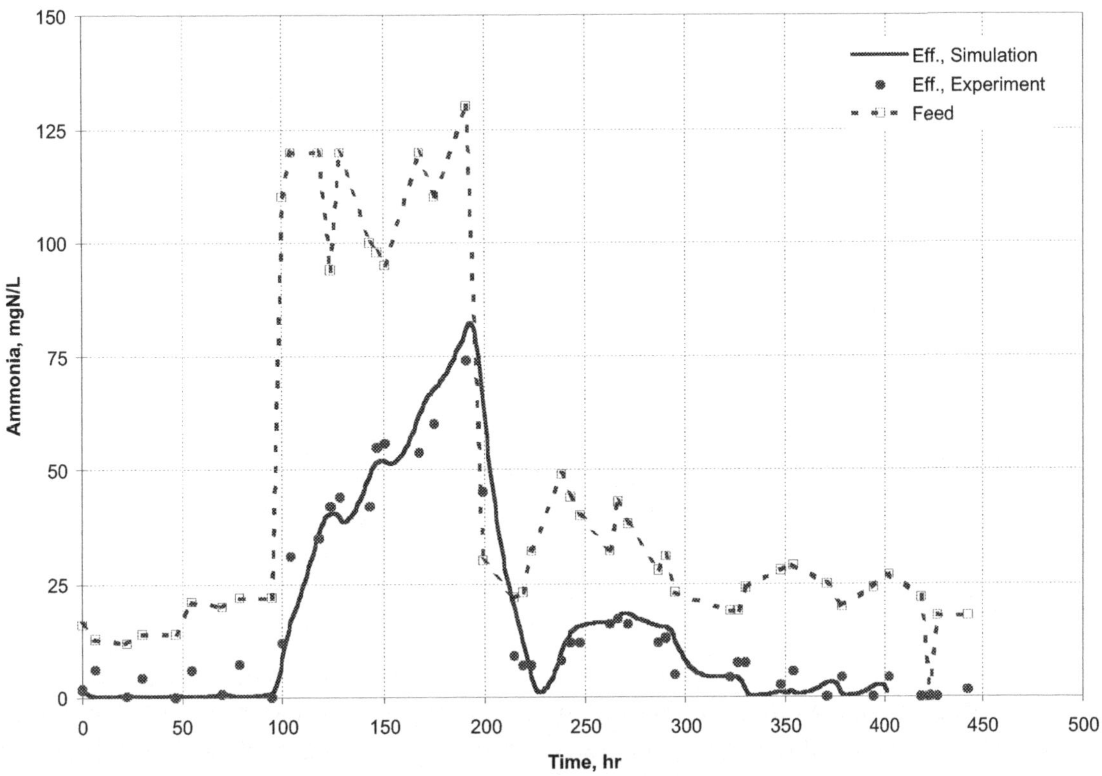

Figure 29-8. Test Results for Ammonia Shock Load Study

Figure 29-9 illustrates the close agreement between the measured and predicted reactor OUR.

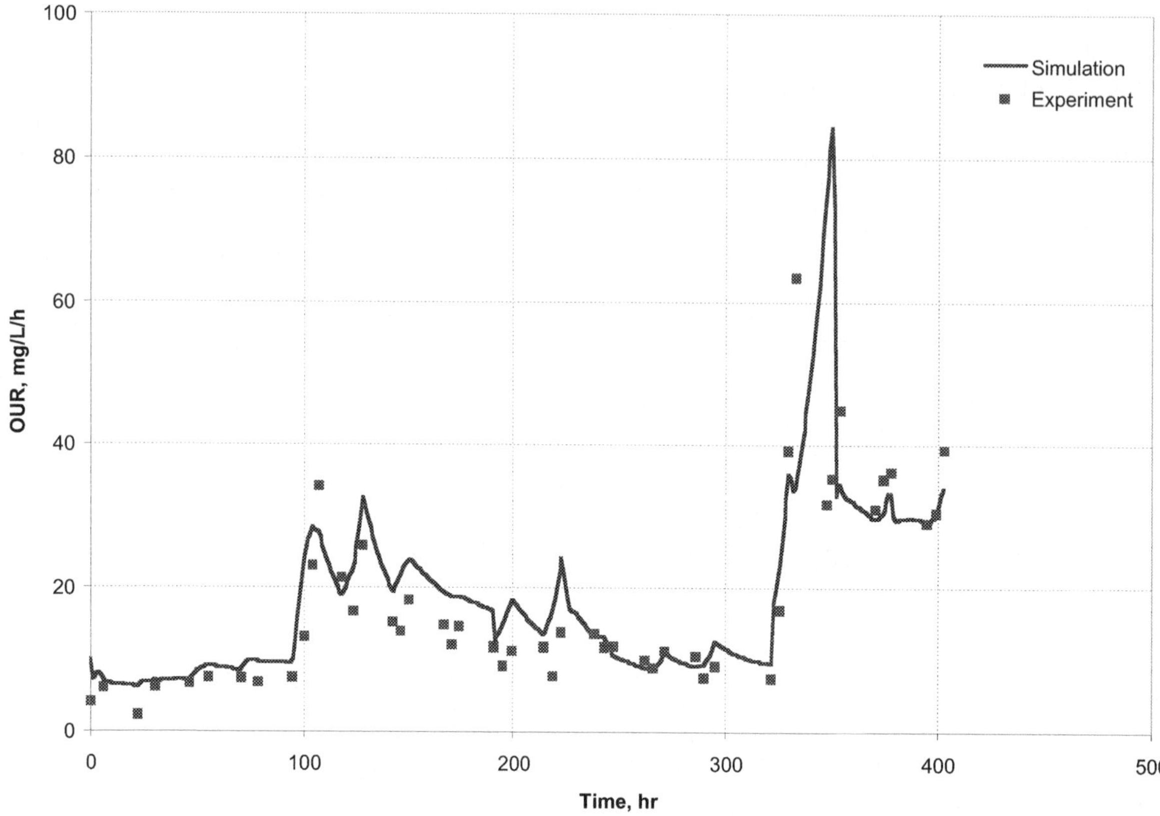

Figure 29-9. The Measured and Simulated OUR for the Pilot-Scale Reactor

29.6 Application

Table 29-6 summarizes the comparison between the long-term full-scale plant performance data with those from model simulations. Based on this comparison, it may be concluded that the model was well calibrated and that it can accurately predict the long-term steady state operation of the plant.

Table 29-6. Comparison of Results and Measurement for Full-Scale, Long Term Operation

Parameters	Simulated	Measured
MLSS (mg/L)	5,356	5,300
MLVSS (mg/L)	3,263	2,933
OUR (mg/L-hr)	14.8	14
Dissolved oxygen (mg/L)	0.96	1
Effluent COD (mg/L)	22	20
Effluent NH_4-N (mg/L)	0.13	<0.1
Effluent NO_3-N (mg/L)	74	62

This calibrated model has been used by the plant operations department to:

1. Design spill/bleed back strategies in the use of the spill diversion tanks;
2. Identify constraints in the treatment plant under various operating conditions;
3. Provide a training tool for operators to allow them to evaluate, through simulation, the impact of different operating strategies; and
4. Provide a tool for diagnosing operational upsets.

29.7 References

Henze, M., W. Gujer, T. Mino, and M. van Loosdrecht. 2000. Activated Sludge Models ASM1, ASM2, ASM2d and ASM3. *IWA Scientific and Technical Report No. 9*. London, UK: International Water Association.

Redmon, D., W.C. Boyle, and L. Ewing. 1983. Oxygen transfer efficiency measurements in mixed liquor using off-gas techniques. *Jour. Water Pollut. Cont. Fed* 55(11):1338-1347.

Ekama, G A., P.L. Dold, and G.v.R. Marais. 1986. Procedures for determining influent COD fractions and the maximum specific growth rate of hetertrophs in activated sludge systems. *Water Sci. Tech.* 18:91-114.

Kappeler, J., and W. Gujer. 1992. Estimation of kinetic parameters of hetertrophic biomass under aerobic conditions and characterization of wastewater for activated sludge modeling. *Water Sci. Tech.* 25(6):125-139.

CHAPTER 30.0

CASE STUDY: MODEL DEVELOPMENT AND CALIBRATION FOR A CHEMICAL WASTEWATER TREATMENT PLANT

30.1 Background

This case study presents an example of modifying the ASM1 model for a chemical industry wastewater treatment plant application and the subsequent calibration of the modified model. The case history concerns a conventional activated sludge process that treats high saline chemical industry wastewater, and which experienced floating sludge problems in the secondary clarifiers (Bury, 2002). The approach taken was to:

- Gather data on the full-scale plant and composition of the influent wastewater,
- Formulate a mechanistic model based on the ASM1 logic, and
- Conduct a series of batch tests using full-scale plant biomass and specific wastewater components to provide data for model calibration.

Note: The nomenclature used in this chapter is unique to this industrial sector, and may not agree with terms listed in Appendix A. However, care has been taken to define variables and terms as they occur.

30.2 Plant Description

The treatment plant consists of three complete mix aeration basins each with a working volume of 13,000 m^3. It is a two-stage system in which flow is directed through one primary reactor followed by two secondary reactors (Figure 30-1).

These tanks are covered, use diffused aeration systems (oxygen in the primary reactor and air in the secondary reactors) and have headspace gas analysis that permits the estimation of oxygen transfer rates. This treatment train receives wastewater from 100 individual chemical process plants, a combination of batch and continuous sources. The train was designed to treat approximately 60,000 kg COD/day. Nitrification was not considered in the design because there was an absence of nitrification in previous laboratory treatability investigations. This was consistent with most of the literature, which indicated that nitrification is inhibited at high salt (greater than seawater) concentrations (Dincer, 1999). Also, ASM1 was not used to develop the plant process design. The influent flow rate varies from 10,000 to 18,000 m^3/day. The system temperature varies from about 30 to 40°C. External cooling is used when the combination of high organic load and high ambient temperatures elevate reactor temperature. The majority of COD removal occurs in the primary reactor.

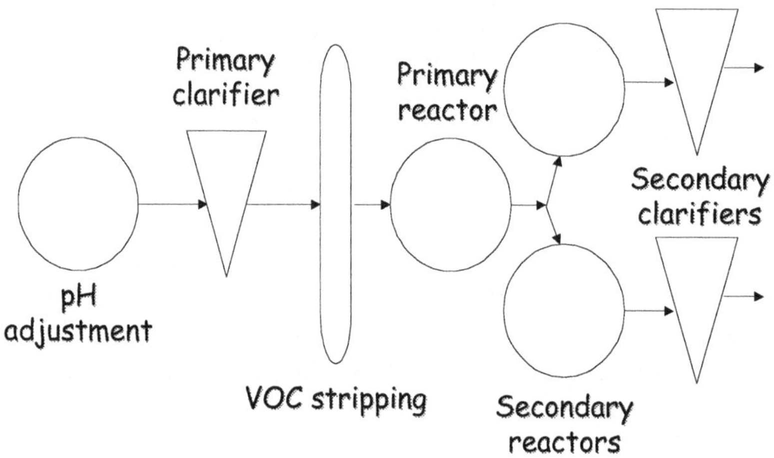

Figure 30-1. Flow Schematic of Chemical Plant Wastewater Treatment Plant

Table 30-1 shows typical influent characterization data.

Table 30-1. Influent Wastewater Characteristics

Parameters	Values		Units
	Low	High	
Flow	10,000	18,000	m^3/day
NaCl	4	7	wt %
Total COD	1,000	4,000	g COD/m^3
BOD_5	330	1,320	g O_2/m^3
BOD_{20}	900	3,600	g O_2/m^3
TKN	0	200	g N/m^3
NH_3-N	0	90	g N/m^3
NO_2-N, NO_3-N	0	0	g N/m^3

The influent has a very high salt content, ranging from 4 percent to 7 percent by weight, and a high organic content with total COD ranging from 1,000 to 4,000 g/m^3. Ammonia constitutes up to 45 percent of the influent nitrogen, the remainder being made up of organic nitrogen sources such as glycol amines. The influent wastewater is pH adjusted and passes through primary clarification for removal of heavy tars and suspended particles. At a high pH of 10-11, it is most unlikely that the influent contains any live biomass. The wastewater next passes through a VOC stripper column. This simplified model development because volatile stripping did not need to be accounted for in the activated sludge system. Further pH adjustment occurs just before the primary reactor. Return activated sludge is introduced at the primary reactor.

30.3 Model Formulation

The general approach to modeling this wastewater treatment plant has been to describe as much as was possible of the observable COD and nitrogen removal phenomena utilizing the standard ASM1 model framework. The need for the model was driven by persistent floating sludge in the secondary clarifiers that occurred about three years after start-up of the plant. It was attributed to excessive denitrification in the secondary clarifiers. The phenomenon was unexpected because denitrification is typically preceded by nitrification; as stated earlier, nitrification was not anticipated to occur in this high salinity wastewater.

Using historical plant performance data and default ASM1 parameters, preliminary simulations suggested that nitrification and denitrification would not occur. However, subsequent data collection demonstrated that significant nitrite and nitrate were present in the reactors. Given that there was no oxidized nitrogen in the feed, it was evident that nitrification had become established. At that time, only ammonia nitrogen but no TKN data were collected, which initially led to an incompatibility between plant data and model predictions. As Table 30-2 indicates, in the absence of TKN data, the preliminary steady state simulation using default ASM1 parameters predicted high levels of readily degradable substrate (S_S) and low effluent NH_3-N levels, which suggested that nitrogen was limiting the reaction, contrary to field observations. Nitrification was not predicted. The only way to match observed removals in the primary reactor was to increase the ammonia or organic nitrogen load.

The more current wastewater characterization that includes TKN data shows that ammonia typically accounts for up to 45 percent of the incoming nitrogen (Table 30-2). The remainder was from organic nitrogen sources like glycol amines. Using this revised nitrogen load gave more reasonable simulation results, although nitrification was still not predicted.

Table 30-2. Preliminary Steady-State Simulation of Effluent Using Temperature-Corrected Default ASM1 Parameters

Reactor	Parameter	Predicted Concentration	Average Field Concentration	Units
Primary	S_S	427	5.0	g COD /m
Primary	S_F	2.28	-	g COD/m^3
Primary	X_S	7.95	-	g COD/m^3
Primary	S_{NH}	0.01	26.6	g N/m^3
Primary	S_{NO}	1e-6	12.0	g N/m^3
Primary	S_{ALK}	20	23	mol/m^3
Secondary	S_S	1.76	2.0	g COD/m^3
Secondary	S_F	1.21	-	g COD/m^3
Secondary	X_S	4.19	-	g m^3
Secondary	S_{NH}	18.65	23.2	g N/m^3
Secondary	S_{NO}	1e-6	38	g N/m^3
Secondary	S_{ALK}	20	21	mol/m^3

This prompted a more comprehensive influent wastewater characterization to be carried out, which focused upon COD, TKN, BOD_5, BOD_{20}, nitrite and nitrate concentration measurements to supplement prior TOC and ammonia data. The analysis of COD had to take into account high salinity interferences. A breakdown of influent COD is given in Table 30-3. The influent COD is entirely soluble. Based on respirometry experiments and BOD_5/BOD_{20} measurements, the influent soluble COD was defined as approximately 85 percent "bulk degradable" consisting of readily degradable (S_s) and complex (S_F) fractions, 10 percent inert (S_I) and 5 percent "specific degradable".

Table 30-3. Influent Wastewater COD Fractions

Soluble COD Fractions		
	Percent	
Bulk degradable	85	
Readily degradable (S_S)		90
Complex (S_F)		10
Specific degradable	5	
Inert (S_I)	10	
Total	**100**	

Four adjustments were made to the ASM1 model:

(i) Temperature-dependent kinetics.
(ii) The observed accumulated levels of nitrite in the reactors led to the incorporation of a two-step nitrification process in place of the single step nitrification typically assumed in nitrifying systems.
(iii) Hydrolysis of the S_F fraction.

(iv) The use of specialized biomass fractions for specific degradable compounds like phenol. These specific degradable compounds are present in the wastewater at low concentrations (<0.5 percent of influent COD_T or typically <20 mg/L in the influent) and the load may be intermittent in nature.

30.4 Calibration

30.4.1 Steady State Calibration

The correct accounting of influent nitrogen improved the model accountability in terms of COD removal and biomass but the simulation still did not predict nitrification, effluent ammonia concentration or oxygen transfer rate correctly. Therefore, the calibration focused on the model parameters that have a role in the estimation of a nitrogen balance and of the oxygen demand. These variables are shown in Table 30-4. A temperature correction coefficient of 1.072 was used for temperatures up to 35°C.

The value for X_{NB} was determined by direct measure of the biomass. Weekly samples were collected and analyzed. The results of five rounds of sampling gave a value of 0.06+/-0.003 g N/g COD.

The heterotrophic yield was determined using nitrification inhibited respirometry. The major influent organic components were measured to be propylene glycol, glycerol and acetate. Respirometric experiments using a range of S_O/X_O ratios for a mixture of the pure organics were carried out and yielded a value of 0.6 g COD/g COD for Y_h.

Table 30-4. Calibrated Model Parameters

Parameter	Description	Calibrated Value
Y_{HET}	Heterotrophic yield	0.6 g COD/g COD
X_{NB}	Nitrogen fraction of biomass	0.06 g N/g COD
b_h	Heterotrophic decay rate	0.62 d^{-1}
μ_{ai}	Ammonia oxidizer maximum growth rate	0.168 d^{-1}
μ_{ao}	Nitrite oxidizer maximum growth rate	0.11 d^{-1}
$\mathbf{K_{nai}}$	Ammonia oxidizer half saturation rate coefficient	20.84 gN-NH3/m^3
$\mathbf{K_{nao}}$	Nitrite oxidizer half saturation rate coefficient	24.72 gN-NO2/m^3

Figure 30-2 shows a typical respirogram from these experiments.

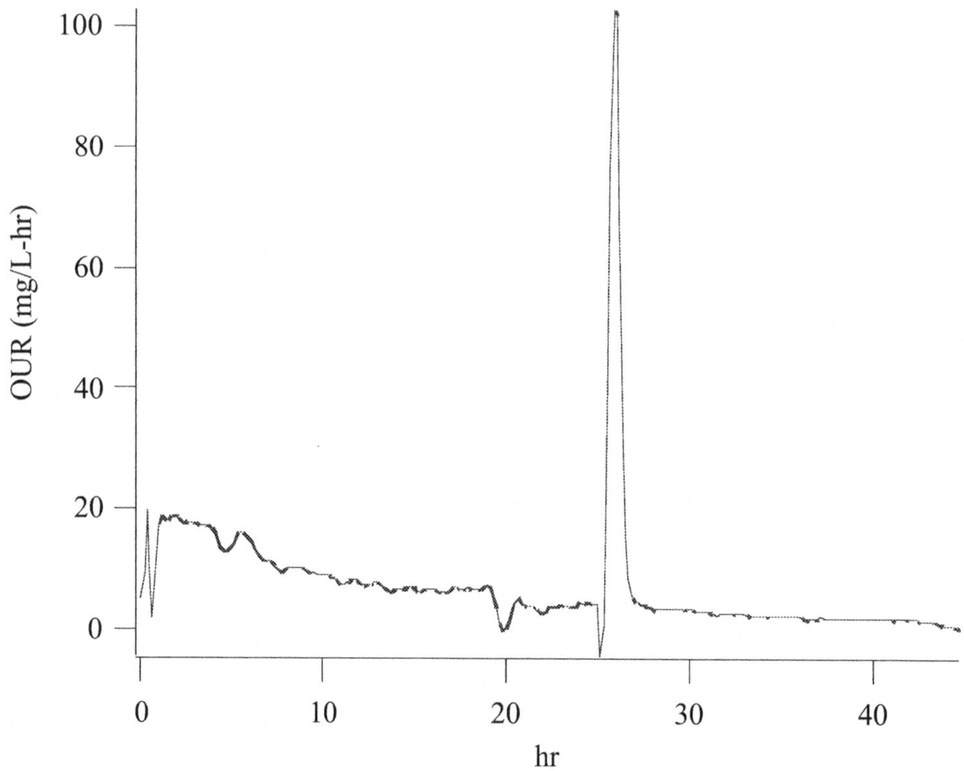

Figure 30-2. A Typical Respirogram for Determining the Heterotrophic Yield Coefficient

The decay coefficient was also determined using respirometry. A 4 L sample of sludge was collected and aerated and kept at 35°C. After 24 hours, a maximum OUR was measured using propylene glycol as a substrate. After a further 8 days, the experiment was repeated and another maximum OUR determined. Using a first order reaction rate expression, the slope of the ln(OUR) gives b_h', which is related to the ASM1 b_h by:

$$b_h = \frac{b_h'}{1 - Y_h(1 - f_u)} \qquad (30.4.1\text{-}1)$$

where f_u is the endogenous fraction, which was assumed to be 0.08. The results of this experiment gave a decay coefficient of 0.62 d^{-1} for b_h.

Plant performance data were then used to calibrate the nitrification parameters using an optimization routine. First the parameters were bounded by manually trying sets until the steady-state simulation provided a good solution. These results are shown in Table 30-5. The accumulation of nitrite in both the primary and secondary reactors is evident. Also, the COD removal and reactor NH_3 levels were consistent with average plant data.

Table 30-5. Steady State Simulation Solution using Experimentally Determined Parameters and Manually Adjusted Nitrification Parameters

Reactor	Species	Predicted Concentration	Average Field Concentration	Units
Primary	S_S	4.04	5.0	g COD/m^3
Primary	S_F	1.8	-	g COD/m^3
Primary	X_S	6.2	-	g COD/m^3
Primary	S_{NH}	17.5	26.6	g N/m^3
Primary	Nitrite-N	16.5	12.0	g N/m^3
Primary	Nitrate-N	3.1	1.2	g N/m^3
Primary	S_{ALK}	20	23	Mol/m^3
Secondary	S_S	1.2	2.0	g COD/m^3
Secondary	S_F	0.8	-	g COD/m^3
Secondary	X_S	3.0	-	g COD/m^3
Secondary	S_{NH}	8.0	23.2	g N/m^3
Secondary	Nitrite-N	35.8	30.0	g N/m^3
Secondary	Nitrate-N	15.1	8.0	g N/m^3
Secondary	S_{ALK}	20	21	Mol/m^3

30.4.2 Dynamic Calibration

Plant performance data for a 48-day period were used for the dynamic calibration of the nitrification parameters. These data included flows, influent concentrations and the reactor and clarifier compositions. The data for flows, COD and TKN are shown in Figure 30-3. An optimization routine was used to vary the specified parameters and minimize the value of the objective function, which consisted of the relative difference between the measured and predicted primary and secondary nitrogen concentrations of ammonia, nitrite and nitrate. An alternative and superior method of seeking optimum parameters is by using a maximum likelihood criterion because of its ability to provide better results than the relative difference method. The final set of parameters is shown in Table 30-5 and the results of the simulation are shown in Figure 30-4.

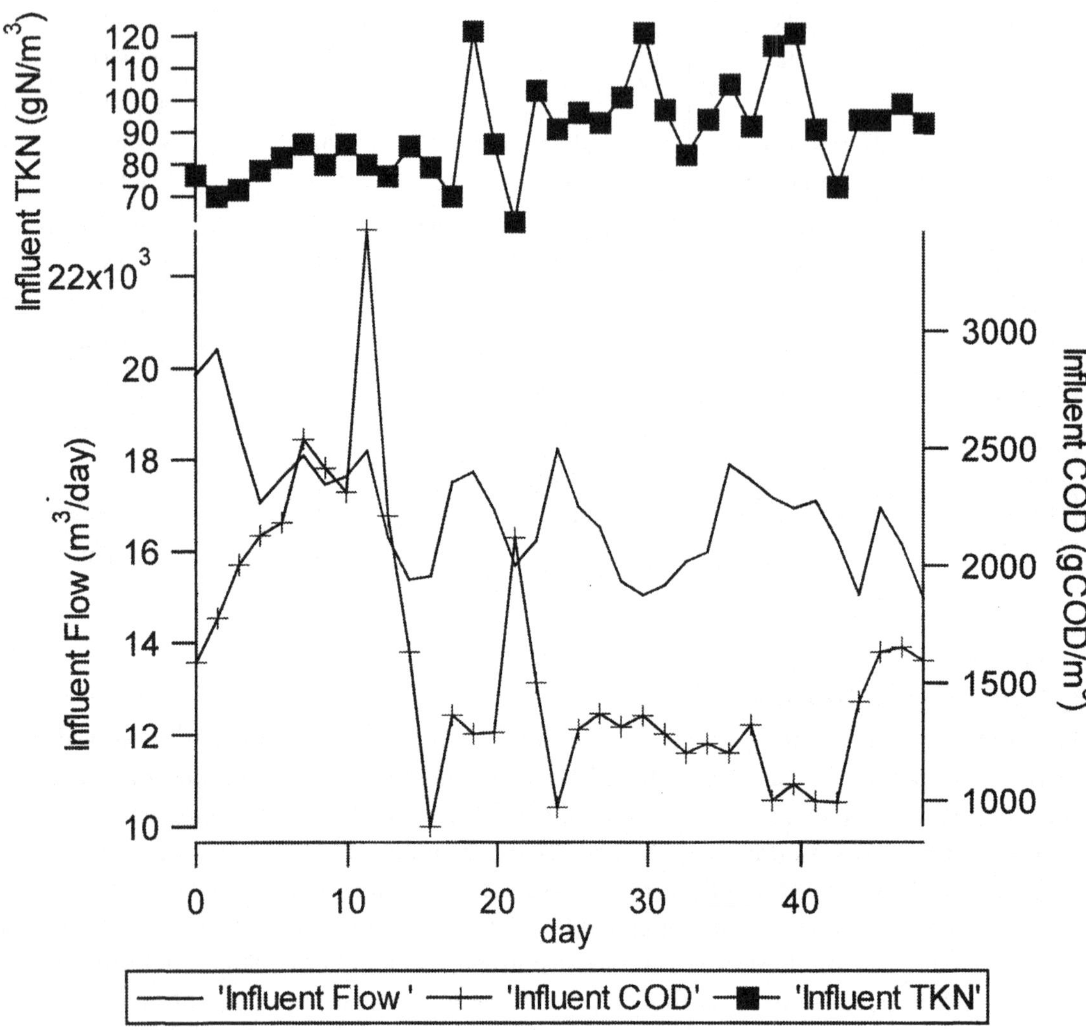

Figure 30-3. Plant Influent Flow, COD and TKN Data Used for Dynamic Calibration

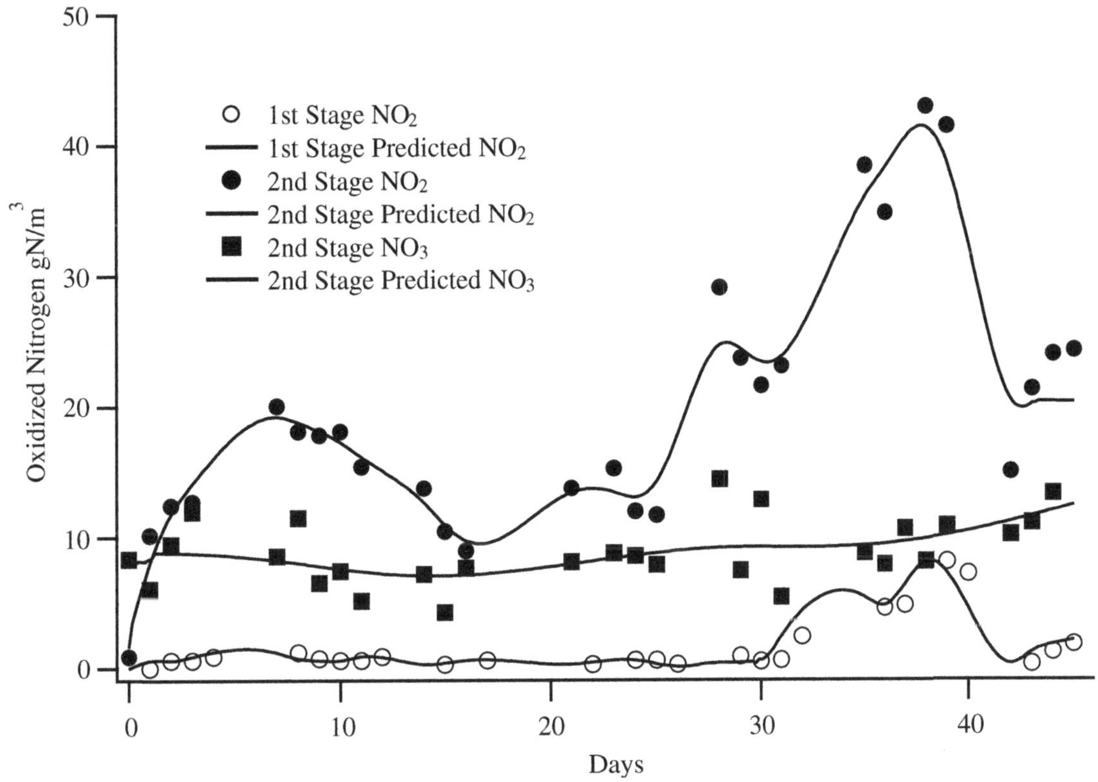

Figure 30-4. Results of the Optimized Simulation for Determination of the Nitrification Parameters

30.5 Application

The use of a modified ASM1 steady state and dynamic model helped to identify limitations in influent wastewater characterization. Calibration provided a tool that could predict effluent relatively well. The knowledge gained from the study guided operational changes to minimize the extent of nitrification under conditions of low COD loadings. In hindsight, had they been utilized in the design phase, up-front simulations and influent characterization as described in the case study would have lead to a different plant design and operation.

30.6 References

Bury, S.J., C.K. Groot, C. Huth, and N. Hardt. 2000. Dynamic simulation of chemical industry wastewater treatment plants. *Water Sci. Tech.* 45(4/5):355-363.

Dincer, A.R., and F. Kargi. 1999. Salt inhibition of nitrification and denitrification in saline wastewater. *Env. Tech.* 20(11):1147-1153.

CHAPTER 31.0

SBR: McMaster University Investigations

A fill-and-draw low F/M SBR system was operated at McMaster University at an early stage in the project. The system was initiated with mixed liquor from the Dundas WWTP and received Dundas raw influent wastewater as feed.

The principal objective of operating the system was to develop an example data set on application of the low F/M SBR protocol for inclusion in the project report and as an example for other facilities where the protocol was to be field tested (Renton and Phoenix).

Results and details of the McMaster SBR study are presented in Appendix D, and are not repeated here. In summary:

- Appendix D outlines the experimental protocol for conducting an SBR study.
- Appendix D provides an example of evaluation of SBR data and calculation of COD and nitrogen mass balances.
- Appendix D illustrates the methods for using SBR data to determine influent wastewater characteristics (COD and TKN fractionation) and nitrifier maximum specific growth rate.
- Elsewhere in earlier chapters, data from the McMaster SBR study are used for purposes of illustrating methods for determining certain wastewater characteristics.

CHAPTER 32.0

SBR: King County Investigations

32.1 Background

This chapter outlines the operation of and data collected for a laboratory-scale fill-and-draw SBR system that was operated with sludge from King County's South Wastewater Treatment Plant in Renton, Wash.

32.2 SBR Operation

For a detailed operation protocol, please refer to Appendix D.

For the Renton system, the SBR was operated for the following conditions:

V_P = 10 L

V_{WW} = 8 L

q_W = 0.625 L/d

θ_X = 10/0.625 = 16 days

Temp. = 20°C

After start-up, the system was operated for a period of 60 days (4 sludge ages) so as to attain a quasi-steady state. At this time, an intensive daily monitoring schedule was commenced for a period of 7 days. Analyses were conducted on samples of influent, waste and decant. Throughout the period of operation, oxygen utilization rate was monitored semi-continuously using an integrated DO controller/meter.

On two days (Days 3 and 5) during the intensive monitoring period, profiles of ammonia and nitrate concentration were monitored over the first 8 to 10 hours of the react period. Analyses were performed on small sample volumes withdrawn from the reactor.

32.2.1 Influent Wastewater

To attain the quasi-steady state SBR response described above, the system received a fixed COD and TKN load each cycle.

32.2.2 Selection of SBR Sludge Age

The target sludge age in the SBR was 16 days. However, due to the unintentional wasting of some solids in the SBR decant, the actual SRT was 14.05 days.

32.2.3 SBR Start-up Period

At start-up, the SBR was seeded with two liters of activated sludge taken from the Renton plant. The system was operated for an acclimation period of 60 days (4 sludge ages). Figure 32-1 shows a plot of the parameters monitored during the start-up period.

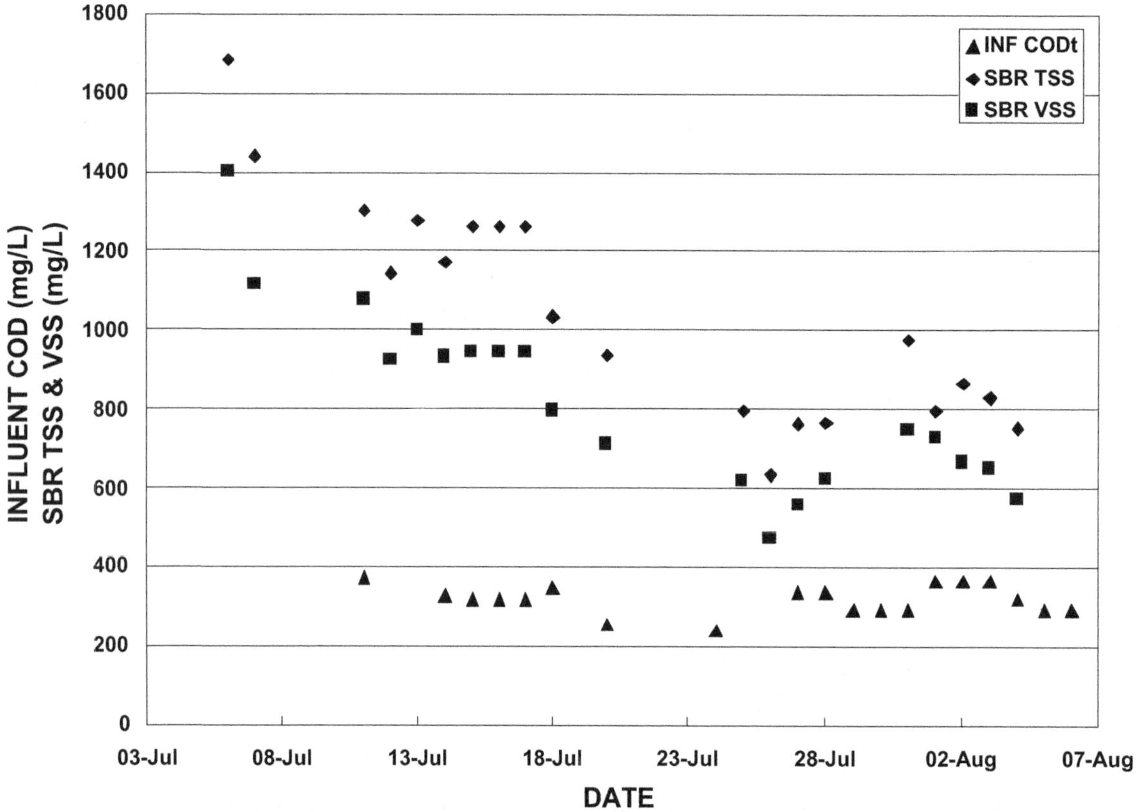

Figure 32-1. Plot of Influent COD, SBR TSS and SBR VSS Prior to Intensive Period

32.2.4 Monitoring during Intensive Testing

Intensive monitoring of the SBR system was performed for a period of 7 days as follows:

SBR Stream	Parameter Monitored
Influent	TSS
	VSS
	Total COD
	Flocculated/filtered COD
	Total TKN
	Ammonia-N
Reactor	TSS
	VSS
	Total COD
	Total TKN
	OUR
	Temperature
Decant	Total COD
	Membrane Filtered COD
	Flocculated/Filtered COD
	Membrane Filtered Total TKN
	Membrane Filtered Ammonia-N
	Membrane Filtered Nitrate-N

Oxygen utilization rate (OUR) in the SBR was monitored semi-continuously using an integrated DO controller/ eter. Results were downloaded to a PC spreadsheet daily. The data were analyzed off-line to calculate the mass of oxygen utilized in the reactor each day (further details may be found in Randall et al. (1991)). This enabled COD mass balances to be performed on the experimental data. Results from the intensive monitoring period are recorded in Table 32-3.

On Days 3 and 5, profiles of ammonia and nitrite/nitrate concentration were monitored over the first 8 to 10 hours of the react period. Analyses were performed on small sample volumes withdrawn from the reactor (and filtered immediately). The ammonia and nitrate responses are shown in Figures 32-2 and 32-3, together with the OUR response. Raw data for the profiles are given in Tables 32-1 and 32-2.

Table 32-1. Ammonia, Nitrate and OUR Profile Data for Day 3

Time (hours)	Nitrate (mgN/L)	Ammonia (mgN/L)	Time (hours)	OUR (mg/L/hr)
0.0	4.28	19.97	1.12	20.7
1.0	6.06	16.35	1.44	20.3
2.0	8.33	15.01	1.71	20.2
3.0	10.71	11.79	1.94	20.2
4.0	13.27	10.21	2.16	20.3
4.5	14.56	8.64	2.36	20.1
5.0	15.55	7.07	2.60	19.6
5.5	16.83	5.50	2.84	19.3
6.0	18.11	4.55	3.06	19.8
6.5	19.50	2.90	3.27	19.9
7.0	20.68	1.80	3.48	19.5
7.5	22.46	0.78	3.68	19.4
8.0	23.45	0.00	3.89	19.1
			4.09	18.8
			4.28	19.3
			4.48	19.0
			4.67	20.0
			4.85	18.7
			5.05	18.3
			5.24	18.5
			5.43	18.4
			5.61	18.3
			5.80	18.4
			5.98	17.7
			6.17	18.0
			6.35	17.9
			6.54	17.5
			6.72	17.5
			6.91	17.7
			7.09	16.8
			7.28	16.4
			7.47	15.7
			7.67	14.1
			7.89	11.3
			8.14	9.1
			8.43	8.4
			8.72	7.7
			9.02	7.2

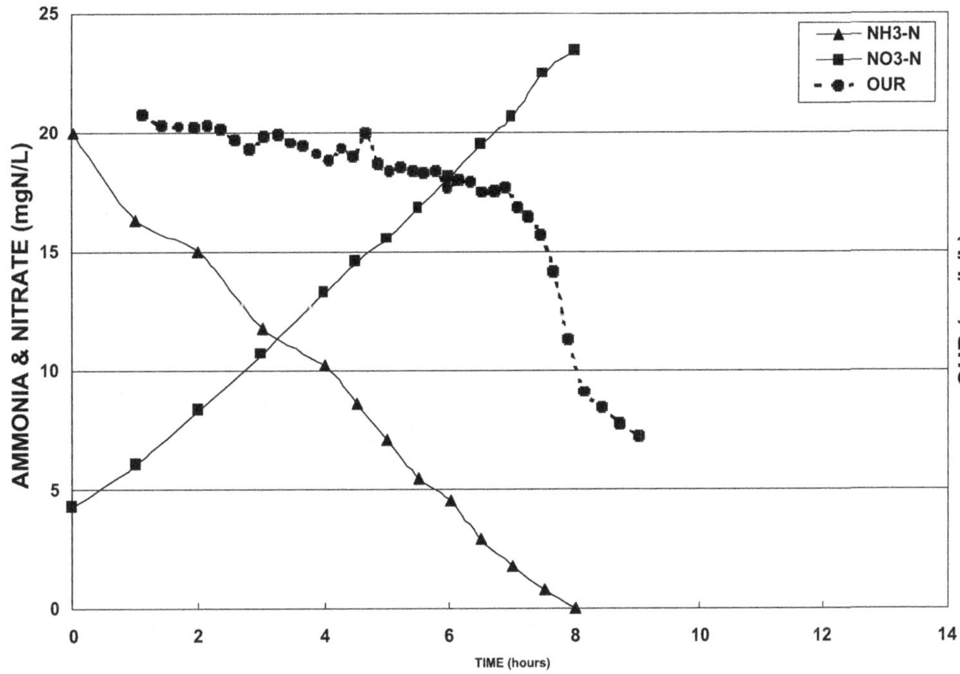

Figure 32-2. Ammonia, Nitrate and OUR Response Over the Initial Period on Day 3

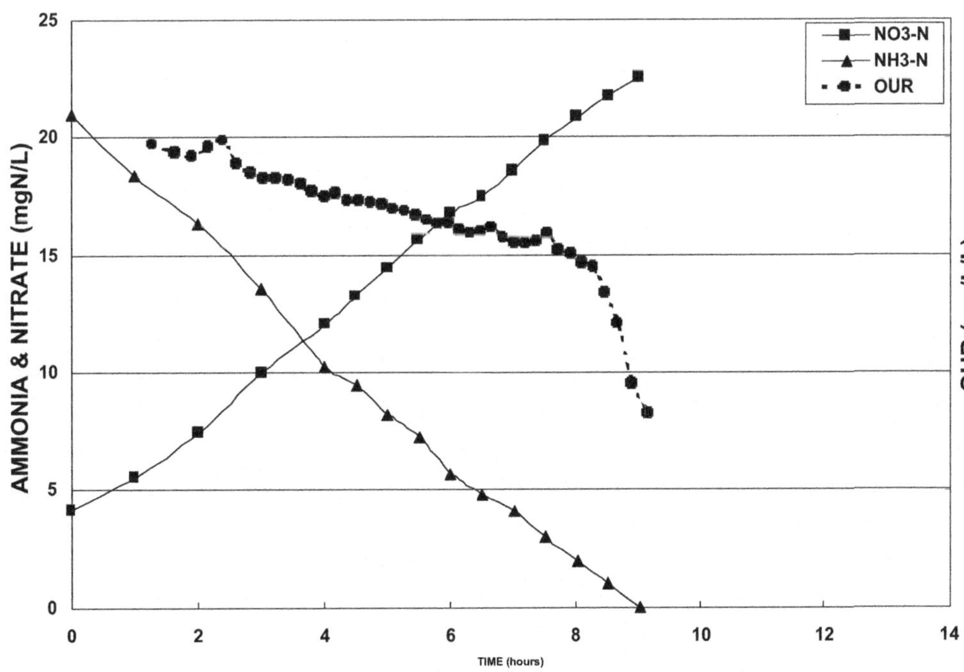

Figure 32-3. Ammonia, Nitrate and OUR Response Over the Initial Period on Day 5

Table 32-2. Ammonia, Nitrate and OUR Profile Data for Day 5

Time (hours)	Nitrate (mgN/L)	Ammonia (mgN/L)	Time (hours)	OUR (mg/L/hr)
0.0	4.19	20.99	1.28	19.7
1.0	5.57	18.39	1.63	19.4
2.0	7.45	16.35	1.91	19.2
3.0	10.01	13.60	2.16	19.6
4.0	12.09	10.29	2.39	19.9
4.5	13.27	9.51	2.61	18.9
5.0	14.46	8.25	2.83	18.5
5.5	15.64	7.30	3.03	18.3
6.0	16.83	5.73	3.23	18.3
6.5	17.52	4.87	3.43	18.2
7.0	18.61	4.16	3.63	18.0
7.5	19.89	2.98	3.81	17.7
8.0	20.88	1.96	4.00	17.5
8.5	21.77	1.01	4.18	17.6
9.0	22.56	0.00	4.37	17.3
			4.55	17.3
			4.73	17.3
			4.92	17.2
			5.10	17.0
			5.27	16.9
			5.45	16.7
			5.63	16.5
			5.80	16.4
			5.97	16.4
			6.15	16.1
			6.32	16.0
			6.49	16.0
			6.66	16.2
			6.84	15.7
			7.02	15.5
			7.20	15.5
			7.37	15.6
			7.55	15.9
			7.72	15.2
			7.90	15.1
			8.08	14.6
			8.27	14.5
			8.46	13.4
			8.66	12.1
			8.89	9.5
			9.14	8.2

Table 32-3. Daily Results from SBR Intensive Testing Period

DATE	DAY	INFLUENT						REACTOR					DECANT						
		TSS	VSS	COD tot.	COD floc./filt.	TKN tot.	NH3	TSS	VSS	COD tot.	TKN tot.	OXYGEN UTILIZED	COD tot.	COD filt.	COD floc./filt.	TKN filt.	NH3	NO3	VSS
		(mg/L)	(mg/L)	(mg/L)	(mg/L)	(mg/L)	(mg/L)	(mg/L)	(mg/L)	(mg/L)	(mg/L)	(mg/d)	(mg/L)	(mg/L)	(mg/L)	(mg/L)	(mg/L)	(mg/L)	(mg/L)
14-Aug	1	95	76	267	44	37.5	24.6	1,000	792	1,139	108	2,281	45	30	18	2.9	0.04	24.6	11.6
15-Aug	2	108	91	334	88	42.0	25.2	1,080	834	1,173	87	2,230	35	28	18	3.0	0.48	24.6	5.8
16-Aug	3	106	84	317	77	38.9	25.0	1,010	788	1,105	87	2,317	37	22	14	2.9	0.07	25.9	11.6
17-Aug	4	113	82	345	90	38.5	25.6	1,088	848	1,090	97	2,255	47	35	16	3.1	0.06	22.9	8.7
18-Aug	5	106	90	321	90	38.9	26.8	1,106	864	1,207	100	2,335	51	36	25	4.2	0.42	24.9	10.9
19-Aug	6	127	108	306	92	41.6	26.8	1,057	825	1,143	96	2,355	63	44	27	4.4	0.08	24.6	9.7
20-Aug	7	114	91	270	65	33.6	23.0	1,057	825	1,143	96	2,195	46	32	19	3.4	0.19	24.6	9.7
AVERAGE		110	89	309	78	38.7	25.3	1,057	825	1,143	95.9	2,281	46	32	19	3.4	0.2	24.6	9.7

32.3 SBR Data Validation: Mass Balances

Mass balances for COD and nitrogen were performed on the experimental data from the 7-day intensive testing period. This provides an overall validation of the experimental data. The balances should be within a few percentage points of 100 percent. If not, it is an indication that a problem exists with the experimental data. The basis for calculating the balances was as follows:

COD Balance:

$$
\begin{aligned}
\textit{Daily Input} &= V_{ww} * \textit{Influent COD} \\
\textit{Daily Output} &= \left(V_{ww} - q_w\right) * \textit{Filtered decant COD} \\
&\quad + q_w * \textit{SBR total COD} \\
&\quad + \textit{total mass of oxygen utilized} \\
&\quad - V_{ww} * \textit{Decant nitrate } N * 4.57 \\
&\quad (\textit{mass of oxygen for nitrification})
\end{aligned}
$$

Nitrogen Balance:

$$
\begin{aligned}
\textit{Daily Input} &= V_{ww} * \textit{Influent TKN} \\
\textit{Daily Output} &= V_{ww} * \textit{Filtered decant TKN} \\
&\quad + V_{ww} * \textit{Decant nitrate } N \\
&\quad + \left(V_{ww} - q_w\right) * \textit{Decant VSS} * N \textit{ Content of VSS} \\
&\quad + q_w * \textit{SBR VSS} * N \textit{ Content of VSS}
\end{aligned}
$$

The results of mass balance calculations are recorded in Table 32-4. An accumulation term was included in the mass balances to account for the difference in the masses of COD and nitrogen in the reactor between the start and end of the 7-day period. The overall COD and nitrogen balances closed to 100 percent and 97 percent, respectively.

Table 32-4. COD and Nitrogen Mass Balances on Data from SBR Intensive Testing Period

DATE	DAY	COD INPUT (mg/d)	COD Eff. (mg/d)	COD WAS (mg/d)	COD OURt (mg/d)	COD OURn (mg/d)	COD OURc (mg/d)	COD OUTPUT (mg/d)	COD Daily %	NITROGEN INPUT (mgN/d)	NITROGEN NO3 out (mgN/d)	NITROGEN TKNs out (mgN/d)	NITROGEN Eff. VSS (mgN/d)	NITROGEN WAS (mgN/d)	NITROGEN OUTPUT (mgN/d)	NITROGEN Daily %
14-Aug	1	2,136	362	712	2,281	899	1,382	2,455	115	300	197	23	10	56	285	95
15-Aug	2	2,672	283	733	2,230	899	1,331	2,347	88	336	197	24	5	59	284	84
16-Aug	3	2,536	298	691	2,317	948	1,369	2,357	93	311	207	23	10	55	296	95
17-Aug	4	2,760	376	681	2,255	838	1,417	2,474	90	308	183	25	7	60	275	89
18-Aug	5	2,568	408	754	2,335	911	1,424	2,586	101	311	199	33	9	61	302	97
19-Aug	6	2,448	502	714	2,355	899	1,456	2,672	109	333	197	35	8	58	298	90
20-Aug	7	2,160	371	714	2,195	899	1,296	2,381	110	269	197	27	8	58	290	108
ACCUMULATION								40	101						37	94
CUMULATIVE TOTAL		17,280	Overall % 100					17,325		2,168	Overall % 97				2,104	

32.4 Characteristics from Direct Measurements

A number of characteristics could be derived from the analyses performed on samples collected during the intensive testing period. These are recorded in Table 32-5. It should be recalled that two different batches of influent were used during this period. As a result, certain characteristics in the influent differed slightly from batch to batch. However, variations from the overall average were not significant.

Influent soluble biodegradable COD fraction (f_{SS}): The average value of 0.19 mg COD/mg COD is typical for a raw municipal wastewater.

Influent soluble unbiodegradable COD fraction (f_{SI}): The average value of 0.06 mg COD/mg COD is typical for a raw municipal wastewater.

Influent inorganic suspended solids concentration (X_{ISS}): The average value was 21 mg/L, which again is typical for a raw municipal wastewater.

Influent TKN:COD ratio: The average value of 0.13 mg N/mg COD is typical for a raw municipal wastewater.

Influent Ammonia/TKN fraction (f_{SNH}): The average value of 0.65 mg N/mg N is typical for a raw municipal wastewater.

Nitrogen content of VSS (f_N): The average value of 0.11 mg N/mg VSS for the nitrogen content of the VSS is typical.

COD/VSS ratio VSS (f_{CV}): The average value of 1.35 mg COD/mg VSS is slightly lower than the expected value of 1.42 mg COD/mg VSS.

Table 32-5. Data Derived from Results of SBR Intensive Testing Period

DATE	DAY	INFLUENT					MIXED LIQUOR				DECANT
		ISS (mg/L)	f_{SS} (mgCOD/ mgCOD)	f_{SI} (mgCOD/ mgCOD)	TKN/COD (mgN/ mgCOD)	f_{SNH} (mg/L)	ISS (mg/L)	Fcv (mgCOD/ mgVSS)	Fn (mgN/ mgVSS)	VSS/ TSS	PRED VSS (mg/L)
14-Aug	1	18	0.10	0.07	0.14	0.66	208	1.40	0.13	0.79	12
15-Aug	2	16	0.21	0.05	0.13	0.60	246	1.37	0.10	0.77	6
16-Aug	3	22	0.20	0.04	0.12	0.64	222	1.37	0.11	0.78	12
17-Aug	4	31	0.22	0.05	0.11	0.66	240	1.24	0.11	0.78	9
18-Aug	5	16	0.20	0.08	0.12	0.69	242	1.35	0.11	0.78	11
19-Aug	6	19	0.21	0.09	0.14	0.64	232	1.33	0.11	0.78	14
20-Aug	7	23	0.17	0.07	0.12	0.68	232	1.35	0.11	0.78	10
	AVERAGE	21	0.19	0.06	0.13	0.65	232	1.35	0.11	0.78	10

32.5 Characteristics from Modeling of SBR Operation

Certain influent wastewater characteristics (and the nitrification rate) cannot be determined by direct measurement. The fraction of the influent COD which is particulate unbiodegradable (f_{XI}) is an example of an influent wastewater characteristic that can be estimated by simulation (this parameter is particularly important with respect to estimating sludge production). These parameters must be estimated iteratively either using an analytical model or through simulation (also based on a model). The analytical approach only is suitable when the system is at, or close to, a true steady state. As noted earlier, in this study the system parameters varied considerably over the testing period; hence the simulation approach was used.

The simulation study of SBR behavior was conducted to model system response over the 7 days of intensive monitoring. The simulator configuration is shown in Figure 32-4.

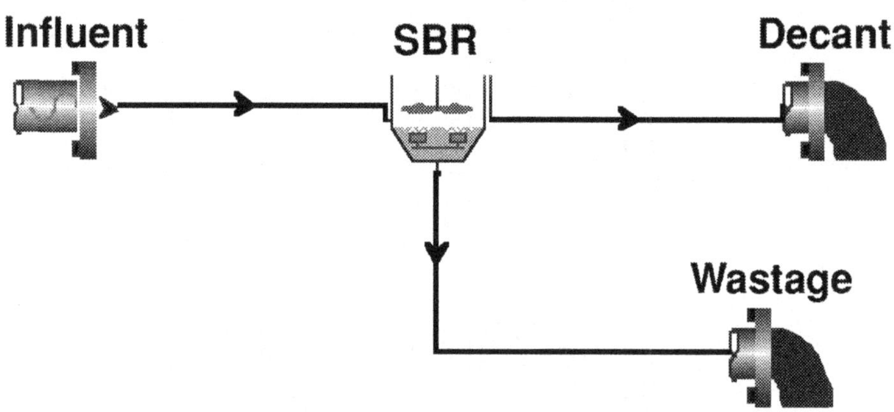

Figure 32-4. Configuration for Simulating the SBR

Operating conditions such as the influent volume and the various periods in the SBR cycle time were set up in the simulator. The influent COD, TKN and X_{ISS} concentrations for each day were specified in the influent element (see Table 32-3), together with the fractional characteristics determined from the direct measurements.

A number of simulation runs were performed, iteratively varying the values of the parameters to be estimated, f_{XI} and μ_{AUT}. The objective was to obtain a reasonable fit of simulated to observed response over the intensive period of operation using a single value for each parameter. In the case of f_{XI}, the primary parameter response to match was VSS concentration. In the case of μ_{AUT}, the data to be matched were the responses of ammonia, nitrate and OUR over the intensive monitoring period.

32.5.1 Particulate Unbiodegradable COD Fraction (f_{XI})

A value for f_{XI} of 0.15 mg COD/mg COD provided a good correspondence between the simulated and observed VSS in the system. This value is typical for a raw municipal wastewater. Figure 32-5 shows a plot of the simulation results over 7 days. It should be noted that the experimental data points correspond to the VSS concentration in the waste sludge drawn from the reactor at 23.5 hours in each daily cycle. Figure 32-6 shows the corresponding plot for TSS concentration.

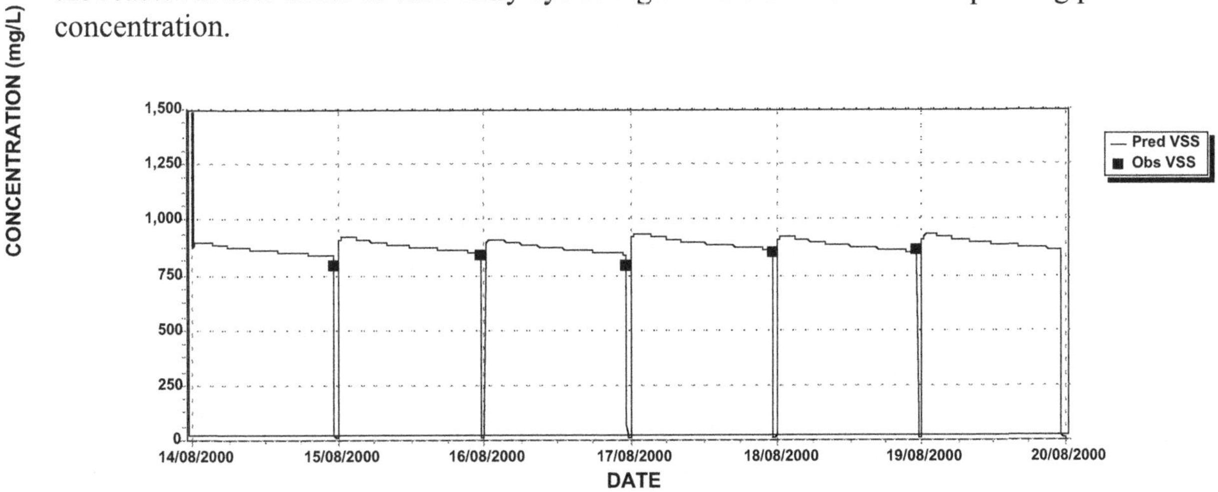

Figure 32-5. Simulated (Solid Line) and Observed (Data Points) VSS in the SBR over 7 days

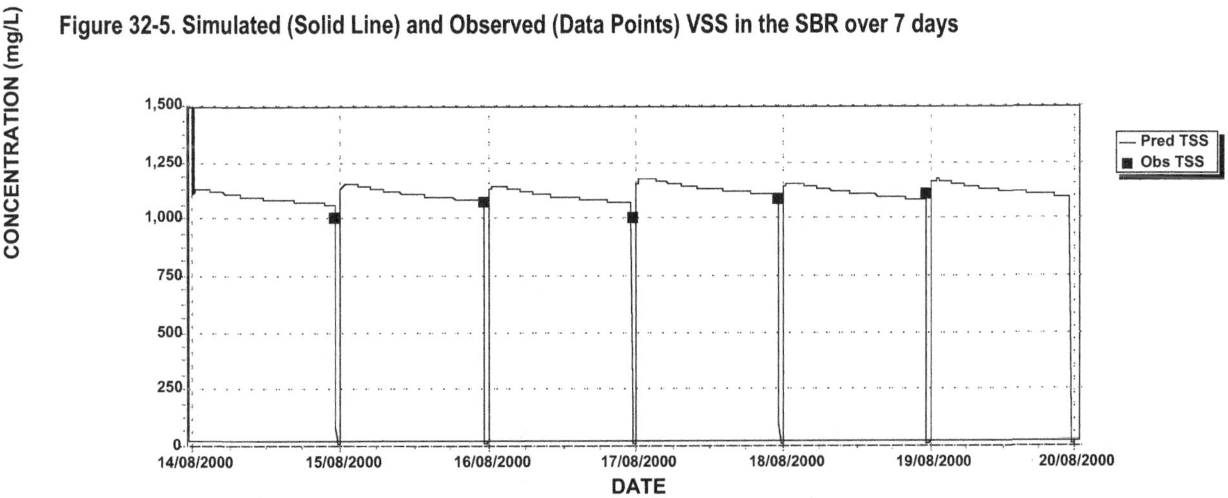

Figure 32-6. Simulated (Solid Line) and Observed (Data Points) TSS in the SBR over 7 days

Throughout the intensive monitoring period, the reactor TSS concentration was slightly over 1,000 mg/L. Prior to the intensive monitoring period, the reactor TSS concentration was lower (i.e., ~ 800 mg/L), despite the fact that the influent COD concentration remained relatively constant over both periods. This difference is attributable to the fact that the reactor effluent solids concentration during the period prior to intensive monitoring was more than twice the value observed during the intensive period.

Note that the simulator predicts a decrease in solids concentration over the course of each day. This is due to normal biological activity, and is expected. At the end of each day, the simulator shows the solids concentration rapidly plunging to zero and then increasing over a very short time period. This solids concentration fluctuation arises when the simulator enters the decant and fill phases.

32.5.2 Nitrification Rate (μ_{AUT})

The nitrification behavior in the system could be simulated accurately with a value for the maximum specific growth rate of the nitrifiers (μ_{AUT}) of 0.80 d^{-1}. (This value is referenced to 20°C, with a temperature dependency coefficient of 1.072. Also, a value of 0.17 d^{-1} (at 20°C) for the nitrifier decay rate (b_{AUT}) was used with a temperature dependency coefficient of 1.029.) Figure 32-7 shows the simulated nitrate response in the SBR over the 7 days of operation. The uppermost data points correspond to the decant samples and the near-vertical clusters of data points are for Days 3 and 5.

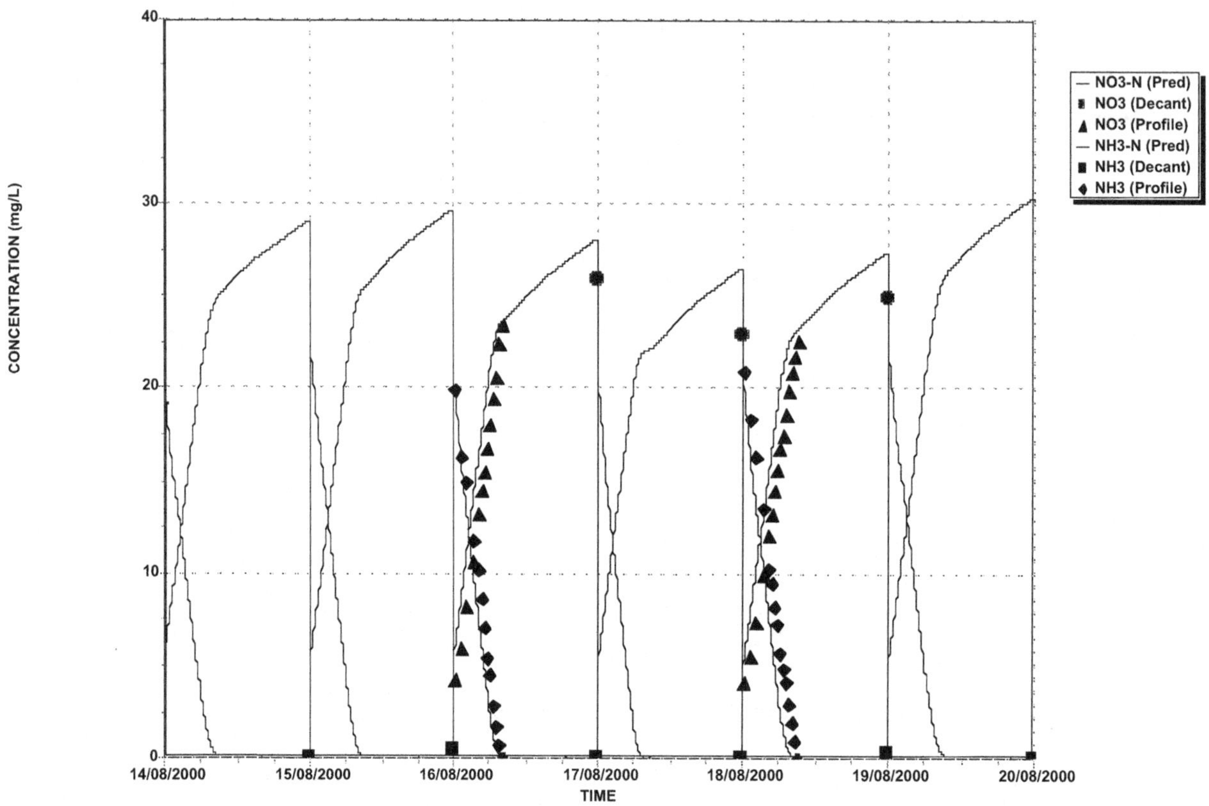

Figure 32-7. Simulated (Solid Line) and Observed (Data Points) Nitrite/Nitrate in the SBR Over 7 days

Data for Days 3 and 5 are shown on an expanded scale in Figure 32-8, together with ammonia concentration.

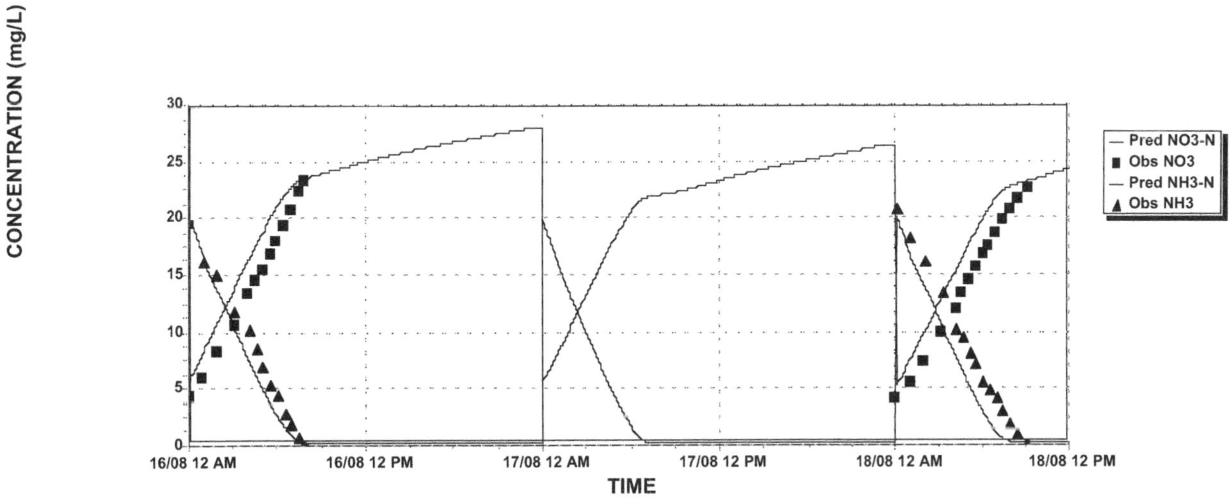

Figure 32-8. Simulated (Solid Line) and Observed (Data Points) Ammonia and Nitrite/Nitrate Response in the SBR over Days 3 and 5

In Figure 32-8, the simulated and observed nitrite/nitrate response over the first several hours are near parallel, indicating that the estimate of nitrifier growth rate is very reasonable.

Oxygen utilization rate (OUR) response is closely linked to nitrification behavior. A further check on the selection of the nitrification rate parameter is provided by comparing the simulated and observed response of OUR over Days 3 and 5 shown in Figure 32-9.

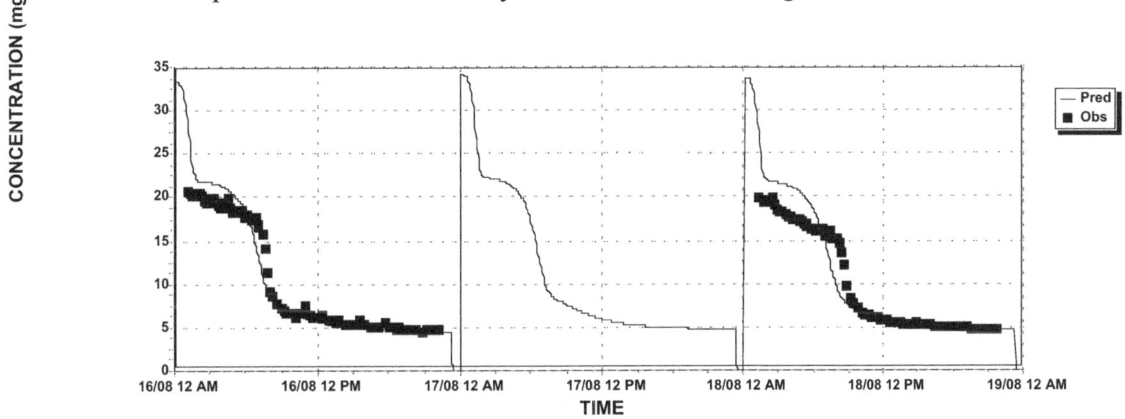

Figure 32-9. Simulated (Solid Line) and Observed (Data Points) OUR Response in the SBR over Days 3 and 5

32.6 References

Randall, E.W., A. Wilkinson, and G.A. Ekama. 1991. An instrument for the direct determination of oxygen utilization rate. *Water SA* 17(1):11-18.

CHAPTER 33.0

SBR: City of Phoenix Investigations

33.1 Background

This chapter outlines the operation of and data collected for a laboratory-scale fill-and-draw SBR system, which was operated with sludge from the 91st Avenue Wastewater Treatment Plant in Phoenix, Ariz.

33.2 SBR Operation

For a detailed operation protocol, please refer to Appendix D.

For the Phoenix system, the SBR was operated for the following conditions:

V_P = 10 L

V_{WW} = 8 L

q_W = 1.2 - 1.5 L/d (average of 1.4 L/d)

θ_X = 10/1.4 = 7 days

Temp. = 22.3°C

After start-up, the system was operated for a period of 47 days (~ 6.5 sludge ages) so as to attain a quasi-steady state. At this time, an intensive daily monitoring schedule was commenced for a period of 9 days. Analyses were conducted on samples of influent, waste and decant. Throughout the period of operation, oxygen utilization rate was monitored semi-continuously using an integrated DO controller/meter.

On two days (Days 2 and 3) during the intensive monitoring period, profiles of ammonia and nitrate concentration were monitored over the first 8 to 10 hours of the react period. Analyses were performed on small sample volumes withdrawn from the reactor.

33.2.1 Influent Wastewater

To attain the quasi-steady state SBR response described above, the system received a fixed COD and TKN load each cycle.

33.2.2 Selection of SBR Sludge Age

The target sludge age in the SBR was 7 days. However, due to the unintentional wasting of some solids in the SBR decant, the actual SRT was 6 days.

33.2.3 SBR Start-up Period

At start-up, the SBR was seeded with two liters of activated sludge taken from the 91st Avenue plant. The system was operated for an acclimation period of 47 days (~ 6.5 sludge ages). Figure 33-1 shows a plot of the parameters monitored during the start-up period.

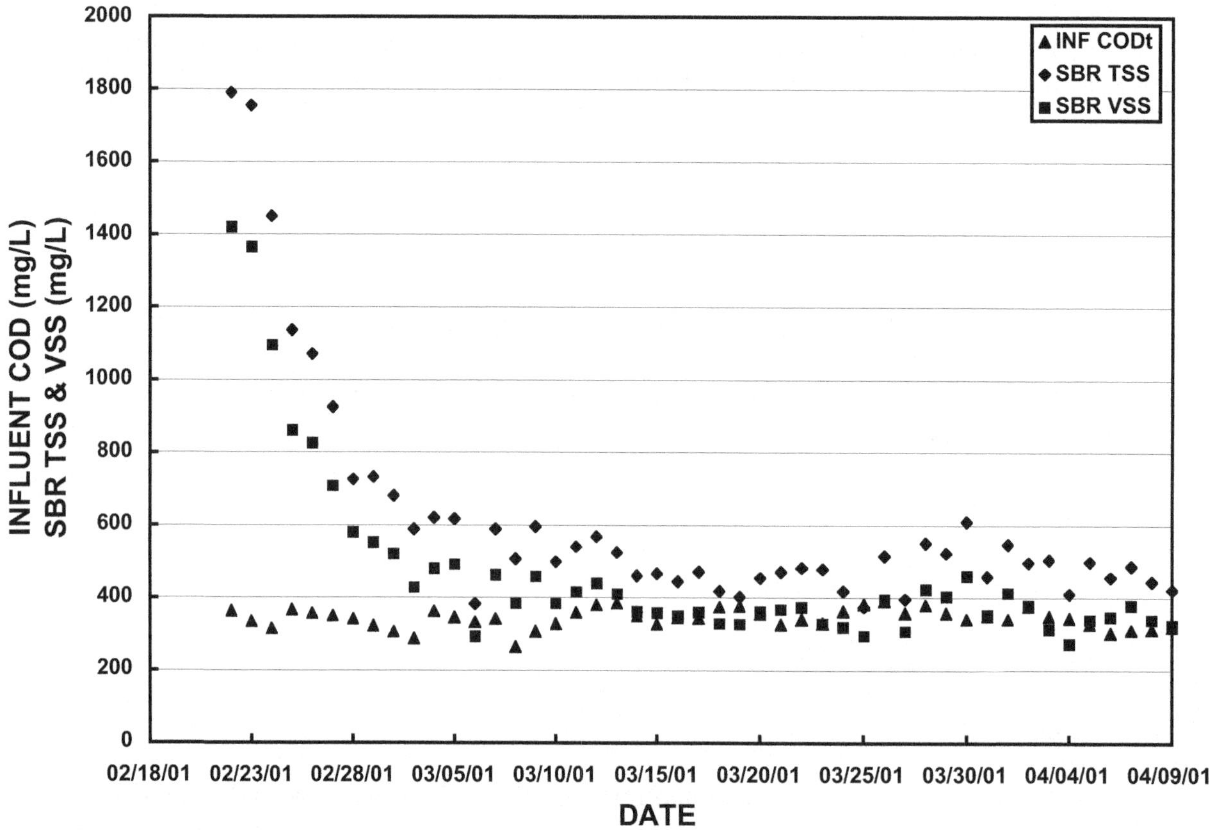

Figure 33-1. Plot of Influent COD, SBR TSS and SBR VSS Prior to Intensive Period

33.2.4 Monitoring during Intensive Testing

Intensive monitoring of the SBR system was performed for a period of 7 days as follows:

SBR Stream	Parameter Monitored
Influent	TSS
	VSS
	Total COD
	Flocculated/filtered COD
	Total TKN
	Ammonia-N
Reactor	TSS
	VSS
	Total COD
	Total TKN
	OUR
	Temperature
Decant	Total COD
	Membrane Filtered COD
	Flocculated/Filtered COD
	Membrane Filtered Total TKN
	Membrane Filtered Ammonia-N
	Membrane Filtered Nitrate-N

COD mass balances could not be performed on the experimental data because oxygen utilization rate (OUR) in the SBR was not monitored continuously. Results from the intensive monitoring period are recorded in Table 33-3.

On Days 2 and 3, profiles of ammonia and nitrite/nitrate concentration were monitored over the first 8 to 10 hours of the react period. Analyses were performed on small sample volumes withdrawn from the reactor (and filtered immediately). The ammonia and nitrate responses are shown in Figures 33-2 and 33-3. Raw data for the profiles are given in Tables 33-1 and 33-2.

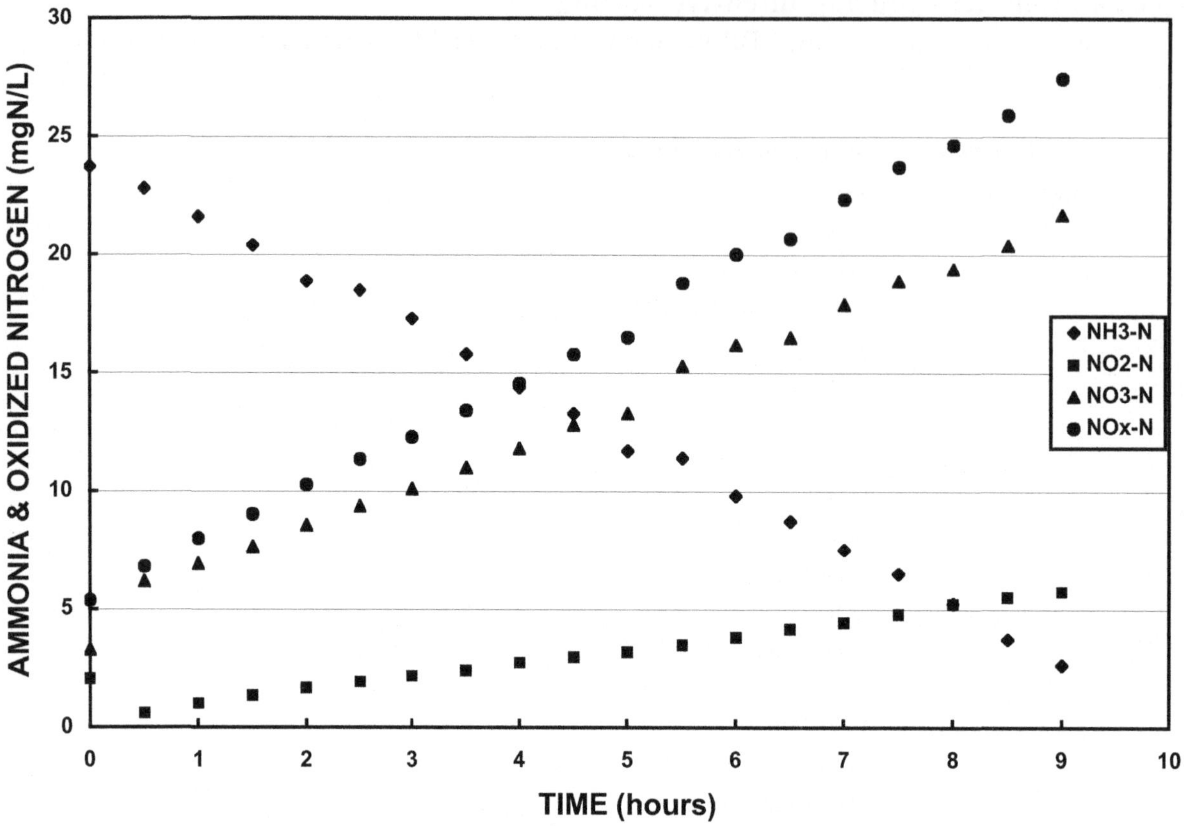

Figure 33-2. Ammonia, Nitrite and Nitrate Response over the Initial Period on Day 2

Table 33-1. Ammonia, Nitrite and Nitrate Profile Data for Day 2

Time (hours)	Nitrite (mgN/L)	Nitrate (mgN/L)	NOx (mgN/L)	Ammonia (mgN/L)
0	2.08	3.30	5.38	23.7
0.5	0.62	6.20	6.82	22.8
1	1.04	6.95	7.99	21.6
1.5	1.37	7.66	9.03	20.4
2	1.70	8.57	10.27	18.9
2.5	1.96	9.39	11.35	18.5
3	2.19	10.1	12.29	17.3
3.5	2.42	11.0	13.42	15.8
4	2.77	11.8	14.57	14.4
4.5	2.99	12.8	15.79	13.3
5	3.21	13.3	16.51	11.7
5.5	3.51	15.3	18.81	11.4
6	3.83	16.2	20.03	9.82
6.5	4.19	16.5	20.69	8.74
7	4.45	17.9	22.35	7.53
7.5	4.81	18.9	23.71	6.51
8	5.24	19.4	24.64	5.25
8.5	5.52	20.4	25.92	3.75
9	5.76	21.7	27.46	2.65

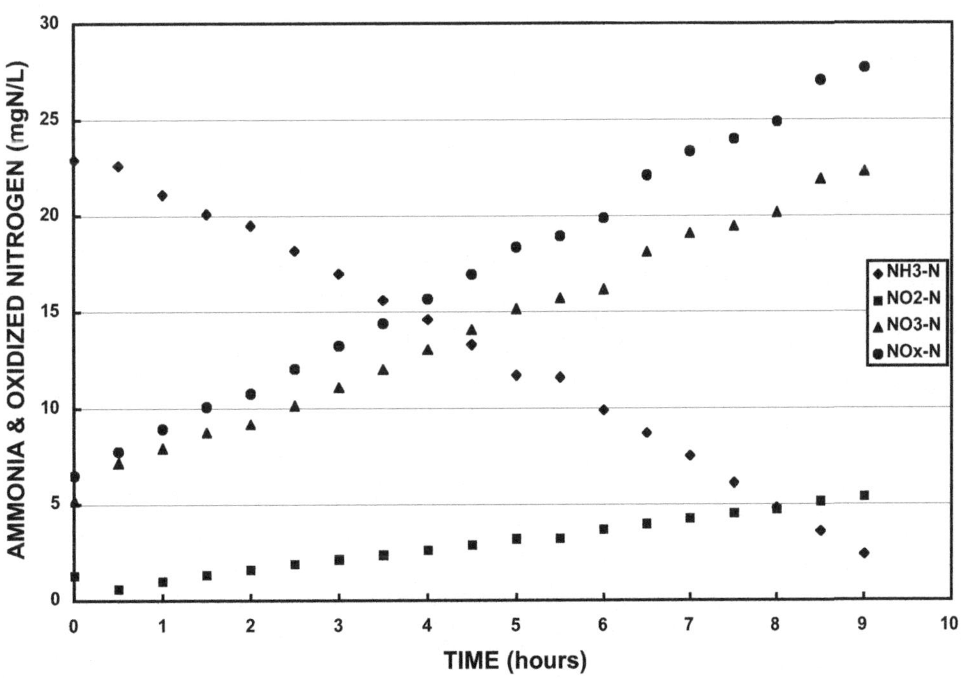

Figure 33-3. Ammonia, Nitrite and Nitrate Response over the Initial Period on Day 3

Table 33-2. Ammonia and Nitrate Profile Data for Day 3

Time (hours)	Nitrite (mgN/L)	Nitrate (mgN/L)	NOx (mgN/L)	Ammonia (mgN/L)
0	1.31	5.19	6.50	22.9
0.5	0.60	7.15	7.75	22.6
1	1.00	7.92	8.92	21.1
1.5	1.34	8.74	10.08	20.1
2	1.61	9.16	10.77	19.5
2.5	1.90	10.2	12.05	18.2
3	2.15	11.1	13.24	17.0
3.5	2.38	12.0	14.40	15.6
4	2.62	13.1	15.67	14.6
4.5	2.89	14.1	16.97	13.3
5	3.21	15.2	18.37	11.7
5.5	3.22	15.7	18.94	11.6
6	3.70	16.2	19.88	9.90
6.5	3.98	18.1	22.09	8.70
7	4.26	19.1	23.35	7.50
7.5	4.53	19.5	23.99	6.10
8	4.72	20.2	24.89	4.80
8.5	5.14	21.9	27.03	3.60
9	5.39	22.3	27.68	2.40

Table 33-3. Daily Results from SBR Intensive Testing Period

DATE	DAY	INFLUENT							REACTOR				DECANT						
		TSS (mg/L)	VSS (mg/L)	COD tot. (mg/L)	COD filt. (mg/L)	COD floc./filt. (mg/L)	TKN tot. (mg/L)	NH_3 (mg/L)	TSS (mg/L)	VSS (mg/L)	COD tot. (mg/L)	TKN tot. (mg/L)	COD tot. (mg/L)	COD filt. (mg/L)	COD floc./filt. (mg/L)	TKN filt. (mg/L)	NH_3 (mg/L)	NO_x (mg/L)	VSS (mg/L)
10-Apr	1	108	96	319	83	61	38.7	29.3	500	390	682	33.0	39	18	12	2.4	0.0	30.2	12
11-Apr	2	114	100	263	80	43	38.9	29.5	480	372	642	34.0	54	20	12	2.8	0.2	32.6	14
12-Apr	3	120	98	306	79	55	39.4	29.9	434	332	584	33.5	47	19	10	2.8	0.0	32.5	9
13-Apr	4	116	92	304	77	56	39.8	30.2	444	332	551	33.5	47	20	10	2.8	0.1	31.2	13
14-Apr	5	102	74	311	78	65	43.7	33.1	444	338	580	33.5	46	21	10	2.8	0.1	30.4	16
15-Apr	6	116	100	366	74	68	34.3	26.0	452	348	645	33.5	48	20	12	2.8	0.1	25.5	15
16-Apr	7	118	96	378	73	61	29.2	22.1	398	298	575	33.5	65	20	10	2.8	0.0	26.4	23
17-Apr	8	108	92	331	82	73	38.4	29.1	418	330	611	33.5	50	15	12	2.8	0.0	26.4	12
18-Apr	9	122	100	341	85	80	35.9	27.2	460	358	641	33.5							
AVERAGE		114	94	324	79	53	37.6	28.5	448	344	612	33.5	50	19	11	2.7	0.1	29.4	14

33.3 SBR Data Validation: Mass Balances

A nitrogen mass balance was performed on the experimental data from the 9-day intensive testing period. As mentioned previously, problems with continuous OUR measurement made COD mass balance calculations impossible. The basis for calculating the nitrogen balance was as follows:

Nitrogen Balance:

$$Daily\ Input = V_{ww} * Influent\ TKN$$

$$\begin{aligned} Daily\ Output = {} & V_{ww} * Filtered\ decant\ TKN \\ & + V_{ww} * Decant\ nitrate\ N \\ & + \left(V_{ww} - q_w\right) * Decant\ VSS * N\ Content\ of\ VSS \\ & + q_w * SBR\ VSS * N\ Content\ of\ VSS \end{aligned}$$

The results of the nitrogen mass balance calculations are recorded in Table 33-4. An accumulation term was included in the mass balance to account for the difference in the mass of nitrogen in the reactor between the start and end of the 9-day period. The overall nitrogen balance closed to 101 percent.

Table 33-4. Nitrogen Mass Balance on Data from SBR Intensive Testing Period

DATE	DAY	NITROGEN						
		INPUT (mgN/d)	NO3 out (mgN/d)	TKNs out (mgN/d)	Eff. VSS (mgN/d)	WAS (mgN/d)	OUTPUT (mgN/d)	Daily %
10-Apr	1	309	242	19	9	49	319	103
11-Apr	2	311	261	22	10	50	344	110
12-Apr	3	316	260	22	7	42	331	105
13-Apr	4	319	250	22	9	45	326	102
14-Apr	5	349	243	22	12	43	320	92
15-Apr	6	274	204	22	11	44	281	102
16-Apr	7	233	211	22	17	35	285	122
17-Apr	8	307	211	22	9	36	278	91
ACCUMULATION							-29	
CUMULATIVE TOTAL		2,419		Overall %	101		2,455	

33.4 Characteristics from Direct Measurements

A number of characteristics could be derived from the analyses performed on samples collected during the intensive testing period. These are recorded in Table 33-5.

Influent soluble biodegradable COD fraction (f_{SS}): The average value of 0.15 mg COD/mg COD is typical for a raw municipal wastewater.

Influent soluble unbiodegradable COD fraction (f_{SI}): The average value of 0.06 mg COD/mg COD is typical for a raw municipal wastewater.

Influent inorganic suspended solids concentration (X_{ISS}): The average value was 20 mg/L, which again is typical for a raw municipal wastewater.

Influent TKN:COD ratio: The average value of 0.12 mg N/mg COD is typical for a raw municipal wastewater.

Influent Ammonia/TKN fraction (f_{SNH}): A plant historical value of 0.76 was used to calculate the influent TKN_{TOT} from measured ammonia data. Therefore, a fraction was not calculated as part of this study.

Nitrogen content of VSS (f_N): The average value of 0.09 mg N/mg VSS for the nitrogen content of the VSS is typical.

COD/VSS ratio VSS (f_{CV}): The average value of 1.72 mg COD/mg VSS is slightly higher than the expected value of 1.42 mg COD / mg VSS.

VSS/TSS ratio: The average value of 0.77 is typical for a plant operating under thc conditions found at the 91st Avenue plant.

Table 33-5. Data Derived from Results of SBR Intensive Testing Period

DATE	DAY	INFLUENT					MIXED LIQUOR				DECANT
		ISS (mg/L)	f_{SS} (mgCOD/ mgCOD)	f_{SI} (mgCOD/ mgCOD)	TKN/COD (mgN/ mgCOD)	f_{SNH} (mg/L)	ISS (mg/L)	Fcv (mgCOD/ mgVSS)	Fn (mgN/ mgVSS)	VSS/ TSS	PRED VSS (mg/L)
10-Apr	1	12	0.15	0.06	0.12		110	1.70	0.078	0.78	14
11-Apr	2	14	0.14	0.08	0.15		108	1.67	0.084	0.78	23
12-Apr	3	22	0.15	0.06	0.13		102	1.70	0.092	0.76	19
13-Apr	4	24	0.15	0.07	0.13		112	1.60	0.092	0.75	18
14-Apr	5	28	0.18	0.07	0.14		106	1.65	0.091	0.76	17
15-Apr	6	16	0.15	0.05	0.09		104	1.80	0.088	0.77	19
16-Apr	7	22	0.13	0.05	0.08		100	1.86	0.103	0.75	30
17-Apr	8	16	0.18	0.05	0.12		88	1.81	0.093	0.79	24
18-Apr	9	22			0.11		102			0.78	
	AVERAGE	20	0.15	0.06	0.12		104	1.72	0.09	0.77	21

33.5 Characteristics from Modeling of SBR Operation

Certain influent wastewater characteristics (and the nitrification rate) cannot be determined by direct measurement. The fraction of the influent COD which is particulate unbiodegradable (f_{XI}) is an example of an influent wastewater characteristic that can be estimated by simulation (this parameter is particularly important with respect to estimating sludge production). These parameters must be estimated iteratively either using an analytical model or through simulation (also based on a model). The analytical approach is only suitable when the system is at, or close to, a true steady state. As noted earlier, in this study the system parameters varied considerably over the testing period; hence the simulation approach was used.

The simulation study of SBR behavior was conducted to model system response over the nine days of intensive monitoring. The simulator configuration is shown in Figure 33-4.

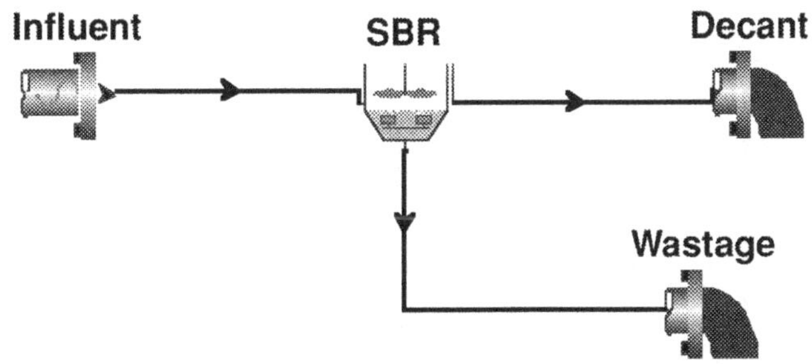

Figure 33-4. Configuration for Simulating the SBR

Operating conditions such as the influent volume and the various periods in the SBR cycle time were set up in the simulator. The influent COD, TKN and X_{ISS} concentrations for each day were specified in the influent element (see Table 33-3), together with the fractional characteristics determined from the direct measurements.

A number of simulation runs were performed, iteratively varying the values of the parameters to be estimated, f_{XI} and μ_{AUT}. The objective was to obtain a reasonable fit of simulated to observed response over the intensive period of operation using a single value for each parameter. In the case of f_{XI}, the primary parameter response to match was VSS concentration. In the case of μ_{AUT}, the data to be matched were the responses of ammonia, nitrate and OUR over the intensive monitoring period.

33.5.1 Particulate Unbiodegradable COD Fraction (f_{XI})

A value for f_{XI} of 0.10 mg COD/mg COD provided a good correspondence between the simulated and observed VSS in the system. This value is typical for a raw municipal wastewater. Figure 33-5 shows a plot of the simulation results over 9 days. It should be noted that the experimental data points correspond to the VSS concentration in the waste sludge drawn from the reactor at 23.5 hours in each daily cycle. Figure 33-6 shows the corresponding plot for TSS concentration.

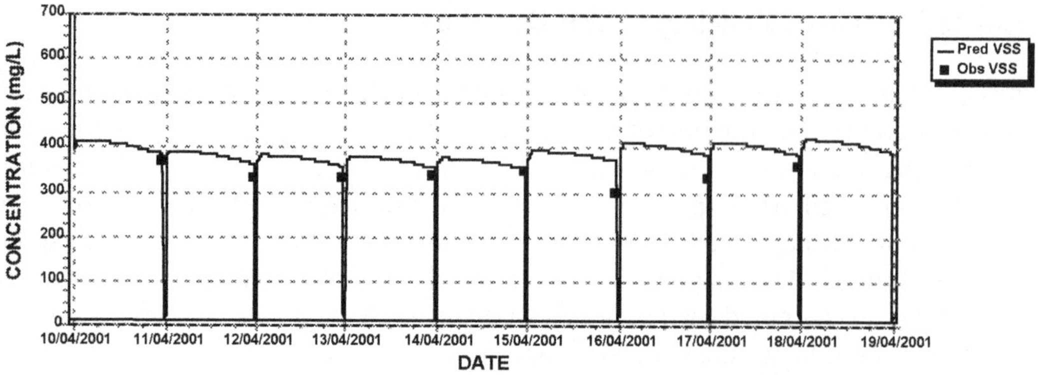

Figure 33-5. Simulated (Solid Line) and Observed (Data Points) VSS in the SBR over 9 Days

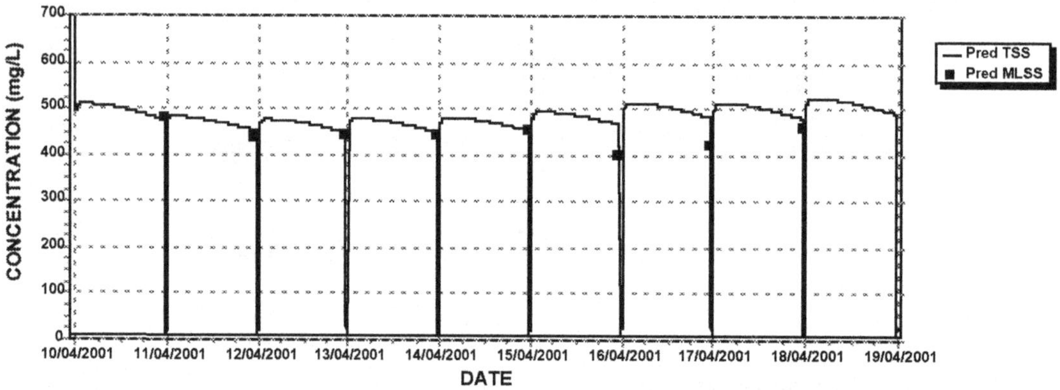

Figure 33-6. Simulated (Solid Line) and Observed (Data Points) TSS in the SBR over 9 days

Note that the simulator predicts a decrease in solids concentration over the course of each day. This is due to normal biological activity and is expected. At the end of each day, the simulator shows the solids concentration rapidly plunging to zero and then increasing over a very short time period. This solids concentration fluctuation arises when the simulator enters the decant and fill phases.

33.5.2 Nitrification Rate (μ_{AUT})

The nitrification behavior in the system could be simulated accurately with a value for the maximum specific growth rate of the nitrifiers (μ_{AUT}) of 0.95 d^{-1} (This value is referenced to 20°C, with a temperature dependency coefficient of 1.072. Also, a value of 0.17 d^{-1} (at 20°C) for the nitrifier decay rate (b_{AUT}) was used with a temperature dependency coefficient of 1.029). Figure 33-7 shows the simulated nitrate response in the SBR over the 9 days of operation. The uppermost data points correspond to the decant samples and the near-vertical clusters of data points are for Days 2 and 3. Data for Days 2 and 3 are shown on an expanded scale in Figure 33-8, together with ammonia concentration.

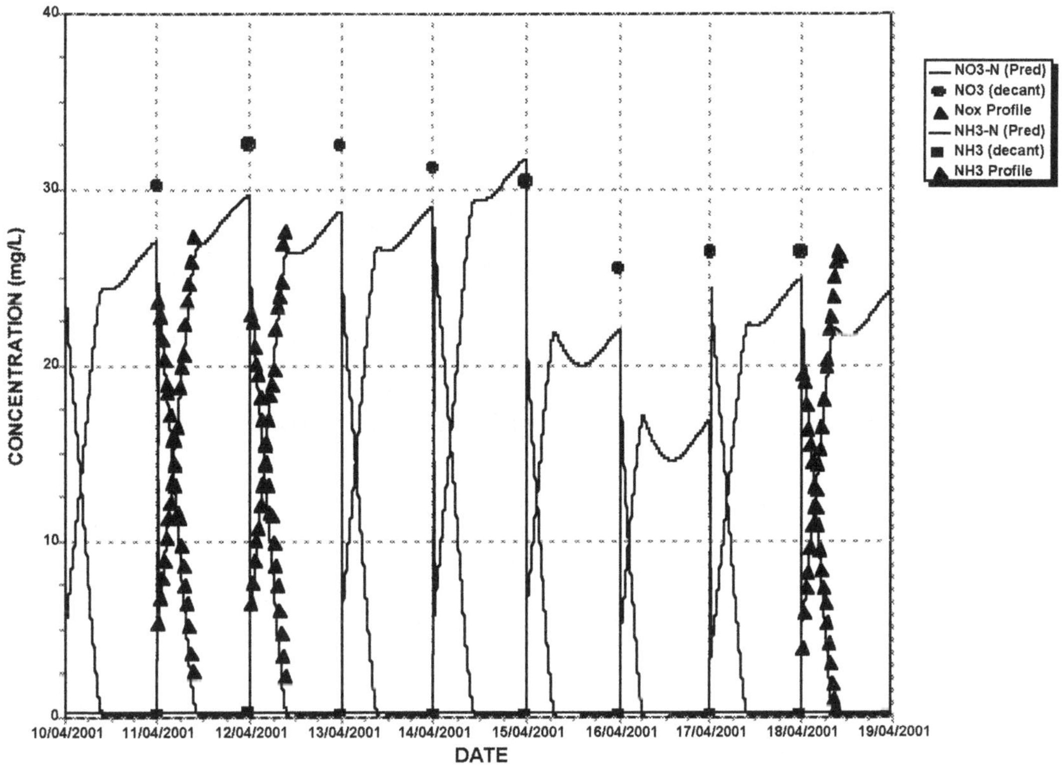

Figure 33-7. Simulated (Solid Line) and Observed (Data Points) Nitrite/Nitrate in the SBR over 9 days

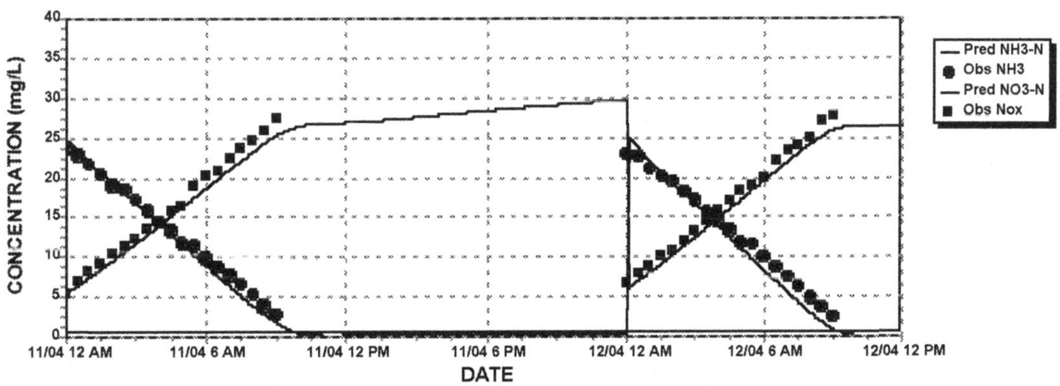

Figure 33-8. Simulated (Solid Line) and Observed (Data Points) Ammonia and Nitrite/Nitrate Response in the SBR over Days 2 and 3

In Figure 33-8, the simulated and observed nitrite/nitrate responses over the first several hours are parallel, indicating that the estimate of nitrifier growth rate is excellent.

Oxygen utilization rate (OUR) response is closely linked to nitrification behavior. A further check on the selection of the nitrification rate parameter is provided by comparing the simulated and observed response of OUR over Day 3 shown in Figure 33-9.

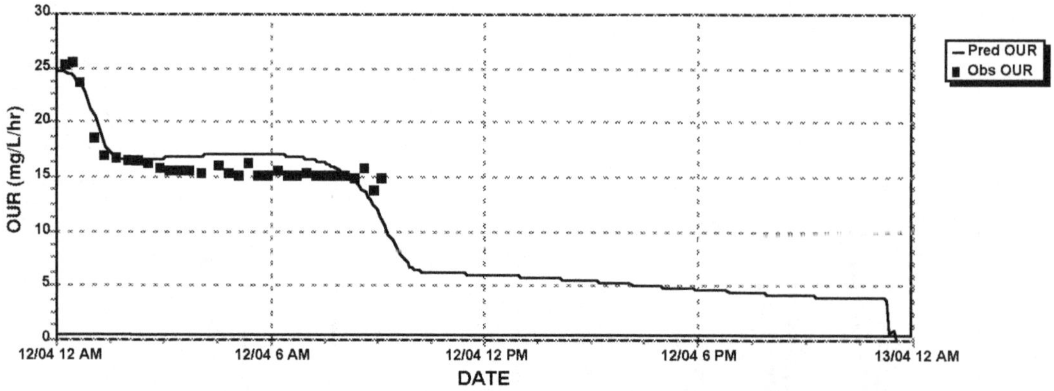

Figure 33-9. Simulated (Solid Line) and Observed (Data Points) OUR Response in the SBR over Day 3

CHAPTER 34.0

LOW F/M SBR METHOD

34.1 Background

A series of investigations were carried out at the McMaster, Renton and Phoenix locations to measure maximum specific nitrifier growth rate using the low F/M SBR method. The method is described in detail in Appendix D. The data collected for the McMaster work were used to illustrate the method of computing wastewater characteristics, as well as the maximum specific nitrifier growth rate (μ_{AUT}), and are reported in Appendix D. Similarly, data collected in low F/M SBR tests at Renton and Phoenix served as supplementary examples of using this method to compute wastewater characteristics and μ_{AUT}, which are reported in Chapters 32.0 and 33.0, respectively. This chapter focuses upon additional low F/M SBR experiments that were conducted at Renton to evaluate the effect of influent feeding regime and temperature on μ_{AUT}.

Note: In assessing the results presented in this chapter, careful attention should be paid to whether estimates of μ_{AUT} values are based on fits to nitrate-only response data or to nitrite+nitrate response data. Estimates based on nitrate-only data will underestimate the μ_{AUT} values. In the early stages of the project tests, nitrite was not monitored. However, during the project it became evident that nitrite accumulation in the various tests may be significant, in particular in high F/M and washout tests, and to a lesser extent in low F/M SBR tests. At that point, nitrite monitoring commenced.

34.2 Impact of Influent Feeding Regime

Four low F/M SBR tests were performed at the Renton location by King County and University Washington staff to evaluate the impact of SBR feeding methods on the observed μ_{AUT}, that is, to determine if the feeding strategy would modify nitrification kinetics brought about by differences in nitrifying populations or enzyme expression that might arise from the different feeding methods.

Three feeding strategies that were thought to provide different growth regimes for nitrifying bacteria were investigated: once-per-day batch feeding (as per the low F/M SBR protocol in

Appendix D), multi-batch feeding, in which a proportion of a day's influent volume was fed every few hours, and semi-continuous feeding, in which small volumes were fed every 15 minutes. In the batch-fed SBR, nitrifying bacteria were exposed to an initially high ammonia concentration, whereas in the semi-continuously fed SBR, they were exposed to a relatively constant low ammonia concentration as would be experienced in a completely-mixed activated sludge reactor. In the multi-batch fed SBR test, the SBR was fed six times per day, so that more wastewater volume could be added to produce a higher mixed liquor concentration, more typical of a full-scale system.

Tests were carried out on three different occasions, designated as Phases I, II and III. Once-per-day batch feeding was carried out in Phase I. Once-per-day feeding was also carried out in Phase II because of a perceived predation problem as discussed below. It is the Phase I batch fed SBR experiment that served as the case in which a large amount of data was collected during an intensive sampling period to facilitate the measurement of wastewater characteristics and maximum specific nitrifier growth rate as described in Chapter 32.0. Semi-continuous feeding was also evaluated in Phase II. Phase III addressed multi-batch feeding.

34.2.1 Operating Conditions

All SBRs received the same nitrifying seed sludge from King County's Vashon Island WWTP oxidation ditch and were fed Renton 24-hour, flow-proportioned composite secondary influent. The composited feed was maintained at 4°C until ready for use, at which time, it was brought to room temperature by immersing and mixing the feed container in warm water over 15 minutes. In Phase II, batch fed and semi-continuous fed SBRs were operated in parallel receiving the same amount of feed each day. Figure 34-1 illustrates the experimental setup during Phase II. These experiments can be carried out using the smaller reactor volume described in Appendix D; in this work, use was made of existing 12 L reactors. Table 34-1 summarizes the operating conditions for each experiment.

Table 34-1. Low F/M SBR Operating Conditions at the Renton Facility

Parameter	Phase I	Phase II		Phase III
	Batch Fed SBR	Batch Fed SBR	Semi-Continuous Fed SBR	Multi-Batch Fed SBR
Feed Volume (L)	8	8	8	2
Reactor Volume (L)	10	10	10	4
Total Feed (L/d)	8	8	8	12
SRT (d)	12	10	10	8
Cycle Time (hr)	24	24	24	4
Aeration Time (hr)	23.5	23.5	23.5	3.4
Settling Time (min)	25	25	25	25
Decant Time (min)	4	4	5	5
Feed Time (min)	1	1	1410	5

Figure 34-1. Consecutive stages of a daily cycle followed in Phase II are shown in A, B, C and D. On the left is the semi-continuous fed SBR and on the right is the batch fed SBR. Each SBR is set on a magnetic stirrer. The timer for controlling feed flow to the semi-continuous fed SBR is shown on the left. The UCT DO monitors are the blue boxes behind each SBR. A shows the start of the cycle when the semi-continuous fed SBR contains 2 L of mixed liquor and the batch fed SBR has received the daily batch of influent. In B, the aeration cycle has been completed and all influent has been fed to the semi-continuous fed SBR. In C, mixed liquor has settled, and decanting is in progress. D shows the residual mixed liquor aerating just before feeding the next batch of influent.

The batch-fed SBRs were fed manually once per day with manual effluent withdrawal. Feed rate was controlled in the semi-continuous and multi-batch fed SBRs with a ChronTrol® timer-controller and Masterflex peristaltic feed pump(s). For both tests, influent was pumped from a container that was continuously mixed with a magnetic stirrer and stir plate. In the semi-continuous fed SBR, the controller initiated the feeding of 86 mL of influent every 15 minutes over a 23.25-hour period. Before the final 15 minutes of the 23.5 hours aeration period, any influent remaining in the feed container was dumped into the SBR, so that it received the same feed volume as the batch fed SBR and to maintain a constant MLSS concentration. The timer turned off the air pump and SBR mixer after 23.5-hour aeration and, allowing settling to occur for 25 minutes. The settled supernatant was manually decanted at the 2 L sample port liquid level.

For the multi-batch fed SBR, the timer controlled the complete operating cycle. The reactor was operated with a total volume of 4 L and received 2 L of feed every 4 hours for a total feed volume of 12 L per day. At the start of each cycle, 12 L of Renton secondary influent feed was placed in the feed container. For each cycle, the timer activated a peristaltic pump, which fed 2 L over a 5 min period. After feeding, the timer turned on the air pump and mixer for a period of 3.5 hours, after which they were turned off and the mixed liquor was allowed to settle for 25 minutes. Then the treated effluent supernatant was withdrawn from the 2 L water level port using another peristaltic pump controlled by the timer, leaving 2 L of settled sludge at the bottom of the reactor. The composite SBR effluent was stored in an effluent container in a refrigerator at approximately 4°C. The SRT was maintained at about 8 days to minimize the time to approach a steady state nitrifier population.

34.2.2 SBR Performance

Table 34-2 summarizes the average influent wastewater characteristics, SBR operating conditions and average effluent concentrations for the period of the experiments. The influent COD and ammonia concentrations were similar for Phases I and II. The higher influent TSS and VSS concentrations and lower ammonia concentration during Phase III were attributed to the rainy season during this test period.

The MLSS and MLVSS concentrations were similar for the SBRs in Phases I and II. The semi-continuous fed SBR in Phase II had slightly lower MLSS and MLVSS concentrations because of the inability to transfer all the solids from the feed container; this constraint was eliminated later in the operating period. With multiple-feeding cycles, the Phase III SBR had much higher MLSS and MLVSS concentrations because of the higher loading rates. The pH was maintained near neutral for all the SBRs.

Low effluent ammonia concentrations were measured for all SBRs. Effluent TSS and VSS concentrations were higher for the Phase II batch fed and semi-continuous fed SBRs, compared to the Phase I batch fed and the multi-batch fed SBR. In general the effluent TSS and VSS concentrations were higher than normally experienced for full-scale facilities.

In Phase I, the SBR contained a higher concentration of *Daphnia*, which were visible in the effluent supernatant. The source was eventually traced to *Daphnia* growth in an aeration basin that was out of service but full of effluent process water that was periodically flushed into the primary effluent. The presence of the high *Daphnia* population raised concerns about their possible predation of nitrifying bacteria with corresponding lower batch nitrification rates and, therefore a lower calculated μ_{AUT} value than for a normal reactor. However, the Phase I and Phase II batch fed SBRs yielded similar μ_{AUT} values (see below), suggesting that the *Daphnia* did not significantly impact the nitrifier population.

Table 34-2. Average Influent and SBR Performance Data for Different SBR Feeding Regimes

Parameter (mg/L except where noted)	Phase I	Phase II		Phase III
	Batch Fed SBR	Batch Fed SBR	Semi-Continuous Fed SBR	Multi-Batch Fed SBR
Days of Operation	45	43	43	27
Feed				
TSS	118 (15.8)	93 (16.2)	93 (16.2)	149 (19.5)
VSS	97 (14.8)	75 (19.5)	75 (19.5)	124 (21.4)
COD	318 (33.7)	279 (31)	279 (31)	-
NH_4-N	30.2 (4.8)	28.3 (4.5)	28.3 (4.5)	24.5 (1.8)
Alkalinity (mg/L $CaCO_3$)	202 (14.3)	260 (57.4)	260 (57.4)	267 (24)
SBR				
SRT, (day)	12.7	10.1 (2.0)	10.1 (1.5)	8.2 (1.0)
Temperature (°C)	21.0 (1.0)	21.0 (1.0)	21.5 (1.0)	20.5 (1.0)
MLSS	1050 (214)	929 (341.4)	878 (411.8)	2553 (505)
MLVSS	822 (177.6)	705 (262.1)	668 (318.8)	2061 (389)
PH	7.1 (0.5)	7.2 (0.4)	7.2 (0.4)	7.2 (0.2)
Solids Yield (gTSS/gCOD)	0.4 (0.08)	0.5 (0.2)	0.4 (0.16)	
Effluent				
TSS	29 (21.3)	49 (45.7)	45 (45.6)	32 (8.9)
VSS	22 (19.7)	33 (31.4)	30 (31.7)	26 (7.6)
TCOD	62 (27.6)	82 (53.2)	79 (49.4)	
NH_4-N	1.9 (6.4)	1.6 (3.4)	0.7 (1.4)	0.5 (0.6)
NO_3-N	20.2 (7.3)	31.5 (4.5)	31.5 (3.0)	25.0 (2.5)
Alkalinity (mg/L $CaCO_3$)	102 (77)	103 (47.8)	100 (40.4)	

(_) – Standard deviation

34.2.3 Nitrification Rate Tests

All SBRs were operated for a time period greater than 3 SRTs before nitrification rate tests were carried out. Up to three tests were conducted on alternate days during a normal SBR feeding cycle. After feeding and starting aeration, 10-mL mixed liquor samples were taken at intervals and immediately filtered through a glass fiber syringe filter. The filtrate was analyzed for ammonia and nitrate nitrogen concentrations using the Hach protocols following storage at 4°C. The sampling time interval was reduced progressively from hourly in Phase I, half-hourly in Phase II, and then every 20 minutes in Phase III. Three nitrification rate tests were completed for the batch fed and semi-continuous fed SBRs and two for the multi-batch fed SBR.

During the multi-batch fed SBR tests, two feed cycles were combined into a batch feed of 4 liters (instead of 2L) and the aeration time was increased from its normal time of 3.5 hours to 7.5 hours.

The measured ammonia and nitrate concentrations for Phases I and II are summarized in Table 34-3, and for Phase III in Table 34-4.

Table 34-3. NH_4-N and NO_3-N Profile Data during Nitrification Rate Tests in Phases I and II

Time (hr)	Phase I		Phase II			
	Batch Fed SBR		Batch Fed SBR		Semi-Continuous Fed SBR	
	NH_4-N	NO_3-N	NH_4-N	NO_3-N	NH_4-N	NO_3-N
	(mg/L for tests 1, 2 and 3, respectively)					
0.0	18.8, 20.0, 21.0	4.5, 4.3, 4.2	20.8, 25.0, 21.5	7.1, 6.7, 6.9	22.4, 24.1, 21.3	6.9, 6.9, 7.5
0.5			20.0, 21.6, 19.9	7.6, 7.2, 7.5	22.1, 22.9, 19.6	7.4, 7.3, 8.4
1.0	16.3, 16.3, 18.4	6.4, 6.1, 5.6	18.5, 20.4, 19.2	8.8, 8.4, 9.5	20.1, 20.8, 19.0	8.1, 8.4,9.6
1.5			17.1, 19.0, 18.7	10.0, 9.7, 10.3	19.1, 19.2, 18.5	9.3, 9.8, 11.2
2.0	14.5, 15.0, 16.3	9.1, 8.3, 7.4	14.5, 15.7, 16.8	11.1, 10.6, 11.9	17.4, 17.2, 17.1	10.4, 10.6, 12.6
2.5			13.5, 14.6, 14.6	12.4, 12.2, 12.9	15.3, 16.3, 14.9	11.5, 11.9, 13.6
3.0	11.9, 11.8, 13.6	11.2, 10.7, 10.0	12.3, 13.9, 13.8	13.3, 13.6, 14.6	13.8, 15.2, 13.8	12.5, 13.1, 15.2
3.5			10.4, 11.8, 12.1	15.0, 14.8, 16.0	11.9, 14.0, 13.0	14.3, 14.9, 16.5
4.0	9.6, 10.2, 10.3	13.9, 13.3, 12.1	9.4, 9.5, 9.9	15.9, 16.4, 17.6	11.2, 10.8, 11.6	15.6, 16.0, 17.9
4.5			7.7, 8.9, 8.1	17.3, 17.4, 19.1	9.3, 10.0, 9.5	16.7, 17.3, 19.5
5.0	6.4, 8.6, 9.5	17.3, 14.5, 13.3	5.8, 7.4, 5.9	18.8, 18.2, 19.8	7.8, 8.9, 7.8	18.1, 18.9, 20.2
5.5			4.4, 4.6, 4.7	20.0, 20.2, 21.3	6.4, 6.2, 6.1	19.7, 20.1, 21.8
6.0	4.2, 7.1, 8.2	20.3, 15.5, 14.4	3.6, 3.5, 3.3	21.2, 21.2, 23.0	4.7, 5.4, 4.8	20.8, 21.4, 23.6
6.5			0.4, 2.2, 1.4	22.8, 23.0, 24.4	2.8, 4.0, 3.5	22.2, 23.0, 25.2
7.0	2.7, 5.5, 7.3	21.3, 16.8, 15.6	0.5, 0.3, 0.5	24.4, 24.4, 25.6	0.5, 2.3, 1.5	24.0, 24.6, 26.5
8.0	1.5, 4.5, 5.7	22.4, 18.1, 16.8				
9.0	0, -, 4.9	24.0, - 17.5				
10.0	0, -, -	25.3, -, -				

Table 34-4. NH_4-N and NO_3-N Profile Data during Nitrification Rate Tests in Phase III

Time (hr)	Phase III	
	Multi-Batch Fed SBR	
	NH_4-N	NO_3-N
	(mg/L)	
0.00	12.3	9.3
0.33	11.6	10.6
0.67	10.1	11.9
1.00	8.6	13.7
1.33	6.6	15.2
1.67	4.7	17.2
2.00	2.6	19.0
2.33	0.0	20.4
2.67	0.1	22.0
3.00	0.0	23.0

Figure 34-2 shows an example of the results obtained from nitrification rate testing in concert with corresponding OUR data. The second plateau for the OUR profile represents the zero-order nitrification rate time period. During this period, a linear increase in NO3-N concentration is observed. These data were regressed using a linear least squares fit for the linear portion of the curve to estimate the nitrification rate. A summary of the nitrification rates measured in all the experiments, given in Table 34-5, shows that they were very similar; the coefficient of variation ranged from 2 to 9 percent.

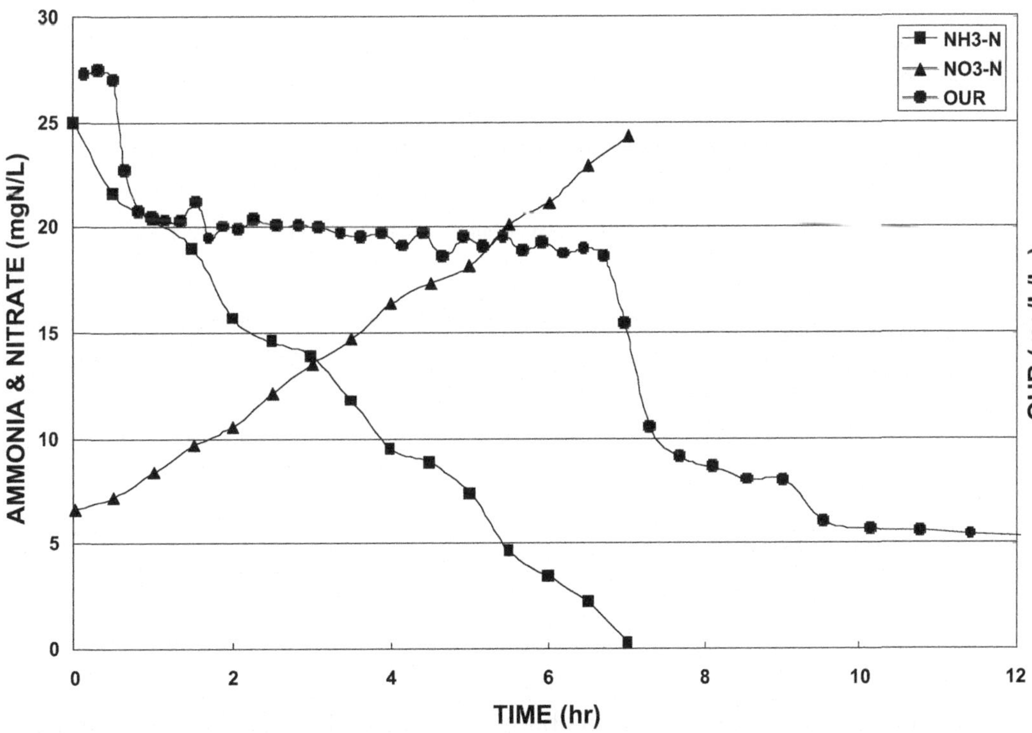

Figure 34-2. Ammonia and Nitrogen Profiles during Test Number 2 in Phase I Batch Fed SBR Experiment

Table 34-5. Summary of Estimated Nitrate Production Rates

Parameter	Phase I	Phase II		Phase III
	Batch Fed SBR	Batch Fed SBR	Semi-Continuous Fed SBR	Multi-Batch Fed SBR
Nitrification Rate (mgN/L-hr)				
Test 1	2.73	2.50	2.50	4.77
Test 2	2.45	2.57	2.59	4.65
Test 3	2.21	2.73	2.73	-
Average	2.46	2.60	2.61	4.71
Standard. Dev	0.21	0.12	0.12	0.08
Temperature during profile tests (°C)				
Test 1	21.0	20.5	20.5	18.7
Test 2	21.0	21.0	21.0	18.3
Test 3	20.5	20.0	20.0	-
Average	20.8	20.5	20.5	18.5

34.2.4 Estimation of μ_{AUT}

Values of μ_{AUT} were estimated using equation 16.4.2-1 in Chapter 16.0. Table 34-6 summarizes values of the pertinent parameters required to solve this equation for each experiment and the corresponding value of μ_{AUT}. Values of the specific nitrifier decay rate and its temperature dependency coefficient used in this calculation were 0.17 d^{-1} at 20°C and 1.029, respectively as described in Chapter 16.0, as well as the μ_{AUT} temperature dependency coefficient of 1.072.

Table 34-6. Summary of Test Results and Maximum Specific Nitrifier Growth Rates

Parameter	Phase I	Phase II		Phase III
	Batch Fed SBR	Batch Fed SBR	Semi-Continuous Fed SBR	Multi-Batch Fed SBR
Average SRT (d)	12.7	10.1	10.1	8.2
Feed Volume (L/d)	8	8	8	12
Reactor Volume (L)	10	10	10	4
Avg. Effluent NO_3-N concentration (mg/L)	23	32	32.1	25
Nitrifier concentration X_{AUT} (mg/L)	10.9	14.3	14.3	39.4
Avg. dN/dt (mgN/L-hr)	2.46	2.60	2.61	7.11
μ_{AUT} at test temp. (d^{-1})	0.81	0.65	0.66	0.65
μ_{AUT} at 20^0C (d^{-1})	0.77	0.63	0.64	0.72

The nearly-identical average μ_{AUT} values obtained in Phase II for the batch-fed and semi-continuous fed SBRs show that similar maximum specific growth rates were obtained, whether the nitrifying bacteria experienced a relatively constant low ammonia feed concentration or a daily feed cycle with a high initial ammonia concentration. The average μ_{AUT} value for the four test conditions was 0.69 with a standard deviation 0.05, which suggests that the μ_{AUT} values are similar regardless of the feeding mode used. It is possible that some variation in nitrification kinetics may be attributed to variations in wastewater characteristics. Comparison of μ_{AUT} values from Phase I and Phase II batch fed experiments suggests that the wastewater in Phase I favored a higher nitrification rate. For all these results, the values of μ_{AUT} are based on nitrate nitrogen concentration measurements only; higher values could be expected if nitrite nitrogen had been present.

34.3 Impact of Temperature

The effect of temperature on nitrification kinetics was evaluated using the low F/M SBR method with Renton wastewater. Two batch fed SBRs were operated in temperature-controlled chambers at the University of Washington (UOW) environmental engineering research laboratory.

34.3.1 Operating Conditions

The SBRs were operated at temperatures of 12°C and 22°C, respectively, each at a 10-day SRT for a period of 40 days (>3 SRTs) until steady state conditions had been established in the reactors. The initial plan was to operate the SBRs at 10°C and 20°C, but nitrification was difficult to maintain at 10°C, so the target temperatures were elevated by 2°C. Because of the large distance between UOW and the South Plant in Renton, feed wastewater was collected as a grab primary effluent sample every five days and stored at 4°C at UOW. Before use, a well-mixed daily feed volume was removed from the refrigerator and warmed to the appropriate temperature before batch feeding the SBRs. Table 34-7 summarizes the operating conditions for the batch fed SBRs.

Table 34-7. Low F/M SBR Operating Conditions at the University of Washington Laboratory

Parameter	Value
Seed Source	Vashon Island WWTP
Feed Source	Renton WWTP Primary Effluent
Reactor Volume (L)	4.0
Feed Volume (L)	3.0
SBR Cycle Time (hr)	24.0
Aeration Time (hr)	23.4
Settling Time (hr)	0.4
Decant Time (hr)	0.1
Batch Fill Time (hr)	0.1
SRT (day)	10.0
Temperature (°C)	12 or 22
DO Range (mg/L)	2.0 to 6.0

After 40 days of operation, two nitrification profile tests were performed for both SBRs. A fresh composite secondary influent sample from the Renton plant was used for each profile determination.

34.3.2 SBR Performance

Table 34-8 summarizes the average influent wastewater characteristics, SBR operating conditions and average effluent concentrations for the period of the experiments. All nitrogen species were monitored in these experiments. Because grab samples of primary effluent were obtained for reactor feeding, the wastewater strength was quite variable, with total CODs ranging from 283 to 540 mg/L. The influent ammonia concentration was less variable and ranged from 23.5 to 31.6 mg/L. During the 5-day intensive testing period in which nitrification profile tests were done, the feed to the SBRs was a daily primary effluent flow proportioned composite sample, which was weaker than the feed grab sample.

Table 34-8. Average Influent and Performance Data for Different Temperature SBRs

Parameter (mg/L except where noted)	n	22^{0}C SBR	12^{0}C SBR
Days of operation		44	44
Influent			
TSS		146 (25)	146 (25)
VSS		107 (17)	107 (17)
COD		365 (74)	365 (74)
NH_4-N		28.6 (2.0)	28.6 (2.0)
Alkalinity (mg/L $CaCO_3$)		290 (12)	290 (12)
SBR			
SRT (day)	44	10.2 (0.9)	10.2 (0.82)
MLSS	44	1,165 (85)	1,287 (102.5)
MLVSS	44	901 (68.2)	986 (86.5)
COD	3	1,127 (55.6)	1,095 (156.1)
TKN	3	83 (1.3)	84 (1.4)
PH	44	7.3 (0.1)	7.3 (0.1)
Solids Yield (gTSS/gCOD)	3	0.54 (0.03)	0.7 (0.05)
Effluent			
TSS	44	46 (13.7)	57 (14.5)
VSS	44	32 (9.4)	39 (11.2)
COD	3	42 (8.6)	83 (10.5)
SCOD	3	20 (5.7)	37 (17)
TKN	3	3.1 (0.5)	6.1 (1.1)
NH_4-N	13	0.2 (0.2)	0.3 (0.44)
NO_3-N	34	27.8 (1.8)	24.9 (2.7)
NO_2-N	34	0.1 (0.2)	0.3 (0.5)
PO_4-P	3	3.7 (0.05)	3.2 (0.01)

( ) – Standard deviation

34.3.3 Nitrification Rate Tests

Three tests were conducted on alternate days during an SBR feeding cycle, using the method given above. In this series of tests, ammonia, nitrate and nitrite nitrogen concentrations were monitored. An example of the nitrogen and OUR profiles are shown for a 22°C SBR test in Figures 34-3 and 34-4.

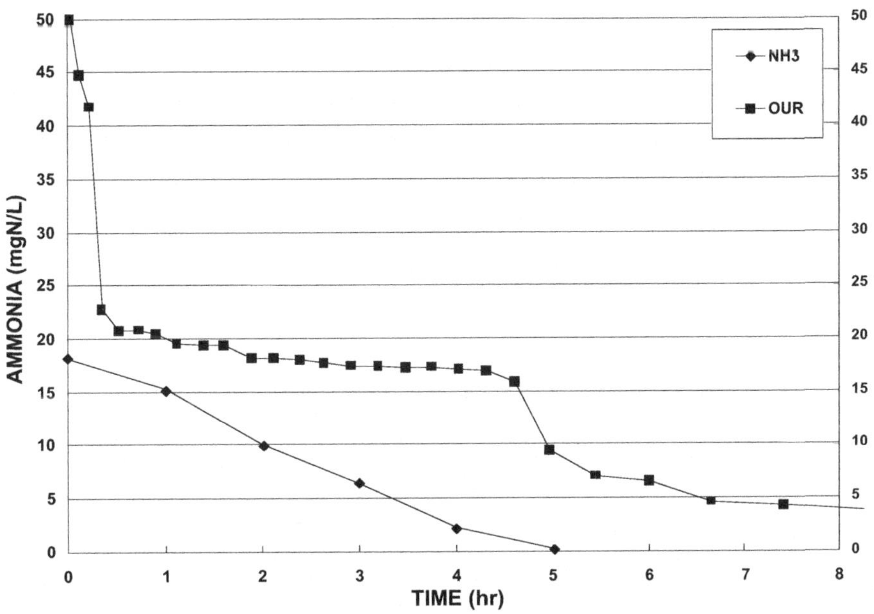

Figure 34-3. Ammonia Concentration and OUR Profiles during a 22°C Nitrification Rate Test

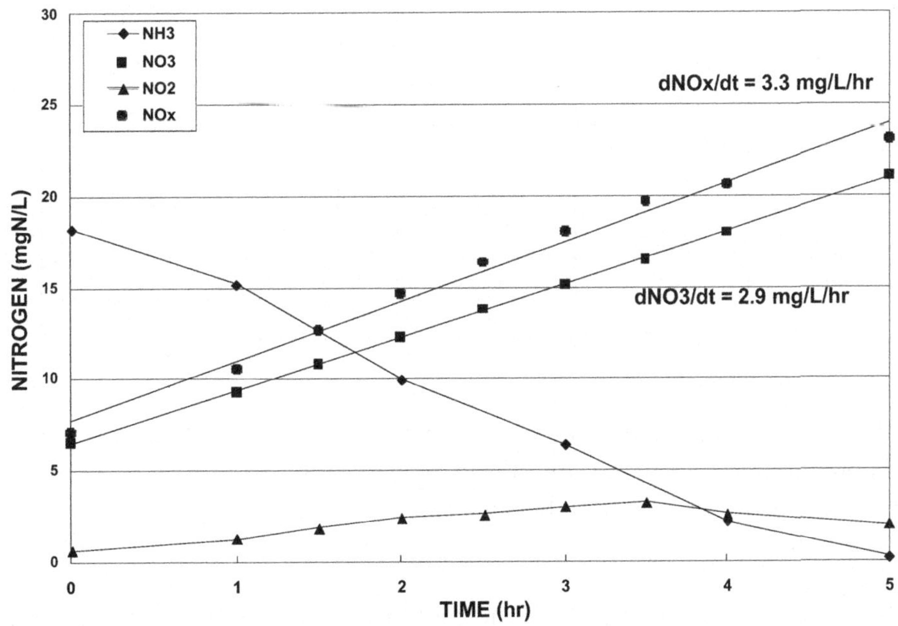

Figure 34-4. Example of Ammonia, Nitrite and Nitrate Nitrogen Concentration Profiles during a 22°C Nitrification Test

Three plateaus typically were observed for the OUR profile. The first lasted less than 30 minutes and represents the sum of the OUR for oxidation of readily biodegradable COD (RBCOD), nitrification and endogenous respiration plus particulate degradation. The second plateau, following depletion of RBCOD, occurred for several hours. Its decline to the final plateau occurred rapidly over a few minutes to indicate the completion of zero-order nitrification. The final plateau represents the oxygen demand for endogenous respiration and particulate degradation.

In Figure 34-4, the increases in NO_X-N and NO_3-N concentrations with time are approximately linear over the first 4 hours. At 4 hours, the NO_2-N concentration had increased to 2.6 mg/L, which was about 13 percent of the NO_X-N concentration increase. In these low F/M tests, the nitrification rate was determined from a linear least squares fit to the increases in NO_X-N or NO_3-N concentration with time, while NH_4-N was still present at a concentration to support a zero-order nitrification rate.

Table 34-9 summarizes the ammonia, nitrite and nitrate concentration collected in the six nitrification rate tests.

Table 34-9. NH_4-N and NO_3-N Data during Nitrification Rate Tests in Phase III

Time (hr)	22°C Batch Fed SBR			12°C Batch Fed SBR		
	NH_4-N	NO_2-N	NO_3-N	NH_4-N	NO_2-N	NO_3-N
	(mg/L for tests 1, 2 and 3, respectively)					
0.0	18.2, 17.2, 22.5	0.6, 0.6, 0.3	6.5, 6.7, 8.2	18.0, 18.7, 20.7	0.4, 0.5,	8.3, 7.8,
0.5		-, -, 0.7	-, -, 9.7		-, -, 0.6	-, -, 9.4
1.0	15.2, -, 17.0	1.3, 0.8, 1.1	9.3, 7.8, 11.1	-, -,	1.3, 0.7, -	11.4, 9.6, -
1.5		1.8, -, 1.4	10.8, 12.9, 12.7	11.4, -, -	1.8, -, -	12.9, -, -
2.0	10.0, 14.6, 11.9	2.4, 1.3, 1.6	12.3, 9.6, 14.5	-, 13.6, 16.3	-, 1.1, 1.2	-, 10.6, 10.5
2.5		2.6, -, 2.3	13.8, -, 15.5	-, 7, -	-, 2.6, -	-, 16.3, -
3.0	6.4, -, 8.2	3.0, 1.6, 2.6	15.1, 11.0, 17.2		3.0, 1.4, 1.6	18.0, 12.1, 11.8
3.5		3.2, -, 3.0	16.5, -, 18.4	2.5, -,	3.5, -, -	19.6, -, -
4.0	2.2, 10.4, 3.3	2.6, 1.9, 3.4	18.0, 12.5, 19.9	-, 11.0, 13.1	3.6, 1.9, 2.1	21.2, 13.7, 13.1
4.5	-, 0.2, -	-, 2.5, 3.9	-, 22.6, 21.7	0.2, -, -	2.5, -, -	22.6, -, -
5.0	0.2, -,-	2.0, 2.1, -	21.1, 14.8, -		-, 2.2, 2.6	-, 15.1, 14.8
6.0	-, 6.7, -	- 2.6, -	-, 16.8, -	-, 6.4, 8.5	-, 2.6, 3.1	-, 16.5, 15.9
7.0		- 2.7, -	-, 18.1, -		-, 3.0, 3.5	-, 18.2, 17.5

A summary of the nitrate production rates estimated from these nitrogen profiles is given in Table 34-10. Rates were estimated based on the sum of nitrite and nitrate nitrogen concentrations (NO_X-N) and compared to rates based on nitrate concentration only. The rates at 12°C were about half of the average rate at 22°C. The standard deviation was about 6 percent to 10 percent of the average nitrification rate.

The nitrification rates based on NO_X-N measurements were about 20 percent higher than the rates based on NO_3-N only measurements. In the low F/M SBR tests, evaluating the effect of the influent feeding regime, nitrite concentrations were not measured and μ_{AUT} estimates were based only on nitrate concentrations. The resultant μ_{AUT} values may have been underestimated by up to 20 percent.

Table 34-10: Summary of Estimated Nitrate Production Rates

	NO_x-N Rate mgN/L-hr	NO_3-N Rate mgN/L-hr	NO_x /NO_3 Rate Ratio
Batch Fed SBR @ 22^{0}C			
	3.27	2.91	1.12
	3.86	3.23	1.20
	3.74	2.96	1.26
Average	*3.62*	*3.03*	*1.19*
Standard dev	0.31	0.17	
Batch Fed SBR @ 12^{0}C			
	2.00	1.69	1.18
	1.82	1.46	1.25
	1.79	1.33	1.35
Average	*1.87*	*1.49*	*1.25*
Standard dev	0.11	0.18	

34.3.4 Estimation of μ_{AUT}

Values of μ_{AUT} were estimated using equation 16.4.2-1 in Chapter 16.0. Table 34-11 summarizes values of the pertinent parameters required to solve this equation for each experiment and the corresponding value of μ_{AUT}. Values of the specific nitrifier decay rate and its temperature dependency coefficient used in this calculation were 0.17 d^{-1} at 20°C and 1.029, respectively, as described in Chapter 16.0. The average μ_{AUT} values at 22 and 12°C were estimated to be 1.06 and 0.52 d^{-1}, respectively. A μ_{AUT} temperature correction factor of 1.072 was estimated from the average μ_{AUT} values determined for the 22 and 12°C SBRs using equation 16.4.1-4 in Chapter 16.0. This gives μ_{AUT} values (corrected to 20°C) for the 22 and 12°C SBRs of 0.92 and 0.91 d^{-1}, respectively.

Table 34-11. Summary of Test Results and Maximum Specific Nitrifier Growth Rates

Temp (^{0}C)	SRT, (day)	Average Effluent NO_x-N (mg/L)	Average dNO_x-N/dt (mgN/L-hr)	Average X_{AUT} (mg/L)	μ_{AUT} (d^{-1})
22	10.2	28.0	3.62	12.3	1.06
12	10.2	25.6	1.87	13.0	0.52

Note: the SBR daily feed rate was 3 L/d and the full reactor volume was 4 L.

CHAPTER 35.0

SUMMARY OF HIGH F/M INVESTIGATIONS

35.1 Background

A series of investigations were carried out at the McMaster, Phoenix and University of Washington locations to measure maximum specific nitrifier growth rate (μ_{AUT}) using the high F/M method. The method used in these experiments is described in detail in Appendix F. The development and background to the method are described in Appendix E. Many experiments were conducted at the McMaster and University of Washington locations to investigate variables that were thought to be important in controlling nitrification reactions and, therefore, the measured value of μ_{AUT}. These included the initial ammonia concentration and its potential for causing inhibition of nitrification, the impact of pH and alkalinity, and variability in wastewater characteristics. Parameter estimation and statistical analysis procedures were developed for determining μ_{AUT} values; these are described in Appendix H. The reproducibility of the test was evaluated by comparing values of μ_{AUT} measured at all three locations. This chapter describes these investigations of the high F/M method.

Note: In assessing the results presented in this chapter, careful attention should be paid to whether estimates of μ_{AUT} values are based on fits to nitrate-only response data or to nitrite+nitrate response data. Estimates based on nitrate-only data will underestimate the μ_{AUT} values. Early in the project, nitrite was not monitored. However, during the project it became evident that nitrite accumulation in the various tests may be significant, in particular in high F/M and washout tests, and to a lesser extent in low F/M SBR tests. At that point nitrite monitoring commenced.

Note: It is evident from a number of the figures below that the data fits to nitrite plus nitrate response data (for determining μ_{AUT}), excluding certain data points towards the end of the experiment. These were excluded on the basis of an examination of plots of residuals (as discussed in Chapter 16.0). In many of these cases, ammonia concentration had decreased sufficiently to exert some rate limiting effect (for further discussion of residuals, see Appendix H).

35.2 University of Washington Investigations

Several runs were conducted at the University of Washington to develop and refine the high F/M method. The results of these runs are summarized in the following sections.

35.2.1 Operating Conditions and Data Collection

The equipment used and the procedure followed in the conduct of these experiments is described in detail in Appendix F. All tests were carried out at 20°C in a temperature-controlled chamber in the University of Washington environmental engineering research laboratory. The experiments were conducted in a 4 L aerated reactor as shown in Fig. 35-1, mixed with a magnetic stirrer and magnetic stir plate.

Before the test started, about 4 L of secondary effluent were collected from the South Wastewater Treatment Plant in Renton, Wash. and warmed to 20°C. The ammonia, nitrate and alkalinity concentrations of the secondary effluent were measured, and sodium bicarbonate and ammonium chloride solution were added to raise the alkalinity to about 250 mg/L as $CaCO_3$ and ammonia concentration to about 120 mg N/L. About 50 mL of nitrifying mixed liquor from one of the low F/M test SBRs or a full-scale plant were added to the reactor containing the prepared secondary effluent. The seed volume was selected to yield an initial nitrification rate of about 1 mg N/L-d based on nitrification rate measured in the low F/M tests.

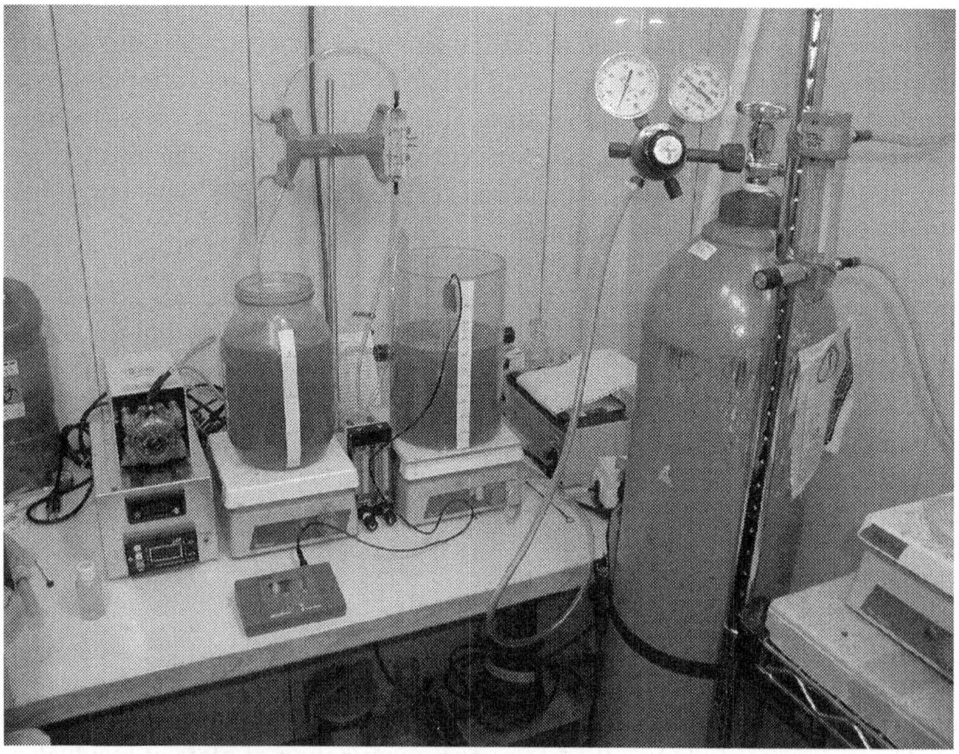

Figure 35-1. Overview of Parallel High F/M Reactors in the University of Washington Laboratory. A Carbon Dioxide Gas Bottle is Shown on the Right Hand Side

Twice per day (morning and afternoon), 10 mL of filtered samples were obtained for measurement of ammonia and nitrate concentrations (and nitrite concentration in later tests). On these occasions, pH, temperature and gas flow rates were measured and recorded and the wall growth on the reactor walls was removed and deposited in the reactor. Whenever the pH was lower than 7.1, some sodium bicarbonate solution was added to bring the pH back to 7.3. Generally, daily additions of 10 to 20 mL of 25.2 g/L sodium bicarbonate solution (15 g/L as $CaCO_3$) were needed after the first two days of the test. Addition of some distilled water was needed to compensate for evaporation losses before sampling to maintain the water level.

Because the test requires continuous aeration with little carbon dioxide generation from heterotrophic bacteria, a high alkalinity concentration at the beginning of the test will likely elevate pH values above the maximum targeted value of 8.0. To improve pH control and maintain it in the range of 7.2 to 7.4, the second method of pH control described in Appendix F was used. Carbon dioxide gas was added to the sparge air. The air and carbon dioxide gas flow rates were measured and controlled by rotameters with needle valves. Air and carbon dioxide flow rates were typically set at 500 and 5 mL/min, respectively (1 percent carbon dioxide by volume). The prepared secondary effluent sample was aerated for at least one hour and the pH was adjusted to about 7.2 by adjusting the carbon dioxide/air flow ratio. The carbon dioxide gas and airflow rates were kept constant during the period of the test.

The resulting data were analyzed by fitting to equation 16.6.1-4 in Chapter 16.0 using the parameter estimating procedure described in Appendix H. A non-linear least squares technique using the Excel solver was applied to equation 16.6.1-4 to adjust the parameters ($\mu_{AUT} - b_{AUT}$), $X_{AUT,0}$, $S_{NO,0}$ in order to minimize the sum of squares of the differences between the measured NO_X and calculated NO_X concentrations. A value of 0.15 for nitrifier yield, Y_{AUT} (Barker and Dold, 1997), was used as were values of 0.17 d^{-1} for b_{AUT} and 1.072 for the μ_{AUT} temperature dependency coefficient.

35.2.2 Evaluation of pH Control Method

The concept of adding carbon dioxide to the air flow to control pH was evaluated by conducting a series of tests using a sparged 2 L flask to determine if carbon dioxide and alkalinity concentrations could significantly affect nitrification rate. These tests were batch experiments using mixed liquor from a low F/M SBR fed Renton WWTP secondary influent. The initial alkalinity was adjusted for different amounts of carbon dioxide/air flow ratios. Ratios of 0, 1.2 and 2.4 percent were applied in incremental steps during a batch nitrification rate test after spiking to a reactor concentration of 30 mg/L ammonia, with corresponding initial alkalinity concentrations of 80, 200 and 400 mg/L, as $CaCO_3$, respectively.

An alkalinity of 80 mg/L, as $CaCO_3$, is in the range of a practical lower limit where a pH close to 7.0 is achievable. During the nitrification rate test, an automated pH controller was used to maintain pH by adding alkalinity to offset the alkalinity consumed by nitrification. Each test was carried out over a 2-hour period before adjusting the alkalinity and carbon dioxide content, and mixed liquor samples were removed and filtered every 30 minutes for measurement of ammonia and nitrate concentrations.

Table 35-1 compares the nitrification rates for the different experimental conditions of alkalinity and carbon dioxide content. As alkalinity and carbon dioxide content were increased, the nitrification rate response was variable; it increased at 200 mg/L alkalinity from 1.9 mg/L-hr to 2.2 mg/L-hr (16 percent increase), but decreased at 400 mg/L alkalinity (16 percent decrease compared to the rate at 80 mg/L alkalinity). These changes were attributed to experimental variability associated with sampling and analytical processes, because there is no fundamental reason to expect the nitrification rate to decrease at the higher alkalinity and carbon dioxide content. The results, in general, suggest that the alkalinity and carbon dioxide levels that would normally be employed in the high F/M test procedure would not affect nitrification kinetics.

Table 35-1. Comparison of Nitrate Production Rates for Evaluation of the Effect of Alkalinity and Carbon Dioxide/Air Ratios

Alkalinity (mg/L as CaCO3)	Volumetric percentage of CO_2 in feed air (percent)	pH	Nitrification production rate (mg/L-hr)	Ratio of N rate to that at 80 mg/L alkalinity
80	0.0	6.9	1.9	1.00
200	1.2	7.0	2.2	1.16
400	2.4	7.0	1.6	0.84

35.2.3 Evaluation of Test Reproducibility

A series of high F/M tests were conducted at the University of Washington to determine consistency in methodology and any variability in measured μ_{AUT} values. Two high F/M reactors were set up to operate in parallel with identical wastewater, seed and operating conditions. Two sets of determinations were made on different occasions. One set (tests 1 and 2 in Table 35-2) were conducted on the same day. A second set (tests 3 and 4) was conducted on consecutive days. This provided information on the reproducibility of the test procedure. The source of wastewater in these tests was secondary effluent from the South Plant in Renton. The source of seed was mixed liquor from the low F/M SBR reactors in the Phase I experiments.

No measurements were made of nitrite concentration in these tests and so the values of $\mu_{AUT,20\ deg\ C}$ presented here are based on nitrate concentration data only.

Table 35-2 shows that there was relatively good agreement between the measured values of $\mu_{AUT,\ 20\ deg\ C}$, that is, good reproducibility between the pairs of tests (0.81 vs. 0.82 for tests 1 and 2, respectively, and 0.72 vs. 0.66 d^{-1} for tests 3 and 4, respectively).

Comparison of $\mu_{AUT,\ 20\ deg\ C}$ values from all four tests for the same source of wastewater (Renton secondary effluent) reveals some variability, with values ranging from 0.66 to 0.81 d^{-1}. This likely is due to the fact that nitrite data were not collected in these tests.

Table 35-2. Comparison of $\mu_{Aut,\ 20\ deg\ C}$ Values for Different Wastewaters

Date	Test No.	Wastewater Source	Seed Source	Initial NH4-N (mg/L)	$\mu_{AUT,\ 20\ deg\ C}$ (d^{-1})*
9/1/00	1	Renton secondary effluent	Low F/M batch fed SBR – Phase I	36	0.81
9/1/00	2			36	0.82
9/14/00	3			28	0.72
9/15/00	4			28	0.66
Average					*0.75*
Standard. Deviation					*0.08*

* Calculated values of $\mu_{AUT,\ 20\ deg\ C}$ based on NO_3-N only data

35.2.4 Evaluation of Initial Ammonia Concentration

Two methods of adding ammonia in the high F/M test (one high initial charge versus periodic additions of ammonia) were evaluated in two sets of experiments using Renton secondary effluent. In both sets, one reactor was spiked with approximately 30 mg/L ammonia nitrogen, and the other with approximately 120 mg/L. The reactor with the lower initial ammonia concentration required intermittent ammonia additions during the test to maintain a zero-order nitrification rate. The advantage of using a high initial concentration is that the ammonia needs to be added only once, simplifying the test procedure. However, the concern over ammonia toxicity at this elevated ammonia concentration and the potential for depressed nitrification kinetics provided further impetus for these tests.

A 70 mL seed was taken from the Phase II high F/M SBR reactor mixed liquor (containing about 1,000 mg TSS/L) was added to the high F/M test reactors containing 2.97 L of Renton secondary effluent. The procedure described in Appendix F was followed. The change in ammonia and nitrate concentrations during the tests is shown in Table 35-3 for both reactors in the pair of tests conducted on 11/20/00. The measured initial ammonia concentrations differed from the target values by the amount of residual ammonia in the secondary effluent samples.

The statistical fit to the model that is described in Chapter 16.0 to determine μ_{AUT} values is shown in Figures 35-2 and 35-3 for the pair of tests reported in Table 35-3, for initial ammonia concentrations of 30 and 120 mg/L, respectively.

Table 35-3. Ammonia and Nitrate Concentrations Response to Changing Initial Ammonia Concentration in High F/M Test

	Initial NH_4N Concentration: 30 mg/L				Initial NH_4N Concentration: 130 mg/L			
Time (days)	NH_4N (mg/L)	pH	Measured NO_3N (mg/L)	Predicted NO_3N (mg/L)	NH_4-N (mg/L)	pH	Measured NO_3N (mg/L)	Predicted NO_3N (mg/L)
0.00	27.5	7.3	0.20	0.54	132.4	7.3	0.2	0.17
1.00		7.2	0.80	0.92		7.2	0.7	0.62
2.00	18.1	7.2	1.50	1.62	121.5	7.2	1.4	1.43
2.92		7.2	3.20	2.75		7.2	2.8	2.72
3.92	17.5	7.2	5.30	4.97	100.6	7.2	5	5.19
4.73		7.1	7.70	8.07		7.1	8.5	8.58
5.08	20.7	7.1	9.40	9.98	82.1	7.1	10.6	10.64
5.73		7.1	13.80	14.70		7	16.2	15.71
6.08	47.9	7.0	19.90	18.19	69.5	7	18.4	19.42
6.69		7.0	28.60	26.25		6.9	29.6	27.91
7.06	17.2	7.0	31.50	32.83	47.9	7	34.6	34.77
7.74		6.9	46.50	49.46		6.8	50.3	51.92
8.13	23.5	6.9	64.20	62.32	16.8	7	66	65.07

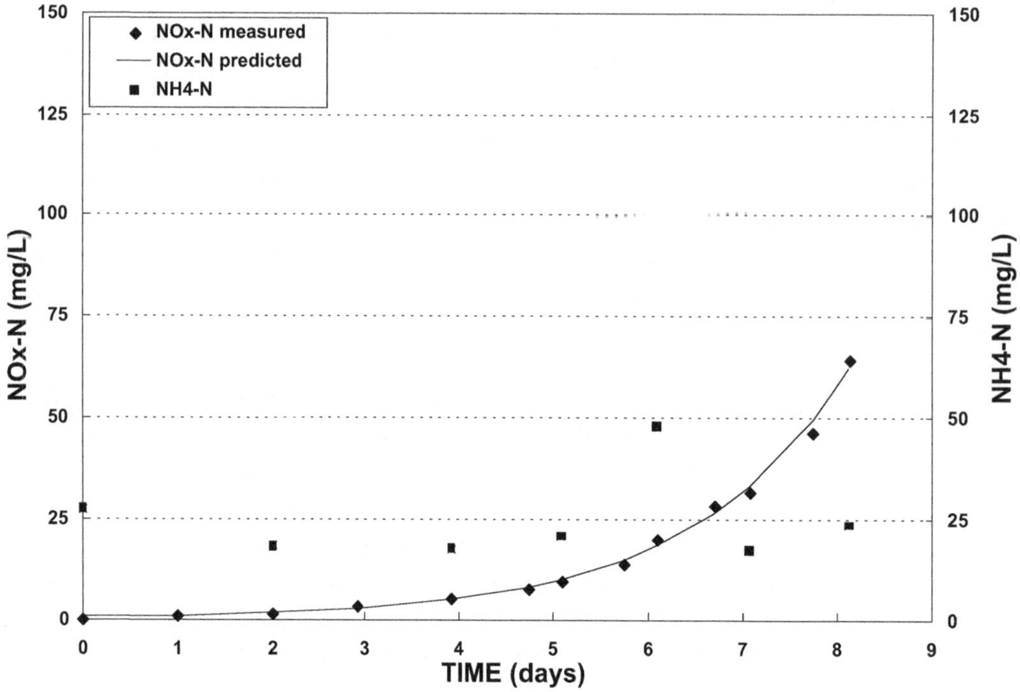

Figure 35-2. Exponential Nitrate Concentration Response in High F/M test on 11/20/00 with Initial Ammonia Concentration of 30 mg N/L

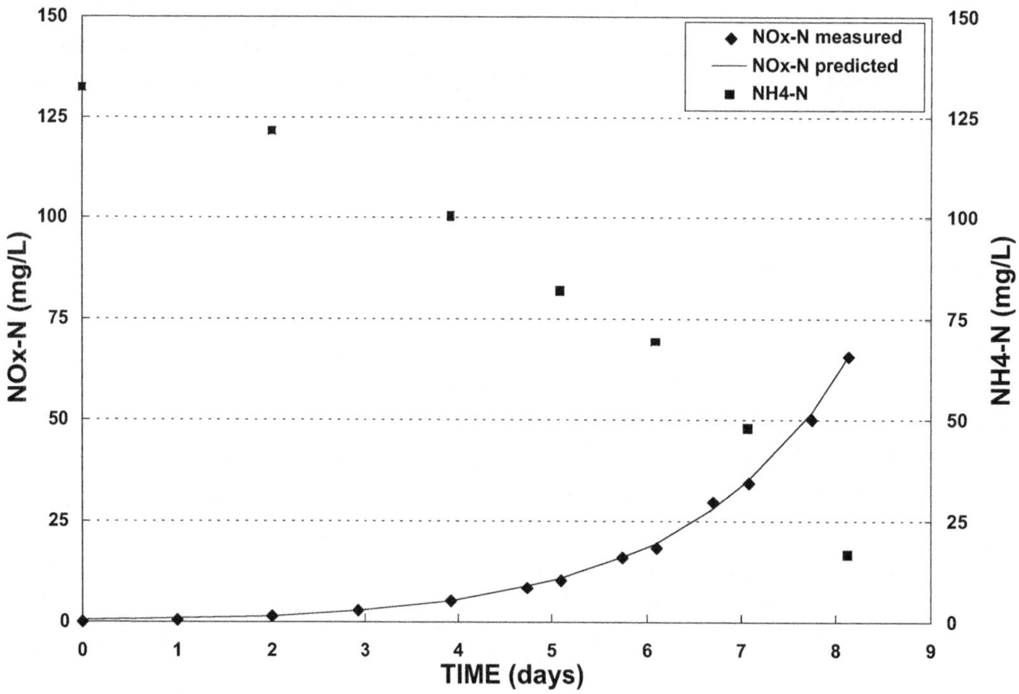

Figure 35-3. Exponential Nitrate Concentration Response in High F/M test on 11/20/00 with Initial Ammonia Concentration of 120 mg N/L

Estimates of $\mu_{AUT, 20\ deg\ C}$ for the two sets of experiments are summarized in Table 35-4. The values are similar for the pairs of tests conducted on the same day, and indicate that the higher initial ammonia concentration did not have any significant effect on the μ_{AUT} values obtained. These results support the use of a single initial ammonia spike, which simplifies the high F/M test method.

Table 35-4. Effect of Initial Ammonia Concentration on $\mu_{AUT, 20\ deg\ C}$ in High F/M Tests

Wastewater Sample	Seed Source	Date	$\mu_{AUT, 20\ deg\ C}$ [1]	
			Initial NH_4-N 30 mg/L	Initial NH_4-N 120 mg/L
Renton secondary effluent	Low F/M batch fed SBR – Phase I	10/06/00	0.64	0.63
Renton secondary effluent	Low F/M batch fed SBR – Phase II	11/20/00	0.73	0.71

[1] Results based on nitrate nitrogen concentration data only

35.2.5 Evaluation of Wastewater Source

Two high F/M tests were carried out to evaluate the response, for a given seed, to changing the wastewater source. A nitrifying mixed liquor seed was taken from an anoxic/aerobic activated sludge wastewater treatment plant in Puyallup, Washington operating with a 12-day SRT. About 100 mL of the mixed liquor was added to Renton secondary influent in one experiment and Puyallup secondary effluent in the second experiment. Only nitrate nitrogen concentration data were collected in the first experiment whereas in the second test, the nitrification rate was determined from measurement of both nitrite and nitrate concentrations.

Estimates of $\mu_{AUT, 20\ deg\ C}$ for the two experiments are summarized in Table 35-5. Based on nitrate concentrations only, almost identical values of $\mu_{AUT, 20\ deg\ C}$ were obtained, that is, values of 0.73 and 0.75 d^{-1} for Renton secondary influent and Puyallup secondary effluent, respectively, when using the same seed.

Table 35-5. Effect of Different Wastewaters on $\mu_{AUT, 20\ deg\ C}$ in High F/M Tests

Wastewater Sample	Seed Ssource	Date	$\mu_{AUT, 20\ deg\ C}$	
			Based on NO_3-N only	Based on NO_2-N & NO_3-N
Renton secondary influent	Puyallup WWTP mixed liquor	2/21/01	0.73	-
Puyallup secondary effluent		6/27/01	0.75	0.86

Figure 35-4 shows the regression of the data from the second experiment.

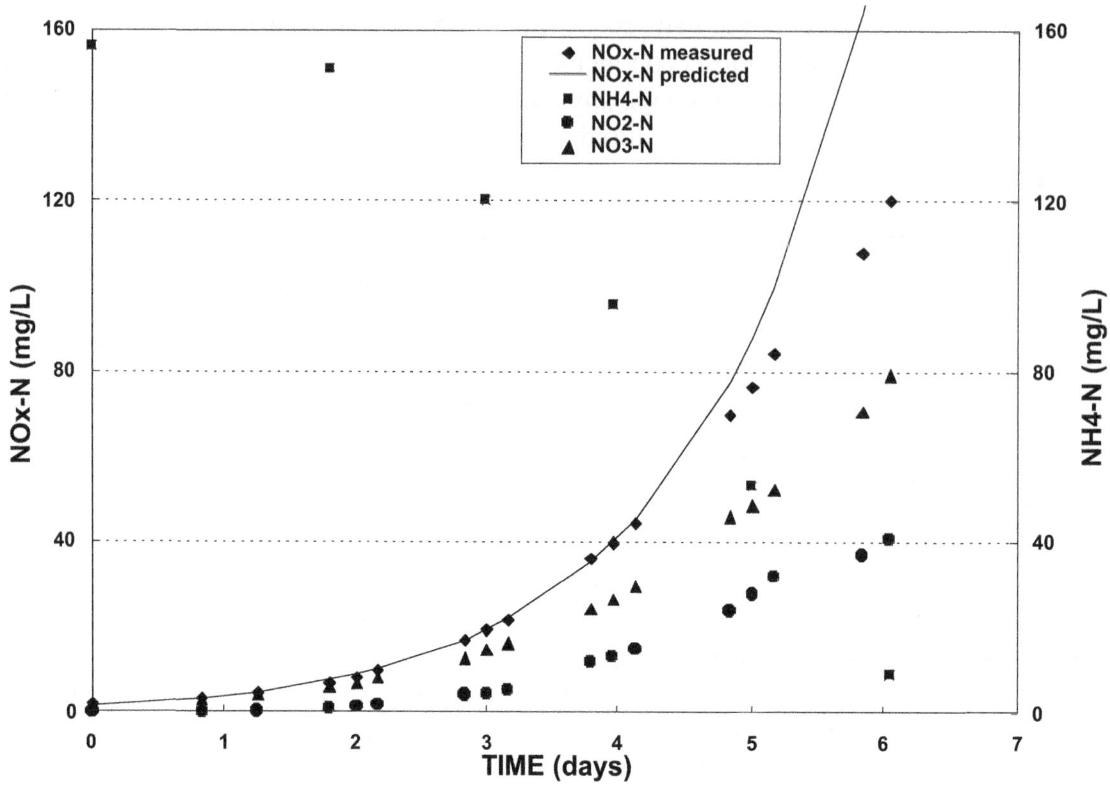

Figure 35-4. Exponential Nitrite and Nitrate Concentration Response in High F/M Test on 6/28/01 with Puyallup WWTP Mixed Liquor Seed and Puyallup Secondary Effluent (Note: Last 5 Points Excluded After Analysis of Residuals)

Figure 35-4 shows that a significant amount of nitrite accumulates during the test. Based on a b_{AUT} value of 0.17 d^{-1} at 20°C, and a temperature dependency coefficient of 1.072 for μ_{AUT}, the estimated $\mu_{AUT, 20\ deg\ C}$ was 0.86 when both nitrite and nitrate data were used. A value of 0.75 d^{-1} was obtained with only the nitrate concentration data, approximately 13 percent lower. The test with Renton secondary effluent and the same seed yielded a $\mu_{AUT, 20\ deg\ C}$ value of 0.73, based on nitrate concentration data only. These experiments showed that for the limited range of wastewaters evaluated, similar values of μ_{AUT} were obtained when using the same seed.

35.3 McMaster University Investigations

Several experiments were conducted at McMaster University to develop and refine the high F/M method. The results of these runs are summarized in the following sections. Each experiment consisted of a pair of high F/M tests. Experimental protocol and equipment follow that outlined in Appendix F (exceptions are noted below). Note that all data for the McMaster University investigations have been analyzed using a value of 0.17 d^{-1} (at 20°C) for the nitrifier decay rate (b_{AUT}) and a temperature dependency coefficient of 1.072 for μ_{AUT}. Nitrite, nitrate and ammonia analyses for the McMaster University investigations were performed with the spectrophotometric methods noted in Appendix F.

35.3.1 Evaluation of Test Reproducibility

Figures 35-5 and 35-6 show the results for two tests conducted in parallel under essentially identical conditions. In each test, 100 mL of SBR sludge was diluted with 2,900 mL of SBR decant (effluent), and spiked with ammonia at the start. Appropriate alkalinity was added at the start of the test and during the test. Comparison of Figures 35-5 and 35-6 illustrates the reproducibility of the method. Each set of data resulted in the same μ_{AUT} estimate within a 95 percent confidence limit (0.91 and 0.90 d^{-1}, respectively).

The importance of pH control in the high F/M test has been mentioned earlier. The pH versus time plot for test number 1 in this run is shown in Figure 35-7. Note that the pH was maintained between 7.4 and 7.8 through intermittent alkalinity addition in the form of sodium bicarbonate.

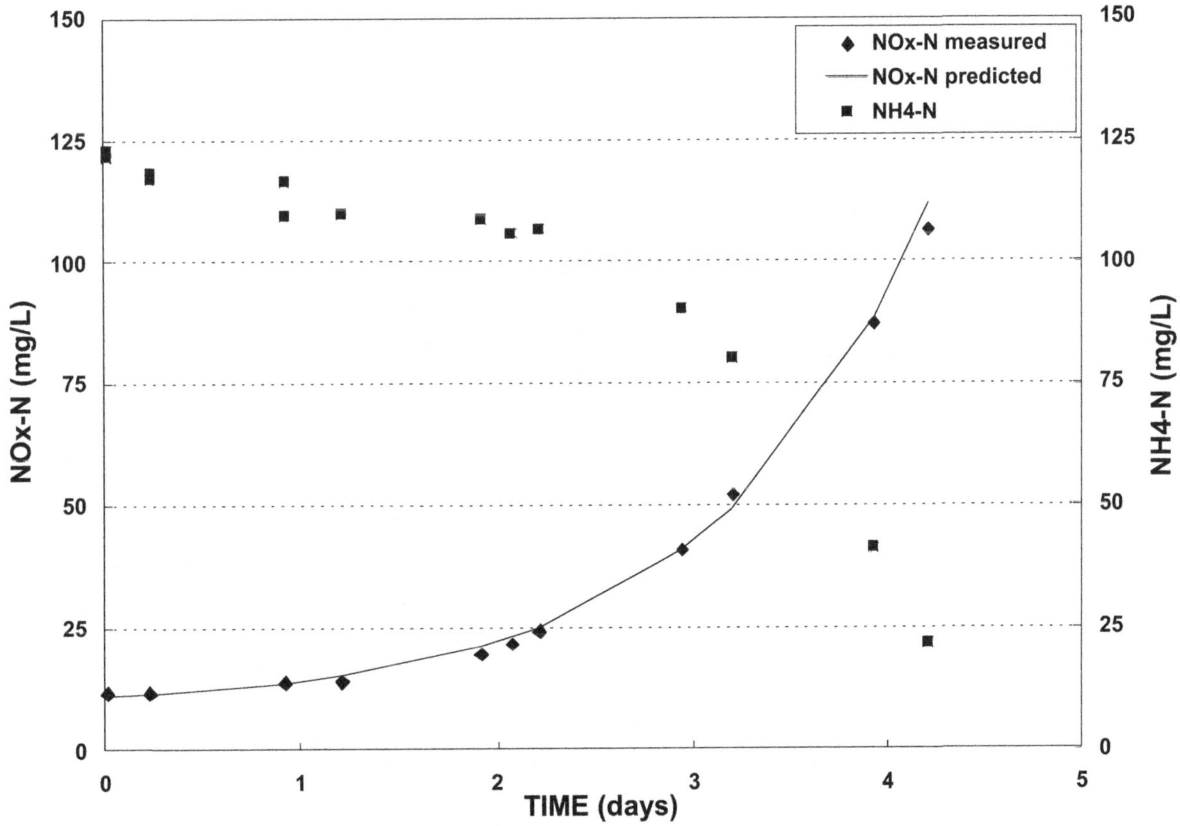

Figure 35-5. Exponential Nitrite / Nitrate Concentration Response in High F/M Test Number 1 on 6/04/01 with McMaster University SBR Mixed Liquor Seed (Originally Obtained from Dundas, Ontario WWTP) and McMaster University SBR Decant (Note: Last Point Excluded After Analysis of Residuals)

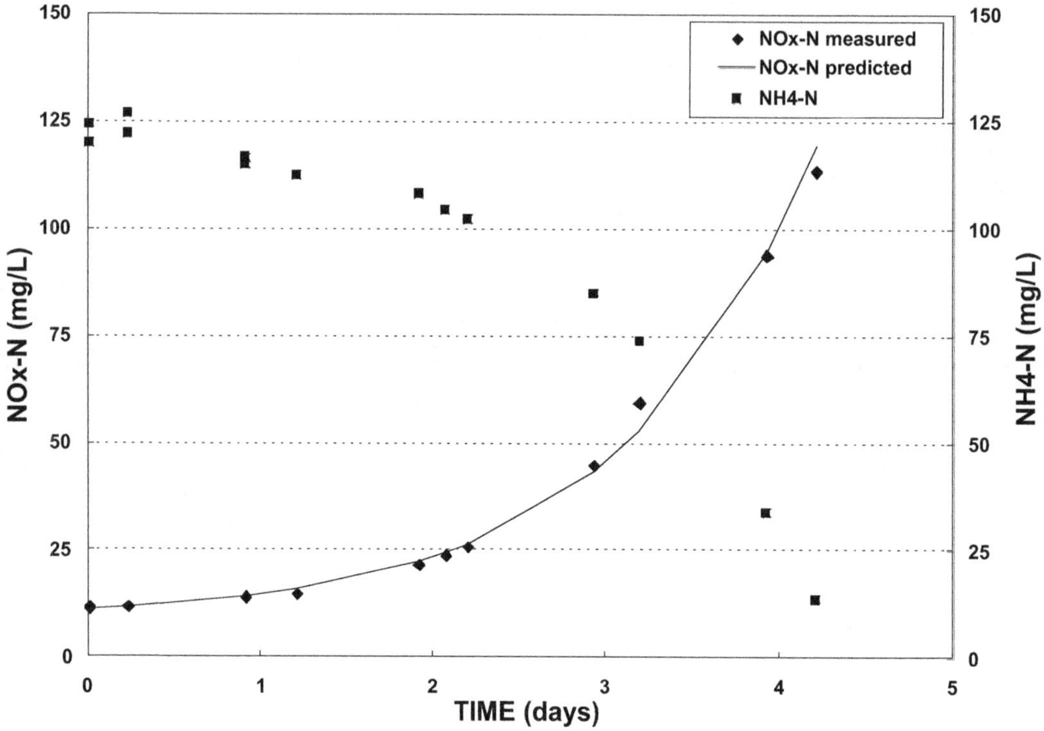

Figure 35-6. Exponential Nitrite/Nitrate Concentration Response in High F/M Test Number 2 on 6/04/01 with McMaster University SBR Mixed Liquor Seed (Originally Obtained from Dundas, Ontario WWTP) and McMaster University SBR Decant (Note: Last Point and Point at 3.2 Days Excluded After Analysis of Residuals)

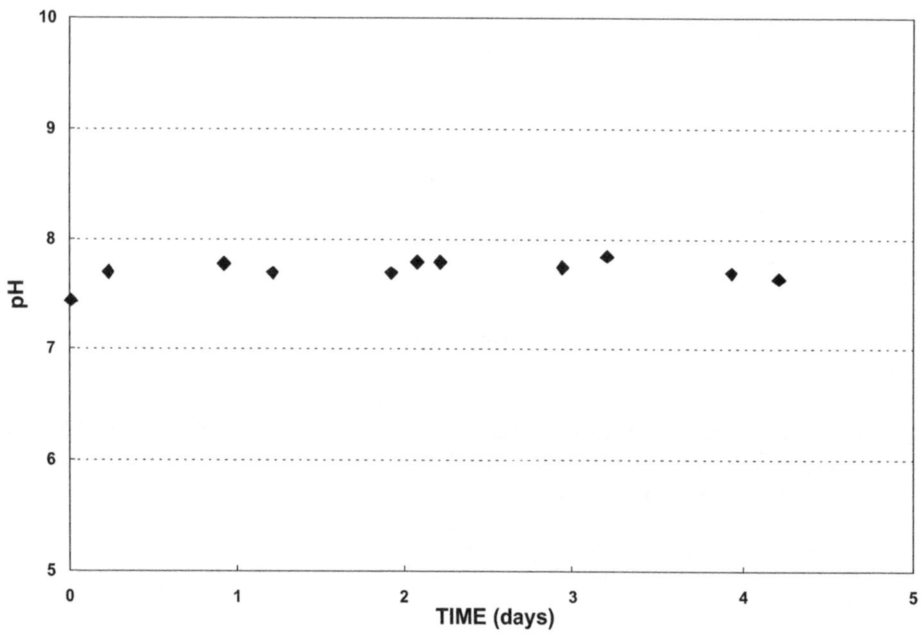

Figure 35-7. pH versus Time in High F/M Test #1 on 6/04/01 with McMaster University SBR Mixed Liquor Seed (Originally Obtained from Dundas, Ontario WWTP) and McMaster University SBR Decant

35.3.2 Evaluation of Influent Inhibition

Figures 35-8 and 35-9 show the results of a second pair of tests conducted in parallel. Each test was started with 100 mL of SBR sludge. In one test, the diluent was raw influent wastewater (Figure 35-8). In the other test, 2,900 mL of SBR decant was used as diluent (Figure 35-9). The two responses were slightly different in this case:

- In the test with decant (effluent) as diluent (Figure 35-9), the initial nitrate concentration was approximately 12 mg/L because the decant contained nitrate. With raw influent as diluent (Fig. 35-8), the initial nitrate concentration was zero.
- The response in the test with decant (effluent) as diluent (Figure 35-9) was slightly more rapid presumably because the decant contained a small concentration of nitrifiers, which elevated the initial nitrifier concentration in the test.

The two tests provided estimates of μ_{AUT} (at 20°C) of 0.82 d^{-1} (raw influent) and 0.89 d^{-1} (effluent).

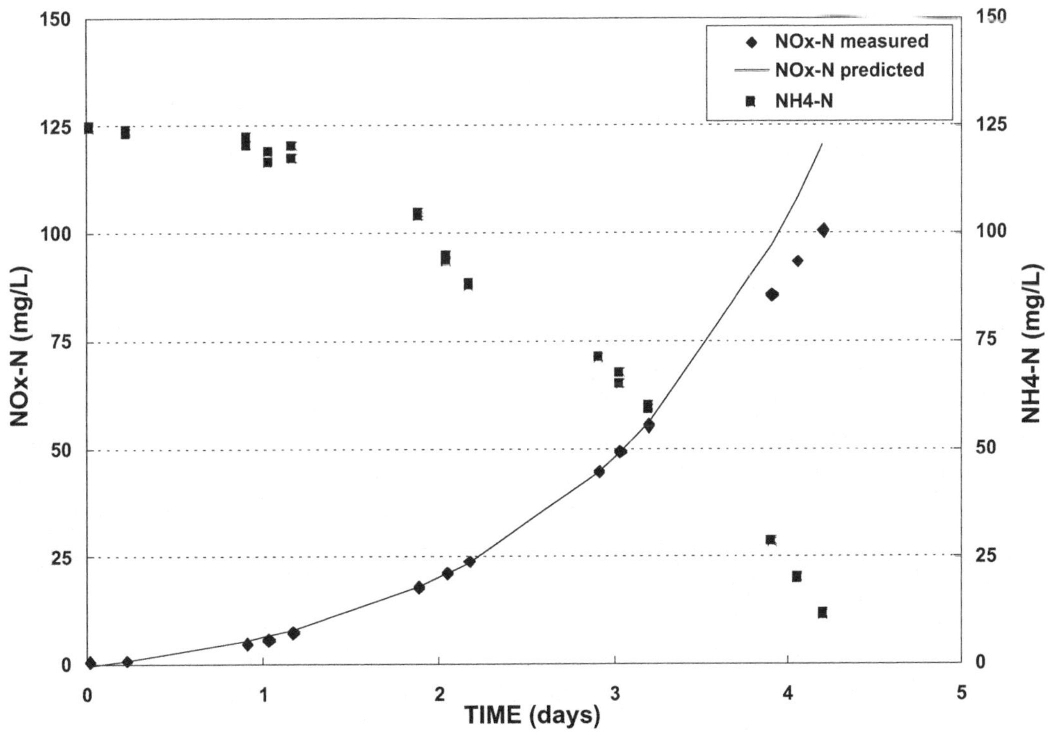

Figure 35-8. Exponential Nitrite/Nitrate Concentration Response in High F/M Test Number 1 on 6/18/01 with McMaster University SBR Mixed Liquor Seed (Originally Obtained from Dundas, Ontario WWTP) and Dundas, Ontario WWTP Raw Influent (Note: Last 3 Points Excluded After Analysis of Residuals)

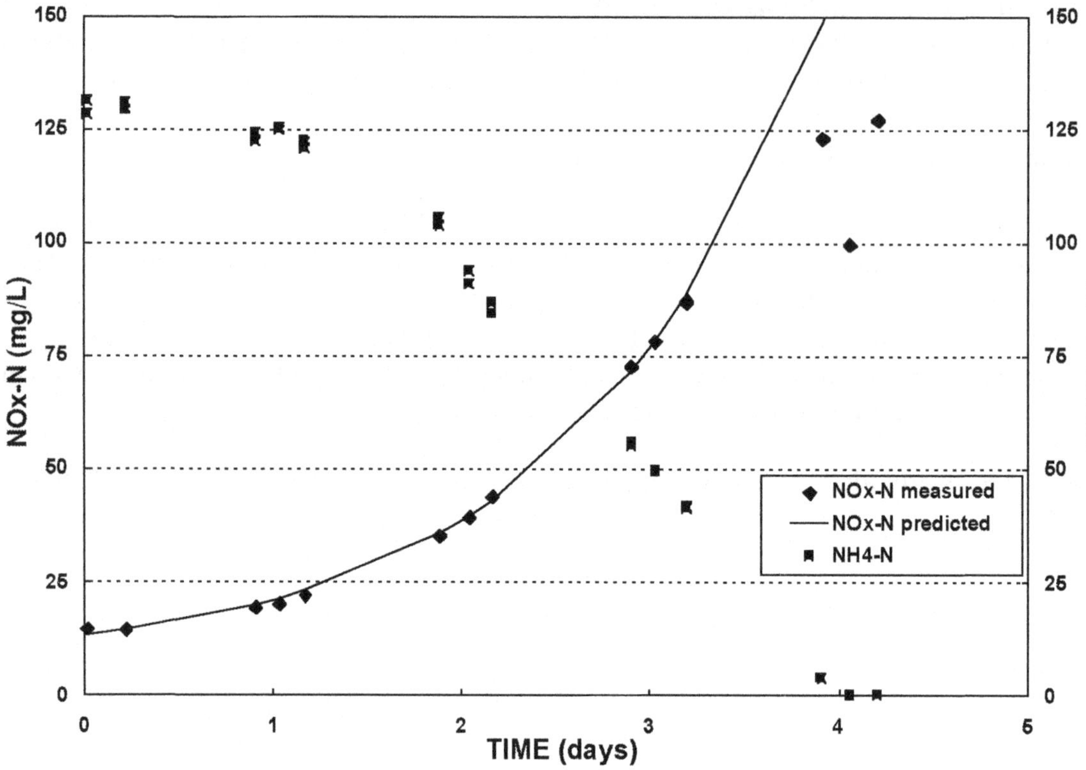

Figure 35-9. Exponential Nitrite/Nitrate Concentration Response in High F/M Test Number 2 on 6/18/01 with McMaster University SBR Mixed Liquor Seed (Originally Obtained from Dundas, Ontario WWTP) and McMaster University SBR Decant (Note: Last 3 Points Excluded After Analysis of Residuals)

35.3.3 Evaluation of Seed Sludge Source

One concern regarding the low F/M SBR protocol is whether the system configuration and operating mode may lead to the growth of nitrifiers that exhibit different characteristics from the nitrifiers in the full-scale plant, even though the systems receive the same influent. The pair of tests commenced on 6/18/01 above was repeated, but using sludge drawn from the full-scale Dundas, Ontario plant. Figures 35-10 and 35-11 show the results. Each test was started with 100 mL of Dundas sludge. In one test, 2,900 mL of raw Dundas influent wastewater were used as diluent (Figure 35-10). In the other test, 2,900 mL of Dundas secondary effluent were used as diluent (Figure 35-11).

The two tests provided estimates of μ_{AUT} (at 20°C) of 0.88 d^{-1} (raw influent) and 0.83 d^{-1} (effluent). (Note that in this case, the rate for the test with raw influent as diluent was slightly higher.)

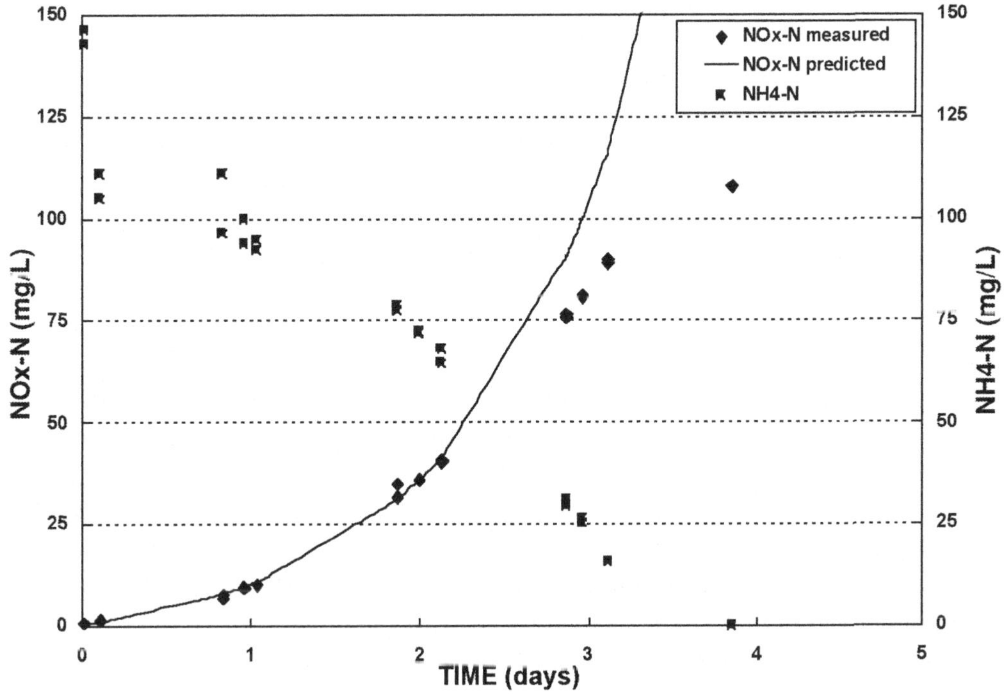

Figure 35-10. Exponential Nitrite/Nitrate Concentration Response in High F/M Test Number 1 on 8/20/01 with Dundas, Ontario WWTP Mixed Liquor Seed and Dundas, Ontario WWTP Raw Influent (Note: Last 4 Points Excluded After Analysis of Residuals)

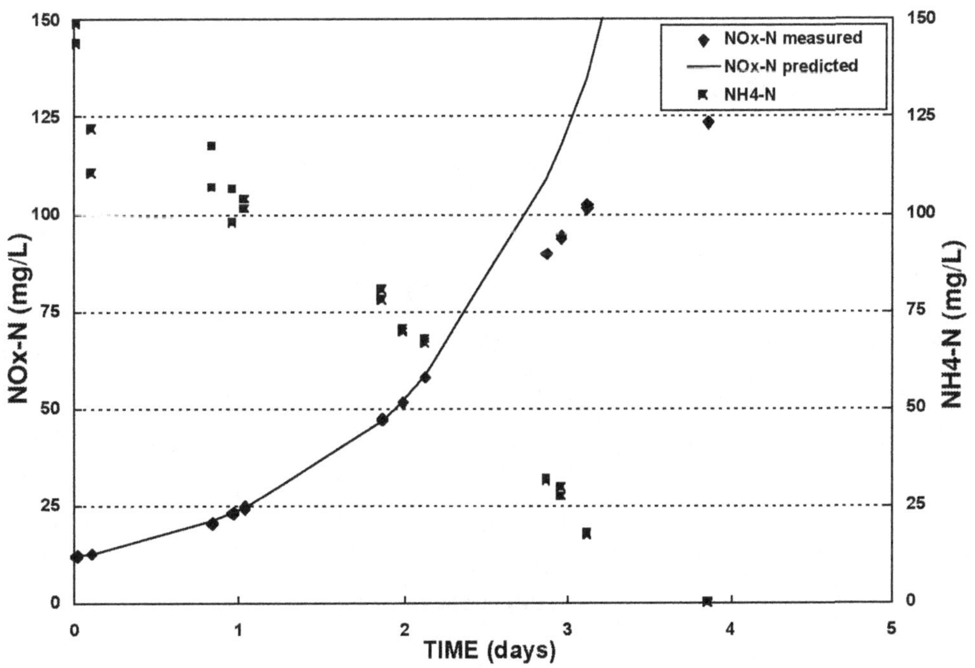

Figure 35-11. Exponential Nitrite/Nitrate Concentration Response in High F/M Test Number 2 on 8/20/01 with Dundas, Ontario WWTP Mixed Liquor Seed and Dundas, Ontario WWTP Secondary Effluent (Note: Last 4 Points Excluded After Analysis of Residuals)

35.3.4 Evaluation of Alkalinity Addition Method (Spike at Beginning or Intermittent)

An area of concern that has been identified with the high F/M technique is maintaining optimal pH conditions throughout the test period. One way to accomplish this is to add a source of alkalinity over the course of the experiment. The alkalinity of previous tests performed at McMaster University was controlled by monitoring pH, and intermittently adding sodium bicarbonate ($NaHCO_3$). One drawback of this method of pH control is increased test complexity. The intermittent alkalinity addition can be automated, but this requires a pH controller, which adds significant cost to the test. If it is done manually (as was the case in the McMaster University investigations), the individual conducting the test must pay close attention to the reactor pH.

A more desirable approach for test simplicity would be to add all the necessary alkalinity at the start of the test. A pair of tests was performed to investigate whether the different alkalinity addition methods impacts μ_{AUT}. Each test was started with 55 mL of Dundas sludge (slightly less was used in this run as the source MLSS was higher). In both tests, 2,945 mL of Dundas secondary effluent were used as diluent. In one test, 4 g of sodium bicarbonate was added at the beginning of the test, and pH was monitored continually (Figure 35-12). In the other test, 4 g of sodium bicarbonate was added in 1 g increments as warranted by periodic pH monitoring (Figure 35-13). Data analysis indicates the two tests provided estimates of μ_{AUT} (at 20°C) of 0.80 d^{-1} (all alkalinity at beginning of test) and 0.75 d^{-1} (intermittent alkalinity).

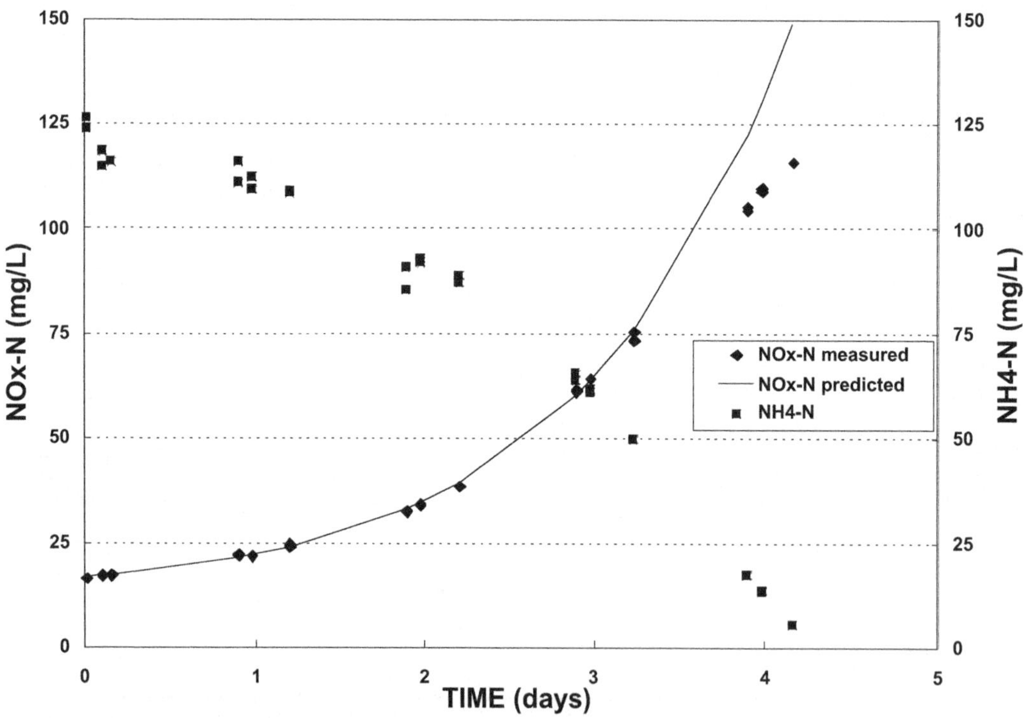

Figure 35-12. Exponential Nitrite/Nitrate Concentration Response in High F/M Test Number 1 on 9/17/01 with Dundas, Ontario WWTP Mixed Liquor Seed and Dundas, Ontario WWTP Secondary Effluent (All Alkalinity Added at Beginning of Test) (Note: Last 3 Points Excluded After Analysis of Residuals)

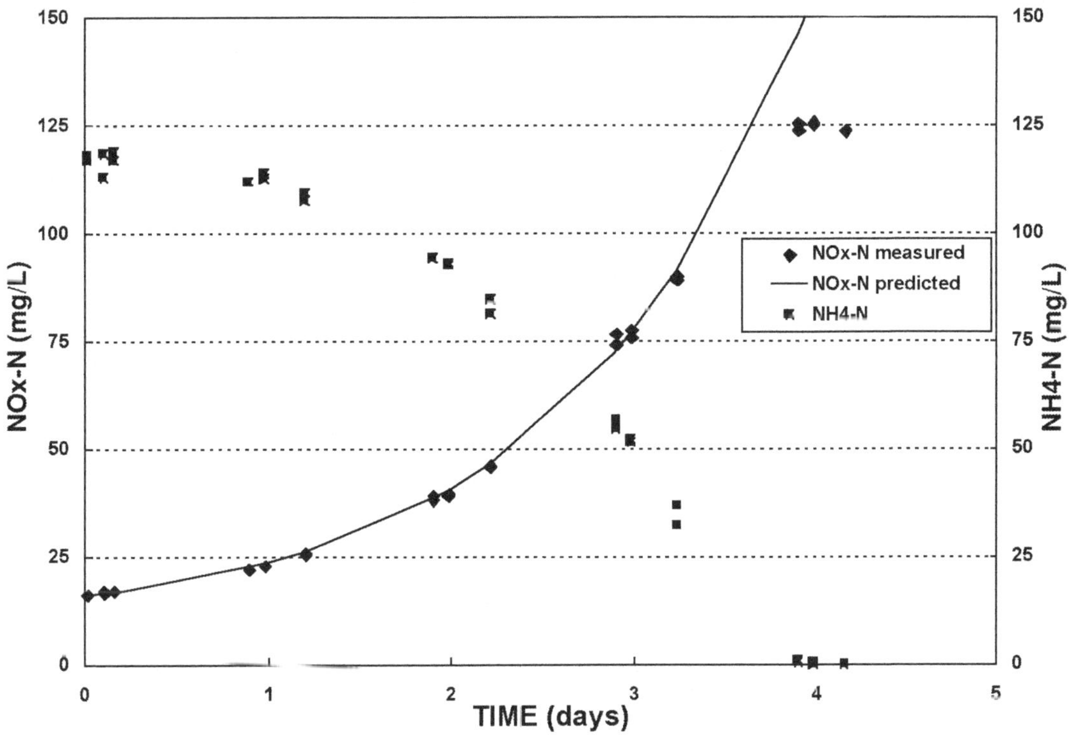

Figure 35-13. Exponential Nitrite/Nitrate Concentration Response in High F/M Test Number 2 on 9/17/01 with Dundas, Ontario WWTP Mixed Liquor Seed and Dundas, Ontario WWTP Secondary Effluent (Intermittent Alkalinity Addition)
(Note: Last 3 Points Excluded After Analysis of Residuals)

35.3.5 Evaluation of Aeration Method

Another area of concern that has been identified with the high F/M technique is the method used for aeration of the test vessel. The aeration method used in previous tests performed at McMaster University used a dissolved oxygen (DO) controller to intermittently switch airflow to the vessel on and off. Two benefits of this approach are:

1. Evaporation of liquid from the test vessel that occurs during aeration is minimized.
2. Stripping of carbon dioxide from the test vessel is minimized—this helps to keep the pH in a suitable range for biological growth.

The main drawback of this approach is that it requires extra equipment and expertise. The high F/M test would be simplified greatly if constant rather than controlled aeration could be used.

A pair of tests was performed to investigate whether the different aeration methods would yield different measured μ_{AUT} values. Each test was started with 75 mL of Dundas sludge (in order to achieve a test VSS of approximately 35 mg/L). In both tests, 2,925 mL of Dundas secondary effluent were used as diluent. One test was aerated with a constant airflow (Figure 35-14). The other test was aerated with intermittent airflow using a DO controller (Figure 35-15).

Data analysis indicates the two tests provided estimates of μ_{AUT} (at 20°C) of 0.50 d^{-1} (constant aeration) and 0.83 d^{-1} (intermittent aeration). Two points are apparent here:

1. There is a statistically significant difference between the measured values.
2. The value with constant aeration (0.50 d-1) is much lower than values previously measured in this study, while the value measured with intermittent aeration (0.83 d-1) is more typical.

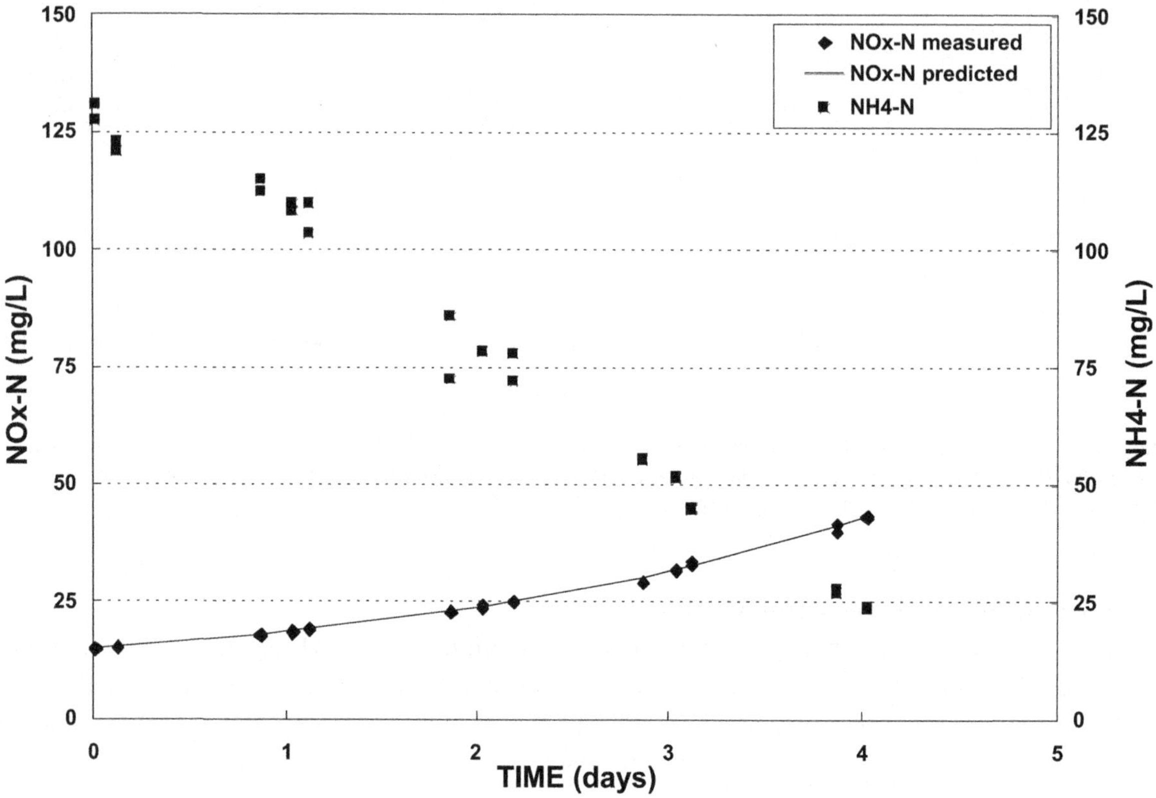

Figure 35-14. Exponential Nitrite/Nitrate Concentration Response in High F/M Test Number 1 on 10/01/01 with Dundas, Ontario WWTP Mixed Liquor Seed and Dundas, Ontario WWTP Secondary Effluent (Constant Aeration)

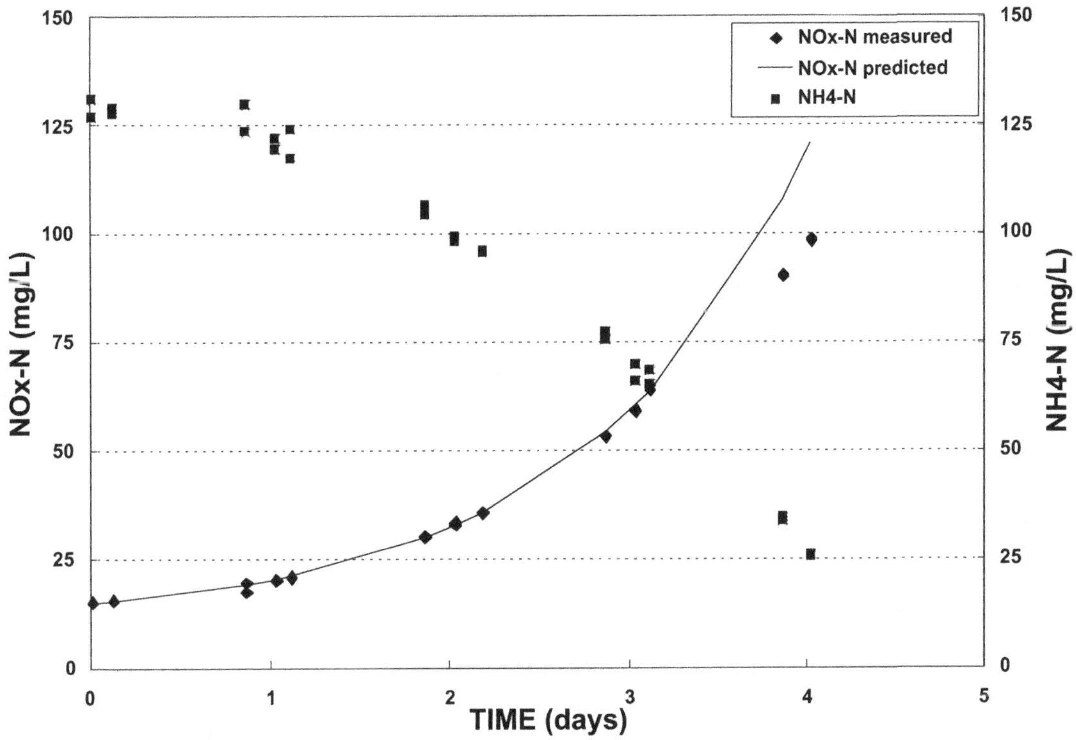

Figure 35-15. Exponential Nitrite/Nitrate Concentration Response in High F/M Test Number 2 on 10/01/01 with Dundas, Ontario WWTP Mixed Liquor Seed and Dundas, Ontario WWTP Secondary Effluent (Intermittent Aeration) (Note: Last 2 Points Excluded After Analysis of Residuals)

Based on this test, it appears that when constant aeration is used, significant stripping of carbon dioxide from the test vessel occurs. The effect of this is to increase the test pH, which has an inhibitory effect on nitrifier growth. Plots showing the nitrate versus time response for the two tests are shown in Figures 35-14 and 35-15. Note the much flatter nitrite/nitrate response for the test using constant aeration. A secondary impact of the high pH is that ammonia is removed from the test vessel via stripping. This is reflected by the linear decline of ammonia with time. A plot of pH versus time for the test with constant aeration is shown in Figure 35-16.

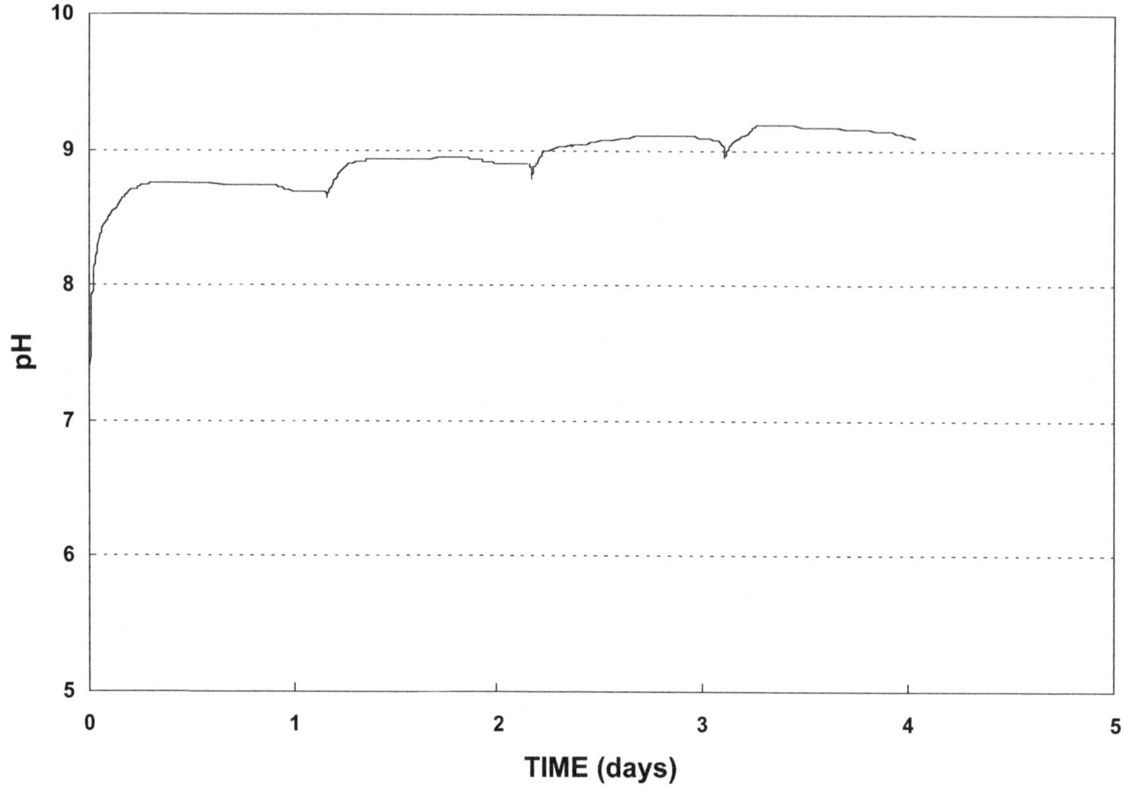

Figure 35-16. pH versus Time in High F/M Test Number 1 on 10/01/01 with Dundas, Ontario WWTP Mixed Liquor Seed and Dundas, Ontario WWTP Secondary Effluent (Constant Aeration)

35.3.6 Summary of McMaster University Results

The results of each McMaster University high F/M test are shown in Fig. 35-17.

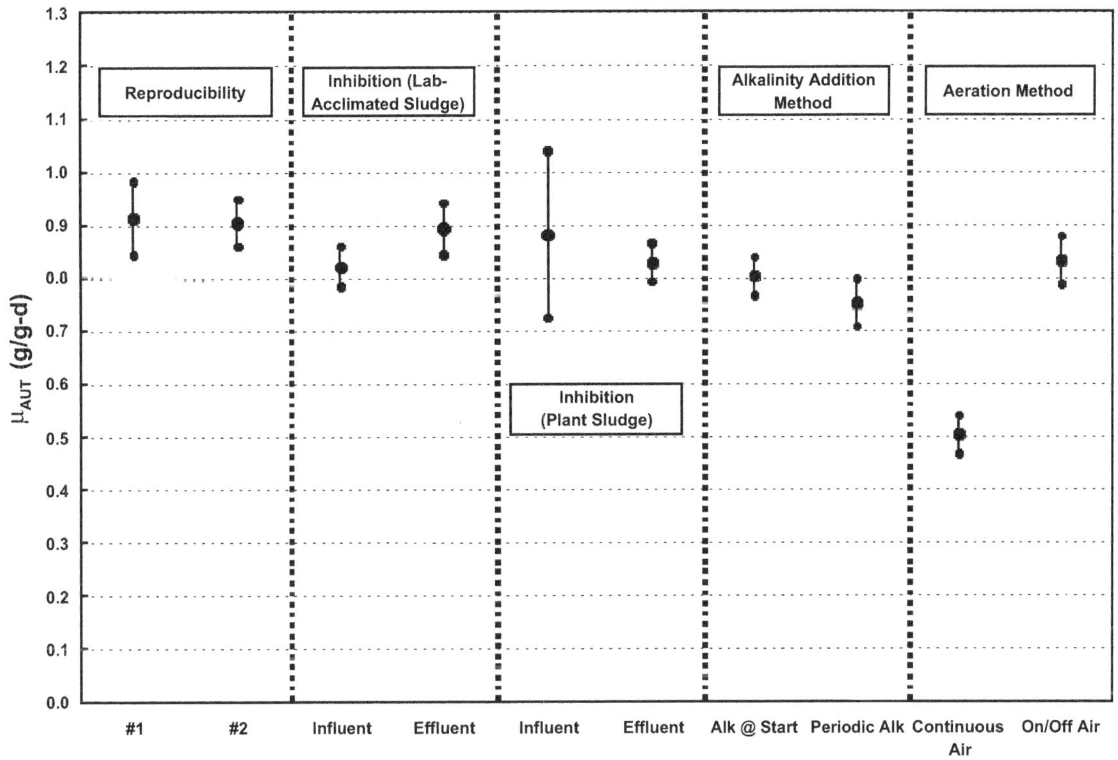

Figure 35-17. Summary of μ_{AUT} Values Estimated for the McMaster University Investigations

In this plot, the results from the various test runs are separated by vertical dashed lines. For each test run, the experimental condition of investigated is noted in a text box (e.g., reproducibility, aeration method). Each test run shows a pair of μ_{AUT} values (i.e., the large middle circle) and their respective 95 percent confidence intervals (i.e., the vertical lines bounded by smaller circles). Important points to note from this plot include:

- The μ_{AUT} values estimated from the first run investigating reproducibility of the test are statistically identical.
- The control test in the run evaluating inhibition effects using laboratory-acclimated seed sludge (i.e., the "effluent") was performed under identical experimental conditions as the first two tests that were run to investigate reproducibility. Note that the μ_{AUT} value estimated from this control test also is statistically identical to the μ_{AUT} values estimated from the first run investigating reproducibility of the test.
- In both experiments evaluating inhibition, there was a small (but statistically significant) difference in the μ_{AUT} estimated values. For the laboratory-acclimated sludge, the μ_{AUT} estimated using raw influent as diluent was the lower of the pair. The opposite was true for the full-scale plant sludge. However, all four values lie within the 0.8 - 0.9 d^{-1} range.
- There was a small (but statistically significant) difference depending on alkalinity addition method. A slightly higher estimate of μ_{AUT} was obtained when all the alkalinity was added at the beginning of the test. It should be noted that the control test in this experiment gave a lower estimate of μ_{AUT} than other control tests (which were operated identically).

- When the high F/M test was conducted with continuous aeration, pH control could not be maintained. It is hypothesized that this was caused by stripping of carbon dioxide from the reactor.

35.4 City of Phoenix Investigations

Three high F/M tests were conducted at the 91st Avenue Wastewater Treatment Plant in Phoenix AZ to investigate the field applicability of the method. The results of these runs are summarized below. Note that all data for the City of Phoenix investigations have been analyzed using a value of 0.17 d^{-1} (at 20°C) for the nitrifier decay rate (b_{AUT}) and a temperature dependency coefficient of 1.072 for μ_{AUT}.

35.4.1 Test #1 – April 18, 2001

This test was started with 50 mL of sludge obtained from the low F/M SBR operated for this study. Secondary effluent (2,950 mL) from the 91st Avenue plant was used as diluent for the test. Alkalinity was added as needed, and airflow was intermittent.

Analysis indicates the test provided an estimate of μ_{AUT} (at 20°C) of 0.48 d^{-1} (Figure 35-18). This low value is attributable to the following factors:

- test pH reached a value of 8.56, which may have resulted in nitrifier inhibition;
- the linear decrease in ammonia indicates that ammonia may have been removed from the test vessel via stripping.

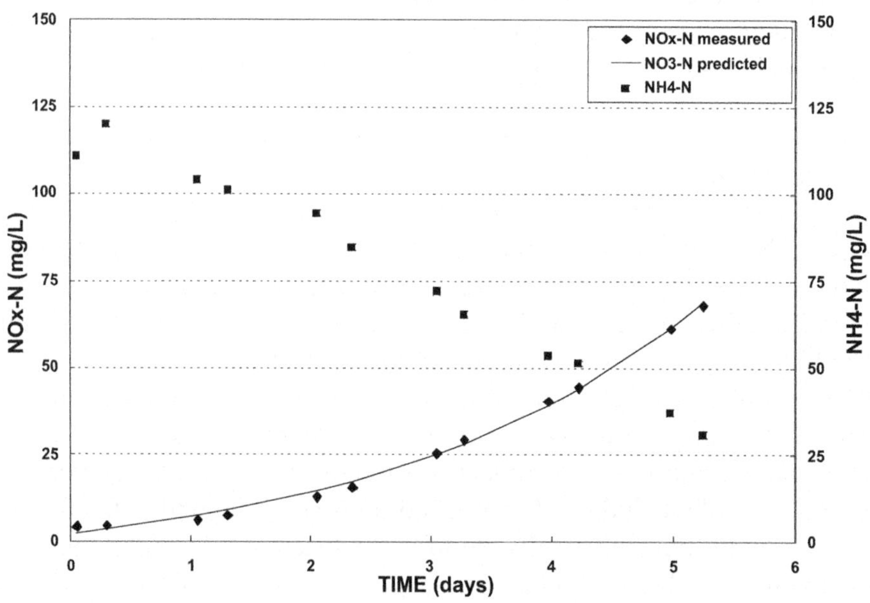

Figure 35-18. Exponential Nitrite/Nitrate Concentration Response in Phoenix High F/M Test on 4/18/01 with Phoenix SBR Mixed Liquor Seed and Phoenix 91st Avenue Secondary Effluent

35.4.2 Test #2 – May 14, 2001

This test was started with 10 mL of sludge obtained from the 91st Avenue plant. Secondary effluent (3,000 mL) from the 91st Avenue plant was used as diluent for the test. Alkalinity was added as needed and airflow was intermittent.

Analysis indicates the test provided an estimate of μ_{AUT} (at 20°C) of 0.73 d^{-1} (Figure 35-19). It should be noted that this value is estimated on a low number of observations due to pH problems that occurred in the later stages of the test.

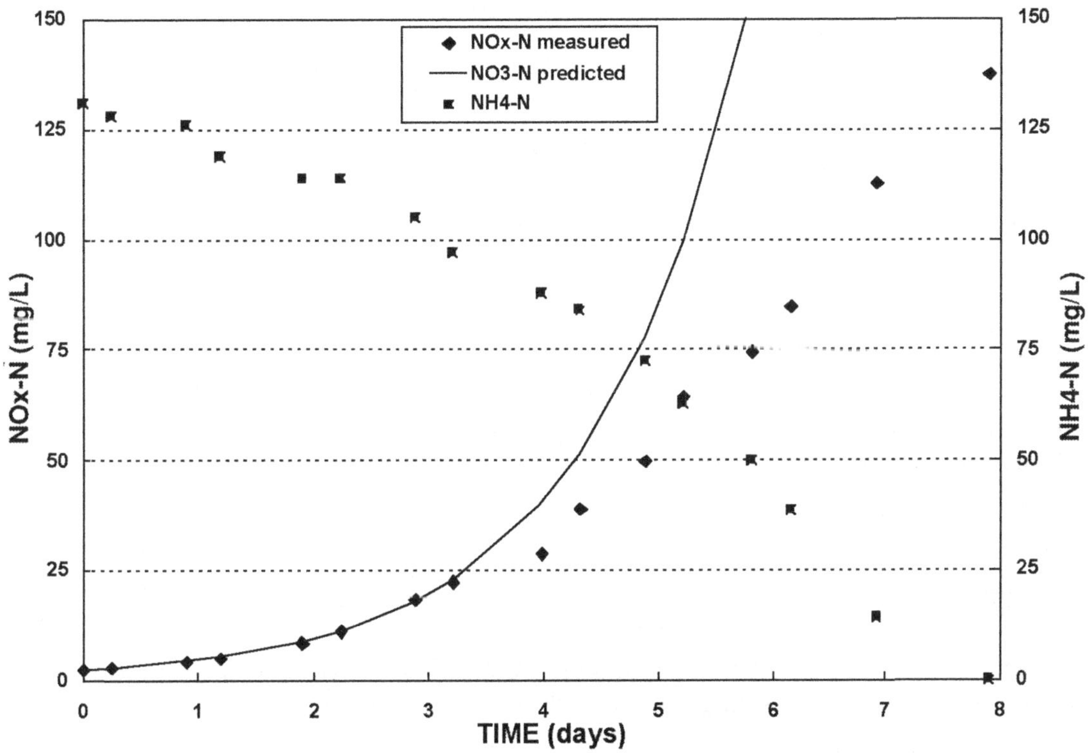

Figure 35-19. Exponential Nitrite/Nitrate Concentration Response in Phoenix High F/M Test on 5/14/01 with Phoenix 91st Avenue Mixed Liquor and Secondary Effluent (Note: Last 9 Points Excluded After Analysis of Residuals)

35.4.3 Test #3 – May 26, 2001

This test was started with 10 mL of sludge obtained from the 91st Avenue plant. Secondary effluent (3,000 mL) from the 91st Avenue plant was used as diluent for the test. Alkalinity was added as needed, and airflow was intermittent.

Analysis indicates the test provided an estimate of μ_{AUT} (at 20°C) of 0.81 d^{-1} (Fig. 35-20). Note that the last three data points have been excluded from the analysis due to dilution errors in sample preparation.

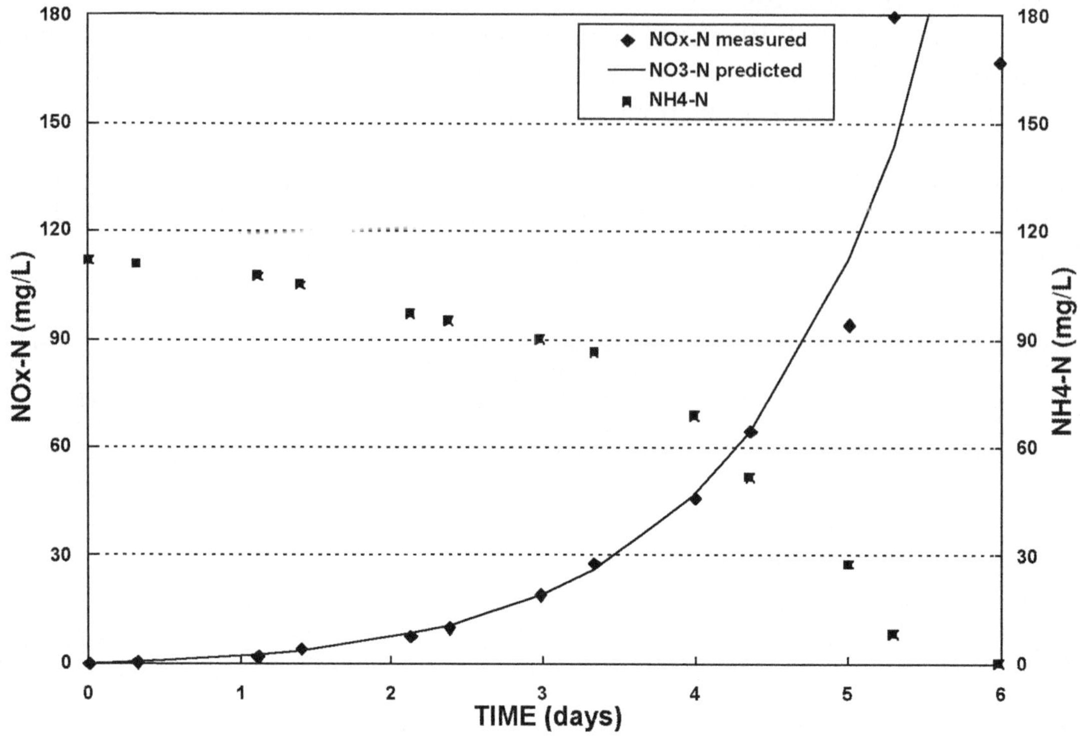

Figure 35-20. Exponential Nitrite/Nitrate Concentration Response in Phoenix High F/M Test on 5/26/01 with Phoenix 91st Avenue Mixed Liquor and Secondary Effluent (Note: Last 3 Points Excluded After Analysis Of Residuals)

35.5 References

Barker, P.S., and P.L. Dold. 1997. General model for biological nutrient removal activated sludge systems: model presentation. *Water Environ.Res.* 69(5):969-984.

CHAPTER 36.0

SUMMARY OF WASHOUT INVESTIGATIONS

36.1 Background

A series of investigations were carried out at the McMaster, Renton and Phoenix locations to measure maximum specific nitrifier growth rate using the washout method. The basis of the method and mathematical background is described in Chapter 16.0. The method is described in detail in Appendix G. This chapter reports on the application of this method to different wastewaters and different seeds at the three test locations.

Note: In assessing the results presented in this chapter, careful attention should be paid to whether estimates of μ_{AUT} values are based on fits to nitrate-only response data or to nitrite+nitrate response data. Estimates based on nitrate-only data will underestimate the μ_{AUT} values. Early in the project nitrite was not monitored. However, during the project it became evident that nitrite accumulation in the various tests may be significant, in particular in high F/M and washout tests, and to a lesser extent in low F/M SBR tests. At that point nitrite monitoring commenced.

36.2 University of Washington Investigations

36.2.1 Apparatus and Operating Conditions

The detailed procedure for this method that is described in Appendix G resulted in part from the experiences reported in this chapter. The method was under development during the experiments described in this section and the following description of the apparatus used at the University of Washington environmental engineering research laboratory serves to illustrate the apparatus that was in use at the time of this portion of the work.

The same reactor used for the low F/M SBR tests at the South Plant in Renton was used for the washout tests. A volume of prepared influent wastewater was stored in a 4 L (approximate 20 cm diameter by 70 cm height) feed container that was placed on a magnetic stirrer. A timer controlled a peristaltic pump to deliver an appropriate volume of wastewater to the reactor every

15 minutes. At the same time, the same volume of excess mixed liquor would overflow from the 4 L level port to another container. Air from an aquarium pump was sparged through a stone diffuser located at the bottom of the reactor, and the rate was controlled by a rotameter and needle valve. A dissolved oxygen/oxygen uptake rate (DO/OUR) meter and DO probe were used to monitor and control the DO at concentrations greater than 4.0 mg/L. A photograph of the experimental set up is shown in Fig. 36-1.

Because of the large distance between the South Plant at Renton (and the Puyallup WWTP) and the University of Washington laboratory, a large enough batch of secondary influent for the duration of the test was collected at the plant, transported to the laboratory and stored at 4°C. Each day of the test, a well-mixed representative feed sample was removed from storage, warmed to the laboratory temperature and added to the reactor feed container.

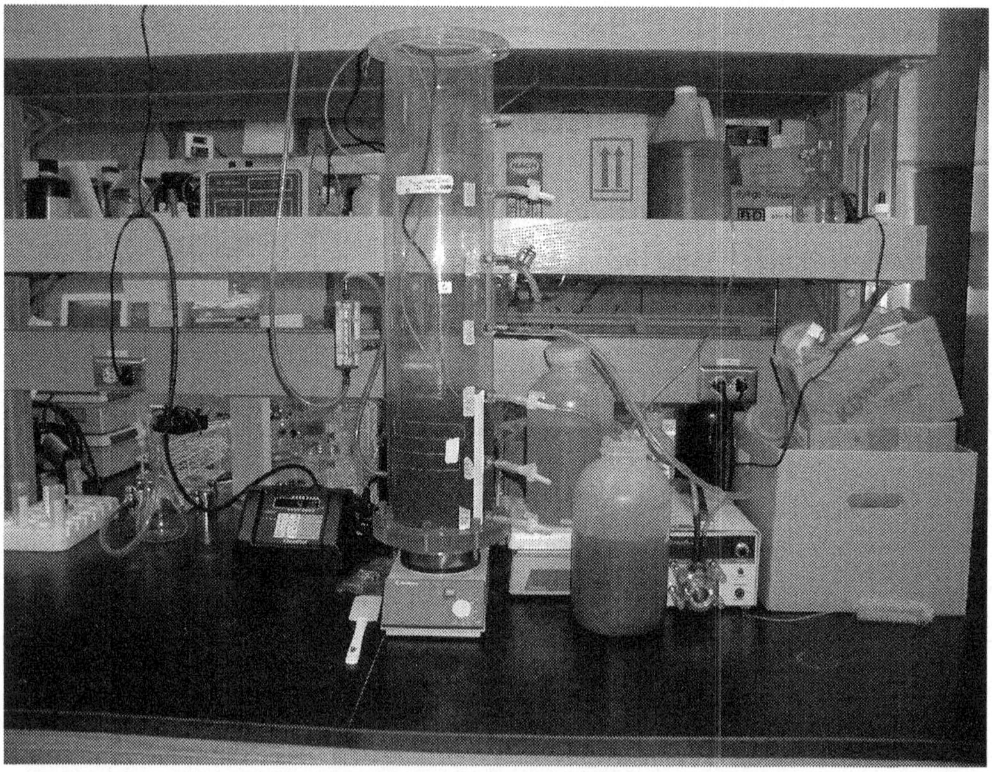

Figure 36-1. Overview of the Washout Apparatus at the University of Washington Laboratory
The reactor is located in the center of the picture with the overflow container to the right of the reactor and the influent feed container between the two. The timer is on the left of the reactor

Ammonia was added to the influent volume at a concentration of approximately 200 mg N/L. Sodium bicarbonate was also added to the influent volume to provide a feed level of approximately 1,400 mg/L alkalinity as $CaCO_3$. This provided a reactor pH of 7.0 to 7.5. As the washout test proceeded, it was sometimes necessary to reduce the ammonia feed concentration to limit the level in the reactor.

Four washout tests were conducted; two of them used mixed liquor from the Phase III low F/M SBR (see Chapter 34.0) with an SRT of 8 days, and two with mixed liquor from the aeration tank of the Puyallup WWTP anoxic/aerobic activated sludge process, with an SRT of 12 days. Test conditions and operating conditions are summarized in Table 36-1.

At test start up, 4 L of nitrifying mixed liquor seed was spiked with 40 mg/L ammonia nitrogen and added to the washout test reactor with aeration and mixing on. Feeding of the spiked wastewater started immediately. At least once or twice each day, the inside wall of the reactor was cleaned to remove wall growth and the material removed retained in the reactor. Table 36-1 summarizes the data collection program during the test.

Table 36-1. Washout Test Operating Conditions at the University of Washington Laboratory

Test No.	Seed Source	Feed Source	Reactor Volume (L)	Feed Volume (L/d)	SRT (days)	Temp (°C)
1	Multi-batch fed SBR-III	Renton WWTP secondary influent	4	2.5	1.6	20
2	Multi-batch fed SBR-III	Renton WWTP secondary influent	4	2.6	1.54	19
3	Puyallup WWTP	Renton WWTP secondary influent	4	2.5	1.6	20
4	Puyallup WWTP	Puyallup WWTP secondary effluent	4	3.8	1.05	25

Table 36-2. Washout Test Data Collection Program

Sample	Parameter	Frequency
Influent	NH_4-N	Daily
Reactor	TSS	Daily
	VSS	Daily
	pH	5/day @ 2-hr intervals
	Temperature	5/day @ 2-hr intervals
Effluent	NH_4-N	Daily
	NO_3-N	5/day @2-hr intervals
	NO_2-N*	5/day @2-hr intervals

* Measured in test no. 4 only

The resulting data were analyzed by fitting to equation 16.7.1-1 in Chapter 16.0 using the parameter estimating procedure described in Appendix H. A non-linear least squares technique using Excel solver was applied to equation 16.7.1-1 to adjust the parameters ($\mu_{AUT} - b_{AUT}$), $X_{AUT,0}$, $N_{NOx,0}$ in order to minimize the sum of squares of the differences between the measured NO_X and calculated NO_X concentrations. A value of 0.15 for nitrifier yield, Y_{AUT} (Barker and Dold, 1997), was used, as were values of 0.17 d^{-1} for b_{AUT} and 1.072 for the μ_{AUT} temperature dependency coefficient.

36.2.2 Washout Test Results

Figure 36-2 shows an example of washout curves for nitrite and nitrate concentrations in test number 4. Within the first two days, the NO_X-N and nitrate concentrations increased and leveled off. Subsequently, they declined from Day 3 as the nitrifying bacteria were washed out of the reactor as a result of the low SRT of 1.1 days. In contrast, nitrite concentrations steadily increased up to Day 4, and then also declined as the ammonia-oxidizing bacteria were being washed out. The increase in nitrite concentration between Days 3 and 4 suggests that nitrite-oxidizing bacteria were also being washed out by the low system SRT.

The model fit yielded a $\mu_{AUT, 20 \text{ deg C}}$ value of 0.75 d^{-1} based on NO_X-N concentration data and a value of 0.71 based on nitrate concentration data only, a difference of 5 percent. $\mu_{AUT, 20 \text{ deg C}}$ values for all four experiments are summarized in Table 36-3. Because nitrite was not measured in the other three experiments, all $\mu_{AUT, 20 \text{ deg C}}$ values presented in Table 36-3 were based on nitrate concentrations only.

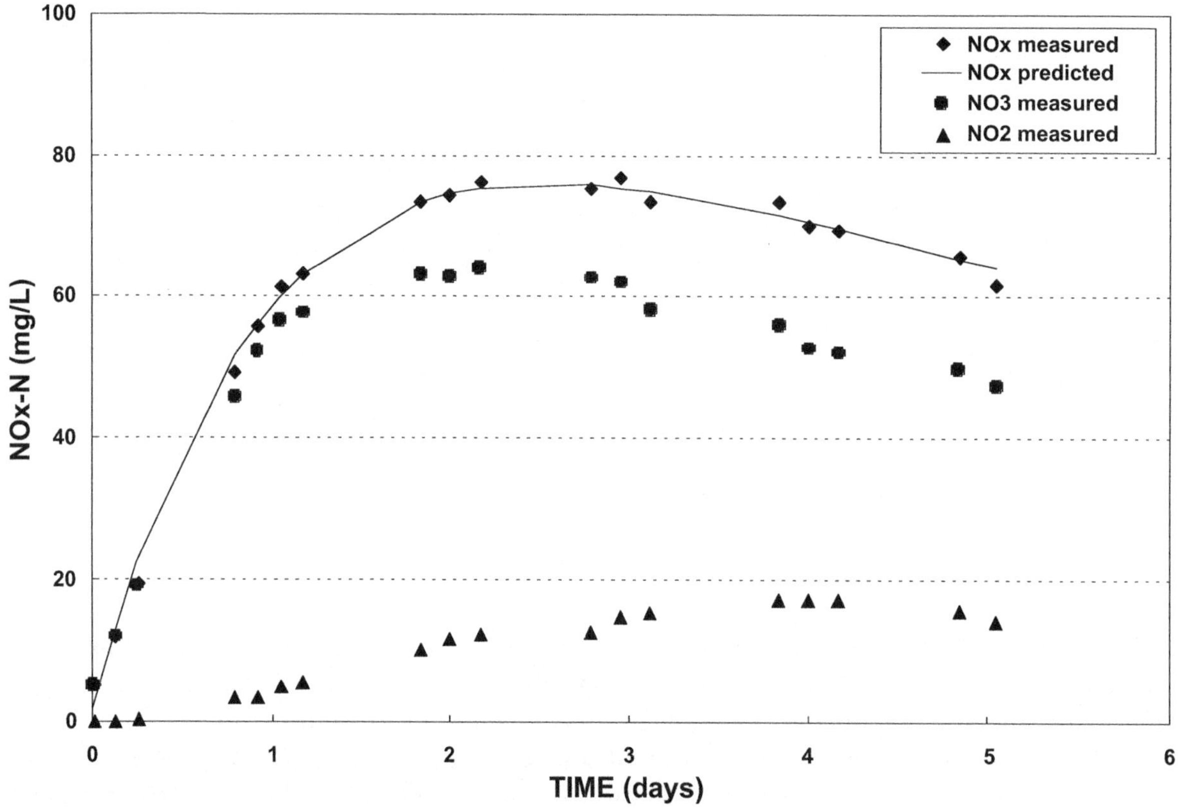

Figure 36-2. Washout Test Data for Test Number 4 with Puyallup WWTP Mixed Liquor and Secondary Effluent, SRT = 1.1 day and Temperature= 25°C

Table 36-3. Summary of Washout Test $\mu_{AUT,\ 20\ deg\ C}$ Values

Test No.	Date	Mixed Liquor Seed Source	Feed Wastewater	SRT (days)	$\mu_{AUT,\ 20\ deg\ C}$ (d^{-1})
1	1/22/01	Low F/M multi-batch fed SBR - Phase III	Renton secondary influent	1.6	0.66
2	2/9/01			1.6	0.61
3	2/21/01	Puyallup WWTP		1.6	0.68
4	6/28/01		Puyallup secondary effluent	1.1	0.71
Average					*0.67*
Std. Dev.					*0.04*

$\mu_{AUT,\ 20\ deg\ C}$ values based on nitrate concentrations only

Values of 0.66 and 0.61 d^{-1} were obtained for identical tests (tests 1 and 2) using seed from the low F/M multi-batch fed SBR mixed liquor and Renton secondary influent as feed wastewater. Almost identical values of 0.68 and 0.71 d^{-1} were measured in tests 3 and 4, with identical seed from Puyallup but different feed wastewater. The $\mu_{AUT,\ 20\ deg\ C}$ values in Table 36-3 were very similar for all tests, with an average value of 0.67 d^{-1} and a standard deviation of 0.04. These results suggest that consistent results can be obtained with the washout test method for different feed wastewaters and different seeds.

36.3 McMaster University Investigations

Two washout tests were conducted at McMaster University to develop and refine the method. The first test was performed on the McMaster University SBR that was run for this study. The second test was performed using seed sludge obtained from the Dundas, OntarioS WWTP. Experimental protocol and equipment follow that outlined in Appendix G (exceptions are noted below). The results of the runs are summarized in the following sections. Note that all data for the McMaster University investigations have been analyzed using a value of 0.17 d^{-1} (at 20°C) for the nitrifier decay rate (b_{AUT}) and a temperature dependency coefficient of 1.072 for μ_{AUT}. Nitrite, nitrate, and ammonia analyses for the McMaster University investigations were performed with the spectrophotometric methods noted in Appendix G.

36.3.1 Washout of the McMaster Low F/M SBR

Figure 36-3 shows the results for the first washout test commenced on 6/25/01. In this test, the McMaster low F/M SBR, which had been operated at steady state conditions for over 120 days, was washed out. Each day, 6 L of primary influent from the Dundas, Ontario WWTP (spiked with ammonia to achieve a concentration of approximately 50 mg/L) were pumped into the 6 L reactor. (Sufficient influent for the duration of the test was collected at the plant, transported to the laboratory and stored at 4°C. Each day of the test, a well-mixed representative feed sample was removed from storage, warmed to the laboratory temperature and added to the reactor feed container). Reactor effluent was pumped out at the same rate. Under these conditions, the reactor essentially was operated as a flow-through system with a 1 day SRT. Appropriate alkalinity was added to each feed batch. Although the fit to the experimental data is good, two problems were identified with this test run:

1. The ammonia response in the reactor differed from the expected theoretical response. It was hypothesized that the nitrifying organisms were substrate-limited at the beginning of the test due to low ammonia concentration in the reactor. It was decided that the reactor should be spiked with ammonia at the start of the next washout test.
2. Examination of Figure 36-3 shows that the NO_X-N response leveled off near the end of the test. It was hypothesized that the operating SRT of 1 day was not low enough. It was decided that the next washout reactor should be operated at a slightly lower SRT.

Analysis of the first washout test data indicates the test provided an estimate of μ_{AUT} (at 20°C) of 0.71 d^{-1}. This value is somewhat lower than μ_{AUT} estimates obtained using other methods (i.e., low F/M SBR, high F/M) in this study. However, this is attributable to the two experimental issues mentioned above.

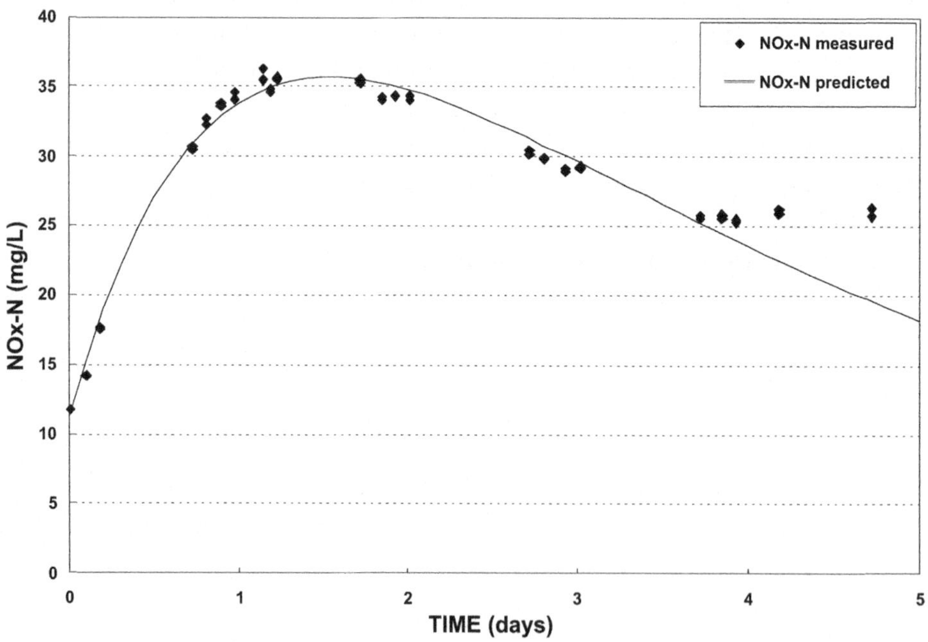

Figure 36-3. Nitrite/Nitrate-Time Response for McMaster University Washout Test Number 1 (Washout of Laboratory-Acclimated Sludge) Commenced on 6/25/01

36.3.2 Washout of Dundas WWTP Sludge

Figure 36-4 shows the results for the second washout test commenced on 9/10/01. In this test, a 6 L reactor containing sludge from the Dundas, Ontario WWTP at a MLSS concentration of approximately 1,500 mg/L was washed out. Each day, 8 L of primary influent from the Dundas, Ontario WWTP (spiked with ammonia to achieve a concentration of approximately 50 mg/L) were pumped into the reactor. (Sufficient influent for the duration of the test was collected at the plant, transported to the laboratory and stored at 4°C. Each day of the test, a well-mixed representative feed sample was removed from storage, warmed to the laboratory temperature and added to the reactor feed container). Reactor effluent was pumped out at the same rate. Under these conditions, the reactor essentially was operated as a flow-through system with a 0.75 day HRT. At the beginning of the test, the reactor was spiked with ammonia to overcome substrate limitation of the nitrifying organisms (as identified in the first McMaster University washout test). Appropriate alkalinity was added to each feed batch. Several points are worth noting for this washout run:

- The addition of ammonia to the reactor at the beginning of the test resulted in the expected reactor ammonia response.
- The NO_X-N response continued to decrease towards the end of the test. This indicates that the lower SRT of 0.75 days in this run was sufficient to properly wash out the nitrifying organisms.
- The decreased SRT in this run results in a more rapid NO_X-N peak response compared to the previous McMaster University washout test.

Analysis of the second washout test data indicates the test provided an estimate of μ_{AUT} (at 20°C) of 0.90 d^{-1}. This value agrees with μ_{AUT} estimates obtained using other methods (i.e., low F/M SBR, high F/M) in this study.

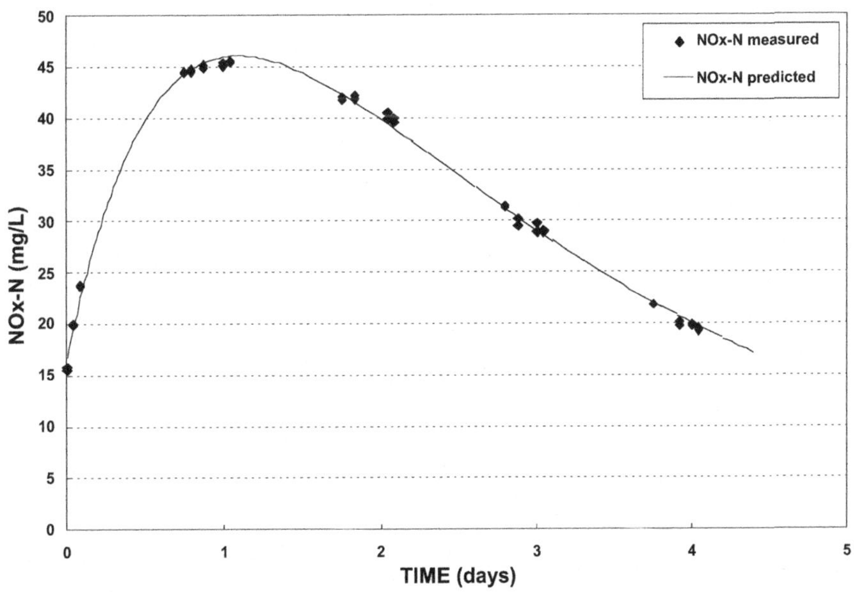

Figure 36-4. Nitrite/Nitrate-Time Response for McMaster University Washout Test Number 2 (Washout of Dundas WWTP Sludge) Commenced on 9/10/01

36.4 City of Phoenix Investigations

Several washout tests were conducted at the 91st Avenue Wastewater Treatment Plant in Phoenix, Ariz. to investigate the field applicability of the method. The results of these runs are summarized below. Note that all data for the City of Phoenix investigations have been analyzed using a value of 0.17 d^{-1} (at 20°C) for the nitrifier decay rate (b_{AUT}) and a temperature dependency coefficient of 1.072 for μ_{AUT}.

36.4.1 May 26, 2001

Figure 36-5 shows the results for the Phoenix washout test commenced on 5/26/01. In this test, a 6 L reactor containing sludge from the 91st Avenue plant at a MLSS concentration of approximately 1,600 mg/L was washed out. Each day, 6 L of primary influent from the 91st Avenue plant (spiked with ammonia to achieve a concentration of approximately 200 mg/L) were pumped into the reactor. Reactor effluent was pumped out at the same rate. Under these conditions, the reactor essentially was operated as a flow-through system with a 1 day HRT. At the beginning of the test, the reactor was spiked with ammonia to overcome substrate limitation of the nitrifying organisms. Appropriate alkalinity was added to each feed batch.

Figure 36-5 shows that there is considerable scatter in the data collected for this run. Also, note that insufficient data were collected around the time of the NO_X-N peak to properly capture the response dynamics.

Analysis of the test data indicates the test provided an estimate of μ_{AUT} (at 20°C) of 0.66 d^{-1}. This value is somewhat lower than other estimates obtained in this study, but likely this is attributable to the poor data quality.

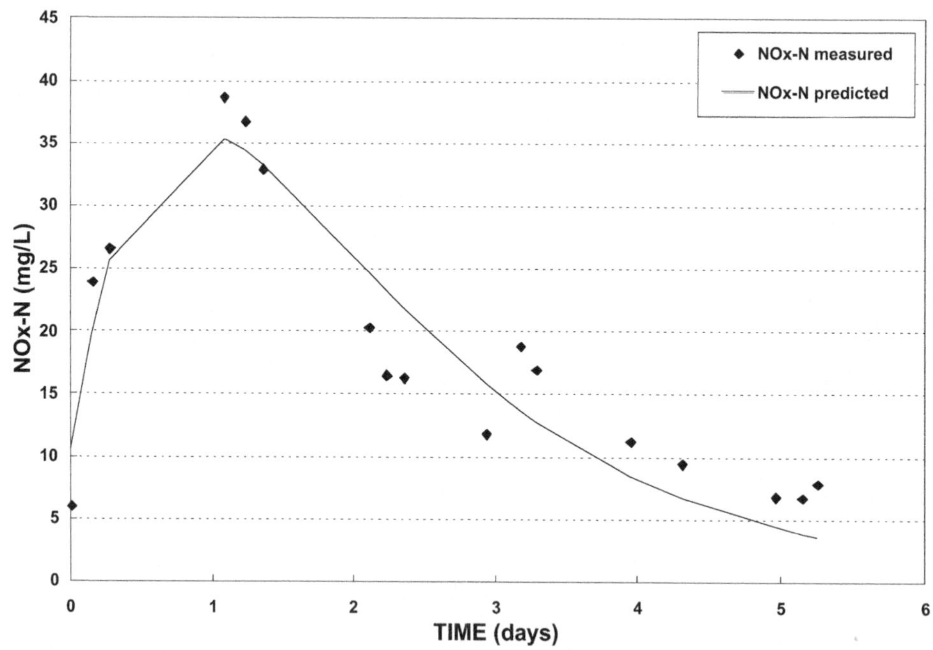

Figure 36-5. Nitrite/Nitrate-Time Response for Phoenix Washout Test Commenced on 5/26/01

36.4.2 September 17, 2001

Figure 36-6 shows the results for the Phoenix washout test commenced on 9/17/01. In this test, a 6 L reactor containing sludge from the 91st Avenue plant at a MLSS concentration of approximately 1,400 mg/L was washed out. Each day, 6.192 L of primary influent from the 91st Avenue plant (spiked with ammonia to achieve a concentration of approximately 40 mg/L) was pumped into the reactor. Reactor effluent was pumped out at the same rate. Under these conditions, the reactor essentially was operated as a flow-through system with a HRT slightly less than 1 day. At the beginning of the test, the reactor was spiked with ammonia to overcome substrate limitation of the nitrifying organisms. However, examination of ammonia concentration versus time data indicates that the reactor ammonia level became negligible after one day and remained low for the remainder of the test. This likely resulted in substrate-limited growth conditions, leading to the lower value (compared to the other two Phoenix washout runs) of μ_{AUT} (at 20°C) of 0.82 d^{-1} estimated for this test.

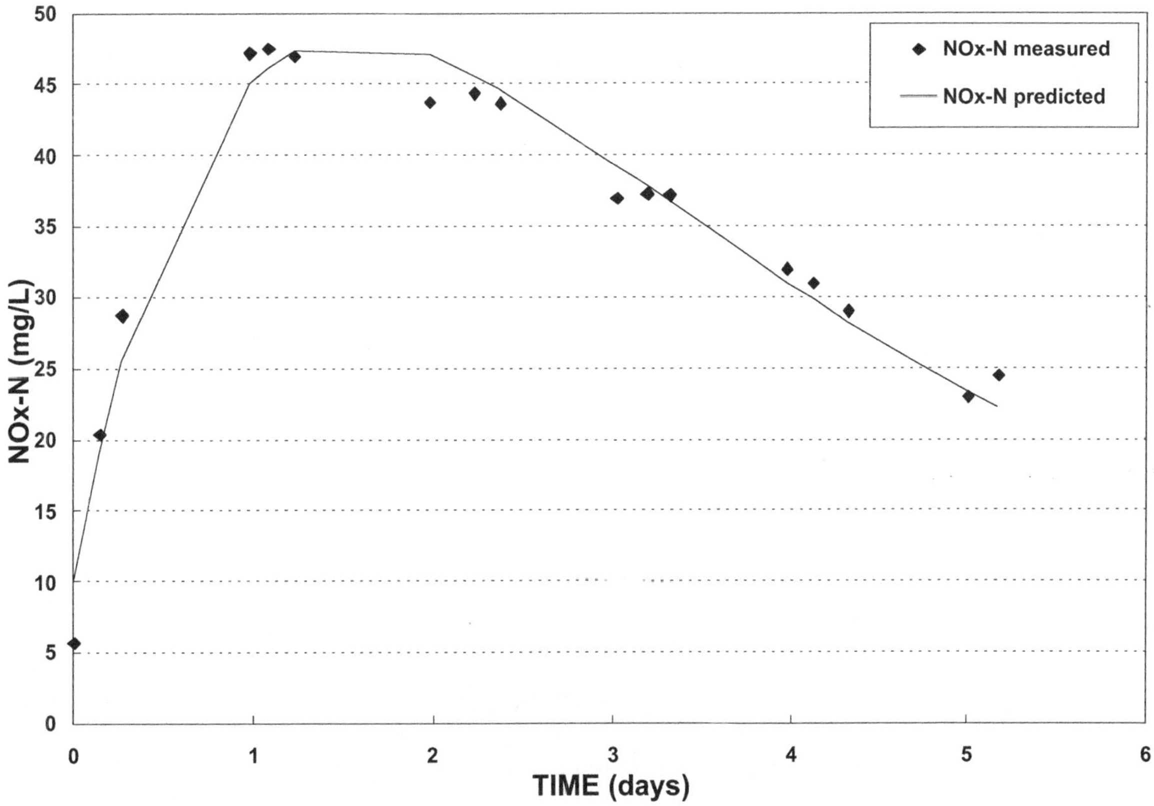

Figure 36-6. Nitrite/Nitrate-Time Response for Phoenix Washout Test Commenced on 9/17/01

36.4.3 September 24, 2001

Figure 36-7 shows the results for the Phoenix washout test commenced on 9/24/01. In this test, a 6 L reactor containing sludge from the 91st Avenue plant at a MLSS concentration of approximately 1,400 mg/L was washed out. Each day, 6.192 L of primary influent from the 91st Avenue plant (spiked with ammonia to achieve a concentration of approximately 40 mg/L) was pumped into the reactor. Reactor effluent was pumped out at the same rate. Under these conditions, the reactor essentially was operated as a flow-through system with a HRT slightly less than 1 day. At the beginning of the test, the reactor was spiked with ammonia to overcome substrate limitation of the nitrifying organisms. Appropriate alkalinity was added to each feed batch.

Analysis of the test data indicates the test provided an estimate of μ_{AUT} (at 20°C) of 1.02 d^{-1}.

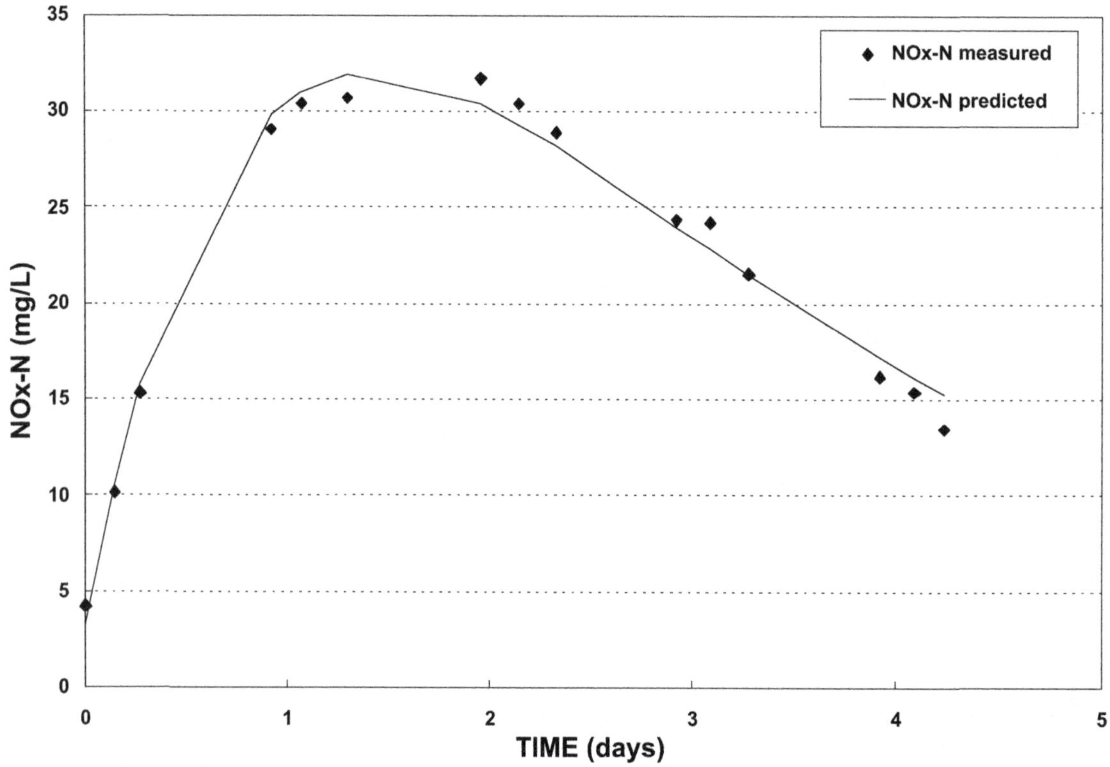

Figure 36-7. Nitrite/Nitrate-Time Response for Phoenix Washout Test Commenced on 9/24/01

36.4.4 October 1, 2001

Figure 36-8 shows the results for the Phoenix washout test commenced on 10/01/01. In this test, a 6 L reactor containing sludge from the 91st Avenue plant at a MLSS concentration of approximately 1,400 mg/L was washed out. Each day, 6.192 L of primary influent from the 91st Avenue plant (spiked with ammonia to achieve a concentration of approximately 40 mg/L) was pumped into the reactor. Reactor effluent was pumped out at the same rate. Under these conditions, the reactor essentially was operated as a flow-through system with a HRT slightly less than 1 day. At the beginning of the test, the reactor was spiked with ammonia to overcome substrate limitation of the nitrifying organisms. Appropriate alkalinity was added to each feed batch.

Analysis of the test data indicates the test provided an estimate of μ_{AUT} (at 20°C) of 1.17 d^{-1}.

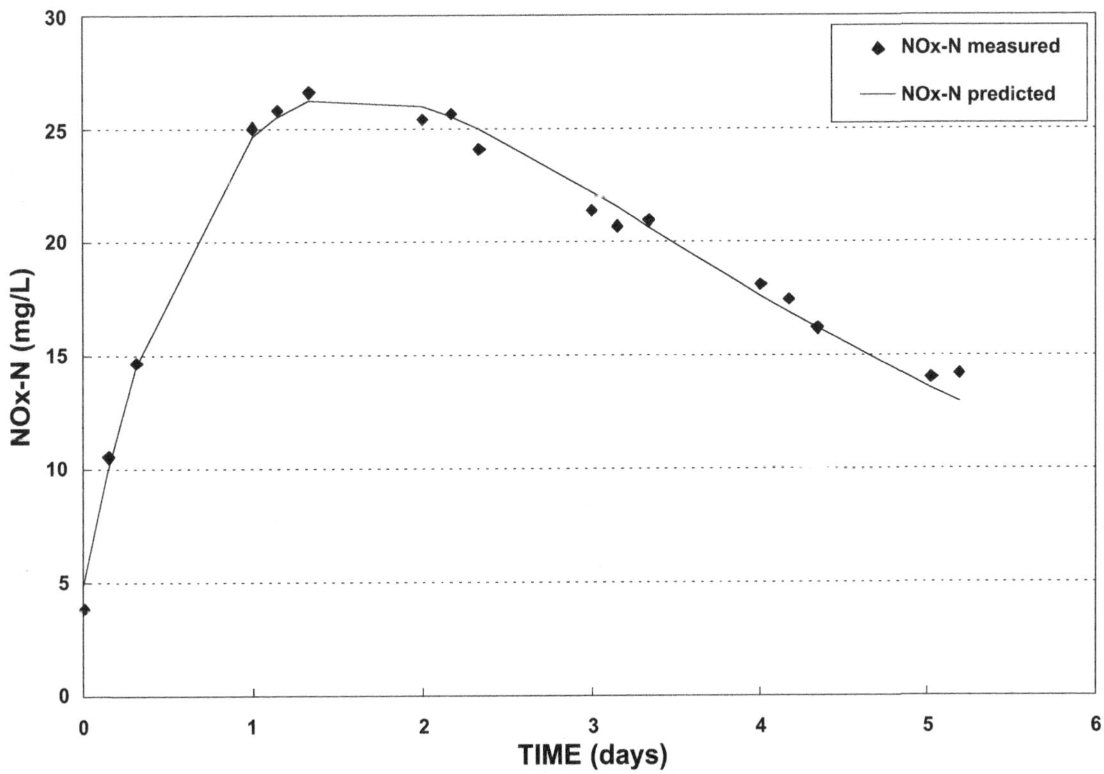

Figure 36-8. Nitrite/Nitrate-Time Response for Phoenix Washout Test Commenced on 10/01/01

36.5 References

Barker, P.S., and P.L. Dold. 1997. General model for biological nutrient removal activated sludge systems: model presentation. *Water Environ.Res.* 69(5), 969-984.

CHAPTER 37.0

NITRIFIER DECAY RATE: UNIVERSITY OF WASHINGTON INVESTIGATIONS

37.1 Background

As the project progressed, it became evident that the nitrifier decay rate had a more significant role in estimating the maximum specific nitrifier growth rate (μ_{AUT}) than had been anticipated. Although not part of the original design of experiments, advantage was taken of a series of experiments that were being conducted at the University of Washington environmental engineering research laboratory to measure the temperature dependency of μ_{AUT} by conducting a measure of nitrifier decay rate using the biomass from the same reactors. This work was conducted before the analysis of nitrifier decay rate described in Chapter 17.0 had been completed and so the method used to measure the nitrifier specific endogenous decay coefficient in this work differs somewhat from the recommendations in Chapter 17.0. The method was a modification of a method proposed by Katehis et al. (2002). This chapter describes the limited examination of nitrifier decay rate that was carried out.

37.2 Apparatus and Operating Conditions

Tests were done concurrently with the lLow F/M SBR nitrification kinetics tests described in Chapter 4.0 to determine the nitrification specific endogenous decay rate at temperatures of 12°C and 22°C. Two once-per-day, batch fed low F/M SBRs with the same 10-day SRT were operated at temperatures of 12°C and 22°C, respectively, to steady state (>3 SRTs). Subsequently, nitrification profile tests were performed to measure the nitrification rate for both SBRs to estimate μ_{AUT} at 12°C and 22°C, respectively. After completing the nitrification rate tests, the operation of the SBRs was changed to allow the reactor nitrifying bacteria concentration to decline by endogenous decay. The SBRs were aerated for five days without any feeding. Each day, a mixed liquor sample was removed to measure the nitrification rate. Nitrification rates declined progressively because of endogenous decay, which decreased the concentration of nitrifying bacteria. In all these tests, nitrification rate measurements were based on the rate of increase in NO_X-N concentration.

Microscopic observation of the reactor mixed liquor was conducted daily during the test to look for evidence of protozoan predation. No significant protozoan population was observed in either of the tests, which suggests that the decline in the nitrifying bacteria population was due solely to endogenous decay.

After each day of aeration, a 250 mL sample of mixed liquor was withdrawn from the reactor to measure its nitrification rate. The sample was placed in an aerated 300 mL flask, and spiked with ammonia to approximately 25 mg N/L. Samples were taken at 30 minute intervals for five hours and immediately filtered and analyzed for ammonia, nitrite and nitrate concentrations. The nitrification rate was calculated from the rate of increase in the sum of nitrite and nitrate nitrogen concentration. The analysis used to determine the decay rate coefficient considered the additional nitrifier bacteria growth occurring in the aerated SBRs due to ammonia released from the decay of heterotrophic bacteria.

37.3 Data Analysis

An estimate of the nitrifying bacteria concentration at the beginning of the decay rate test aeration period was determined from the Low F/M SBR operating data using equation 16.4.2-1 in Chapter 16.0 as described in Chapter 34.0.

The increase in nitrifying bacteria due to growth from ammonia release can be calculated from the amount of NO_X produced each day and an average net decay rate.

$$\Delta X_{AUT,t} = \frac{Y_{AUT} \cdot \Delta NO_{x,t}}{1 + b_{AUT} \cdot (dt/2)} \qquad (37.3\text{-}1)$$

where:

$\Delta NO_{x,t}$ = $NO_{x,t} - NO_{x,t-1}$ = increase in NO_x over time dt (mg/L)

Y_{AUT} = nitrifying bacteria yield coefficient (0.12 g VSS/g NO_x produced)

ΔX_{AUT} = increase in nitrifying bacteria concentration over time dt (mg/L)

b_{AUT} = endogenous decay rate coefficient, d^{-1}

dt = time interval between each batch rate test (~1 day)

By combining equations 16.4.2-1 and 37.3-1, the following expression for the nitrifier bacteria concentration at time t is obtained:

$$X_{AUT,t} = X_{AUT,t-1} e^{-b_{AUT} \cdot dt} + \frac{Y_{AUT} \cdot \Delta NO_{x,t}}{1 + b_{AUT} \cdot (dt/2)} \qquad (37.3\text{-}2)$$

Dividing both side of the equation by $X_{AUT,t-1}$

$$\frac{X_{AUT,t}}{X_{AUT,t-1}} = e^{-b_{AUT} \cdot dt} + \frac{Y_{AUT} \cdot \Delta NOx,t}{(1+b_{AUT} \cdot (dt/2)) \cdot X_{AUT,t-1}} \quad (37.3\text{-}3)$$

The ratio of the nitrification rates for spiked mixed liquor samples at different times should be directly proportional to the nitrifying bacteria concentration ratio in Eq. 37.3-3.

$$\frac{NPR_t}{NPR_{t-1}} = \frac{X_{AUT,t}}{X_{AUT,t-1}} \quad (37.3\text{-}4)$$

where:

NPR_t = Nitrification rate measured at time t

hence:

$$\frac{NPR_t}{NPR_{t-1}} = \frac{X_{AUT,t}}{X_{AUT,t-1}} = e^{-b_{AUT} \cdot dt} + \frac{Y_{AUT} \cdot \Delta NOx,t}{(1+b_{AUT} \cdot (dt/2)) \cdot X_{AUT,t-1}} \quad (37.3\text{-}5)$$

The difference between the nitrification rate ratio and the right side of Eq. 37.3-5 is equal to a factor R_t, which would be zero if a correct value for b_{AUT} is known.

$$R_t = \frac{NPR_t}{NPR_{t-1}} - e^{-b_{AUT} \cdot dt} - \frac{Y_{AUT} \cdot \Delta NO_{x,t}}{(1+b_{AUT} \cdot (dt/2)) \cdot X_{AUT,t-1}} \approx 0 \quad (37.3\text{-}6)$$

By applying the Excel solver function, the decay rate b_{AUT} can be determined from a best fit that minimizes the sum of the squares of the value R_t for selected b_{AUT} values. A value of 0.15 for Y_{AUT} (Barker and Dold, 1997) was used to calculate b_{AUT}. The b_{AUT} values at 12°C and 22°C were used to develop a temperature coefficient for the endogenous decay rate as follows:

$$b_{AUT,12^\circ C} = b_{AUT,22^\circ C} \cdot (\theta)^{12-22} \quad (37.3\text{-}7)$$

where:

θ = temperature correction coefficient

37.4 Washout Test Results

37.4.1 Estimate of b_{AUT} Accounting for Supplementary Nitrifier Growth

Table 37-1 summarizes the nitrification rates in the daily batch tests using mixed liquor from the Low F/M SBRs operated at 12°C and 22°C. The steady decline in both reactors is indicative of a steady loss of nitrifier bacteria activity. The daily increases in the SBR NO_x-N concentrations are also shown, indicating the magnitude of ammonia release used for new nitrifier growth. b_{AUT} values at 12°C and 22°C were calculated using equation 37.3-6 and the data analysis procedure described above, which yielded values of 0.14 and 0.19 d^{-1}, respectively.

Table 37-2 shows the application of the 12°C test data to fit b_{AUT}. The fit to equation 37.3-6 includes the effect of new growth due to hydrolysis and ammonia release in the aerated reactors.

37.4.2 Estimate of Temperature Dependency for b_{AUT}

Using the 12°C and 22°C b_{AUT} values and equation 37.3-7, the specific endogenous decay temperature coefficient was estimated to be 1.03 and b_{AUT} referred to 20°C was 0.17 d^{-1}.

Table 37-1. Summary of Nitrifier Decay Rate Test Results

SBR temp. (°C)	Aerated Time (days)	Daily batch nitrification rate (NPR*) (mg N/L-hr)	SBR		
			NO_x-N	MLSS	MLVSS
			Concentration (mg/L)		
	0.0	3.33	12.9	1247	941
	1.0	3.28	15.4	1290	965
	2.0	2.66	20.4	1255	885
	2.9	2.19	25.5	1195	850
	3.9	1.89	29.3	1235	845
22	4.9	1.80	34.6	1155	785
	0	2.32	17.5	1130	850
	0.96	2.31	20.2	1160	845
	1.96	2.10	21.8	1195	815
	2.96	1.78	22.2	1155	815
	3.96	1.53	25.2	1190	820
12	4.96	1.31	29.4	1100	800

Table 37-2. Estimation of Nitrifer Decay Rate for 12°C Test Data

Day (t)	dt	NPR_t (mg/L-hr)	NPR_t/ NPR_{t-1}	NO_x-N_t (mg/L)	$\Delta NO_{x,t}$	$X_{AUT,t}$	$b_{AUT, 12 deg C}$	R_t	$(R_t)^2$
0	0.00	2.32	1.000	17.5	0.000	11.915	0.143		
0.96	0.96	2.31	0.996	20.2	2.700	10.771		0.092	0.008
2	1.00	2.10	0.909	21.8	1.600	9.563		0.021	0.000
3	1.00	1.78	0.848	22.2	0.400	8.347		-0.025	0.001
4	1.00	1.53	0.860	25.2	3.000	7.656		-0.058	0.003
5	1.00	1.31	0.856	29.4	4.200	7.226		-0.088	0.008
Sum:									*0.021*

37.4.3 Estimate of b_{AUT} Using Method by Katehis et al.

Using the method by Katehis et al. (2001), the nitrifier specific endogenous decay rate is determined from the slope of the logarithmic nitrification rate plot and the resultant b_{AUT} values are 0.15 d^{-1} at 22°C and 0.12 d^{-1} at 12°C. The semi-log plots are shown in Figures 37-1 and 37-2. This method does not consider new growth and thus predicts a lower endogenous decay rate by 15 percent to 20 percent.

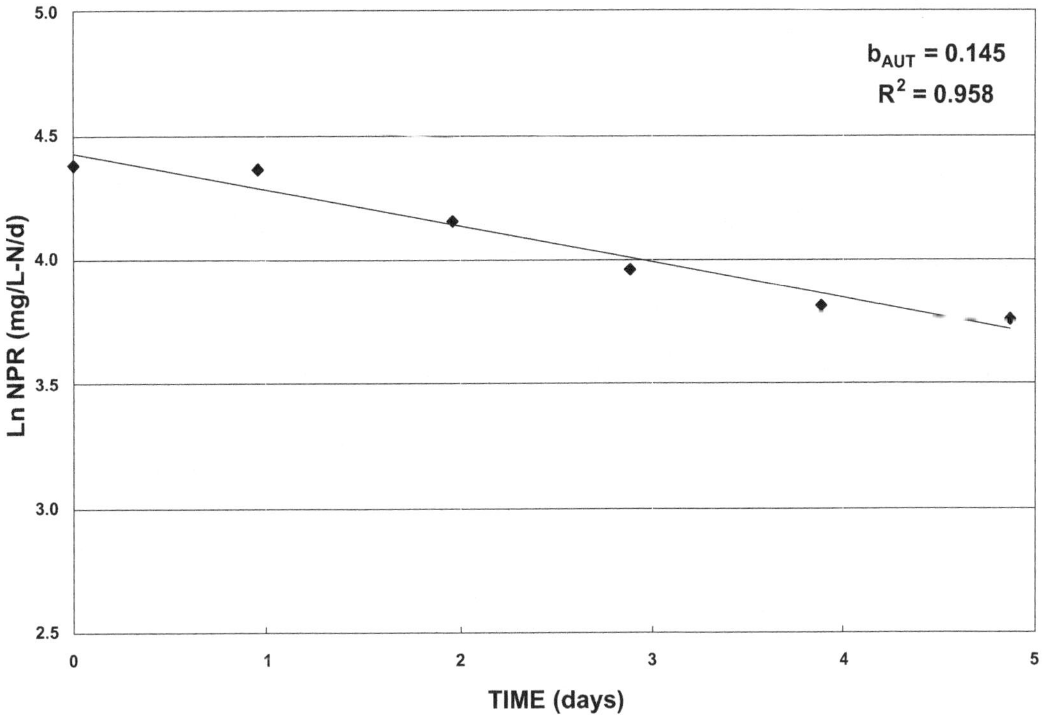

Figure 37-1. Semi-Log Plot of Specific Endogenous Nitrifier Decay Rate at 22°C

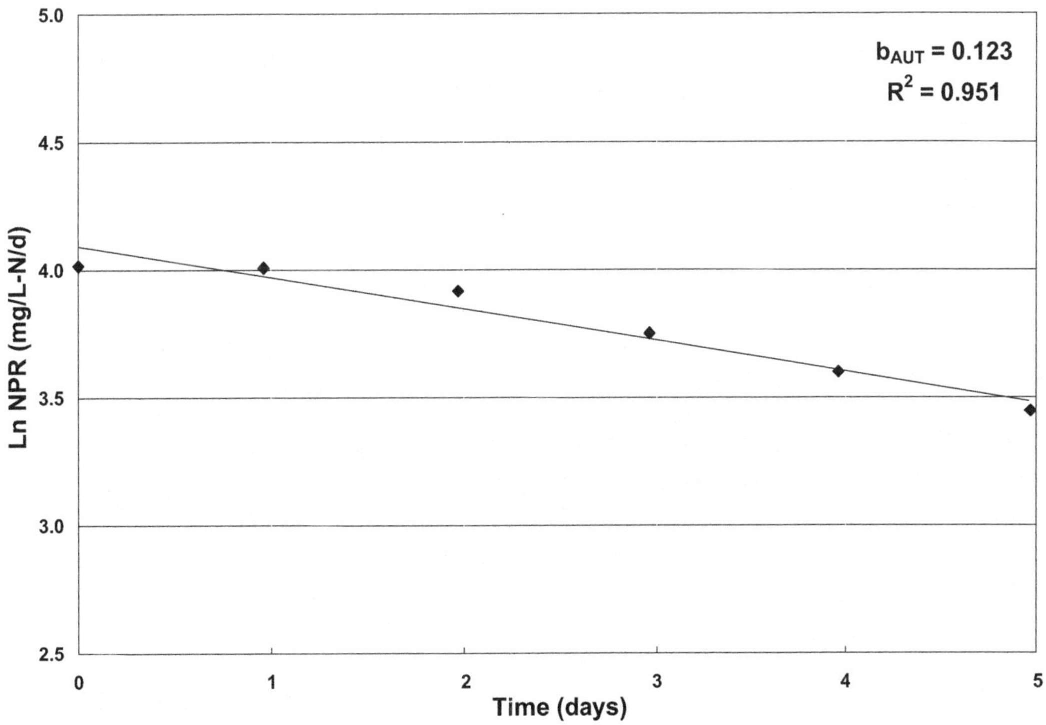

Figure 37-2. Semi-Log Plot of Specific Endogenous Nitrifier Decay Rate at 12°C

37.4.4 Comparison with Established Literature Values

The specific endogenous decay rate coefficient values obtained in these tests are higher than normally used in nitrifying activated sludge design, but the temperature coefficient is similar to the default value in IWA activated sludge models.

The default b_{AUT} value at 0.04 d^{-1} and the EPA *Nitrogen Control Manual* (1993) recommends that the decay values be ignored. However, other reports from nitrification kinetic testing support b_{AUT} values of this magnitude. Copp and Murphy (1995) reported $b_{AUT,\ 20\ deg\ C}$ values in the range of 0.14 to 0.17 d^{-1} for studies with nitrifier enrichments. Katehis et al. (2000), using a similar experimental procedure as this one to study nitrifying bacteria endogenous decay, reported variable b_{AUT} values ranging from 0.09 to 0.11 d^{-1} for bench-scale reactors operated at different SRTs and for aerobic only versus anoxic/aerobic process conditions. This issue is discussed in detail in Chapter 17.0.

37.5 Impact of b_{AUT} on μ_{AUT}

The importance of b_{AUT} is illustrated in Table 37-3, which compares μ_{AUT} values calculated for data collected in the low F/M SBR experiments described in Chapter 4.0 using the b_{AUT} default value of 0.04 d^{-1} versus the 0.17 d^{-1} obtained in this work. Much lower μ_{AUT} values are obtained when a lower b_{AUT} value is assumed. At lower b_{AUT} values, a higher nitrifier biomass concentration is predicted for a given operating SRT in the low F/M SBR tests. Higher biomass levels result in the prediction of a lower specific nitrification rate and thus a lower μ_{AUT} value.

Table 37-3. Comparison of Calculated Average $\mu_{AUT,\ 20\ deg\ C}$ Values Using Default and Experimentally Determined Nitrifier Decay Rates

SBR Feed Method	SRT (day)	$\mu_{AUT,20\ deg\ C}$ (d^{-1}) for b_{AUT} = 0.04 /d	$\mu_{AUT,20\ deg\ C}$ (d^{-1}) for b_{AUT} = 0.17 /d
Batch fed – Phase I	12.7	0.36	0.75
Batch fed – Phase II	10.1	0.34	0.66
Semi-continuous fed – Phase II	10.1	0.34	0.66
Multi-batch fed – Phase III	8.2	0.39	0.71
Average		0.36	0.70

Previous literature values on nitrification kinetics that report only μ_{AUT} values, that is, without corresponding b_{AUT} values, do not provide useful data to accurately predict nitrification performance. Because few b_{AUT} values have been provided in published literature on μ_{AUT} values, the usefulness of the existing literature database on nitrification kinetics may be limited. Both values are necessary when reporting nitrification kinetics and must be applied together in process design.

The following example illustrates the effect of b_{AUT} value on process performance prediction for a completely mixed activated sludge (CMAS) process. The ammonia concentration can be calculated as a function of SRT, b_{AUT} and μ_{AUT} values from:

$$S_{NH,eff} = \frac{K_{NH}(1+b_{AUT}\theta)}{(\mu_{AUT}+b_{AUT})\theta - 1} \quad (37.5\text{-}1)$$

where:

K_{NH} = half-saturation concentration (mg/L)

$S_{NH,eff}$ = CMAS effluent ammonia concentration (mg/L)

Table 37-4 summarizes the calculated effluent ammonia concentrations for three sets of b_{AUT} and μ_{AUT} values. The first two columns show the effect of using b_{AUT} values of 0.17 or 0.04 d^{-1} with their respective μ_{AUT} values calculated from this work, as shown in Table 37-3. The third column uses the recommended b_{AUT} and μ_{AUT} values in the EPA *Nitrogen Control Manual* (1993). A K_{NH} value of 0.75 mg/L (Barker and Dold, 1997) was used in these calculations.

Table 37-4. Predicted Effluent Ammonia Concentrations versus SRT for a CMAS Process as a Function of b_{AUT} and μ_{AUT} Values at 20°C

	Effluent NH4-N (mg/L)		
SRT (day)	b_{AUT} = 0.04 d^{-1} μ_{AUT} = 0.36 d^{-1}	b_{AUT} =0.17 d^{-1} μ_{AUT} = 0.70 d^{-1}	b_{AUT} = 0 d^{-1} μ_{AUT} = 0.5 d^{-1}
2.0	Washout	16.8	Washout
3.0	Washout	1.9	1.5
3.5	7.1	1.4	1.0
4.0	3.1	1.1	0.8
4.5	2.0	1.0	0.6
5.0	1.5	0.8	0.5
6.0	1.0	0.7	0.4
10.0	0.48	0.47	0.19
20.0	0.25	0.34	0.08

Table 37-4 shows that, for a given effluent ammonia concentration, use of the higher b_{AUT} value, which was obtained experimentally in this work, allows a lower design SRT or predicts a greater nitrification capacity for SRT values below 10 days. At an SRT of about 10 days, the SRT used in the low F/M SBR experiments, the effect of b_{AUT} is negligible. Thus, if the low F/M SBR test reactor is operated at the same design SRT expected for the full-scale plant, the error in selecting an incorrect b_{AUT} value is not significant. However, the kinetic data obtained have limited application for evaluating a range of SRT operation for a full-scale system. The table also shows that at SRTs greater than 10 days, a larger nitrification capacity is predicted for the system design based on using the lower b_{AUT} value. In this case, the error is not very significant.

The third column in Table 37-4 shows the CMAS effluent ammonia concentration using b_{AUT} and μ_{AUT} kinetic values in the EPA *Nitrogen Control Manual* (1993). In this case, the nitrification capacity predicted is greater than that predicted by the kinetics derived in the low F/M SBR tests.

37.6 References

Barker, P.S., and P.L. Dold. 1997. General model for biological nutrient removal activated sludge systems: model presentation. *Water Environ. Res.* 69(5):969-984.

Copp, J.B., and K.L. Murphy. 1995. Estimation of the Active nitrifying biomass in activated sludge. *Water Res.* 29:1855-1862.

Katehis, D., J. Fillos, and L.A. Carrio. 2002. Comparison of bench scale testing methods for nitrifier growth rate measurement. *Water Sci. Tech.* 46(1-2):289-295.

U.S. EPA. 1993. *Manual-Nitrogen Control.* EPA/625/R-93/010. Washington, DC: Office of Technology Transfer, U.S. Environmental Protection Agency.

CHAPTER 38.0

COMPARISON OF μ_{AUT} RESULTS FROM DIFFERENT METHODS

38.1 Introduction

In this study, nitrifier maximum specific growth rate (μ_{AUT}) was estimated using three different methods:

- Low F/M SBR,
- High F/M, and
- Washout.

A series of investigations were carried out at the McMaster, Phoenix and University of Washington locations to develop and refine these methods (especially the high F/M and washout methods). The purpose of this chapter is to compare estimated μ_{AUT} values across methods for each location.

Note: In assessing the results presented in this chapter, careful attention should be paid to whether estimates of μ_{AUT} values are based on fits to nitrate-only response data or to nitrite+nitrate response data. Estimates based on nitrate-only data will underestimate the μ_{AUT} values. Early in the project, nitrite was not monitored. However, during the project it became evident that nitrite accumulation in the various tests may be significant, in particular in high F/M and washout tests, and to a lesser extent in low F/M SBR tests. At that point nitrite monitoring commenced.

38.2 University of Washington μ_{AUT} Values

Low F/M SBR experiments were conducted under several different operating regimes and used to estimate μ_{AUT} values. Ten high F/M tests were conducted at the University of Washington to aid in development and refinement of the method. Four washout tests also were conducted. The μ_{AUT} values estimated from the various experiments are summarized in the following sections.

38.2.1 μ_{AUT} Estimated from Low F/M SBR Method

Values of μ_{AUT} were estimated using equation 16.4.2-1 in Chapter 16.0. Table 38-1 summarizes μ_{AUT} values from the different experiments. The key experimental condition also is noted for each estimate. Note that values of the specific nitrifier decay rate and its temperature dependency coefficient used in this calculation were 0.17 d^{-1} at 20°C and 1.029, respectively as described in Chapter 16.0, as well as the μ_{AUT} temperature dependency coefficient of 1.072.

Table 38-1. Summary of μ_{AUT} Estimates (Corrected to 20°C) Obtained from University of Washington Low F/M SBR Experiments

Experimental Condition	μ_{AUT} Estimate (d^{-1})
Phase I Batch-Fed SBR	0.77
Phase II Batch-Fed SBR	0.63
Phase II Semi-Continuous-Fed SBR	0.64
Phase III Multi-Batch-Fed SBR	0.72
SBR Temperature of 22°C	0.92*
SBR Temperature of 12°C	0.91*

* – tests based on nitrite and nitrate monitoring

The average μ_{AUT} values at 22°C and 12°C were estimated to be 1.06 and 0.52 d^{-1}, respectively. A μ_{AUT} temperature correction factor of 1.072 was estimated from the average μ_{AUT} values determined for the 22°C and 12°C SBRs using Equation 16.4.1-4 in Chapter 16.0. This gives μ_{AUT} values (corrected to 20°C) for the 22°C and 12°C SBRs of 0.92 and 0.91 d^{-1}, respectively.

38.2.2 μ_{AUT} Estimated from High F/M Method

Values of μ_{AUT} were estimated by fitting to equation 16.6.1-4 in Chapter 16.0 using the parameter estimating procedure described in Appendix H. Table 38-2 summarizes μ_{AUT} values from the different experiments. The key experimental condition also is noted for each estimate. Note that a value of 0.15 for nitrifier yield (Y_{AUT}) was used in the parameter fitting exercise as were values of 0.17 d^{-1} for b_{AUT} and 1.072 for the μ_{AUT} temperature dependency coefficient.

An important point to note regarding the University of Washington high F/M experimental results is that nine of the ten experiments do not account for any nitrite that may have accumulated in the tests. The importance of measuring nitrite became evident during the McMaster University experiments, which took place after the University of Washington experiments. This likely is the reason that μ_{AUT} values estimated in the University of Washington high F/M experiments generally tend to be lower than those estimated at other locations (and with other methods).

Table 38-2. Summary of μ_{AUT} Estimates (Corrected to 20°C) Obtained from University of Washington High F/M Experiments

Seed Source	Objective	Diluent	μ_{AUT} Estimate (d^{-1}) (fit with NO_3 only)	μ_{AUT} Estimate (d^{-1}) (fit with NO_3+NO_2)
Batch Fed SBR-I	Evaluation of test reproducibility	9/1/00 Renton 2nd Effluent	0.81	
		9/1/00 Renton 2nd Effluent	0.82	
	Evaluation of test reproducibility (same seed source but different sample date)	9/14/00 Renton 2nd Effluent	0.72	
		9/15/00 Renton 2nd Effluent	0.66	
	Evaluation of initial ammonia concentration with the same seed and diluent	10/6/00 Renton 2nd Effluent	0.64	
		10/6/00 Renton 2nd Effluent	0.63	
Batch Fed SBR-II	Evaluation of Hi/Lo initial NH4-N with the same seed and diluent	11/20/00 Renton 2nd Effluent	0.73	
		11/20/00 Renton 2nd Effluent	0.71	
Puyallup Plant	Evaluation of seed source from a full scale plant	2/21/01 Renton 2nd Influent	0.73	
	Evaluation of nitrite production measurement on _AUT value observed	6/27/01 Puyallup 2nd Effluent	0.75	0.86

38.2.3 μ_{AUT} Estimated from Washout Method

Values of μ_{AUT} were estimated by fitting to equation 16.7.1-2 in Chapter 16.0 using the parameter estimating procedure described in Appendix H. Table 38-3 summarizes μ_{AUT} values from the different experiments. Seed sludge source and feed wastewater also are noted for each estimate. Note that a value of 0.15 for nitrifier yield (Y_{AUT}) was used in the parameter fitting exercise as were values of 0.17 d^{-1} for b_{AUT} and 1.072 for the μ_{AUT} temperature dependency coefficient.

Table 38-3. Summary of μ_{AUT} Estimates (Corrected to 20°C) Obtained from University of Washington Washout Experiments

Seed Source	Test Start Date	Wastewater Feed Source	μ_{AUT} Estimate (d^{-1}) (fit with NO_3 only)	μ_{AUT} Estlmate (d^{-1}) (fit with NO_3+NO_2)
Low F/M multi-batch fed SBR - Phase III	1/22/01	Renton secondary influent	0.66	-
	2/9/01	Renton secondary influent	0.61	-
Puyallup Plant	2/21/01	Renton secondary influent	0.69	-
	6/28/01	Puyallup secondary effluent	0.71	0.75

An important point to note regarding the University of Washington washout results is that three of the four experiments do not account for any nitrite that may have accumulated in the tests. The importance of measuring nitrite became evident during the McMaster University experiments, which took place after the University of Washington experiments. This likely is the reason that μ_{AUT} values estimated in the University of Washington washout experiments generally tend to be lower than those estimated at other locations (and with other methods).

38.3 McMaster University μ_{AUT} Values

One low F/M SBR experiment was conducted to establish a baseline estimate for μ_{AUT}. Ten high F/M tests were conducted to aid in development and refinement of the method. Two washout tests also were conducted. The μ_{AUT} values estimated from the various experiments are summarized in the following sections.

38.3.1 μ_{AUT} Estimated from Low F/M SBR Method

A value of μ_{AUT} was estimated using the low F/M SBR protocol outlined in Appendix D. The μ_{AUT} value estimated was 0.90 d^{-1} (corrected to 20°C).

38.3.2 μ_{AUT} Estimated from High F/M Method

Values of μ_{AUT} were estimated by fitting to equation 16.6.1-4 in Chapter 16.0 using the parameter estimating procedure described in Appendix H. Table 38-4 summarizes μ_{AUT} values from the different experiments. The key experimental condition also is noted for each estimate. Note that a value of 0.15 for nitrifier yield (Y_{AUT}) was used in the parameter fitting exercise as were values of 0.17 d^{-1} for b_{AUT} and 1.072 for the μ_{AUT} temperature dependency coefficient.

Table 38-4. Summary of μ_{aut} Estimates (Corrected to 20°C) Obtained from McMaster University High F/M Experiments

Seed Source	Objective	Diluent	μ_{AUT} Estimate (d^{-1})
McMaster SBR	Evaluation of test reproducibility	McMaster SBR decant	0.91 0.90
McMaster SBR	Evaluation of influent inhibition	Dundas WWTP raw influent McMaster SBR decant	0.82 0.89
Dundas WWTP	Evaluation of seed sludge source	Dundas WWTP raw influent Dundas WWTP effluent	0.88 0.83
Dundas WWTP	Evaluation of alkalinity addition method	Dundas WWTP effluent	0.80 (spike) 0.75 (intermittent)
Dundas WWTP	Evaluation of aeration method	Dundas WWTP effluent	0.50 (constant) 0.83 (on/off)

All of the high F/M experiments at McMaster University accounted for both nitrite and nitrate accumulation during the tests. The majority of the μ_{AUT} values estimated in McMaster University high F/M experiments lie within the 0.80 - 0.90 d^{-1} range. (The one low value of 0.50 d^{-1} was observed in the test with pH control problems.)

38.3.3 μ_{AUT} Estimated from Washout Method

Values of μ_{AUT} were estimated by fitting to equation 16.7.1-2 in Chapter 16.0 using the parameter estimating procedure described in Appendix H. Table 38-5 summarizes μ_{AUT} values from the different experiments. Seed sludge source and feed wastewater also is noted for each estimate. Note that a value of 0.15 for nitrifier yield (Y_{AUT}) was used in the parameter fitting exercise as were values of 0.17 d^{-1} for b_{AUT} and 1.072 for the μ_{AUT} temperature dependency coefficient.

Table 38-5. Summary of μ_{AUT} Estimates (Corrected to 20°C) Obtained from McMaster University Washout Experiments

Seed Source	Test Start Date	Wastewater Feed Source	μ_{AUT} Estimate (d^{-1})
McMaster SBR	6/25/01	Dundas WWTP raw influent (ammonia limitation problem)	0.71
Dundas WWTP	9/10/01	Dundas WWTP raw influent	0.90

All of the washout experiments at McMaster University accounted for both nitrite and nitrate accumulation during the tests. Note that in the first McMaster University washout test, two experimental issues were identified as causes for the low estimate of μ_{AUT}:

1. Low ammonia concentration in the reactor at the start of the test resulted in substrate limited growth conditions.
2. The operating SRT was not sufficiently low.

38.4 City of Phoenix μ_{AUT} Values

One low F/M SBR experiment was conducted to estimate μ_{AUT}. Several high F/M tests were conducted to ascertain field applicability of the method; however, only one (conducted on May 26, 2001) yielded satisfactory results (data collection problems occurred in other tests conducted at Phoenix). Several washout tests also were conducted; however, only three yielded satisfactory results due to data collection problems. The μ_{AUT} values estimated from the various experiments are summarized in the sections below.

38.4.1 μ_{AUT} Estimated from Low F/M SBR Method

A value of μ_{AUT} was estimated using the low F/M SBR protocol outlined in Appendix D. Data and analysis details may be found in Chapter 33.0. The μ_{AUT} value estimated was 0.95 d^{-1} (corrected to 20°C).

38.4.2 μ_{AUT} Estimated from High F/M Method

A value of μ_{AUT} was estimated by fitting data from the good Phoenix high F/M test (conducted on May 26, 2001) to equation 16.6.1-4 in Chapter 16.0 using the parameter estimating procedure described in Appendix H. Note that a value of 0.15 for nitrifier yield (Y_{AUT}) was used in the parameter fitting exercise as were values of 0.17 d^{-1} for b_{AUT} and 1.072 for the μ_{AUT} temperature dependency coefficient. The μ_{AUT} value estimated was 0.81 d^{-1} (corrected to 20°C).

38.4.3 μ_{AUT} Estimated from Washout Method

Values of μ_{AUT} were estimated by fitting to equation 16.7.1-2 in Chapter 16.0 using the parameter estimating procedure described in Appendix H. Table 38-6 summarizes μ_{AUT} values from the different experiments. Seed sludge source and feed wastewater also is noted for each estimate. Note that a value of 0.15 for nitrifier yield (Y_{AUT}) was used in the parameter fitting exercise as were values of 0.17 d^{-1} for b_{AUT} and 1.072 for the μ_{AUT} temperature dependency coefficient.

Table 38-6. Summary of μ_{AUT} Estimates (Corrected to 20°C) Obtained from Phoenix Washout Experiments

Seed Source	Test Start Date	Wastewater Feed Source	μ_{AUT} Estimate (d^{-1})
Phoenix 91st Ave. WWTP	9/17/01	Phoenix 91st Ave. WWTP	0.82
Phoenix 91st Ave. WWTP	9/24/01	Phoenix 91st Ave. WWTP	1.02
Phoenix 91st Ave. WWTP	10/01/01	Phoenix 91st Ave. WWTP	1.17

All of the washout experiments at Phoenix accounted for both nitrite and nitrate accumulation during the tests. As discussed in the Summary of Washout Investigations chapter, the low value of 0.82 d^{-1} likely is due to the low ammonia concentrations reached during the test. The runs conducted on 9/24/01 and 10/01/01 resulted in estimates that were statistically different. This may be due to a time-dependent difference in the plant influent, however, at this stage this is speculative.

CHAPTER 39.0

BACKGROUND ON MEASURING RBCOD

39.1 Background

The biodegradable organic material in municipal wastewater is divided into a readily biodegradable fraction and a slowly biodegradable fraction. The subdivision is based on the dynamic response observed in activated sludge systems (Dold et al., 1980) and relates to the bi-substrate modeling approach that identifies there is approximately an order of magnitude difference between the degradation rates of the two fractions. That is, the division is a biokinetic one.

The readily biodegradable fraction is hypothesized to consist of low molecular weight material that can be absorbed readily by the organism and metabolized for energy and synthesis. Organic compounds that can be metabolized directly are low molecular weight compounds such as volatile fatty acids (VFAs), carbohydrates, alcohols, peptones and amino acids. Table 39-1 contains an estimate of the typical content of these compounds in European raw municipal wastewater (adapted from Henze, 1992).

Table 39-1. Example of Breakdown of RBCOD Components in European Raw Municipal Wastewater

Compound	Concentration (mg/L)	N content (gN/g COD)
Acetic acid	25	0.0
Higher VFAs	10	0.0
Alcohols (ethanol, methanol)	5	0.0
Lower amino acids	10	0.14
Simple carbohydrates	10	0.0

In contrast to the RBCOD, the slowly biodegradable fraction is assumed to consist of particulate/colloidal material and complex organic molecules that require extracellular enzymatic breakdown prior to absorption and utilization.

The division between readily and slowly biodegradable COD has a major influence on process dynamic behavior (for example, oxygen demand and denitrification capacity) and is of crucial importance in the design of biological nitrogen and phosphorus removal systems.

Chapter 7.0 outlines methods for measuring the readily biodegradable COD content of influent wastewaters. This chapter provides background information on the rationale for the bioassay or oxygen uptake rate (OUR)-based respirometric procedures and physicochemical methods.

39.2 OUR-Based Respirometric Methods

The first methods proposed for measuring RBCOD content of influent wastewaters were based on bioassay techniques, and typically involved measuring oxygen utilization rate (or nitrate utilization rate) in tests combining influent wastewater with activated sludge (e.g., Ekama et al., 1986) either in continuous flow or batch tests. The early continuous flow activated sludge method developed by Ekama and Marais (1979) produced good measurements of RBCOD but was criticized for its cost and difficulty of operation. Aerobic or anoxic batch procedures require that acclimated sludge be obtained from laboratory-scale continuous flow reactors (Ekama et al., 1986; Sollfrank and Gujer, 1989; Kappeler and Gujer, 1992) or from a full-scale plant (Nichols et al., 1985). Using sludges from biological phosphorus removal systems can produce erroneous estimates of RBCOD due to the phenomenon of RBCOD uptake and storage with phosphorus release by polyP organisms under anoxic or anaerobic conditions (Wentzel et al., 1995).

A simpler aerobic batch method of measuring RBCOD was presented by Wentzel et al. (1995). This procedure is conducted on unsettled wastewater samples without activated sludge seed. The OUR is monitored continuously with an automated apparatus (Randall et al., 1991), following the respiration of the organisms that were indigenous to the wastewater. The simplicity of this test is appealing since no mixed liquor source is required, and no independent determination of unbiodegradable COD is required.

Xu and Hasselblad (1996) proposed a modification to the aerobic batch test method. A single OUR test is used, where the DO consumed for readily biodegradable COD is determined by a change in the slope of the DO concentration versus time curve. However, there was a requirement to generate a calibration curve between oxygen consumption and readily biodegradable COD in the tests. It also seemed that some repeats of the test would be required to determine the proper dilution to ensure that not all of the dissolved oxygen is consumed during the test. Because of these additional requirements it does not seem that this method offers a great advantage over the aerobic batch method.

Other authors (Shulan and Hultman, 1996) have looked at dividing soluble biodegradable COD into readily biodegradable and rapidly hydrolysable. The readily biodegradable fraction is determined from the first slope of a single OUR test. The rapidly hydrolysable fraction is determined by subtracting the readily biodegradable from the biodegradable soluble as determined by the floccCODsol test of Mamais et al. (1993).

The more complex respirometric procedures developed by Spanjers and Vanrolleghem (1995) and Brouwer et al. (1998) were used by Witteborg et al. (1996) to measure RBCOD. This method was reported to overestimate RBCOD by about 15 percent when compared to the classical batch methods discussed above. Mathieu and Etienne (2000) demonstrated a simpler respirometric procedure in which they used a combination of two simple clearly defined kinetic experiments at high and low S_O/X_O ratios to evaluate the structural and practical identifiability properties of the system. Comparison with estimates of RBCOD using the classic continuous flow Ekama et al. (1986) method showed excellent agreement.

39.3 Physicochemical Methods

The experimental protocols for the bioassay methods require close attention. Obtaining reliable results requires considerable experience with experimental activated sludge systems and associated measurement techniques. Because of these difficulties, attention has been directed at devising alternative methods using physicochemical means.

The terms "readily" and "slowly" biodegradable often have been used synonymously with the terms "soluble" and "particulate." Therefore, it would appear obvious that RBCOD (S_S) can be measured by separating out the particulate material from an influent wastewater sample (by, say, filtration) and measuring the residual COD. A complication is that the influent wastewater also contains soluble unbiodegradable COD (S_I), and S_S and S_I would group together in a physicochemical separation. Therefore, some means is required to distinguish between S_S and S_I (see Chapters 6.0, 7.0 and 8.0). This problem has been addressed by assuming the amount of RBCOD in the effluent from a properly operating activated sludge system is negligible, and that the unbiodegradable soluble *influent* COD (S_I) is equal to the *effluent* soluble COD. Hence S_I can be estimated by measuring the filtered effluent COD from a continuous flow activated sludge system (Dold et al., 1986; Bortone et al., 1993; Mamais et al., 1993) or sequencing batch reactor (Torrijos et al., 1993) or from measurements of filtered COD over at least 10 days in batch tests (Lesouef et al., 1992). That is, the influent RBCOD can be estimated by difference from:

$$RBCOD = Influent\ soluble\ COD\ -\ Effluent\ soluble\ COD \qquad (39.3\text{-}1)$$

A potential problem with this approach is that part of the slowly biodegradable influent COD consists of colloidal material that likely spans a wide range of molecular sizes and weights. For example, Dold et al. (1980) found that, by using 0.45 μm filters on municipal wastewater, a significant portion of the slowly biodegradable colloidal COD passed through the filter causing overestimation of the RBCOD. Therefore the physicochemical separation technique should exclude this colloidal material. Two approaches for resolving this issue have been proposed:

- Filtration using low molecular weight cut-off membranes (Dold et al., 1986; Lesouef et al., 1992; Bortone et al., 1993; Mamais et al., 1993; Torrijos et al., 1993). The ultrafiltration work of Dold et al. (1986) showed that, for municipal wastewater, membranes with a cutoff < 10,000 molecular weight gave RBCOD values that correlated closely with those determined by conventional bioassay methods. Bortone et al. (1993) found that with a textile wastewater, membranes with a cutoff < 10,000 molecular weight gave lower RBCOD values (13 percent of total COD) than measured in bioassay batch tests (20 percent of total COD). The issue of molecular weight cut-off membrane selection is discussed later.
- Flocculation of the colloidal material prior to filtration through 0.45 μm filters. Mamais et al. (1993) demonstrated that flocculation of a sample with a zinc sulfate solution removed all colloidal material such that subsequent filtration through a 0.45 μm filter

gave RBCOD values similar to those determined by conventional bioassay methods for four different types of wastewater (raw wastewater, two sources of primary effluent and primary effluent mixed with digester supernatant).

The physicochemical methods for measuring RBCOD incorporate the assumption that the concentration of unbiodegradable soluble COD in the *effluent* equals the concentration in the *influent*. This, in turn, implies that generation of unbiodegradable soluble COD within the process is negligible. This issue is now examined.

39.4 Constitution of the Soluble Effluent COD

The filterable or "soluble" portion of the effluent COD from an activated sludge system is made up of a wide range of organic compounds. Because the effluent COD often is the basis for evaluating plant performance, and because of the non-specificity of the "lumped parameter" COD determination, there has always been interest in characterization of the individual compounds (or chemical groups) and identification of the contributing sources. Identifying the sources contributing to the effluent carbonaceous content would be the first step in attempting to reduce the effluent soluble COD so as to improve plant performance.

A wide range of experimental techniques have been utilized in the characterization and fractionation of secondary effluent. Amongst these methods (with examples of researchers) are: sedimentation and centrifugation (Hunter and Heukelekian, 1960); dialysis membranes (Bunch et al., 1961); ultrafiltration membranes (Saunders and Dick, 1981); gel permeation chromatography (DeWalle and Chian, 1974). Comparison of the information from the many studies is difficult because of the wide range of experiment systems, techniques and substrates utilized. However, the data confirm the presence of substances ranging from low molecular weight volatile acids to high molecular weight polymeric compounds.

From the various experimental investigations three principal sources contributing to the soluble effluent COD have been identified and/or proposed: (1) the influent COD; (2) intermediates and end products of various metabolic pathways; and (3) material from cell lysis and death.

39.4.1 Influent COD

In the treatment of municipal wastewaters compounds or groups of compounds which are known to be resistant to biodegradation have been identified in the influent and therefore will appear as soluble effluent COD (e.g., Painter, 1973). To classify all of these compounds as "unbiodegradable" perhaps is not strictly correct because the compounds, although possibly biodegradable, may not be degraded for a number of reasons; for example:

1. The reaction rates may be very low and the compounds are not degraded within the retention time of the process.
2. The compounds may not be degraded because the enzymes required to degrade them may not be present in the specific system.

3. The presence of certain co-substances may cause repression or inhibition of enzymes resulting in non-degradation of a particular substrate.

Reasons such as these are likely to influence the behavior in pure culture systems or systems treating very specific wastes. However, in the context of municipal wastewater treatment, the organism population will be very diverse and should adapt to the usual constitution of the influent. Therefore, it is reasonable to refer to a fraction of the influent material as "unbiodegradable." For the purposes of developing the separation technique for measurement of the readily biodegradable COD (RBCOD), the unbiodegradable influent portion is not central to the present discussion because it does not involve generation of effluent COD per se. However, it is of interest to inquire into the magnitude of this portion in relation to the other contributors to the effluent COD. Typically it has been reported that between approximately 5 and 10 percent of the raw wastewater influent COD is soluble unbiodegradable matter. That is, for a total influent COD of 500 mg/L, between 25 and 50 mg/L of the soluble effluent COD is derived from the influent (Water Research Commission, 1984).

39.4.2 Intermediates and End Products

Products of the metabolic activities in a system will be present within the organisms, and may accumulate to appreciable levels (as a result of, for example, all functions in the control mechanisms). Because the accumulation products will diffuse out of cells at a rate proportional to the concentration gradient across the cell membrane, these materials, although possibly degradable, will be present in the liquid phase and therefore also in the effluent - the magnitude of the residual amount will depend primarily on the rate of degradation.

A number of researchers have concluded that primary and secondary metabolites are the major contributor to the effluent COD; that is, generation is within the system as opposed to being introduced via the influent. (For example, Chudoba et al. (1969), Grady and Williams (1975) and Daigger and Grady (1977), among others). Molecular weight fractionation work of Leidner et al. (1984) provides strongest evidence demonstrating the presence of intermediates and end products in the effluent. Low-pressure gel chromatography was used to measure the molecular weight distribution (MWD) of the dissolved organic carbon (DOC) in the influent and effluent from a laboratory unit treating a mixture of glucose and glutamic acid. Component-selective tests for glucose and glutamic acid in the effluent were negative, indicating that the DOC was derived from microbial sources only. Parallel tests were conducted on cell-free extracts of the biomass from the system. That is, samples of sludge were centrifuged, washed and broken-up to release the liquid contents (extract); the particulate matter then was removed by filtration. MWD fractionation of the cell-free extract showed only two peaks for compounds at the extremes of the range. These peaks appear to correspond to peaks in the effluent MWD, strongly supporting the proposal that products of the metabolic activities in a system will be present in the effluent. These results are corroborated by Grady et al. (1984), who identified two peaks in the effluent MWD using ultrafiltration fractionation and COD measurement on effluent.

39.4.3 Cell Lysis Products

Researchers have also recognized that residual matter from cell lysis and death will add to the soluble COD from microbial origin (Saunders and Dick, 1981; Gloor et al. (1981); Leidner et al., 1984; Chudoba, 1985; Namkung and Rittman, 1986; Germili et al., 1991; Eremektar et al. 1998; Orhon and Cokgor, 1997). (It should be noted, however, that the charge characteristics and polar nature of functional groups in these compounds (lipoproteins, complex carbohydrates) result in a large degree of adsorption and enmeshment in the sludge matrix, thereby reducing the amounts in the effluent.) Again, confirmatory evidence is provided by the results of Leidner et al. (1984). The MWD for the cell-free extract exhibits only two peaks whereas the effluent MWD identifies an indistinct distribution of compounds between the two peaks. This would suggest that both lysis products and particulate cellular debris produced during operation are subject to some degree of hydrolysis and degradation.

39.4.4 Quantifying the "Internal" Generation of Soluble Effluent COD

The magnitude of the contribution to the effluent COD from the two microbial sources above will determine the viability of the proposed physicochemical separation methods for RBCOD measurement. If the contribution is large, then the methods will be flawed because the assumption that the effluent COD is nearly equal to the effluent S_I will be incorrect. However, if the generation within the system is small compared to the influent S_S and S_I then the error introduced in the test result will be small, and probably less significant than the errors encountered with the bioassay technique.

Generally, data on the microbial generation of effluent COD has been presented in the form of a linear relation between the effluent soluble COD (0.45μm filtrates usually) and the influent COD. For example, experimental work of Chudoba (1985) on both batch and continuous flow laboratory systems has shown that the effluent COD is approximately 1 per cent of the influent COD; that is, for a typical influent COD of 500 mg/L, the portion of the effluent COD originating from microbial activity would be approximately 5 mg/L.

A generation of 1 percent would seem surprisingly small considering the attention given to this aspect in the sanitary engineering literature. Perhaps this attention has arisen partially because these observations are based on experimentation using only synthetic substrates such as glucose where all of the influent substrate is biodegradable. In this case all of the effluent, COD is derived from the microbial source, and assumes a certain importance in terms of percentages. However, in the case of a typical 500 mgCOD/L municipal wastewater, an amount of 5 mg/L is relatively small in a soluble effluent COD of 25 to 50 mg/L, and is certainly small compared to a typical influent RBCOD of 120 mg/L for the same influent. In the case of municipal wastewater treatment, an additional factor is likely to minimize the microbial contribution, namely, the diversity of organisms. With the wide range of substrates present in a municipal wastewater the mixed organism population should be extremely diverse; this diversity should increase the possibility of degrading the complex molecules arising from microbial activity compared to either a pure culture or a culture grown on a limited synthetic substrate.

The contribution from microbial sources to the effluent COD has been related principally to the influent concentration. However, limited research has demonstrated that system operating parameters, and in particular the sludge age or solids retention time, also will influence the production. Grady and Williams (1975), on the basis of results from completely-mixed reactor systems (without recycle) receiving a mixed synthetic substrate, proposed that the soluble effluent COD is related inter alia inversely to the sludge age and can be modeled by an expression of the form:

$$S = K'' \cdot S_0 + \frac{K' \cdot S_0}{SRT} \qquad (39.4.4\text{-}1)$$

where:

S = effluent soluble COD concentration

S_0 = influent soluble COD concentration

SRT = solids retention time (sludge age)

The data of Grady and Williams (1975) were collected over a range of operating conditions very different to those encountered in municipal wastewater treatment practice, namely, solids retention times ranging from 0.17 to 0.46 days with average suspended solids concentrations between 148 and 797 mg/L. It is possible that, at these low reactor solids concentrations, the influence of adsorption and enmeshment of microbially-generated matter would be less evident than in full-scale treatment plants where the solids concentration is in the range 3,000 mg/L to 5,000 mg/L. In this respect, the data of Saunders and Dick (1981) is more pertinent to this discussion. In that study, laboratory plants were operated over a range of sludge ages from 0.8 to 19.4 days, with a 1:1 mixture of raw sewage and synthetic substrate. The soluble (<0.45μm) effluent COD samples were fractionated into three divisions shown in Table 39-2.

Table 39-2. Example of Molecular Weight Fractionation of Effluent COD

Fraction	Approximate MW range
High molecular weight	> 100,000
Low molecular weight	500 – 100,000
Dissolved	< 500

Note: Results similar to those of Saunders and Dick (1981) have been observed by Burke, Dold and Marais (1986) in low sludge age biological phosphorus removal plants. As the aerobic sludge age decreased below 2 days there was a marked increase in the 0.45μm effluent COD; concomitantly there was an increase in effluent turbidity.

The effect of sludge age on the 0.45μm COD was marked with the average value decreasing from 57 to 20 mg/L in an inverse manner for sludge ages from 0.8 to 19.4 days with an influent COD of approximately 450 mg/L; in contrast the dissolved fraction (< 500 MW) was near constant. These results show that the contribution of microbial material with molecular weight in excess of 500 increases with decreasing sludge age; an observation in conformity with the interpretation of data by Chudoba (1985). These researchers have proposed different explanations for the phenomenon: Saunders and Dick (1981) proposed that as sludge age decreases certain organisms will be "washed out" depending on the maximum specific growth rates; therefore it is likely that at low sludge ages the culture no longer contains organisms capable of degrading all of the higher molecular weight material from the microbial source.

Chudoba (1985), on the other hand, proposed that at low sludge ages the rate of cell replication is increased and the rate of production of refractory material will be increased.

39.4.5 Summary

The assumption that generation of unbiodegradable soluble COD is negligible would appear reasonable as the filtered effluent COD from systems receiving the same influent COD concentration (from the same waste source) does not increase for sludge ages ranging from 3 to 30 days (Marais and Ekama, 1976). (For biological phosphorus removal systems there may be some generation of unbiodegradable soluble material from the decay processes (Wentzel et al., 1989); however, the quantity of material generated should be small enough to be considered negligible in the estimation of the influent unbiodegradable soluble COD.)

In summary, it would appear from data in the literature that for sludge ages greater than, say, 2 or 3 days the microbial contribution is minimal. Therefore, providing the system sludge age is greater than about 3 days it likely can be accepted that the generation of soluble effluent COD is sufficiently small to be neglected for the purposes of measuring the RBCOD by the physicochemical methods.

39.5 Selection of Filtration Medium

Note: Many variables will influence the separation of organic molecules by ultrafiltration, for example, pore size distribution adsorptive characteristics of membrane, charge of organic compounds being filtered. Therefore, the designated molecular mass cut-off value, which is based on only a limited spectrum of molecular species, should be seen as only approximate.

An issue with the filtration method for RBCOD determination is selection of the appropriate filter. Because the method requires isolating the small-molecule low molecular weight material in the influent and the effluent the obvious approach is by ultrafiltration. This technique can be carried out readily without the availability of extensive laboratory facilities using a proprietary ultrafiltration cell; membranes are available in approximate molecular mass cut-off divisions of, for example, 500, 1,000, 10,000 and 100,000. However, the soluble material in a municipal wastewater spans a wide range of molecular sizes and weights. Therefore, the problem is one of electing an appropriate nominal molecular mass cut-off ultrafiltration membrane, which will allow passage of small- molecular readily biodegradable COD but which will retain molecules that require extracellular hydrolysis. This selection problem will not be a trivial one if the soluble influent COD is present in a continuum of molecular weights and sizes.

Very little information on the range of molecular sizes and/or weights of the organic material in municipal wastewaters is available. However, Gloor et al. (1981) presented one set of data in the form of a molecular weight distribution chromatogram of the dissolved organic carbon in the influent and effluent from a full- scale activated sludge system. While it should be conceded that the data are restricted to a single wastewater and treatment plant, information very pertinent to this study is provided in the MWD chromatogram:

- The major portion of the influent DOC is in the low molecular weight fraction (< 200).
- The portion of the DOC, which is removed in the treatment process, is almost entirely in this low molecular weight fraction; this observation serves to support the supposition that the S_{BS} is comprised of small molecules.
- It would appear that almost the entire DOC of molecular mass greater than approximately 1,000 is "unbiodegradable" because the influent and effluent profiles are essentially identical in this region.
- The very close similarity between the influent and effluent profiles (except for S_S) lends strong support to the proposal that the contribution to the effluent soluble effluent COD is closely equal to the influent unbiodegradable COD, S_I.

On the basis of these results, it would appear that there is a large degree of flexibility in selection of an ultrafiltration membrane with the suitable nominal molecular weight cut-off. For example, a cut-off of, say, 1,000, 10,000 or even 100,000 probably would be equally suitable for isolating the readily biodegradable influent COD fraction in this wastewater. As proposed, filtration of influent and effluent samples with the same cut-off membrane should yield a close estimate of RBCOD by difference.

39.6 References

Bortone, G., J.S. Chech, F. Germirli, R. Bianchi, and A. Tilche. 1993. Experimental approaches for the characterization of a nitrification/denitrification process on industrial wastewater. In *Proc.of 1st International Specialty Conference on Microorganisms in Activated Sludge and Biofilm Processes* (pp.129-136), Paris, France. London UK: International Water Association.

Brouwer, H., A. Klapwijk, and K.J. Keesman. 1998. Identification of activated sludge and wastewater characteristics using respirometric batch experiments. *Water Res.* 32(4):1240-1254.

Bunch, R.L., E.F Barth, and M.B. Ettinger. 1961. Organic materials in secondary effluents. *Jour. WPCF* 33(2):122-126.

Burke, R., P.L. Dold, and G.v.R. Marais. 1986. *Phosphorus removal in short sludge age plants.* University of Cape Town Report No. W58. Rondebosch, South Africa: Department of Civil Engineering, University of Cape Town.

Chudoba, J. 1985. Quantitative estimation in COD units of refractory organic compounds produced by activated sludge microorganisms. *Water Res.* 19(1):37-43.

Chudoba, J., J. Cervenka, and M. Zima. 1969. Residual organic matter in activated sludge process effluents: IV – Treatment of municipal sewage. *Technol. Water* 14:5.

Daigger, G.T., and C.P.L. Grady. 1977. A model for the bio-oxidation process based on product formation concepts. *Water Res.* 11:1049-1057.

DeWalle, F.B., and E.s.K. Chian. 1974. Removal of organic matter by activated carbon columns. *Jour. Env. Eng. Div.*, ASCE 100(EE5):1089-1104.

Dold, P.L., G.A. Ekama, and G.v.R. Marais. 1980. A general model for the activated sludge process. *Prog. in Water Tech.* 12(6):47-77.

Dold, P.L., W.K. Bagg and G.v.R Marais (1986) *Measurement of readily biodegradable COD fraction in municipal wastewater by ultrafiltration.* University of Cape Town Report No. W57. Rondebosch, South Africa: Department of Civil Engineering, University of Cape Town.

Ekama, G.A., and G.v.R. Marais. 1979. Dynamic behaviour of the activated sludge process. *J. Water Pollut. Control Fed.* 51:534-556.

Ekama, G.A., P.L. Dold, and G.v.R. Marais. 1986. Procedures for determining influent COD fractions and the maximum specific growth rate of heterotrophs in activated sludge systems. *Water Sci. Tech.* 18(6):91-114.

Eremektar, G., F. G. Babuna, and D. Orhon. 1998. Inert COD fractions in two-stage treatment of a pulp and paper mill effluent. *J. Chem. Technol. Biotechnol*. 72:7-14.

Germirli, F., D. Orhon, and N. Artan. 1991. Assessment of the initial inert soluble COD in industrial wastewaters. *Water Sci. Tech.* 23:1077-1086.

Gloor, R., H. Leidner, K. Wuhrmann, and T. Fleischmann. 1981. Exclusion chromatography with carbon detection. A tool for further characterisation of dissolved organic carbon. *Water Res.* 15(4):457-462.

Grady, C.P.L., E.J. Kirsch, M.K. Koczwara, B. Trgovich, and R.D. Watt (1984) Molecular weight distributions in activated sludge effluents. *Water Res.* 18(2):239-246.

Grady, C.P.L., and D.R Williams. 1975. Effects of influent substrate concentration on the kinetics of natural microbial populations in continuous culture. *Water Res.* 9(2):171-180.

Henze, M. 1992. Characterization of wastewater for modeling of activated sludge processes. *Water Sci. Tech.* 25(6):1-15.

Henze, M., W. Gujer, T. Mino, T. Matsuo, M. Wentzel, and G.v.R. Marais. 1994. Activated Sludge Model No. 2. *IAWQ Scientific and Technical Report No. 3*, London, UK: International Water Association.

Kappeler, J., and W. Gujer. 1992. Estimation of kinetic parameters of heterotrophic biomass under aerobic conditions and characterization of wastewater for activated sludge modeling. *Water Sci. Tech.* 25(6):125-139.

Leidner, H.A., T. Fleischmann, and G. Hamer. 1984. Molecular weight fractionation for the study of complex biodegradation processes. *Analytica Chimica Acta* 163:35-42.

Lesouef, A., M. Payraudeau, F. Rogalla, and B. Kleiber. 1992. Optimizing nitrogen removal reactor configuration by on-site calibration of the IAWPRC activated sludge model. *Water Sci. Tech.* 25(6):105-123.

Mamais, D., D. Jenkins, and P. Pitt. 1993. A rapid physical-chemical method for the determination of readily biodegradable soluble COD in municipal wastewater. *Water Res.* 27(1):195-197.

Marais, G.v.R., and G.A. Ekama. 1976. The activated sludge process: part 1—steady state behaviour. *Water SA* 2:163-200.

Mathieu, S., and P. Etienne. 2000. Estimation of wastewater biodegradable COD fractions by combining respirometric experiments in various S_o/X_o ratios. *Water Res.* 34(4):1233-1246.

Namkung, E., and B.E. Rittman. 1986. Soluble microbial products (SMP) formation kinetics by biofilms. *Water Res.* 20(6):795-806.

Nichols, H.A., A.R. Pitman, and D.W. Osborn. 1985. The readily biodegradable fraction of sewage: its influence on phosphorus removal and measurement. *Water Sci. Tech.* 17:73-87.

Orhon, D., and U.C. Cokgor. 1997. COD fractionation in wastewater characterization—the state of the art. *J. Chem. Technol. Biotechnol.* 68:283-293.

Painter, H.A. 1973. Organic compounds in solution in sewage effluents. *Chem. Ind.* 818-822.

Randall, E.W., A. Wilkinson, and G.A. Ekama. 1991. An instrument for the direct determination of oxygen utilization rate. *Water SA* 17(1):11-18.

Saunders, F.M., and R.I. Dick. 1981. Effect of mean-cell residence time on organic composition of activated sludge effluents. *J. Water Poll. Control Fed.* 53(2):201-215.

Shulan, X., and B. Hultman. 1996. Experiences in wastewater characterization and model calibration for the activated sludge process. *Water Sci. Tech.* 33(12):89-98.

Sollfrank, U., and W. Gujer. 1991. Characterization of domestic wastewater for mathematical modeling of the activated sludge process. *Water Sci. Tech.* 23(4-6):1057-1066.

Spanjers, H., and P. Vanrolleghem. 1995. Respirometry as a tool for rapid characterization of wastewater and activated sludge. *Water Sci. Tech.* 31(2):105-114.

Torrijos, M., R.M. Cerro, B. Capdeville, S. Zeghal, M. Payraudeau, and A. Lesouef. 1993. Sequencing batch reactor: a tool for wastewater characterization for the IAWPRC model. In *Proc. 1st International Specialty Conference on Microorganisms in Activated sludge and Biofilm Processes* (pp. 81-90), Paris, France. London, UK: International Water Association.

Wentzel, M.C., P.L. Dold, G.A. Ekama, and G.v.R. Marais. 1989. Enhanced polyphosphate organism cultures in activated sludge systems. Part III: Kinetic Model. *Water SA* 15:89-102.

Wentzel, M.C., A. Mbewe, and G.A. Ekama. 1995. Batch test for measurement of readily biodegradable COD and active organism concentrations in municipal wastewaters. *Water SA* 21(2):117-124.

Witteborg, A., A. van der Last, R. Hamming, and L. Hemmers. 1996. Respirometry for determination of the influent S_S concentration. *Water Sci. Tech.* 33(1):311-323.

Water Research Commission. 1984. *Theory, design and operation of nutrient removal activated sludge processes*. P.O. Box 824, Pretoria, South Africa: Water Research Commission.

Xu, S., and S. Hasselblad. 1996. A simple biological method to estimate the readily biodegradable organic matter in wastewater. *Water Res.* 30(4):1023-1025.

CHAPTER 40.0

READILY BIODEGRADABLE COD: UNIVERSITY OF WASHINGTON INVESTIGATIONS

40.1 Background

A limited series of investigations were carried out at the at the University of Washington environmental engineering research laboratory to compare the different methods of measuring readily biodegradable COD for their reproducibility, ease of application and recovery of COD. This work was limited because the project focus was directed at developing and verifying methods of measuring maximum specific nitrifier growth rate.

40.2 Experimental Methods

In the two physicochemical methods (flocculation/filtration and ultrafiltration) suspended solids, colloidal solids and high molecular weight compounds are theoretically removed from a sample, and the remaining dissolved COD is measured. For influent wastewater, the dissolved COD remaining will consist of biodegradable and non-biodegradable compounds. For wastewater after biological treatment, the dissolved COD remaining will consist of non-biodegradable COD. RBCOD is determined by the difference in dissolved COD between the influent and the biologically treated effluent.

The amount of the readily biodegradable COD remaining in the effluent should be negligible. Therefore, the methods should be applied only to systems with sludge ages in excess of about 3 days so as to validate this assumption.

The assertion that the effluent filtrate COD (representing the unbiodegradable soluble portion of the effluent) is nearly equal to the influent unbiodegradable soluble COD assumes that the generation of inert soluble material within the process is negligible.

40.2.1 Flocculation/Filtration COD Method

The procedure for the flocculation/filtration (FFCOD) method is as follows:

1. 1 mL of 100 g/L zinc sulfate solution is added to 100 mL of wastewater.
2. The sample is mixed vigorously for approximately 1 minute.
3. The sample pH is adjusted to approximately 10.5 using 2 M sodium hydroxide solution with slow mixing for 5 minutes.
4. The sample is then allowed to settle for a few minutes and a volume of clear supernatant is withdrawn for filtration through a 0.45 μm membrane filter.
5. The filtrate COD is measured.

The procedure was conducted on both the raw wastewater feed to, and effluent samples from, the SBR. Filtrate COD was measured using Hach spectrophotometric procedures. The test was repeated using aliquots from the same sample to assess test reproducibility. The RBCOD concentration was calculated as follows:

$$RBCOD = Influent\ ffCOD\ -\ Effluent\ ffCOD \qquad (40.2.1\text{-}1)$$

40.2.2 Ultrafiltration COD Method

Ultrafiltration membranes with a specific molecular weight separation are used to filter the wastewater and treated effluent samples. The inherent assumption is that all the RBCOD compounds are at a molecular weight that is below the membrane separation. The RBCOD values derived from this method are compared to that from the respirometric and FFCOD methods to determine if this assumption is reasonable. The 10,000 and 100,000 molecular weight cut-off (MWCO) membrane filters were used for ultrafiltration COD measurements.

A pre-filtration procedure was used to protect ultrafiltration membranes from being blinded by colloidal and particulate matter. A procedure proposed by Dold and Bagg (1986) was used and is described as follows:

1. Samples are centrifuged at 2,200 rpm for 20 minutes and centrate removed.
2. The centrate sample is subjected to two stages of pre-filtration: filtration through a Whatman GF/C glass fiber filter paper followed by filtration through a 0.45μm membrane filter. At this point all suspended solids and most colloidal solids are removed.
3. 50 mL of the 0.45μm membrane filtrate is passed through an ultrafiltration membrane in a closed, stirred sample apparatus (Amicon Corp. Model 8050 50-ml cell) under nitrogen gas pressure of approximately 100 Kpa. A magnetic stir plate rotates a paddle in the sample cell above the membrane to help prevent binding of the membrane.

Filtrate COD was measured using Hach spectrophotometric procedures. The test was repeated using aliquots from the same sample to assess test reproducibility. The RBCOD concentration was based on analysis of wastewater and SBR effluent samples:

$$RBCOD = Influent\ Ultrafiltrate\ COD\ -\ Effluent\ Ultrafiltrate\ COD \qquad (40.2.2\text{-}1)$$

40.2.3 Respirometric Test Method

The respirometric tests were conducted during the operation of a 4 L SBR that was fed Renton primary effluent once per day and operated at a 10-day SRT in a 20°C temperature chamber at the University of Washington environmental engineering laboratory. This was the same apparatus used for low F/M tests described in Chapter 34.0. The change in oxygen uptake rate (OUR) with time was monitored after a daily batch feeding of wastewater. Because of the nature of the SBR operation, replicates were not conducted for the respirometric method.

40.2.3.1 Description of SBR test reactor and operation

The reactor used had a total volume of 5 L and was constructed of 0.2 cm thick Plexiglass with a 16 cm inside diameter, and was operated with a 4 L liquid volume. Sample taps were located at 1 L increments and an on/off valve located at the 1 L liquid level was used to decant settled supernatant. A stone diffuser was located at the bottom of the reactor, and air was provided by an aquarium pump through a rotameter with a needle valve and controlled at about a 600 mL/min flow rate. Each reactor was mixed with a stirring bar and a magnetic stir plate. A dissolved oxygen/oxygen uptake rate (DO/OUR) meter and DO probe were used to monitor and control DO concentration, as well as provide OUR data throughout the reactor aeration period. The DO probe was immersed in the reactor with the probe sensor side up to prevent air bubbles impinging on the sensor and affecting the DO measurement. At the high DO set point (6 mg/L), the DO meter/controller switched off the air pump, and at the low DO set point (3 mg/L), the DO controller turned on the air pump. As the DO concentration declined with time with the air off, the meter calculated and logged the OUR. The OUR and temperature data were stored by the meter and subsequently downloaded later to a PC.

The seed for the SBR was drawn from nitrifying mixed liquor at King County's Vashon Island, Wash. WWTP. Every 5 days, a 30 L grab sample of primary effluent was taken from the Renton WWTP to serve as the SBR feed source. The feed was stored in a 4°C refrigerator, and each day 3 L of a well-mixed representative feed sample was warmed to the test temperature before reactor feeding. Sodium bicarbonate solution was added to the feed sample so that the initial alkalinity was about 250 mg/L as CaCO3, to ensure that each reactor had enough alkalinity for complete nitrification and resulted in a pH in the range of 7.0 to 7.5 during aeration. The pH was measured at the first hour and at the end of the aeration period.

The SRT was controlled at 10 days by wasting a certain volume of mixed liquor near the end of the aeration period, based on daily effluent TSS and reactor MLSS measurements. The MLSS measurement sample was taken at the full SBR volume. Previous day's samples were used to calculate the next day wasting. At least once per day, the reactor's wall growth was brushed off and returned to the reactor to maintain the integrity of the solids mass balance inventory.

40.2.3.2 Calculation of RBCOD from the SBR OUR response

After rapid batch feeding of 3 L of primary effluent to the SBR containing a 1 L volume of settled mixed liquor, OUR measurements were made at intervals of a few minutes for a period of approximately 2 to 3 hours. To obtain the initial OUR quickly, the feed sample was pre-aerated to provide an initial high DO concentration.

Estimation of the RBCOD concentration was achieved through analysis of the OUR response over the first part of the aeration period. Figure 40-1 illustrates a typical OUR response observed during the SBR aeration period. The area under the OUR curve is divided into four sections: Area 1 indicates the mass of oxygen utilized for oxidation of RBCOD; Area 2 indicates the mass utilized for nitrification; Area3 denotes the mass consumed for oxidation of slowly biodegradable COD (rate limited by hydrolysis of the enmeshed COD); and Area 4 indicates the oxygen associated with endogenous respiration.

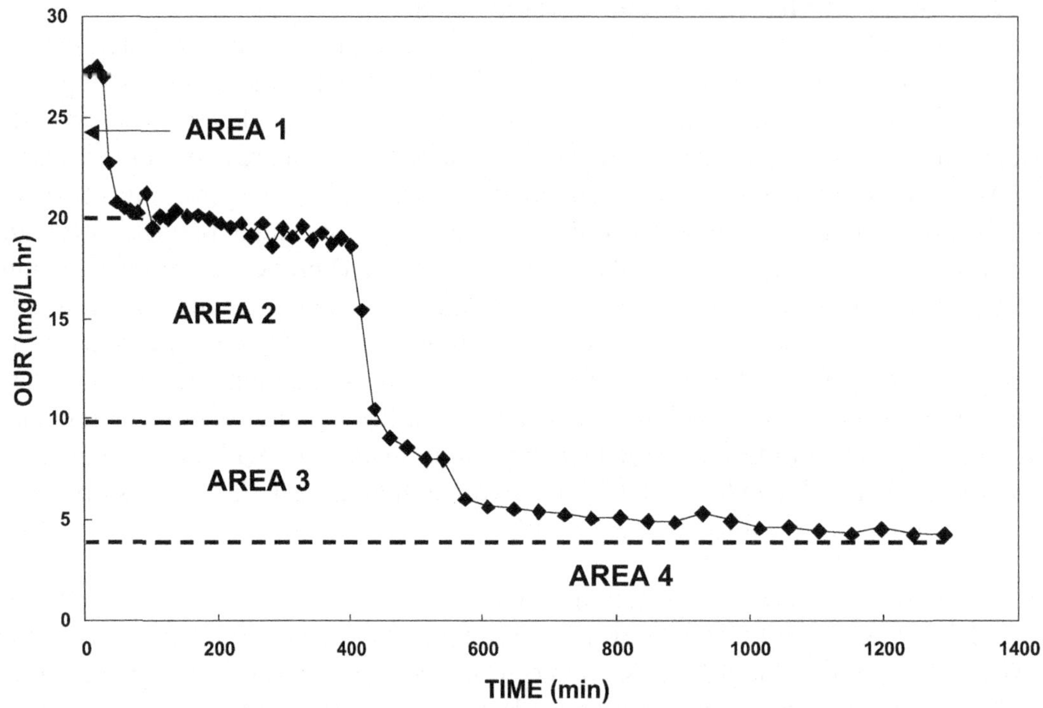

Figure 40-1. Example of OUR Response Observed during the SBR Aeration Period

The RBCOD concentration is estimated based on measuring the area under the OUR CURVE (Area 1 shown in Figure 40-1), and applying equation 40.2.3.2-1 below:

$$RBCOD = \frac{M_O \cdot V_m}{(1 - Y_{HET}) \cdot V_f} \qquad (40.2.3.2\text{-}1)$$

where:

M_O = mass of oxygen consumed (Area 1) per liter of the SBR reactor volume (mgO_2/L)

Y_{HET} = heterotrophic yield coefficient (0.66 mg cell COD/mg substrate COD)

V_m = SBR reactor volume during OUR test (4 L)

V_f = daily feed wastewater volume (3 L)

40.3 Results and Discussion

The test reproducibility and acetate recovery results are presented first. Then the RBCOD results from the different test methods are compared.

40.3.1 Test Reproducibility for Flocculation/Filtration and Ultrafiltration Methods

The COD of the filtrate for each of the flocculation/filtration and ultrafiltration procedures were measured for a sample of secondary influent and a sample of secondary effluent from the Renton WWTP in up to seven replicate determinations to measure the reproducibility of the methods. Only the 100K MWCO membrane was subjected to reproducibility trials in the ultrafiltration method.

The data in Table 40-1 shows excellent reproducibility for both methods. The coefficient of variation was minimal ranging from 2 to 5 percent.

Table 40-1. Influent and Effluent Filtrate COD for Renton WWTP Wastewater

Method	Number of replicates	Filtrate COD		
		Average (mg/L)	Standard Deviation	Coefficient of variation (%)
Renton secondary influent				
Flocculation / Filtration	7	75.6	2.8	3.7
Ultrafiltration - 100K MWCO	6	107.3	1.9	1.8
Renton secondary effluent				
Flocculation / Filtration	6	35.2	1.5	5.3
Ultrafiltration - 100K MWCO	6	58.1	1.6	3.0

Note: The raw data from which these averages are drawn are located in the Appendix to this chapter.

40.3.2 Recovery of Soluble COD

Wastewater samples were supplemented with acetate to measure the recovery of soluble COD in the filtrate from the flocculation/filtration and ultrafiltration procedures. Acetate, present as a 20 mg COD/L solution, was added at to a wastewater sample, which was then split and subjected to the flocculation/filtration and ultrafiltration methods. This was carried out on a total of 10 days. Acetate recovery data are shown in Table 40-2. A recovery of 90 percent was measured for influent samples by the flocculation/filtration method, and 92 percent by the ultrafiltration method. This was considered to be within acceptable tolerances for the COD test method and sampling errors. The reasonable recovery of acetate in influent samples confirms previous work by Narkis et al. (1980). They measured almost complete acetate recovery and found that volatile fatty acids were not removed to any significant amount by lime or ferric hydroxide chemical treatment of wastewater.

A lower recovery of only about 80 percent was measured for the effluent samples by both methods. The reason for the lower recovery could not be explained. This finding raises concerns about the laboratory technique or it could be due to some other factor not recognized at this time. More evaluations of effluent samples are needed to resolve this issue.

Table 40-2. Acetate Recovery in the Flocculaton/Filtration and Ultrafiltration Methods

	Filtrate COD					
	Influent (mg/L)	Influent + acetate (mg/L)	Recovery (%)	Effluent (mg/L)	Effluent + acetate (mg/L)	Recovery (%)
	Flocculation/filtration					
Average	75.9	91.6	89.9	35.7	48.9	79.4
Std. dev.	12.7	14.3	11.3	9.2	10.5	13.5
	Ultrafiltration 100K MWCO					
Average	106.2	122.4	92.0	49.0	62.8	83.1
Std. dev.	11.0	13.0	12.1	9.7	9.4	8.2

Note: The raw data from which these averages are drawn are located in the Appendix to this chapter

In the respirometric method, OURs were measured after spiking the reactor with acetate at 100 mg/L. The recovery results are shown in Table 40-3. Using a value for the yield coefficient, Y_{HET}, of 0.66 g cell COD/g COD removed, the RBCOD recovery was 80 to 90 percent and averaged 87 percent. The possible causes for the difference between the RBCOD added and measured are: 1) experimental errors in the acetate measurement and addition, 2) errors in the interpretation of the OUR curve to calculate the area where the rapid COD removal occurs, and 3) error in the assumed yield value. Of these, the error in the assumed yield value seems least likely. The yield would need to be 0.71 g cell COD/g COD removed to fully account for the acetate added, which is high compared to previously reported yields with readily degradable substrates. Some inaccuracies in the test procedure can be expected. Based on these results, an error of at least 10 percent in RBCOD may be expected for the respirometric method.

Table 40-3. Acetate COD Recovery in the Respirometric Method

Date	Acetate Added (mg COD/L)	RBCOD for Y_{HET}=0.666 (mg COD/L)	Recovery (%)
9/26/01	100	80	80.2
9/27/01	100	90	89.8
10/4/01	100	90	89.6
Average	100	87	86.5

40.3.3 Comparison of Flocculation/Filtration and Ultrafiltration Methods

A number of tests were conducted with Renton and Puyallup wastewaters to compare the RBCOD determined by the flocculation/filtration method to the ultrafiltration method using the 100K molecular weight cut off (MWCO) membranes. Figure 40-2 shows that the RBCOD by the flocculation/filtration test method is consistently lower than that for the ultrafiltration method using the 100K MWCO membrane.

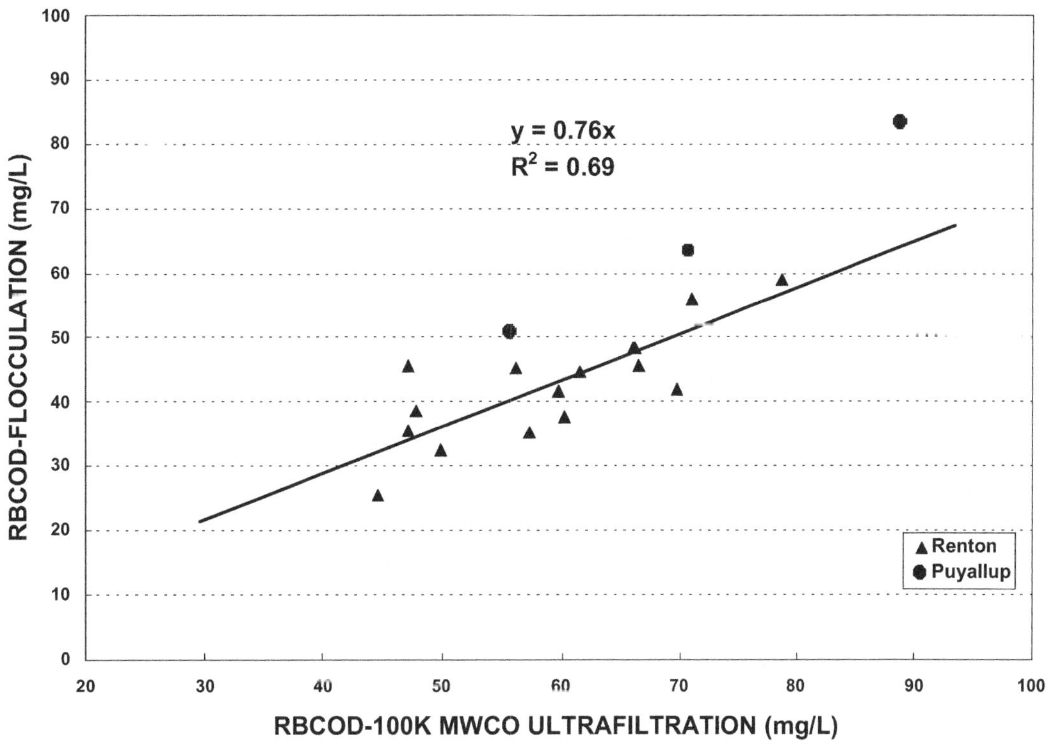

Figure 40-2. Flocculation/Filtration RBCOD versus 100K MWCO Ultrafiltration RBCOD

A smaller subset of six experiments were conducted in which the flocculation/filtration method was compared to the ultrafiltration method for both 10 K and 100K MWCO membranes using Puyallup and Renton WWTP wastewaters. Table 40-4 summarizes the average measured values of RBCOD for these experiments.

Table 40-4. RBCOD by Flocculation/Filtration versus Ultrafiltration

Wastewater Source	Flocculation Filtration	10K MWCO Ultrafiltration	100K MWCO Ultrafiltration
Puyallup			
Influent	92.2	82.4	105.8
Effluent	26.3	28.5	34.1
RBCOD	65.9	53.9	71.7
Renton			
Influent	89.9	86.2	107.3
Effluent	36.3	31.7	38.7
RBCOD	53.6	54.5	68.6

Note: The raw data from which these averages are drawn are located in the Appendix to this chapter

For both wastewater sources, the RBCOD was higher for the 100 K MWCO membrane than for the 10 K MWCO membrane, suggesting that the 100 K MWCO membrane pore size is too large and may be allowing the passage of colloidal COD to the filtrate. The Renton RBCOD was almost identical for both filtration/flocculation and ultrafiltration (10 K MWOC membrane) methods.

In the case of the Puyallup wastewater, however, the RBCOD by the ultrafiltration (10 K MWOC membrane) method was lower than by the flocculation/filtration method. It is possible that the Puyallup wastewater contained soluble COD that was removed by the 10 K MWCO membrane, but not in the flocculation/filtration procedure. Examination of Figure 40-2 shows that the highest flocculation/filtration derived RBCOD values were for the Puyallup WWTP. It is possible that a comparison of these physical-chemical methods might be compromised by the wastewater source. More comprehensive analysis of several wastewaters would be needed to confirm this supposition.

40.3.4 Comparison of Flocculation/Filtration and Respirometric Methods

In a later period, comparison was made between the respirometric and flocculation/filtration methods for split samples of Renton wastewater. Figure 40-3 shows that the measured RBCOD values based on the respirometric method were consistently lower than that those measured by the flocculation/filtration method by nearly 30 percent. Welch (1995) reported similar results for a similar OUR respirometer procedure for more than 20 influent wastewater samples from the Rya, Sweden wastewater treatment facility. The wastewater RBCOD fraction for the respirometric method was about half of that for the flocculation filtration method (11.4 versus 21.6 percent).

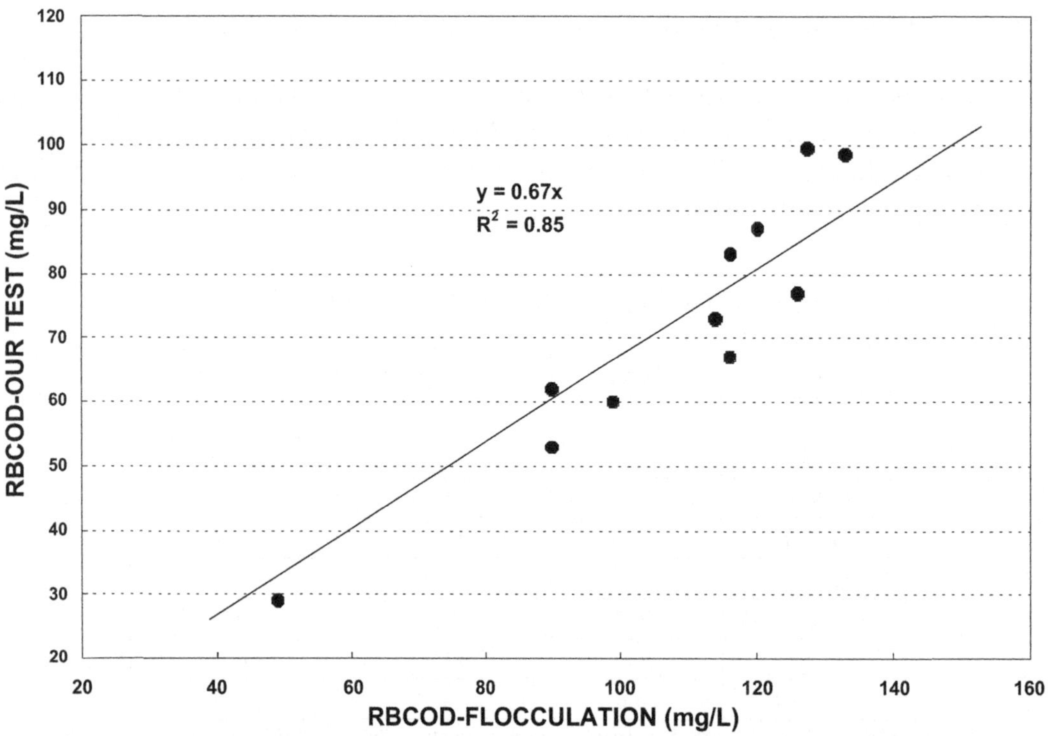

Figure 40-3. Respirometric RBCOD versus Flocculation/Filtration RBCOD

In contrast, Mamais et al. (1993) obtained almost identical RBCOD results between an activated sludge OUR respirometric method and the flocculation/filtration method. The reason for better correlation of results in the work by Mamais et al. could be related to their use of an activated sludge reactor with a separate clarifier and the method proposed by Ekama et al.

(1984), which involves 12 hours of feeding and 12 hours with no feeding. The step change in the OUR when the feed is turned off is used to determine the oxygen used to degrade the RBCOD. The period of aeration time used for the RBCOD measurement is much longer than the time used to observe RBCOD removal in the batch test method used here. It may be possible that absorption of some of the RBCOD substrate occurs during the batch-feeding period and that substrate is oxidized later. With the 12 hour feeding method, absorption may be less likely as the bacteria are operating at a less starved condition.

40.3.5 Flocculation/Filtration After Ultrafiltration

The flocculation/filtration method was used to treat influent and secondary wastewater filtrates following ultrafiltration in order to determine if any soluble but precipitable COD remained after ultrafiltration. Table 40-5 compares the filtrate COD values following flocculation/filtration of the ultra-filtered filtrates.

Table 40-5. COD Removal after Ultrafiltration Followed by Flocculation/Filtration

Wastewater Source	10K MWCO Ultrafiltration			100K MWCO Ultrafiltration		
	COD (mg/L)	COD removed by floc/filt (mg/L)	Loss (%)	COD (mg/L)	COD removed by floc/filt (mg/L)	Loss (%)
Puyallup						
Influent	82.4	2.6	3.1	105.8	16.8	15.9
Effluent	28.5	5.6	19.6	34.1	4.7	13.7
Renton						
Influent	86.2	5.5	6.4	107.3	19.0	17.7
Effluent	31.7	0	0	38.7	8.3	21.4

As noted in Table 40-4, the 100 K MWCO membrane passes more colloidal COD than the 10 K MWOC membrane. Table 40-5 shows that a significant amount of COD can be removed from the 100 K MWOC filtrate. However, the results with 10K MWCO membrane are less conclusive. With the exception of the Puyallup effluent sample, the amount of COD removed by the flocculation/filtration method is not significant and not discernible from normal errors in the COD test. The lack of COD removal by the flocculation/filtration of the 10K filtrate is important for two reasons. First, it suggests that the flocculation/filtration method is not adsorbing soluble COD, and second, it suggests that no significant colloidal COD is passing through the 10K ultrafiltration process. At a low molecular weight of 10,000, very little, if any, colloidal material should be expected in the filtrate, so the results are likely more indicative of the lack of COD adsorption by the flocculation/filtration test. However, because of the significant level of COD removal for the Puyallup effluent sample, it cannot be concluded that the flocculation/filtration test does not adsorb COD in all cases. The higher COD removal for the Puyallup effluent may be related to sampling and analysis (the COD difference was only 5.6 mg/L) or due to some unique feature of that wastewater sample. Because not all wastewater samples can be expected to have the same chemical compound composition, extensive testing would be needed with numerous samples to develop conclusions about adsorption with the flocculation/filtration procedure. The results from these limited tests are at least encouraging in terms of suggesting that adsorption of RBCOD during the flocculation filtration test may be minimal.

40.4 Ease of Use

The flocculation filtration procedure was the easiest method to use for the wastewater evaluation. Preparation of individual COD determinations required 15 to 30 minutes to complete followed by another 2 hours for the Hach COD measurement. A group of 10 samples would take half a day to complete. In contrast, the ultrafiltration method was much more labor intensive and time consuming due to the repetitive treatment of the sample via centrifugation and double filtration before membrane filtration at high pressure. Special equipment and a greater degree of skill are required than for the flocculation/filtration procedure. A group of ten samples could take a day to complete. Although in this case the respirometric method was relatively simple, with respect to simplicity and labor intensiveness, the application of this method benefited from the fact that the SBR used for the testing was operated for other test goals, mainly to determine nitrification kinetics. If a laboratory scale activated sludge system were to be set up specifically for the respirometric method, this would require the most labor and time of the three methods to operate and maintain. A bench scale activated sludge system can take at least three weeks to set up and reach equilibrium before carrying the respirometric test.

The flocculation filtration method can be more readily done by less experienced personnel and in less time than the other methods.

40.5 References

Dold, P.L., W.K. Bagg, and G.v.R Marais. 1986. *Measurement of readily biodegradable COD fraction in municipal wastewater by ultrafiltration.* University of Cape Town Report No. W57. Rondebosch, South Africa: Department of Civil Engineering, University of Cape Town.

Ekama, G.A., P.L. Dold, and G.v.R. Marais. 1986. Procedures for determining influent COD fractions and the maximum specific growth rate of heterotrophs in activated sludge systems. *Water Sci. Tech.* 18(6):91-114.

Mamais, D., D. Jenkins, and P. Pitt. 1993. A rapid physical-chemical method for the determination of readily biodegradable soluble COD in municipal wastewater. *Water Res.* 27(1):195-197.

Narkis, N., S. Henefeld-Forrier, and M. Rebhun. 1980. Volatile organic acids in raw wastewater and in physico-chemical treatment. *Water Res.* 14(9):1215-1223.

Welch, D. 1995. An evaluation of the EFOR model's ability to predict the performance of an activated sludge nitrogen removal system for Goteborg, Sweden. Masters Thesis, Department of Civil Engineering, University of Washington, Seattle, WA.

Appendix – Experimental Data

Renton Wastewater RBCOD Test Reproducibility

			Flocculation			
Date	**Sample**	**Test 1**	**Test 2**	**Average**	**St.Dev.**	**St.dev/Average, %**
3/17/2001	Influent	90.2	87.2	88.7	2.12	2.4%
3/19/2001	Influent	78.5	76.5	77.5	1.41	1.8%
3/20/2001	Influent	82	87.7	84.85	4.03	4.8%
4/6/2001	Influent	74.6	70.8	72.7	2.69	3.7%
4/10/2001	Influent	71.5	78.2	74.85	4.74	6.3%
4/12/2001	Influent	53.8	56.5	55.15	1.91	3.5%
	Average			75.6	2.8	3.7%
3/17/2001	Effluent	42.4	43.5	42.95	0.78	1.8%
3/19/2001	Effluent	44.5	45.5	45	0.71	1.6%
3/20/2001	Effluent	47.1	45.5	46.3	1.13	2.4%
4/6/2001	Effluent	31.6	30.2	30.9	0.99	3.2%
4/10/2001	Effluent	29.4	23.3	26.35	4.31	16.4%
4/12/2001	Effluent	19	20.8	19.9	1.27	6.4%
	Average			35.2	1.5	5.3%

			Ultrafiltration			
Date	**Sample**	**Test 1**	**Test 2**	**Average**	**St.Dev.**	**St.dev/Average, %**
3/17/2001	Influent	117.6	115.8	116.7	1.27	1.1%
3/19/2001	Influent	109.2	106.8	108	1.70	1.6%
3/20/2001	Influent	110.4	111.8	111.1	0.99	0.9%
4/6/2001	Influent	105.3	110.8	108.05	3.89	3.6%
4/10/2001	Influent	109.7	108.5	109.1	0.85	0.8%
4/12/2001	Influent	86.5	90.3	88.4	2.69	3.0%
4/14/2001	Influent	108.5	111.1	109.8	1.84	1.7%
	Average			107.31	1.89	1.8%
3/17/2001	Effluent	58.3	52.3	55.3	4.24	7.7%
3/19/2001	Effluent	57.5	58.7	58.1	0.85	1.5%
3/20/2001	Effluent	63.8	62.9	63.35	0.64	1.0%
4/6/2001	Effluent	49	48	48.5	0.71	1.5%
4/10/2001	Effluent	41.3	44.7	43	2.40	5.6%
4/12/2001	Effluent	30.7	31.6	31.15	0.64	2.0%
	Average			58.10	1.62	3.0%

Renton Wastewater Acetate Recovery/Variability

Method			Flocculation				RBCOD
Sample Date	Influent	Influent +20	% Recovery	Effluent	Effluent +20	% Recovery	mg/L
20 mg/L		24.1	120.5%				
3/14/2001	88	105.1	85.5%	42.3	62.8	102.5%	45.7
3/15/2001	90.9	109.3	92.0%	42.5	59	82.5%	48.4
3/16/2001	89.7	108.7	95.0%	44.1	57.4	66.5%	45.6
3/17/2001	87.8			43			44.8
3/19/2001	77.5			45			32.5
3/20/2001	84.9			46.3			38.6
4/6/2001	72.7	88.9	81.0%	30.9	47.4	82.5%	41.8
4/9/2001	56.1	69.5	67.0%	30.6	44.7	70.5%	25.5
4/10/2001	74.9	96.3	107.0%	26.4	39.5	65.5%	48.5
4/11/2001	59.9	76.8	84.5%	24.1	39.3	76.0%	35.8
4/12/2001	55.2	74.5	96.5%	19.9	33.8	69.5%	35.3
4/13/2001	69.8	90.1	101.5%	27.9	43	75.5%	41.9
4/14/2001	78.8	96.5	88.5%	41.1	61.6	102.5%	37.7
Average	75.9	91.6	89.9%	35.7	48.9	79.4%	40.2

Method			Ultrafiltration	100K MWCO			RBCOD
Sample Date	Influent	Influent +20	% Recovery	Effluent	Effluent +20	% Recovery	mg/L
20 mg/L		22.2	111.0%				
3/14/2001	109.1	126.2	85.5%	62	79.2	86.0%	47.1
3/15/2001	116.3	136.2	99.5%	50.4	68.7	91.5%	65.9
3/16/2001	121.1	140.2	95.5%	54.6	70.4	79.0%	66.5
3/17/2001	116.7			55.3			61.4
3/19/2001	108			58.1			49.9
3/20/2001	111.1			63.35			47.75
4/6/2001	108.1	122.9	74.0%	48.5	64.6	80.5%	59.6
4/9/2001	87.4	100.3	64.5%	42.8	59.9	85.5%	44.6
4/10/2001	109.1	129.6	102.5%	43	57.3	71.5%	66.1
4/11/2001	89.1	107.9	94.0%	42	56.9	74.5%	47.1
4/12/2001	88.4	108.1	98.5%	31.15	47.6	82.3%	57.25
4/13/2001	105.9	125	95.5%	36.3	53.9	88.0%	69.6
4/14/2001	109.8	128.05	91.3%	49.7	69.6	99.5%	60.1
Average	106.16	122.45	92.0%	49.02	62.81	83.1%	57.15
St.Dev	11.02	12.98	12.13%	9.72	9.40	8.20%	8.81

Respirometric Data

Sample	TCOD		FLOCCULATION		RESPIRATION		OUR / Flocculate
Date	Influent	Influent ffCOD	Effluent ffCOD	RBCOD	RBCOD	Aeration	RBCOD Ratio
9/9/2001	361	122	23	99	60	Air only	61%
9/13/2001	283	64	15	49	29	Air only	59%
9/18/2001	352	140	20	120	87	Pure Oxygen	73%
9/19/2001	352	147	21	126	77	Pure Oxygen	61%
9/22/2001	287	103	13	90	62	Pure Oxygen	69%
9/24/2001	287	108	18	90	53	Air + Oxygen	59%
9/25/2001	287	132	18	114	73	Air + Oxygen	64%
9/30/2001	435	130	14	116	67	Air + Oxygen	58%
10/4/2001	442	133	17	116	83	Air + Oxygen	72%
10/10	540	145	17.7	127	99.5	Air + Oxygen	78%
10/11	540	147	14.1	133	98.6	Air + Oxygen	74%
Average	378.7	124.6	17.3	107.3	71.7		66.1%

RBCOD OUR Test (9/9)

Time (min)	OUR	Ro
0	47.29	1.35
4.7	47.48	2.48
13.2	42.8	1.81
21.7	43.52	1.74
29.4	44.47	1.81
36.9	45.33	1.61
43.2	45.21	2.05
51.3	44.72	1.72
58.3	31.9	0.24
66	30.36	0.05
74.3	30.24	0.03
82.6	29.69	SUM =
90.7	30	14.90
98.8	30.27	
106.9	30.09	

RBCOD OUR Test (9/22)

Time (min)	OUR	Ro
0	**56.00**	**3.46**
6.7	53.32	1.51
9.9	47.33	1.67
14.4	47.51	1.65
18.8	47.61	2.19
24.6	48.16	2.55
31.2	38.39	1.52
38	29.41	0.73
47.9	25.81	SUM =
57.8	25.48	15.28
68.1	25.35	
78.6	25.02	
88.2	26.44	
101.5	31.19	
120.4	31.3	
128.4	31.28	
136.5	31.28	
144.4	31.22	
152	31.49	
159.8	31.49	
167.5	31.23	
175.2	31.19	
183	30.83	

RBCOD OUR Test (9/30)

Time (min)	OUR	Ro
0	71.13	1.84
2.9	69.87	1.90
6	69.54	2.19
9.6	69.70	2.02
12.9	69.87	2.21
16.5	69.82	2.27
20.2	67.18	2.45
24.5	45.31	0.94
29.1	36.90	0.41
35.4	36.43	0.29
40.4	35.46	0.18
44.8	33.48	SUM =
52.9	34.68	16.71
58.1	33.61	
65.8	34.5	
73.2	33.92	
80.4	33.66	
86	32.58	
96.4	33.25	
103.6	33.09	
110.7	33.4	
117.7	33.22	
124.8	33.14	

Flocculation/Filtration of Ultra-filtered Filtrates

Renton

	Method	Flocculation	10K MWCO	Ultrafiltration	100K MWCO	Ultrafiltration
Test		Flocculation	Ultrafiltration	UF => Floc	Ultrafiltration	UF => Floc
Date	Sample	COD, mg/L	COD, mg/L	ffCOD, mg/L	COD, mg/L	ffCOD, mg/L
	Influent	81.3	77.1	73.3	96.3	79.4
5/11/2001	Effluent	36	31.2	27.8	40.1	29.2
	RBCOD	45.3	45.9	45.5	56.2	50.2
	Influent	89.1	89.8	84.5	112.2	93.7
5/17/2001	Effluent	29.9	30	22.8	33.6	29.5
	RBCOD	59.2	59.8	61.7	78.6	64.2
	Influent	99.2	91.7	81.2	113.4	91.7
5/18/2001	Effluent	43	34	29	42.4	32.6
	RBCOD	56.2	57.7	52.2	71	59.1

Puyallup

	Method	Flocculation	10K MWCO	Ultrafiltration	100K MWCO	Ultrafiltration
Test		Flocculation	Ultrafiltration	UF => Floc	Ultrafiltration	UF => Floc
Date	Sample	COD, mg/L	COD, mg/L	ffCOD, mg/L	COD, mg/L	ffCOD, mg/L
	Influent	109	95.6	92	117.3	89.6
6/25/2001	Effluent	25.6	29.6	27	28.5	22.8
	RBCOD	83.4	66	65	88.8	66.8
	Influent	89.5	81.4	78.5	101.9	79.8
6/26/2001	Effluent	26	27.1	23	31.2	29.8
	RBCOD	63.5	54.3	55.5	70.7	50
	Influent	78	70.3	69	98.2	97.5
6/27/2001	Effluent	27.2	28.8	18.8	42.5	35.5
	RBCOD	50.8	41.5	50.2	55.7	62

APPENDIX A

NOMENCLATURE

A.1 Discussion

This appendix summarizes the nomenclature used throughout the manual. For model state variables, there has been an attempt to follow the system of nomenclature used in IWA publications, where the key principle is that soluble components are represented by "$S_{subscript}$" and particulate components by "$X_{subscript}$". When there is a generally accepted name for a combined variable (i.e., COD) this name has been used. There are some combined variables that are specifically required in a measurement method (e.g., SBCOD), and these have beed defined appropriately.

Note: Additional subscripts INF and EFF are used to denote influent and effluent parameters, respectively.

Table A-1. Nomenclature

Symbol	Model State Variable	Units
S_S	Soluble readily biodegradable COD concentration	(mg COD/L)
S_A	Soluble readily biodegradable volatile fatty acid COD concentration	(mg COD/L)
S_F	Soluble readily biodegradable complex COD concentration	(mg COD/L)
S_I	Soluble unbiodegradable COD concentration	(mg COD/L)
S_{COL}	Slowly biodegradable colloidal COD concentration	(mg COD/L)
S_O	Dissolved Oxygen when used as a model state variable	(mgO/L)
X_S	Slowly biodegradable COD concentration	(mg COD/L)
X_I	Particulate unbiodegradable COD concentration	(mg COD/L)
X_{PAO}	Active polyP organism (or PAO) concentration	(mg COD/L)
X_{AUT}	Active autotrophic organism concentration	(mg COD/L)
X_{HET}	Active heterotrophic organism concentration	(mg COD/L)
X_{ISS}	Inorganic suspended solids	
S_{NB}	Soluble biodegradable organic nitrogen	(mg N/L)

Table A-1. Nomenclature (cont.)

Symbol	Model State Variable	Units
S_{NH}	Ammonia nitrogen	(mg N/L)
S_{NO}	Nitrate + nitrite nitrogen	(mg N/L)
S_{NI}	Soluble unbiodegradable organic nitrogen	(mg N/L)
X_{NB}	Particulate biodegradable organic nitrogen	(mg N/L)
X_{NI}	Particulate unbiodegradable organic nitrogen	(mg N/L)
S_{PB}	Soluble biodegradable organic phosphorus	(mg P/L)
S_{PO4}	Orthophosphate phosphorus	(mg P/L)
S_{PI}	Soluble unbiodegradable organic phosphorus	(mg P/L)
X_{PB}	Particulate biodegradable organic phosphorus	(mg P/L)
X_{PI}	Particulate unbiodegradable organic phosphorus	(mg P/L)

Symbol	Combined Variable	Units
COD_T	Total COD	(mg COD/L)
TKN	Total Kjeldahl Nitrogen	(mg N/L)
TP	Total Phosphorus	(mg P/L)
N_{OB}	Biodegradable organic nitrogen (soluble + particulate)	(mg N/L)
P_{OB}	Biodegradable organic phosphorus (soluble + particulate)	(mg P/L)
S_{ffCOD}	Flocculated and filtered COD	(mg COD/L)
RBCOD	Readily biodegradable COD concentration	(mg COD/L)
SBCOD	Slowly biodegradable COD concentration	(mg COD/L)
BOD	Biochemical oxygen demand	(mg/L)
X_{VSS}	Volatile Suspended Solids	(mg/L)

Symbol	Kinetic/Stoichiometric Parameters – Heterotrophs	Units
μ_{HET}	Maximum specific growth rate for heterotrophs	(/d)
K_{SS}	Half-saturation coefficient for heterotrophs	(mg N/L)
b_{HET}	Heterotroph decay rate	(/d)
Y_{HET}	Heterotroph yield coefficient	(mg COD/mg COD)
η_g	ratio of growth rate with nitrate versus that for oxygen	

Symbol	Kinetic/Stoichiometric Parameters – Autotrophs (Nitrifiers)	Units
μ_{AUT}	Maximum specific growth rate for autotrophs	(/d)
K_{NH}	Half-saturation coefficient for autotrophs	(mg N/L)
b_{AUT}	Autotroph decay rate	(/d)
Y_{AUT}	Autotroph yield coefficient	(mg COD/mg N)

Table A-1. Nomenclature (cont.)

Symbol	Influent Wastewater Fractions	Units
f_{SS}	Fraction of total influent COD which is readily biodegradable	(mg COD/mg COD)
f_{XI}	Fraction of total influent COD which is particulate unbiodegradable	(mg COD/mg COD)
f_{SI}	Fraction of total influent COD which is soluble unbiodegradable	(mg COD/mg COD)
f_{XS}	Fraction of slowly biodegradable COD that is particulate	(mg COD/mg COD)
f_{XHET}	Fraction of total influent COD that is active hetorotrophic organisms	(mg COD/mg COD)
f_{SA}	Fraction of influent readily biodegradable COD which is VFA	(mg COD/mg COD)
f_{SNH}	Fraction of influent TKN that is free and saline ammonia	(mg N/mg N)
f_{SNI}	Fraction of influent TKN that is soluble unbiodegradable	(mg N/mg N)
f_{XNB}	Fraction of biodegradable organic TKN that is particulate	(mg N/mg N)
f_{XNI}	N content of the particulate unbiodegradable influent COD	(mg N/mg COD)
f_{SPO4}	Fraction of influent TP that is orthophosphate	(mg P/mg P)
f_{SPI}	Fraction of influent TP that is soluble unbiodegradable	(mg P/mg P)
f_{XPB}	Fraction of biodegradable organic TKN that is particulate	(mg P/mg P)
f_{XPI}	P content of the particulate unbiodegradable influent COD	(mg P/mg COD)

Symbol	Process Variables	Units
M_O	Mass of oxygen consumed / litre batch test volume in utilization of RBCOD	(mgO_2/L)
OUR	Oxygen utilization rate	($mgO_2/L/h$)
NUR	nitrate-nitrogen uptake rate	(mgN/L/h)
Q	Average influent flowrate	(volume/time)
V	Volume	(volume)
q_w	Average wastage flowrate	(volume/time)
V_P	Volume of reactor	(volume)
V_{sl}	Volume of activated sludge mixed liquor	(volume)
V_{ww}	Volume of wastewater	(volume)
SRT	Solids retention time (sludge age)	(days)
θ_x	Sludge age	(days)
VSS	Volatile suspended solids	(mg/L)
TSS	Total suspended solids	(mg/L)
f_N	Nitrogen content of waste sludge	(mg N/mg VSS)
f_{CV}	Sludge COD/VSS ratio	(mg COD/mg VSS)

Symbol	Miscellaneous
EBPR	Excess biological phosphorus removal
SBR	Sequencing batch reactor
IWA	International Water Association
BNR	Biological Nutrient Removal
ASM	Activated Sludge Model (specifically those published by the IWA)

APPENDIX B

BOD CALCULATIONS

B.1 Background

In the IWA-based models, characterization of the carbonaceous material in municipal wastewater influent is in terms of the chemical oxygen demand (COD). This selection is based on a number of factors, but primarily because COD provides a consistent basis for description of the activated sludge process, and for quantifying sludge production, oxygen demand, etc. The rationale for preferring COD over other parameters such as biochemical oxygen demand (BOD) or total organic carbon (TOC) is discussed briefly later.

In many instances, data on treatment plant operation are recorded in terms of the carbonaceous biochemical oxygen demand (BOD) for both the influent and effluent, usually reported as the 5-day value (BOD_5). (Certain European communities use the BOD_7 parameter.)

Observation on a large number of municipal wastewaters indicates that there is a near-constant ratio between COD and BOD for a particular wastewater, viz.:

$$\frac{COD}{BOD_5} = F$$

The factor or ratio, F, usually is in the range of 1.9 to 2.2 mg COD/mg BOD_5. For a particular treatment plant where there is an historical BOD database, and no COD data, a limited number of parallel measurements of the two parameters likely will establish a reasonably accurate value for F. Thereafter, the F factor can be used to estimate influent COD values for modeling, based on measurements of BOD_5.

The following section presents a method for empirically calculating carbonaceous BOD of an influent wastewater based on the total COD and the various wastewater characteristic fractions.

Table B-1. Additional Definition of Terms for This Appendix

Symbol	Parameter
b_{HET}	Endogenous decay rate constant ($\cong$ 0.24 d^{-1})
BOD_E	Biochemical oxygen demand due to endogenous respiration
BOD_S	Soluble component of biochemical oxygen demand
BOD_{XSP}	Particulate slowly biodegradable component of biochemical oxygen demand
BOD_{XHET}	Active biomass component of biochemical oxygen demand
BOD_T	Total biochemical oxygen demand
F	Fraction of active mass remaining as endogenous residue ($\cong$ 0.20)
K	First order rate constant for X_{SP} degradation ($\cong$ 0.40 d^{-1})
$(MO_2)_G$	Mass of oxygen utilized for growth on soluble substrate
$(MO_2)_E$	Mass of oxygen utilized for endogenous metabolism
OUR	Oxygen utilization rate (mg/L/day)
OUR_E	Oxygen utilization rate due to endogenous metabolism
OUR_G	Oxygen utilization rate for substrate utilization (growth)
T	Time (d)
X_{HET}	Active heterotrophic organism concentration at time zero (mg COD/L)
$X_{S,0}$	Slowly biodegradable particulate COD concentration at time zero (mg COD/L)
Y_{HET}	Yield of active organisms ($\cong$ 0.666 mg cell COD mg COD^{-1})

B.2 Basis for BOD Calculations

Note: Setting up the equations in a spreadsheet to calculate soluble and total BOD based on total COD and the various wastewater characteristics provides a method for estimating f_{XI} through comparing predicted and observed COD/BOD ratios for different f_{XI} estimates.

The objective here is to demonstrate a method for calculating filtered and total carbonaceous BOD concentrations based on wastewater characteristics in terms of COD, and show how the equations were developed. Understanding development of the equations presented below is facilitated by Figure 4-1, which illustrates the division of a wastewater stream into the different components.

Calculation of BOD is based on differing rates of degradation of the different components (e.g., influent biodegradable material (readily and slowly biodegradable), active organism masses which exert an endogenous oxygen demand and hence a BOD).

The approach for calculating BOD is to distinguish three components, and to calculate the BOD contribution for each component independently. These components are as follows:

1. BOD associated with utilization of soluble COD (both readily biodegradable and colloidal slowly biodegradable) and the biomass generated from this utilization;
2. BOD associated with utilization of slowly biodegradable particulate COD and the biomass generated from this utilization; and

3. BOD exerted by active biomass present in the sample. That is, biomass initially present; not biomass generated through utilization referred to in 1 and 2.

B.3 BOD Associated with Soluble Biodegradable COD (S_B):

For the purpose of BOD calculations, the "soluble" biodegradable COD (S_B) can be considered to be the sum of the readily biodegradable COD (S_S) and the biodegradable colloidal COD (S_{COL}), i.e.,

$$S_B = S_S + S_{COL} \qquad \text{(B.3-1)}$$

The oxidation of this material occurs rapidly, and for the purpose of these calculations is assumed to occur instantaneously (i.e., at t = 0). Assuming that the soluble biodegradable COD is oxidized instantly to form new cells:

$$S_B \rightarrow X_{HET} = Y_{HET} \cdot S_B \quad \text{(at time t = 0)} \qquad \text{(B.3-2)}$$

then the mass of oxygen consumed for growth of these organisms $(MO_2)_G$ is:

$$(MO_2)_G = (1 - Y_{HET}) \cdot S_B \qquad \text{(B.3-3)}$$

Assuming that the rate of change in organism concentration due to endogenous decay is first order with respect to the active organism concentration, an expression for active organism concentration can be obtained, i.e.,

$$\frac{dX_{HET}}{dt} = -b_{HET} \cdot X_{HET} \qquad \text{(B.3-4)}$$

$$\Rightarrow X_{HET} = X_{HET,0} \cdot e^{-b_{HET}\,t} \qquad \text{(B.3-5)}$$

The BOD due to endogenous metabolism by the organisms generated from growth on S_S can then be calculated using either of the methods described below.

B.3.1 Method 1

From equation B.3-2 and B.3-5, the endogenous mass loss at time t can be calculated:

$$\begin{aligned}\Delta X_{HET} &= X_{HET,0} - X_{HET} \\ &= X_{HET,0} \cdot (1 - e^{-b_{HET}\,t}) \\ &= Y_{HET} \cdot S_B \cdot (1 - e^{-b_{HET}\,t})\end{aligned} \qquad \text{(B.3.1-1)}$$

and the mass of oxygen consumed for endogenous metabolism:

$$\begin{aligned}(MO_2)_E &= (1 - f) \cdot \Delta X_{HET} \\ &= (1 - f) \cdot Y_{HET} \cdot S_B \cdot (1 - e^{-b_{HET}\,t})\end{aligned} \qquad \text{(B.3.1-2)}$$

This is equal to the BOD due to endogenous decay, i.e.,

$$\begin{aligned} BOD_E &= (1-f)\cdot X_{HET,0}\cdot(1-e^{-b_{HET}\,t}) \\ &= (1-f)\cdot Y_{HET}\cdot S_B\cdot(1-e^{-b_{HET}\,t}) \end{aligned} \qquad \text{(B.3.1-3)}$$

The soluble BOD component can now be calculated by summing Eqs. B.3-3 and B.3.1-3, i.e.,

$$\begin{aligned} BOD_S &= (MO_2)_G + (MO_2)_E \\ &= (1-Y_{HET})\cdot S_B + (1-f)\cdot Y_{HET}\cdot S_B\cdot(1-e^{-b_{HET}\,t}) \end{aligned} \qquad \text{(B.3.1-4)}$$

B.3.2 Method 2

The OUR due to endogenous metabolism can also be related to the rate of change in active organism concentration (taking into account that a fraction of the active mass becomes endogenous residue), i.e.,

$$\begin{aligned} OUR_E &= -(1-f)\cdot\frac{dX_{HET}}{dt} \\ &= b_{HET}\cdot(1-f)\cdot X_{HET} \\ &= b_{HET}\cdot(1-f)\cdot X_{HET,0}\cdot e^{-b_{HET}\,t} \end{aligned} \qquad \text{(B.3.2-1)}$$

The BOD due to endogenous decay is then the cumulative mass of oxygen used over time, i.e.,

$$\begin{aligned} BOD_E &= \int_0^t OUR_E\,dt \\ &= b_{HET}\cdot(1-f)\cdot X_{HET,0}\int_0^t e^{-b_{HET}\,t}dt \\ &= b_{HET}\cdot(1-f)\cdot X_{HET,0}\cdot\left(-\frac{1}{b_{HET}}e^{-b_{HET}\,t}\right)+C \\ &= -(1-f)\cdot X_{HET,0}\cdot e^{-b_{HET}\,t}+C \end{aligned} \qquad \text{(B.3.2-2)}$$

and since at t = 0, $BOD_E = 0$,

$$\Rightarrow C = (1+f)\cdot X_{HET,0}$$

Substituting the expression obtained for $X_{HET,0}$ (equation B.3-2) into the above gives an expression for the BOD due to endogenous decay:

$$\begin{aligned} BOD_E &= -(1-f)\cdot X_{HET,0}\cdot e^{-b_{HET}\,t}+(1-f)\cdot X_{HET,0} \\ &= (1-f)\cdot X_{HET,0}\cdot(1-e^{-b_{HET}\,t}) \\ &= (1-f)\cdot Y_{HET}\cdot S_B\cdot(1-e^{-b_{HET}\,t}) \end{aligned} \qquad \text{(B.3.2-3)}$$

This is the same expression obtained using Method 1 (equation B.3.1-3). The total soluble BOD portion is given by equation B.3.1-4.

B.4 BOD Associated with Particulate Slowly Biodegradable COD (X_S):

The biochemical oxygen demand related to the particulate slowly biodegradable material can be calculated from the cumulative oxygen consumption for growth on X_S and endogenous respiration exerted by organisms from this growth. The OUR is the sum of two components:

$$\begin{aligned} OUR &= OUR_G + OUR_E \\ &= (1-Y_{HET}) \cdot \left(-\frac{dX_S}{dt}\right) + (1-f) \cdot \left(-\frac{dX_{HET}}{dt}\right)_E \\ &= (1-Y_{HET}) \cdot \left(-\frac{dX_S}{dt}\right) + (1-f) \cdot \left(b_{HET} X_{HET}\right) \end{aligned} \qquad \text{(B.4-1)}$$

The rate of change in particulate slowly biodegradable substrate concentration is assumed to be first order with respect to substrate concentration, i.e.,

$$\frac{dX_S}{dt} = -k \cdot X_S$$

$$\Rightarrow X_S = X_{S,0} \cdot e^{-kt} \qquad \text{(B.4-2)}$$

The active organism concentration in equation B.4-1 is obtained by integrating an expression for the rate of change of X_{HET}. This rate is a consequence of an increase due to growth on X_S, minus a decrease due to endogenous metabolism, i.e.,

$$\frac{dX_{HET}}{dt} = Y_{HET} \cdot \left(-\frac{dX_S}{dt}\right) - b_{HET} \cdot X_{HET} \qquad \text{(B.4-3a)}$$

Substituting equation B.4-2 into the above results in a linear differential equation for X_{HET} as a function of time, i.e.,

$$\begin{aligned} \frac{dX_{HET}}{dt} &= -Y_{HET} \cdot \frac{dX_S}{dt} - b \cdot X_{HET} \\ &= -Y_{HET} \frac{d}{dt}\left(X_{S,0} \cdot e^{-kt}\right) - b_{HET} \cdot X_{HET} \\ &= kY_{HET} X_{S,0} \cdot e^{-kt} - b_{HET} \cdot X_{HET} \end{aligned} \qquad \text{(B.4-3b)}$$

Or equivalently,

$$\frac{dX_{HET}}{dt} + b_{HET} \cdot X_{HET} = kY_{HET} X_{S,0} \cdot e^{-kt} \qquad \text{(B.4-3c)}$$

An integration factor of $e^{b_{HET} t}$ can be used to solve the differential equation. Multiplying both sides of the equation by $e^{b_{HET} t}$ results in a simplified expression:

$$e^{b_{HET}t} \cdot \frac{dX_{HET}}{dt} + b \cdot e^{b_{HET}t} \cdot X_{HET} = e^{b_{HET}t} \cdot kY_{HET}X_{S,0} \cdot e^{-kt}$$

$$\frac{d}{dt}(e^{b_{HET}t} \cdot X_{HET}) = kY_{HET}X_{S,0} \cdot e^{(b_{HET}-k)t} \qquad \text{(B.4-4)}$$

Integrating both sides of equation B.4-4:

$$(e^{b_{HET}t} \cdot X_{HET}) = \int kY_{HET}X_{S,0} \cdot e^{(b_{HET}-k)t}\,dt$$

$$(e^{b_{HET}t} \cdot X_{HET}) = \frac{kY_{HET}X_{S,0}}{(b_{HET}-k)} \cdot e^{(b_{HET}-k)t} + C \qquad \text{(B.4-5)}$$

$$X_{HET} = \left(\frac{kY_{HET}X_{S,0}}{b_{HET}-k}\right) \cdot e^{-kt} + Ce^{-b_{HET}t}$$

At t = 0, $X_{HET} = X_{HET,0}$, therefore solving for C gives:

$$X_{HET} = \left(\frac{kY_{HET}X_{S,0}}{b_{HET}-k}\right) \cdot e^{-kt} + \left(X_{HET,0} - \frac{kY_{HET}X_{S,0}}{b_{HET}-k}\right) \cdot e^{-} \qquad \text{(B.4-6)}$$

Returning to equation B.4-1 for the oxygen utilization rate and substituting for X_{HET}:

$$\begin{aligned} OUR &= (1-Y_{HET}) \cdot \left(-\frac{dX_S}{dt}\right) + (1-f) \cdot \left(b_{HET}X_{HET}\right) \\ &= (1-Y_{HET}) \cdot kX_S + (1-f) \cdot b_{HET}X_{HET} \\ &= (1-Y_{HET}) \cdot kX_{S,0} \cdot e^{-kt} + (1-f) \cdot b_{HET} \cdot \left[\left(\frac{kY_{HET}X_{S,0}}{b_{HET}-k}\right) \cdot e^{-kt} + \left(X_{HET,0} - \frac{kY_{HET}X_{S,0}}{b_{HET}-k}\right) \cdot e^{-b_{HET}t}\right] \end{aligned} \qquad \text{(B.4-7)}$$

Integrating equation B.4-7 gives an expression for the BOD of the slowly biodegradable material as a function of time, i.e.,

$$\begin{aligned} BOD_{XS} &= \int_0^t OUR\,dt \\ &= \int_0^t (1-Y_{HET}) \cdot kX_{S,0} \cdot e^{-kt} + (1-f) \cdot b_{HET} \cdot \left[\left(\frac{kY_{HET}X_{S,0}}{b_{HET}-k}\right) \cdot e^{-kt} + \left(X_{HET,0} - \frac{kY_{HET}X_{S,0}}{b_{HET}-k}\right) \cdot e^{-b_{HET}t}\right] dt \\ &= -(1-Y_{HET}) \cdot X_{S,0} \cdot e^{-kt} - \frac{(1-f) \cdot b_{HET}Y_{HET}X_{S,0}}{(b_{HET}-k)} \cdot e^{-kt} - (1-f) \cdot \left(X_{HET,0} - \frac{kY_{HET}X_{S,0}}{b_{HET}-k}\right) \cdot e^{-b_{HET}t} \\ &\quad + (1-Y_{HET}) \cdot X_{S,0} + \frac{(1-f) \cdot b_{HET}Y_{HET}X_{S,0}}{(b_{HET}-k)} + (1-f) \cdot \left(X_{HET,0} - \frac{kY_{HET}X_{S,0}}{b_{HET}-k}\right) \end{aligned}$$

Assuming $X_{HET,0} \approx 0$ and re-arranging the above expression,

$$BOD_{XS} = X_{S,0} \cdot \left[\left((1 - Y_{HET}) + \frac{(1-f) \cdot b_{HET} Y_{HET}}{(b_{HET} - k)} \right) \cdot \left(1 - e^{-kt}\right) - \left(\frac{(1-f) \cdot Y_{HET} k}{(b_{HET} - k)} \right) \cdot \left(1 - e^{-b_{HET} t}\right) \right] \tag{B.4-8}$$

The total BOD (if no biomass is present) is then the sum of BOD_S and BOD_{XS},. (Eqs. B.3.1-4 and B.4-8), i.e.,

$$BOD_T = X_{S,0} \cdot \left[\left((1 - Y_{HET}) + \frac{(1-f) \cdot b_{HET} Y_{HET}}{(b_{HET} - k)} \right) \cdot \left(1 - e^{-kt}\right) - \left(\frac{(1-f) \cdot Y_{HET} k}{(b_{HET} - k)} \right) \cdot \left(1 - e^{-b_{HET} t}\right) \right] \tag{B.4-9}$$

$$+(1 - Y_{HET}) \cdot S_B + (1-f) \cdot Y_{HET} \cdot S_B \cdot (1 - e^{-b_{HET} t})$$

B.5 BOD Associated with Active Biomass

In cases where there are active organisms present, the BOD exerted by the organisms can be calculated from the rate of change in active organisms due to endogenous metabolism, i.e.,

$$BOD_{XHET} = (1-f) \cdot X_{HET} \cdot (1 - e^{-b_{HET} t}) \tag{B.5-1}$$

This must then be added to the soluble and slowly degradable BODs to obtain the total BOD (i.e., sum equations B.4-9, and B.5-1).

B.6 Example

This example demonstrates the BOD calculation procedure for a municipal influent wastewater stream with the following characteristics:

COD_T = 500 mg COD/L

f_{SI} = 0.10

f_{XI} = 0.08

f_{SS} = 0.20

f_{XHET} = 0.00

f_{XS} = 0.75

The influent "soluble" and slowly biodegradable concentrations are calculated using the influent fractions:

$$S_S = f_{SS}\, COD_T$$

$$= (0.20)(500)$$

$$= 100\ mg\ COD/L$$

$$SBCOD = (1 - f_{SI} - f_{XI} - f_{SS} - f_{XHET})\, COD_T$$

$$= (1 - 0.10 - 0.08 - 0.20 - 0.0)(500)$$

$$= 310\ mg\ COD/L$$

$$S_{COL} = (1 - f_{XS})\, X_S$$

$$= (1 - 0.75)(310)$$

$$= 77.5\ mg\ COD/L$$

$$X_S = (f_{XS})\, SBCOD$$

$$= (0.75)(310)$$

$$= 232.5\ mg\ COD/L$$

$$S_B = S_S + S_{COL}$$
$$= 100 + 77.5$$
$$= 177.5\ mg\ COD/L$$

Note: The parameter values for f and b are for endogenous respiration, and should not be confused with the corresponding terms used in the death-regeneration approach for modeling biomass decay in the IWA-type models.

The soluble BOD_5 can then be calculated using Eq. B.3.1-4, and the parameters listed below:

f = 0.20

Y_{HET} = 0.666 mg COD/mg COD

b_{HET} = 0.24 d^{-1}

k = 0.40 d^{-1}

$$BOD_S = (1 - Y_{HET}) \cdot S_B + (1 - f) \cdot Y_{HET} \cdot S_B \cdot (1 - e^{-b_{HET}t})$$

$$= (1 - 0.666)(177.5) + (1 - 0.20)(0.666)(177.5)(1 - e^{-(0.24)(5)})$$

$$= 125.4 \ mg \ BOD / L$$

Similarly, the slowly degradable BOD_5 can be calculated using equation B.4-8:

$$BOD_{XS} = X_{S,0} \cdot \left[\left((1 - Y_{HET}) + \frac{(1 - f) \cdot b_{HET} Y_{HET}}{(b_{HET} - k)} \right) \cdot \left(1 - e^{-kt}\right) - \left(\frac{(1 - f) \cdot Y_{HET} k}{(b_{HET} - k)} \right) \cdot \left(1 - e^{-b_{HET}t}\right) \right]$$

$$= (232.5) \left[\left((1 - 0.666) + \frac{(1 - 0.20)(0.24)(0.666)}{(0.24 - 0.40)} \right) \cdot (1 - e^{-(0.40)(5)}) - \left(\frac{(1 - 0.20)(0.666)(0.40)}{(0.24 - 0.40)} \right) \cdot (1 - e^{-(0.24)(5)}) \right]$$

$$= 122.9 \ mg \ BOD / L$$

In this example the active organism fraction is zero; therefore the total BOD_5 is the sum of the BOD_S and the BOD_{XS}, i.e.,

$$BOD_T = BOD_S + BOD_{XSP}$$
$$= 125.4 + 122.9$$
$$= 248.3 \ mg \ BOD / L$$

B.7 COD versus BOD as a Modeling Parameter

The advantage of selecting COD as the parameter for quantifying the "strength" of organic material in the influent, as opposed to BOD or TOC, is that it provides a consistent basis for description of the activated sludge process. Marais and Dold (1985) outlined the rationale which makes COD the appropriate parameter. It is worth reviewing this rationale briefly as selection of the COD parameter is fundamental to the application of the models. Briefly, the suitability of COD is established by considering utilization of organic substrate. In the process of metabolism the organic substrate serves two functions for the organisms as shown schematically in Figure B-1:

1. A portion of the organic material is oxidized to CO_2 and water, providing energy for maintaining the homeostatic balance for existing cell mass (osmotic pressure, ionic balance, membrane potential, etc.) and for (2) below. The energy is provided by transferring electrons from the organic substrate through the electron transport chain to the terminal electron acceptor (oxygen in the case of an aerobic system, or nitrate under anoxic conditions). Under the substrate limited conditions usually encountered in activated sludge systems, the organisms utilize a relatively fixed fraction of the energy available from oxidation for the energy consuming processes.

2. The remaining portion of the organic material is converted into new heterotrophic cell mass, utilizing energy available from process (1).

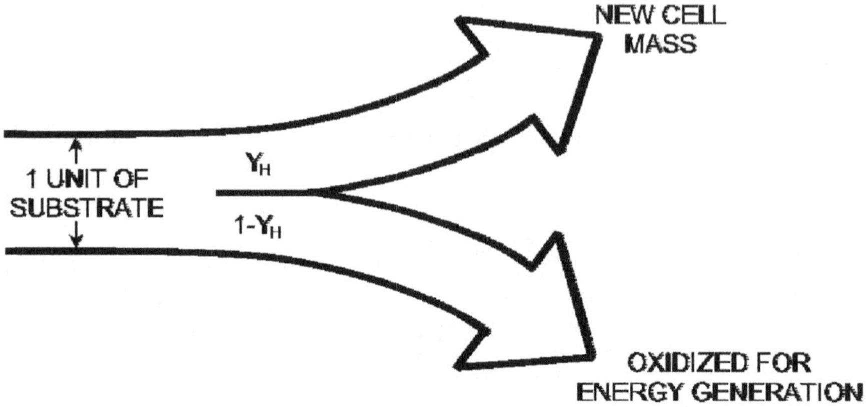

Figure B-1. Schematic Representation of the Utilization of Substrate by Heterotrophic Organisms Showing the Division between Substrate Oxidized for Energy Generation and Growth of New Cell Mass

Regarding the relative proportions allocated between (1) and (2), this is quantified by the ratio

$$Y_H = \frac{cell\ mass\ formed}{substrate\ utilized}$$

termed as the yield coefficient. Therefore, if the electron donating potential of the organic substrate in the activated sludge system influent (consisting of a broad spectrum of compounds) is measured it is possible to quantify sludge production (from Y_H) and oxygen demand (from $(1-Y_H)$). This observation provides the rationale for selecting the COD parameter.

The electron donating capacity of organic material is measured in the COD test. In the test each mole of oxygen (O_2) accepts four electron equivalents (e^- eq); therefore, the COD is a direct measure of the electron donating potential. The link between electron equivalents (COD) of the substrate, the near constant yield of organism mass per unit substrate COD, and the corresponding fixed oxygen requirement per unit substrate COD metabolized, makes the COD a fundamental parameter in the analysis of activated sludge behavior.

Note: Regarding the COD of the waste sludge, this may be measured directly by the COD test or be estimated from the VSS measurement. Because cell mass has an approximately constant composition and is made up of an essentially fixed number of electron equivalents (e^- eq) per unit mass, the COD/VSS ratio is near constant (approximately 1.48 mg COD/mg VSS).

A factor that adds impetus to the selection is that mass balances can be made in terms of COD. As electrons cannot be created or destroyed in a biosystem operated at steady state, the mass of COD entering with the influent per unit time must equal the sum of (1) the mass of COD leaving in the effluent, (2) the COD of the wasted sludge, and (3) the oxygen consumed in the utilization of the organic material (from the oxygen utilization rate measurement); that is, a mass balance is possible.

Characteristics of the COD as a measure of the "strength" of a wastewater are not realized by either the BOD or the TOC parameters. The BOD measures only that portion of the e^- eq in the substrate utilized for energy generation and excludes the portion of the substrate e^- eq transformed into new cell mass. Therefore, BOD cannot be used as the basis for a mass balance. The TOC is deficient in that the ratio of carbon/e^- eq differs between organic compounds and therefore TOC is an inappropriate parameter when dealing with the mixed substrate influents to wastewater treatment plants.

B.8 References

Marais, G.v.R., and P.L. Dold. 1985. Biological removal of carbon, nitrogen and phosphorus in single sludge systems. In *Proc. Advances in Biological Wastewater Treatment Seminar* (pp . 207-233). Rome Italy: Instituto di Ricerca Sulle Acque (IRSA).

APPENDIX C

OPERATING A BENCH-SCALE CONTINUOUS FLOW SYSTEM

C.1 Background

Accurate estimation of a number of influent wastewater characteristics can be achieved through the analysis of data gathered in the operation of a bench-scale single reactor aerobic activated sludge system operated under steady state conditions at a fixed sludge age. These include:

- Unbiodegradable COD fractions (f_{SI} and f_{XI})
- Ammonia fraction (f_{SNH})
- Soluble unbiodegradable organic nitrogen (f_{SNI})
- Division of biodegradable organic nitrogen into soluble and particulate portions (f_{SNB} and f_{XNB}, respectively)

C.2 Operation

Steady state operation is achieved by feeding the system a fixed daily volume of influent (Q) of a specified COD concentration (COD_T) at a constant rate over the day. Maintaining the fixed influent loading is facilitated by storing batches of wastewater sufficient for 7 to 10 days' operation at 4°C. Each day the feed is made up by diluting wastewater to the selected influent COD_T concentration with tap water. The selected COD_T should be slightly lower than the minimum anticipated COD_T of the collected batches. Care should be taken to ensure that there are no losses of COD in the feed system. The feed container should be cleaned thoroughly on a daily basis to avoid loss of COD through biological growth. Maintaining the container at a low temperature will aid in avoiding COD loss. The feed should be mixed gently on a continuous basis to keep particulate material in suspension and maintain a uniform influent composition.

The sludge age of the system must be maintained at a constant value in excess of approximately 8 days. This is best attained through hydraulic control of sludge age, where a fixed volume of mixed liquor is removed from the reactor each day. The system sludge age (θ_X)

given by the mass of sludge in the system divided by the mass removed per day. In a laboratory unit where the mass of sludge in the settling tank is minimal, the sludge age is given by:

$$\theta_X = \frac{V_P \cdot X_{VSS}}{q_W \cdot X_{VSS}} = \frac{V_P}{q_W} \qquad \text{(C.2-1)}$$

where:

X_{VSS} = mixed liquor VSS concentration (mg VSS/L)

V_P = reactor volume (L)

q_W = influent flow rate (L/d)

For instance, to fix the system sludge age at 10 days, the volume of mixed liquor withdrawn directly from the reactor should be 1/10th of the reactor volume.

Under this mode of operation, with a constant influent loading and operating at a fixed sludge age, the system will approach a steady state condition where the monitored parameters do not vary significantly from day to day. Once at steady state, the parameters listed in Table C-1 should be measured on a daily basis. Measurements of the reactor unfiltered COD and TKN should be made periodically to assess the values of the mixed liquor solids COD/VSS ratio (f_{CV}) and nitrogen content (f_N), respectively. A 0.45 μm membrane filter should be used to remove particulate material for the filtered sample analyses. Care should be taken to wash 0.45 mm membrane filters before using them to measure effluent filtered COD. Organic matter leaching from the filters can give artificially high values. Monitoring of the system should continue until acceptable steady state average values are observed. This will require operation for at least three sludge ages.

Table C-1. Daily Data Collection Protocol for Continuous Flow Laboratory-Scale System

Parameter	Influent	Effluent	Reactor
Unfiltered COD	✔	✔	✔ (1)
Filtered COD	✔	✔	
ffCOD	✔	✔	
Unfiltered TKN	✔	✔	✔ (1)
Filtered TKN		✔	
Ammonia	✔	✔	
Nitrate/Nitrate		✔	
VSS	✔		✔
TSS	✔		✔
OUR			✔

1- Measured periodically to estimate f_{CV} and f_N.

A check on the acceptability of the experimental data should be made by conducting mass balances for COD and nitrogen (N). Satisfactory balances should account for the masses of COD and N in the influent within 95 to 105 percent (Barker and Dold, 1995). Poorer balances indicate that errors have been made either in the analyses or in the operation of the system.

C.2.1 COD Balance

Under steady state conditions, the daily mass of COD in the influent should equal the sum of the unfiltered effluent COD, the COD of the waste sludge and the oxygen utilized for oxidation of organic material. The oxygen component is obtained by deducting oxygen utilized for nitrification from the total oxygen utilization, noting that 4.57 mg O_2 is utilized per mg S_{NO} formed (assuming that ammonia is oxidized fully to nitrate). For a mass balance on COD in the single reactor *aerobic* activated sludge system:

$$Q \cdot COD_T = Q \cdot COD_{T,EFF} + q_W \cdot X_{VSS} \cdot f_{CV} + OUR \cdot V_P \cdot 24 - 4.57 \cdot S_{NO} \cdot Q \qquad \text{(C.2.1-1)}$$

where:

COD_T = total influent COD concentration (mg COD/L)

$COD_{T,EFF}$ = total effluent COD concentration (mg COD/L)

OUR = oxygen utilization rate (mg/L/hour)

S_{NO} = effluent nitrate concentration (mg NO_3-N/L)

f_{CV} = COD/VSS ratio of waste solids (≅ 1.48 mg COD/mg VSS)

C.2.2 Nitrogen Balance

Under steady state conditions, the daily mass of TKN in the influent should equal the sum of the unfiltered effluent TKN, the effluent nitrate and the TKN in the waste sludge. For a mass balance on nitrogen in the single reactor *aerobic* activated sludge system:

$$Q \cdot TKN = Q \cdot TKN_{EFF} + Q \cdot S_{NO} + q_W \cdot X_V \cdot f_N \qquad \text{(C.2.2-1)}$$

where:

TKN = influent TKN concentration (mg N/L)

TKN_{EFF} = effluent TKN concentration (mg N/L)

f_N = TKN/VSS ratio of waste solids (≅ 0.1 mgN/mgVSS)

Having established the acceptability of the experimental data, the wastewater characteristics can be estimated from the measured data using the methods outlined in the earlier chapters.

C.2.3 System Sludge Age (SRT)

Note: The low effluent VSS concentration cannot be measured directly, and must be estimated.

Before attempting to estimate the particulate unbiodegradable COD fraction (f_{XI}) the sludge age should be recalculated to account for any loss of solids in the effluent. Noting that the concentration of VSS in the effluent is given by the difference of the unfiltered and filtered effluent COD concentrations divided by the COD/VSS ratio, the correct system sludge age can be calculated as follows:

$$\theta_X = \frac{V_P \cdot X_{VSS}}{q_W \cdot X_{VSS} + Q \cdot \dfrac{(COD_{T,EFF} - COD_{T,EFF,FILT})}{f_{CV}}} \qquad \text{(C.2.3-1)}$$

where:

$COD_{T,EFF,FILT}$ = filtered effluent COD concentration (mg COD/L)

C.2.4 Note on Estimating f_{XI}

The magnitude of f_{XI} is estimated by trial using a model. The reactor configuration and operating parameters are specified as inputs to the model, together with the measured values of the other wastewater characteristics (f_{SI}, f_{SNH}, f_{XNB}); initially, f_{SNI} can be set to zero. Selecting an initial value for f_{XI} of say 0.10, the model is used to predict the response of the system. Incremental adjustments are made to the estimate of f_{XI} until good correspondence is obtained between the predicted and observed VSS, oxygen utilization rate, effluent TKN and nitrate concentrations. It should be noted that, at each iteration, the magnitude of f_{XNB} should be adjusted according to equation 10.8, according to the estimate of f_{XI}; this will have a minor influence on the results.

C.3 Example of Continuous Flow System

The experimental protocol outlined above was followed using unsettled (raw) wastewater drawn from the influent to the full-scale treatment plant. A 6.3 L single-reactor activated sludge system was operated at a nominal sludge age of 10.6 days; sludge age was established by wasting a volume of 0.60 L of mixed liquor from the reactor each day (see equation C.2-1). Temperature in the laboratory was maintained at 20°C. The system was operated for a period of 50 days at a constant daily loading (Q = 15 L/d, influent COD ≅ 310 mg COD/L), using 5 different batches of wastewater. Differences in the wastewater constitution from batch to batch (TKN/COD ratio, readily biodegradable COD, etc.) were relatively small. Once a steady state was attained, values of the parameters monitored daily were averaged for operation with each batch of wastewater. A typical data set was as follows:

$Unfiltered\ \ influent\ COD = 310\ \ mg/L$
$Unfiltered\ \ influent\ TKN = 36.0\ \ mgN/L$
$Filtered\ \ influent\ TKN = 28.8\ \ mgN/L$
$Influent\ NH_3 - N = 25.2\ mg/L$
$Unfiltered\ \ effluent\ COD = 71\ \ mg/L$
$Filtered\ \ effluent\ COD = 41\ \ mg/L$
$Unfiltered\ \ effluent\ TKN = 4.2\ \ mgN/L$
$Filtered\ \ effluent\ TKN = 2.2\ \ mgN/L$
$Effluent\ NH_3 - N = 0.3\ mg/L$
$Effluent\ NO_3\ \text{-}\ N = 25.2\ \ mgN/L$
$Mixed\ liquor\ VSS = 1784\ \ mg/L$
$Reactor\ OUR = 24.1\ \ mg/L/hour$

C.3.1 COD Balance

$$Q \cdot COD_T = Q \cdot COD_{T,EFF} + q_W \cdot X_{VSS} \cdot f_{CV} + OUR \cdot V_P \cdot 24 - 4.57 \cdot S_{NO} \cdot Q$$
$$Q \cdot COD_T = 15 \cdot 310 = 4650\ mg/day$$
$$Q \cdot COD_{T,EFF} + q_W \cdot X_{VSS} \cdot f_{CV} + OUR \cdot V_P \cdot 24 - 4.57 \cdot S_{NO} \cdot Q -$$
$$15 \cdot 71 + 0.6 \cdot 1784 \cdot 1.48 + 24.1 \cdot 6.3 \cdot 24 - 4.57 \cdot 25.2 \cdot 15 = 4566\ mg/day$$
$$\%\ Recovery\ = (4566/4650) \cdot 100 = 98.2\%$$

C.3.2 Nitrogen Balance

$$Q \cdot N_T = Q \cdot N_{TE} + Q \cdot N_{NO3} + q_W \cdot X_V \cdot f_N$$
$$Q \cdot N_T = 15 \cdot 36.0 = 540\ mg/day$$
$$Q \cdot N_{TE} + Q \cdot N_{NO3} + q_W \cdot X_V \cdot f_N =$$
$$15 \cdot 4.2 + 15 \cdot 25.2 + 0.6 \cdot 1784 \cdot 0.1 = 548\ mg/day$$
$$\%\ Recovery = (548/540) \cdot 100 = 101.4\%$$

C.3.3 Unbiodegradable Soluble COD (f_{SI})

$Unfiltered\ \ influent\ COD = 310\ \ mg/L$
$Filtered\ \ effluent\ COD = 41\ \ mg/L$
$S_{SI} = 41\ \ mg/L$
$$f_{SI} = \frac{41\ \ mg/L}{310\ \ mg/L} = 0.13$$

C.3.4 Ammonia Nitrogen (f_{SNH})

$$Unfiltered\ influent\ TKN = 36.0\ mg/L$$
$$Influent\ NH_3 - N\ (N_{NH3}) = 25.2\ mg/L$$
$$f_{SNH} = \frac{25.2\ mg/L}{36.0\ mg/L} = 0.70$$

C.3.5 Unbiodegradable Soluble Organic Nitrogen (f_{SNI})

$$S_{NI} \approx\ Filtered\ effluent\ TKN\ -\ N_{NH3} - 0.4\ mgN/L$$
$$\approx\ 2.2\ -\ 0.3 - 0.4\ = 1.5\ mgN/L$$
$$f_{SNI} = \frac{1.5}{36.0} = 0.042$$

C.3.6 Division of Biodegradable Organic Nitrogen into Soluble and Particulate: f_{XNB}

$$f_{XNB} = \frac{X_{NB}}{S_{NB} + X_{NB}}$$
$$= \frac{36.0 - 28.8 - 0.07 \cdot 0.22 \cdot 310}{(1 - 0.70 - 0.042)\ 36.0 - 0.07 \cdot 0.22 \cdot 310}$$
$$= 0.54$$

C.3.7 System Sludge Age (SRT)

$$\theta_X = \frac{V_P \cdot X_{VSS}}{q_W \cdot X_{VSS} + Q \cdot \frac{(COD_{T,EFF} - COD_{T,EFF,FILT})}{f_{CV}}}$$
$$= \frac{6.3 \cdot 1784}{0.6 \cdot 1784 + 15 \cdot \frac{(71 - 41)}{1.48}}$$
$$= 8.24\ days$$

C.3.8 Particulate Unbiodegradable COD (f_{XI})

With this sludge age, together with the other operating conditions and measured influent parameters as inputs, a simulator was used to estimate f_{XI} iteratively. No changes were made to the default values of the kinetic and stoichiometric parameters in the model. Table C-2 shows a comparison of the observed VSS, OUR and effluent nitrate concentration to the predicted values for different estimates of f_{XI}. The strong influence of f_{XI} on volatile solids production even at this relatively low sludge age is evident. From the simulation results, a value of 0.22 gives a good compromise between the observed and predicted values. It should be noted that this value is higher than that usually encountered for a raw municipal wastewater; a value of 0.11 to 0.13 would be more typical.

Table C-2. Comparison of Observed and Predicted VSS, OUR and Effluent Nitrate for Different Estimates of Influent Particulate Unbiodegradable COD Fraction

Parameter	Observed Value	Predicted Value for Different f_{XI}			
		0.10	0.15	0.20	0.25
VSS (mg/L)	1784	1478	1613	1748	1883
OUR (mgO_2/L/h)	24.1	28.5	27.2	25.8	24.4
S_{NO} (mgN/L)	25.2	26.3	25.6	25.0	24.3

Application of equation C.2-1 (the simplified steady state equation) using an f_{XI} value of 0.22 yields:

$$X_{VSS} = \frac{15 \cdot 310 \cdot 8.24}{6.3}\left\{\frac{(1-0.13-0.22)0.45}{1+0.24 \cdot 8.24}(1+0.2 \cdot 0.24 \cdot 8.24) + \frac{0.22}{1.48}\right\}$$

$$= 1738\ mg/L$$

C.4 References

Barker, P.S., and P.L. Dold. 1995. COD and nitrogen mass balances in activated sludge systems. *Water Res.* 29(2):633-643.

APPENDIX D

OPERATING A BENCH-SCALE SBR SYSTEM

D.1 Background

This appendix outlines the operation of a laboratory-scale fill-and-draw SBR system for wastewater characterization. The procedure is most readily described by using the low F/M SBR that was operated at McMaster University as part of the nitrifier maximum specific growth rate work performed for this study.

D.2 SBR Operation

In the protocol for wastewater characterization a single cycle in SBR operation consists of five operating modes or periods. The periods are fill, react, waste, settle and draw (decant), in sequence. The sequence in a cycle is illustrated in Table D-1. The SBR is operated on the basis of a 24-hour cycle, with a selected maximum volume (V_P). The volume of supernatant (effluent) withdrawn after the settle period is equal to the volume of wastewater added at each cycle (V_{WW}), less that wasted (q_W).

At start up, the system should be seeded with a mass of microorganisms from an activated sludge system. Quasi-steady state conditions are achieved by repeating the following 24-hour cycle of actions over a period of approximately three sludge ages:

1. At the start of each cycle, a fixed volume of wastewater (V_{WW}) of a specified COD concentration ($COD_{T,INF}$) is added to the reactor (FILL).
2. After the "instantaneous" fill, the contents of the reactor are aerated and mixed for a period of $23^1/_2$ hours (REACT). At this point, any liquid volume lost through evaporation is replaced with distilled, de-ionized water.
3. At the end of the react period, a fixed volume (q_W) of mixed liquor is wasted from the reactor to maintain a constant sludge age for the system (WASTAGE). For example, to maintain a 15 day sludge age, 1/15th of the reactor volume is wasted directly from the reactor. The mixed liquor wastage provides sample volumes for analysis.
4. After wastage, the air supply and mixer are turned off and the sludge is allowed to settle for a period of approximately 30 minutes (SETTLE).

5. At the end of the settle period, the treated effluent (supernatant) is withdrawn from the reactor (DRAW), leaving the settled sludge at the bottom of the reactor. The volume of supernatant drawn off is equal to the volume of wastewater added initially, less the volume of mixed liquor wasted ($V_{WW} - q_W$).

Table D-1. Bench Scale SBR Operating Conditions

Phase	Time	Reactor Condition	Action
FILL	T = 0 (Instantaneous fill)	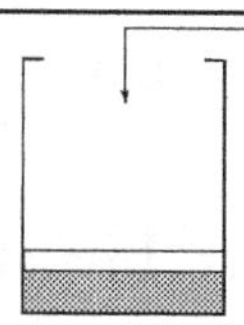	V_{WW} of wastewater is added to reactor containing V_P - V_{WW} of mixed liquor, i.e. reactor is filled to the maximum volume (V_P).
REACT	T = 0 - 23_ hours	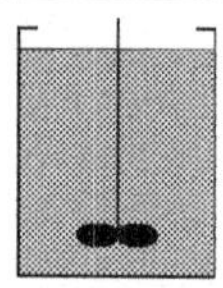	Reactor volume constant (V_P). Mixing and aeration on.
WASTE	T = 23_ hours (Instantaneous wastage)	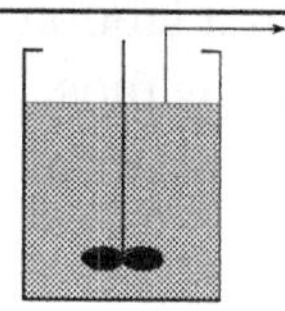	Mixing and aeration on. Withdraw volume of mixed liquor for wastage (q_W). $q_W = V_P/\theta_X$.
SETTLE	T = 23_ - 24 hours	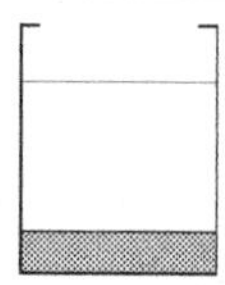	No mixing or aeration. Allow sludge to settle.
DRAW	T = 24 hours (Instantaneous draw of treated water)	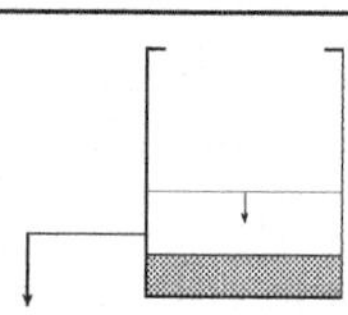	No mixing or aeration. Decant off supernatant (effluent) volume of (V_{WW} - q_W), leaving a volume of $V_P - V_{WW}$ in the reactor.
Parameters for this study: $V_P = 6$ L, $V_{WW} = 5$ L, $q_W = 0.4$ L (i.e. $\theta_X = 6/0.4 = 15$ days)			

For the McMaster system, the SBR was operated for the following conditions:

V_P = 6 L

V_{WW} = 5 L

q_W = 0.4 L/d

θ_X = 6/0.4 = 15 days

Temp. = 21°C

After start-up, the system was operated for a period of 80 days (five sludge ages) so as to attain a quasi-steady state. At this time, an intensive daily monitoring schedule was commenced for a period of 12 days. Analyses were conducted on samples of influent, waste and decant.

Water Environment Research Foundation

Throughout the period of operation oxygen utilization rate was monitored semi-continuously using a UCT DO meter.

On two days (Days 2 and 11) during the intensive monitoring period, profiles of ammonia and nitrate concentration were monitored over the first six hours of the react period. Analyses were performed on small sample volumes withdrawn from the reactor at intervals of approximately 15 minutes.

D.2.1 Influent Wastewater

To attain the quasi-steady state SBR response described above, the system should receive a fixed COD and TKN load each cycle. Attaining uniform influent waste composition is facilitated by storing batches of wastewater sufficient for several days' operation at 4°C. The concentration of the influent feed to the system is kept as near to a target value as possible by diluting successive batches of wastewater with tap water.

In practice, if the SBR method were being applied at a wastewater treatment plant, it may not be feasible to store batches of wastewater as described above. As a reasonable alternative, it is suggested that each day, at the same time, the daily feed volume (V_{WW}) is collected from the influent to the plant and added directly to the reactor at the start of each cycle (FILL). The influent wastewater flow rate and concentrations of COD and TKN to the plant usually follow a daily cyclic pattern that is repeated with limited day-to-day variation. As a result, the concentrations of COD and TKN in the feed volume (V_{WW}) collected daily should be reasonably constant from day to day.

The SBR study used in the example was not conducted on site so it was not feasible to collect samples of SBR influent daily. Rather, 50 L batches of wastewater were collected at weekly intervals. These batches were stored in plastic containers in a cold room at 4°C to avoid degradation.

D.2.2 Selection of SBR Sludge Age

One requirement for the proposed SBR operating procedure is that good solid-liquid separation is achieved during the settle period. This allows decanting of effluent with minimal solids loss. For effective settling, the total suspended solids (TSS) concentration in the SBR should be at least about 1,200 mg/L. Under these circumstances, the sludge should flocculate well and exhibit zone settling behavior.

In most cases, the influent COD to the SBR will be in the region of 300 mg/L to 400 mg/L. With the suggested operating procedure, the SBR receives influent only once per day and the feed volume necessarily is less than the reactor volume, i.e., the effective hydraulic retention time in the SBR exceeds one day. In this situation, with typical wastewater characteristics, the SBR sludge age necessarily should exceed approximately 15 days to obtain the required TSS concentration.

An advantage of operating at a high sludge age is that the unbiodegradable particulate fraction of the VSS becomes more significant as the sludge age increases. As a result, determination of the influent f_{XI} fraction should be more accurate than that determined at a lower sludge age. A disadvantage of operating at a high sludge age, however, is that the SBR must be operated over a long period of time (at least 2 to 3 sludge ages) in order to attain steady state conditions.

D.2.3 SBR Start-up Period

At start-up, the SBR was seeded with one liter of activated sludge taken from the Dundas, Ontario WWTP. The system was operated for an acclimation period of 80 days (five sludge ages) so as to attain a stable operating sludge age (or SRT).

No monitoring was conducted for the first 40 days of operation. For the next 20 days, SBR and decant total suspended solids concentrations were monitored. For the remainder of the start-up period the following parameters were measured daily: (1) influent COD concentration, (2) SBR total suspended solids concentration (TSS) and (3) SBR decant total suspended solids.

Figure D-1 shows a plot of the monitored parameters. The influent COD exhibited some fluctuations during the start-up period, but batches collected the week prior to the intensive period showed more consistency. The two batches of influent used during intensive testing (Days 81 to 92) were similar in strength, with an average of COD of approximately 330 mg/L.

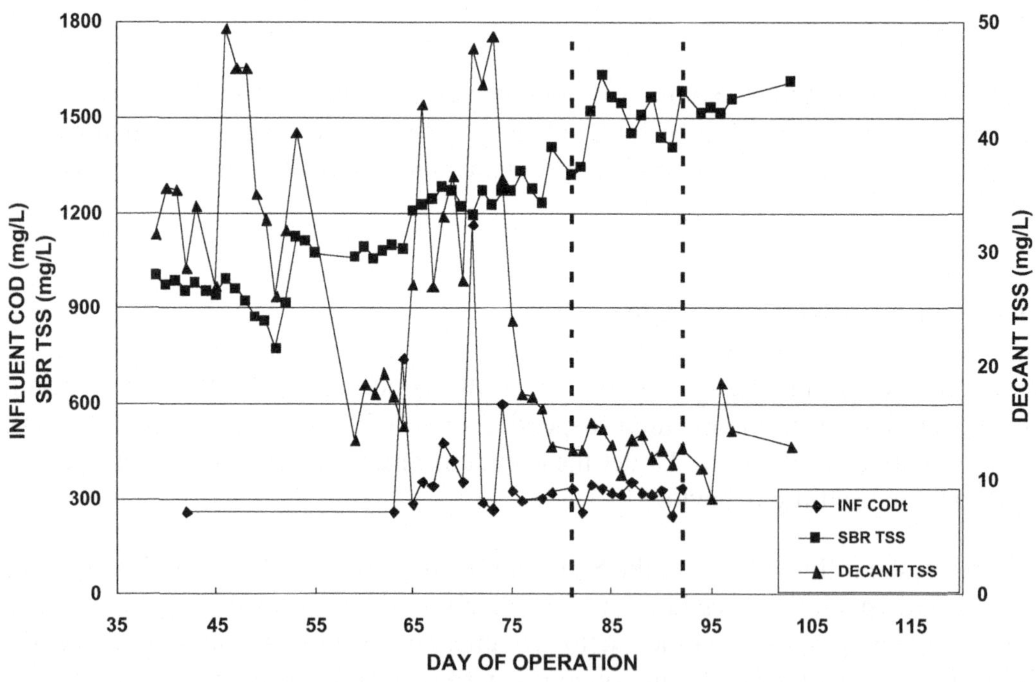

Figure D-1. Plot of Influent COD, SBR TSS and Decant TSS

D.2.4 Monitoring during 12 Day Intensive Testing

Intensive monitoring of the SBR system was performed for a period of 12 days from Day 81 to Day 92 as follows:

SBR Stream	Parameter Monitored
Influent	TSS
	VSS
	Total COD
	Filtered COD (on VSS/TSS filtrate)
	Flocculated/filtered COD
	Total Nitrogen
	Ammonia-N
Reactor	TSS
	VSS
	Total COD
	Total Nitrogen
	OUR
	Temperature
Decant	TSS
	VSS
	Total COD
	Membrane Filtered COD
	Membrane Filtered Total Nitrogen
	Membrane Filtered Ammonia-N
	Membrane Filtered Nitrate-N

Oxygen utilization rate (OUR) in the SBR was monitored semi-continuously using a DO/OUR meter. Results were downloaded to a PC spreadsheet daily. The data were analyzed off-line to calculate the mass of oxygen utilized in the reactor each day (further details may be found in Randall et al. (1991)). This enabled COD mass balances to be performed on the experimental data.

Results from the intensive monitoring period are recorded in Table D-2. The horizontal dashed line indicates when there was a change to a different batch of influent wastewater.

On Day 11 profiles of ammonia and nitrite/nitrate concentration were monitored over the first 6 hours of the react period. Analyses were performed on small sample volumes withdrawn from the reactor (and filtered immediately) at intervals of approximately 15 minutes. The ammonia and nitrate responses are shown in Figure D-2, together with the OUR response.

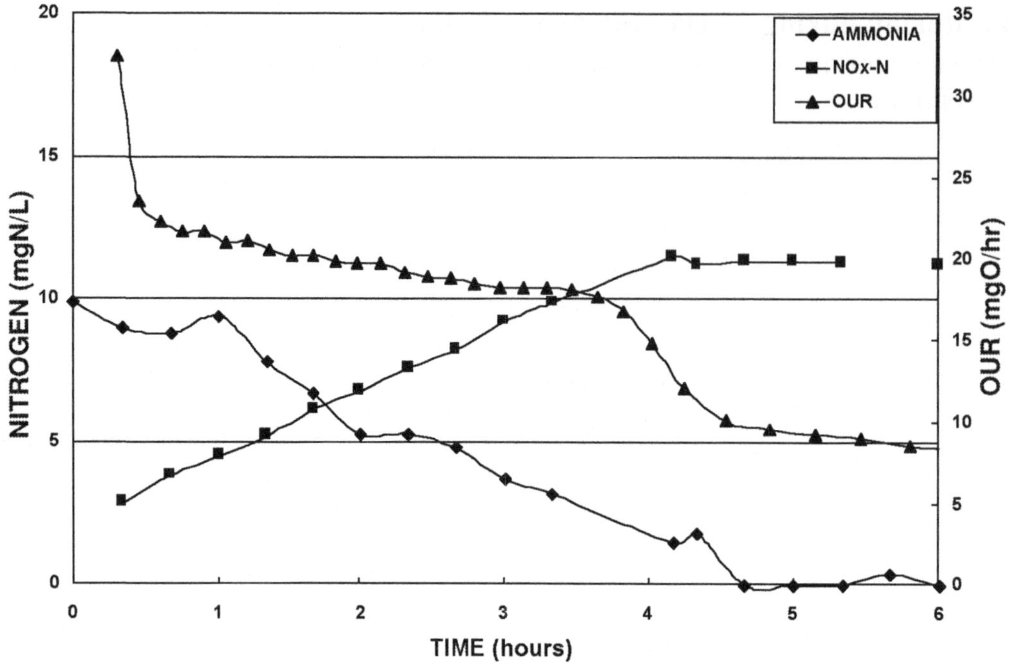

Figure D-2. Ammonia, Nitrite/Nitrate and OUR Response over the Initial Period on Day 11

Table D-2. Daily Results from SBR Intensive Testing Period

DATE	DAY	INFLUENT						REACTOR					DECANT						
		TSS (mg/L)	VSS (mg/L)	COD tot. (mg/L)	COD filt. (mg/L)	TN tot. (mg/L)	NH3 (mg/L)	TSS (mg/L)	VSS (mg/L)	COD tot. (mg/L)	TN tot. (mg/L)	OXYGEN UTILIZED (mg/d)	COD tot. (mg/L)	COD filt. (mg/L)	TN filt. (mg/L)	NH3 (mg/L)	NO3 (mg/L)	TSS (mg/L)	VSS (mg/L)
14-May	1	194	130	333	110	19.3	10.7	1,319	847	1,040	64.1	1,057	43	25	12.1	0.0	11.6	13	9
15-May	2	121	81	341	166	19.9	15.5	1,345	827	1,219	77.5	1,111	42	35	15.7	0.0	13.8	13	9
16-May	3	211	138	351	101	21.7	16.5	1,521	942	1,279	76.8	1,365	41	29	15.3	0.2	13.6	15	9
17-May	4	219	142	338	169	24.3	15.2	1,632	976	1,569	78.2	1,401	43	31	14.9	0.7	14.5	15	10
18-May	5	118	92	312	169	23.1	17.7	1,567	957	1,464	81.9	1,283	39	28	18.7	0.7	16.0	13	9
19-May	6	129	102	318	154	24.8	17.0	1,544	927	1,337	101.1	1,313	45	20	19.8	1.1	15.9	11	7
20-May	7	153	123	349	147	27.1	16.3	1,452	908	1,251	92.9	1,315	21	7	21.4	1.3	16.4	14	9
21-May	8	169	138	326	130	27.3	15.6	1,512	955	1,240	93.5	1,284	26	12	20.8	1.2	16.7	14	9
22-May	9	100	77	314	184	30.5	18.6	1,568	979	1,253	80.4	1,443	29	18	20.4	1.2	16.2	12	8
23-May	10	130	93	330	147	25.1	18.3	1,438	899	1,251	83.3	1,371	32	17	14.9	0.3	15.7	13	9
24-May	11	168	92	327	112	24.6	14.0	1,412	866	1,213	72.4	1,236	35	22	18.0	0.0	14.5	11	8
25-May	12	191	111	337	157	24.7	12.3	1,586	936	1,187	76.0	1,126	31	25	17.4	0.3	14.5	13	9
AVERAGE		159	110	332	145	24.4	15.6	1,491	918	1,275	81.5	1,275	36	22	17.5	0.6	15.0	13	9

D.3 SBR Data Validation: Mass Balances

Mass balances for COD and nitrogen were performed on the experimental data from the 12 day intensive testing period. This provides an overall validation of the experimental data. The balances should be within a few percentage points of 100 percent. If not, it is an indication that a problem exists with the experimental data. The basis for calculating the balances was as follows:

D.3.1 COD Balance

$$\begin{aligned}
\textit{Daily Input} &= V_{ww} * \textit{Influent COD} \\
\textit{Daily Output} &= \left(V_{ww} - q_w\right) * \textit{Filtered decant COD} \\
&\quad + q_w * \textit{SBR total COD} \\
&\quad + \textit{total mass of oxygen utilized} \\
&\quad - V_{ww} * \textit{Decant nitrate N} * 4.57 \\
&\quad (\textit{mass of oxygen for nitrification})
\end{aligned}$$

D.3.2 Nitrogen Balance

$$\begin{aligned}
\textit{Daily Input} &= V_{ww} * \textit{Influent TKN} \\
\textit{Daily Output} &= V_{ww} * \textit{Filtered decant TKN} \\
&\quad + V_{ww} * \textit{Decant nitrate N} \\
&\quad + \left(V_{ww} - q_w\right) * \textit{Decant VSS} * \textit{N Content of VSS} \\
&\quad + q_w * \textit{SBR VSS} * \textit{N Content of VSS}
\end{aligned}$$

The results of mass balance calculations are recorded in Table D-3. An accumulation term was included in the mass balances to account for the difference in the masses of COD and nitrogen in the reactor between the start and end of the 15 day period. The overall COD and nitrogen balances closed to 101 and 98 percent, respectively.

Table D-3. COD and Nitrogen Mass Balances on Data from SBR Intensive Testing Period

DATE	DAY	COD								NITROGEN						
		INPUT (mg/d)	Eff. (mg/d)	WAS (mg/d)	OURt (mg/d)	OURn (mg/d)	OURc (mg/d)	OUTPUT (mg/d)	Daily %	INPUT (mgN/d)	NO3 out (mgN/d)	TKNs out (mgN/d)	Eff. VSS (mgN/d)	WAS (mgN/d)	OUTPUT (mgN/d)	Daily %
14-May	1	1,665	199	416	1,057	266	791	1,407	84	96	58	2	3	24	87	90
15-May	2	1,705	195	488	1,111	316	795	1,477	87	99	69	9	3	23	104	105
16-May	3	1,753	189	512	1,365	311	1054	1,755	100	109	68	9	3	26	106	97
17-May	4	1,690	199	627	1,401	331	1071	1,897	112	121	72	2	3	27	105	87
18-May	5	1,562	179	586	1,283	366	917	1,682	108	116	80	14	3	27	123	106
19-May	6	1,588	205	535	1,313	363	949	1,690	106	124	80	20	2	26	127	103
20-May	7	1,747	98	500	1,315	375	940	1,538	88	136	82	25	3	25	135	100
21-May	8	1,630	118	496	1,284	382	902	1,516	93	137	84	20	3	27	133	98
22-May	9	1,572	133	501	1,443	370	1073	1,707	109	153	81	21	3	27	132	86
23-May	10	1,650	147	500	1,371	360	1011	1,659	101	126	79	0	3	25	107	85
24-May	11	1,635	163	485	1,236	331	904	1,552	95	123	73	18	3	24	117	95
25-May	12	1,683	144	475	1,126	332	794	1,413	84	123	73	14	3	26	116	94
ACCUMULATION								882							37	
CUMULATIVE TOTAL		19,880		Overall %	101			20,177		1,462		Overall %	98		1,429	

D.4 Characteristics from Direct Measurements

A number of characteristics could be derived from the analyses performed on samples collected during the intensive testing period. These are recorded in Table D-4. It should be recalled that two different batches of influent were used during this period. As a result certain characteristics in the influent differed slightly from batch to batch. However, variations from the overall average were not significant.

Influent soluble unbiodegradable COD fraction (f_{SI}): The average value of 0.07 mg COD/mg COD is typical for a raw municipal wastewater.

Influent inorganic suspended solids concentration (X_{ISS}): The average value was 49 mg/L, which again is typical for a raw municipal wastewater.

Influent Ammonia/TKN fraction (f_{SNH}): The average value of 0.65 mg N/mg N is typical for a raw municipal wastewater where this treatment plant is located.

Nitrogen content of VSS (f_N): The average value of 0.07 mg N/mg VSS for the nitrogen content of the VSS is slightly lower than the expected value of 0.10 mg N/mg VSS.

COD/VSS ratio VSS (f_{CV}): The average value of 1.36 mg COD/mg VSS is slightly lower than the expected value of 1.42 mg COD/mg VSS.

Table D-4. Data Derived from Results of SBR Intensive Testing Period

DATE	DAY	INFLUENT					MIXED LIQUOR				DECANT	
		X_{ISS} (mg/L)	f_{SI} (mgCOD/ mgCOD)	COD coll. (mg/L)	f_{cv} (mgCOD/ mgVSS)	f_{SNH} (mg/L)	X_{ISS} (mg/L)	f_{cv} (mgCOD/ mgVSS)	f_n (mgN/ mgVSS)	VSS/ TSS	Calc'd S_{NB} mg/L	PRED VSS (mg/L)
14-May	1	64	0.08	110	1.72	0.55	472	1.20	0.061	0.64	0.47	12
15-May	2	40	0.10	166	2.16	0.78	518	1.43	0.075	0.61	1.83	5
16-May	3	74	0.08	101	1.81	0.76	579	1.33	0.065	0.62	1.53	8
17-May	4	77	0.09	169	1.19	0.63	656	1.58	0.065	0.60	0.00	9
18-May	5	25	0.09	169	1.56	0.76	609	1.50	0.066	0.61	1.97	7
19-May	6	27	0.06	154	1.60	0.69	617	1.42	0.088	0.60	2.77	17
20-May	7	30	0.02	147	1.64	0.60	544	1.37	0.079	0.63	3.67	10
21-May	8	31	0.04	130	1.42	0.57	557	1.29	0.076	0.63	2.83	9
22-May	9	23	0.06	184	1.69	0.61	590	1.26	0.061	0.62	3.00	7
23-May	10	37	0.05	147	1.97	0.73	539	1.37	0.076	0.63	0.00	10
24-May	11	77	0.07	112	2.34	0.57	545	1.37	0.063	0.61	3.50	9
25-May	12	80	0.07	157	1.63	0.50	650	1.24	0.063	0.59	2.53	4
	AVERAGE	49	0.068	145	1.73	0.65	573	1.36	0.070	0.62	2.01	9

D.5 Characteristics from Modelling of SBR Operation

Certain influent wastewater characteristics (and the nitrification rate) cannot be determined by direct measurement. These are (1) the fraction of the influent COD, which is particulate unbiodegradable (fXI), and (2) the fraction of the influent TKN, which is soluble unbiodegradable (f_{SI}). (The former is particularly important with respect to estimating sludge production.) These must be estimated iteratively either using an analytical model or through simulation (also based on a model). The analytical approach is only suitable when the system is at, or close to, a true steady state. As noted earlier, in this study, the system parameters varied considerably over the testing period; hence the simulation approach was used.

The simulation study of SBR behavior was conducted to model system response over the 12 days of intensive monitoring. The configuration for simulation is shown in Figure D-3.

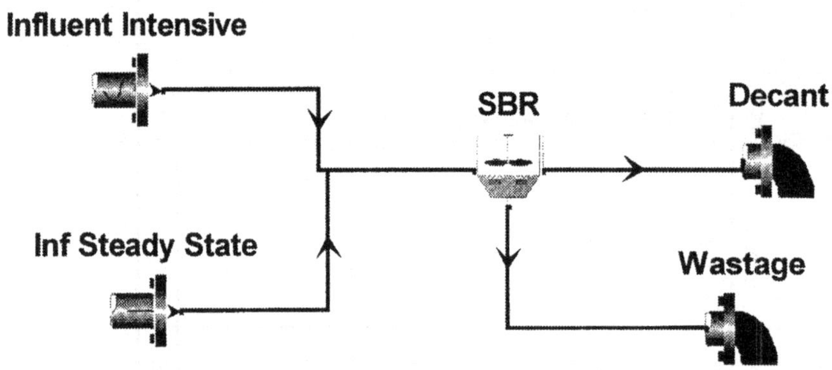

Figure D-3. Simulation Configuration

Operating conditions such as the influent volume and the various periods in the SBR cycle time were set-up in the simulator. The influent COD, TKN and X_{ISS} concentrations for each day were specified in the influent element (see Table D-2), together with the fractional characteristics determined from the direct measurements.

A number of simulation runs were performed, iteratively varying the values of the parameters to be estimated, f_{XI} and f_{SNI}. The objective was to obtain a reasonable fit of simulated to observed response over the fifteen days of operation using a single value for each parameter. In the case of f_{XI}, the primary parameter response to match was VSS concentration. For f_{SNI} the most important parameter was decant filtered TN.

The maximum specific growth rate of the nitrifiers (μ_{AUT}) was determined in a similar fashion. However, in this case the data to be matched were the responses of ammonia, nitrate and OUR over six hours on Day 11 (shown in Figure D-2), as well as the longer term response over 12 days.

To perform the simulations it was necessary to assign initial concentration values to model components at time zero, the start of the simulation. These concentrations were determined by simulating system response over a period of 60 days (i.e., four SRTs) using the intensive monitoring period influent information as model input. As is evident from the plots which follow, this provided reasonable estimates of the system status at the start of the simulation period.

D.5.1 Particulate Unbiodegradable COD Fraction (f_{XI})

A value for f_{XI} of 0.16 mg COD/mg COD provided a good correspondence between the simulated and observed VSS in the system. This value is typical for a raw municipal wastewater. Figure D-4 shows a plot of the simulation results over 12 days. It should be noted that the experimental data points correspond to the VSS concentration in the waste sludge drawn from the reactor at 23.5 hours in each daily cycle. Figure D-5 shows the corresponding plot for TSS concentration.

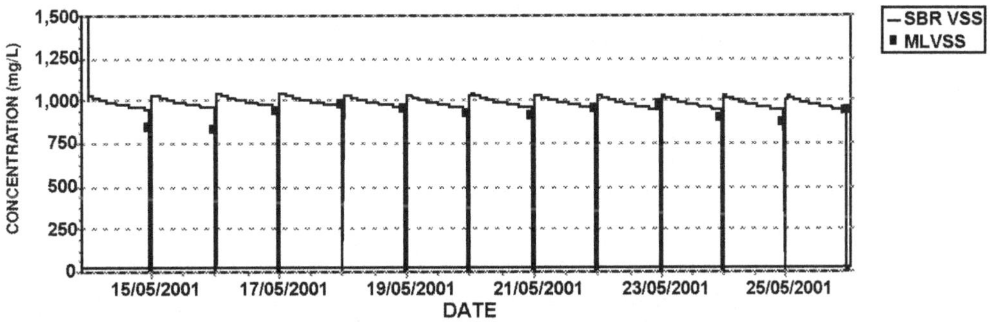

Figure D-4. Simulated (Solid Line) and Observed (Data Points) VSS in the SBR over 12 Days

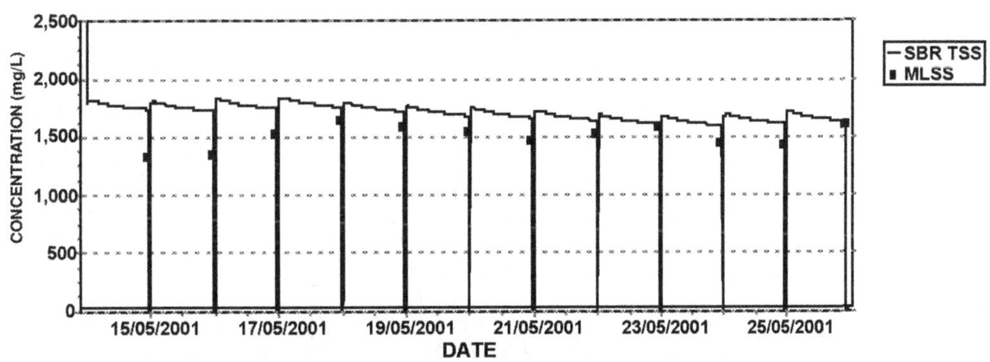

Figure D-5. Simulated (Solid Line) and Observed (Data Points) TSS in the SBR over 12 Days

Note that the simulator predicts a decrease in solids concentration over the course of each day. This is due to normal biological activity, and is expected. At the end of each day, the simulator shows the solids concentration rapidly plunging to zero and then increasing over a very short time period. This solids concentration fluctuation arises when the simulator enters the decant and fill phases.

D.5.2 Soluble Unbiodegradable TKN Fraction (f_{SNI})

A value for f_{SNI} of 0.04 mg N/mg N provided a good correspondence between the simulated and observed decant filtered TKN from the system. Figure D-6 shows a plot of the simulation results over 12 days. It should be noted that the data correspond to the concentration in the decant drawn from the reactor at the end of the settle period so the experimental data points should lie at the top of each simulated "spike."

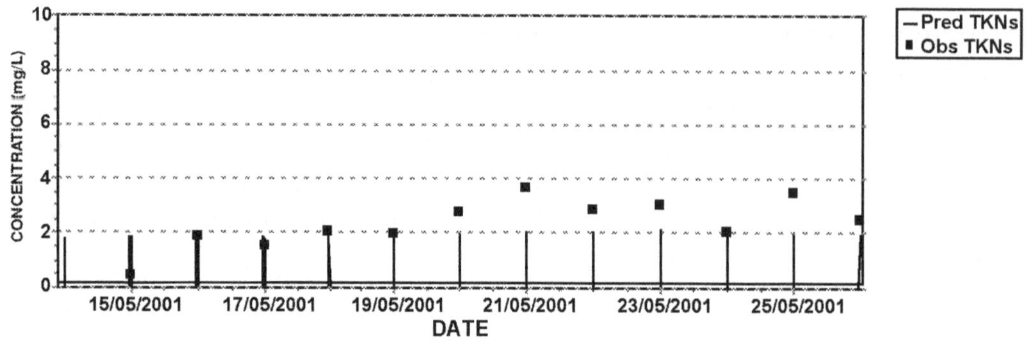

Figure D-6. Simulated (Solid Line) and Observed (Data Points) Decant Filtered TKN over 12 Days

D.5.3 Nitrification Rate (μ_{AUT})

The nitrification behavior in the system could be simulated accurately with a value for the maximum specific growth rate of the nitrifiers (μ_{AUT}) of 0.90 d^{-1} (referenced to 20°C, with a temperature dependency coefficient of 1.072). Figure D-7 shows the simulated nitrate response in the SBR over the 12 days of operation. The uppermost data points correspond to the decant samples and the near-vertical cluster of data points is for Day 11. Data for Day 11 are shown on an expanded scale in Figure D-8, together with ammonia concentration.

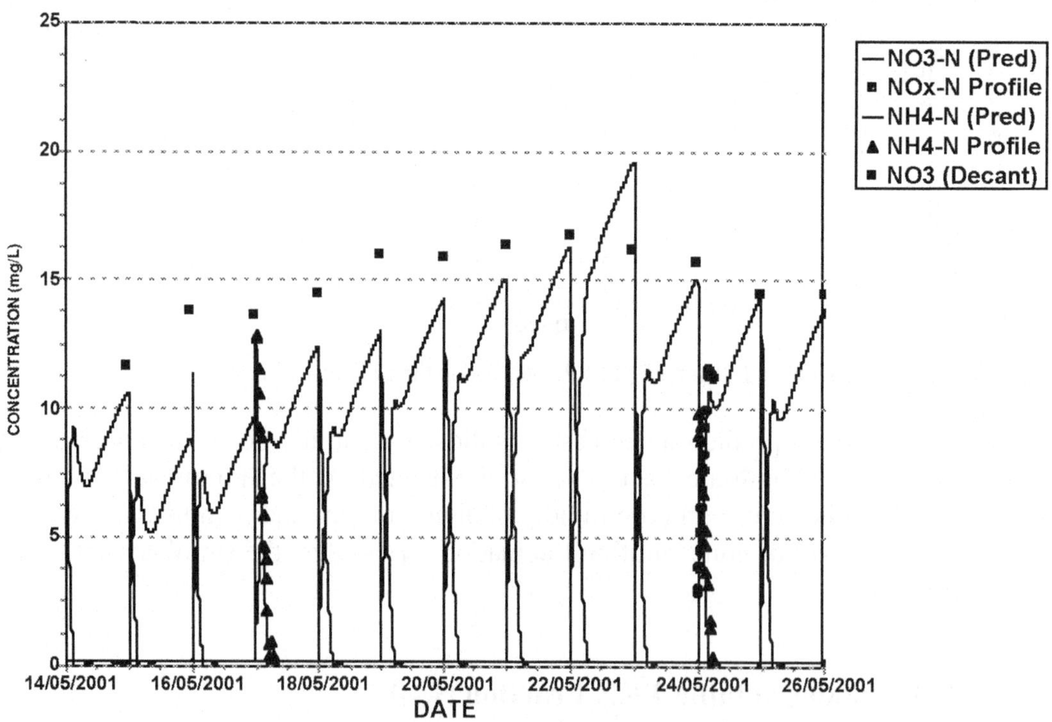

Figure D-7. Simulated (Solid Line) and Observed (Data Points) Ammonia and Nitrite/Nitrate in the SBR over 12 Days

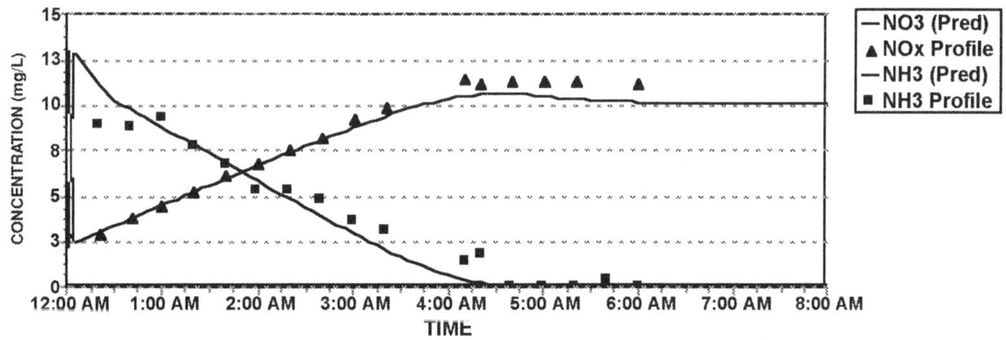

Figure D-8. Simulated (Solid Line) and Observed (Data Points) Ammonia and Nitrite/Nitrate Response in the SBR over Day 11

In Figure D-8, the simulated and observed nitrite/nitrate responses over the first six hours are near parallel, indicating that the estimate of nitrifier growth rate is very reasonable.

Oxygen utilization rate (OUR) response is closely linked to nitrification behavior. A further check on the selection of the nitrification rate parameter is provided by comparing the simulated and observed response of OUR over Day 11 shown in Figure D-9.

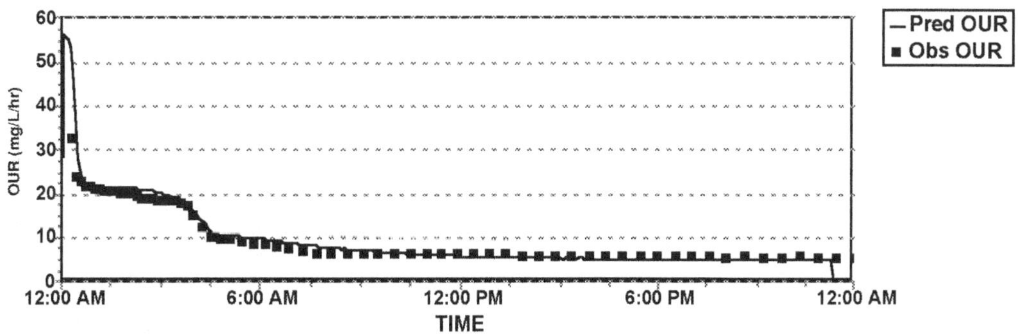

Figure D-9. Simulated (Solid Line) and Observed (Data Points) OUR Response in the SBR over Day 11

D.6 Estimated Labor Requirements

The following table summarizes the test labor requirements.

Table D-5. Estimated Labor Requirements

Time For Setup and Cleanup	1 day
Time For Data Collection	4 to 6 weeks
Analytical Load	Low for first month, high for last 1-2 weeks (see below)
Manpower Required	3 to 4 hr/day (may vary with time, see below)

As indicated in the above table, the test analytical load varies with time. If analyses are to be performed using Hach™ spectrophotometric methods by the person running the test, there will be a concommittant time variation in manpower requirement. For example, during the start-up period, only the following analyses are required daily:

1. Influent total COD
2. SBR TSS
3. SBR decant TSS

Due to this low analytical load, the manpower requirement during the initial period is low (i.e., likely 2 hours per day).

During the intensive monitoring period, the analyses defined in the "Monitoring during 12-day intensive testing" section above are required daily, which will increase the manpower requirement to about 7 hours per day.

D.7 References

Randall, E.W., A. Wilkinson, and G.A. Ekama. 1991. An Instrument for the direct determination of oxygen utilization rate. *Water SA* 17(1):11-18.

APPENDIX E

DEVELOPMENT OF HIGH F/M METHOD

E.1 Background

Two papers in the literature refer to application of the high F/M batch test method for estimating the maximum specific growth rate of nitrifiers (μ_{AUT}): Antoniou et al. (1990) and Sözen et al. (1996). Lesouef et al. (1992) proposed a similar approach, except that instead of utilizing data over the whole period of exponential growth, the nitrification rate at only two points (24 hour interval) is considered. Both the test methodology and the estimation technique developed in this study, as outlined in Chapter 16.0 and Appendices F and H, are a modification of what was proposed in these references. This chapter provides background on why the methodology and estimation techniques were modified. Certain material from Chapter 16.0 and Appendices F and H is repeated here for purposes of clarity, although this may be repetitious.

E.2 Basis for the High F/M Method

In the high F/M bioassay, a relatively low concentration of nitrifying mixed liquor is spiked with ammonia, and the increase in nitrite/nitrate concentration with time is monitored. This batch test method for estimating the maximum specific growth rate of the nitrifiers (μ_{AUT}) is based on a few simple principles:

- The ammonia concentration in the batch test is high enough (relative to the half-saturation coefficient, K_{NH}) to ensure that the nitrification rate is at the maximum. That is:

$$\mu = \mu_{AUT}\frac{S_{NH}}{K_{NH}+S_{NH}} \approx \mu_{AUT} \qquad \text{(E.2-1)}$$

- The initial nitrifier concentration is small, and the rate of growth of nitrifiers during the test is exponential.
- Dissolved oxygen concentration, pH and alkalinity should be such that growth is not limited by these factors.
- Experimental error in the measured nitrite/nitrate values should be small relative to the change in nitrite/nitrate concentration during the test.

The principal advantages of the high F/M method over the low F/M methods are that:

- Computation of the maximum specific growth rate (μ_{AUT}) does not depend on an estimate of the mass of nitrifiers in the test
- There is not a requirement to operate a laboratory-scale system for an extended period

E.3 Nitrite/Nitrate Response in the High F/M Method

At the start of the batch test, with a non-limiting ammonia concentration, the nitrite/nitrate production rate is:

$$\frac{dS_{NO}}{dt} = \frac{\mu_{AUT}}{Y_{AUT}} \cdot X_{AUT} \qquad \text{(E.3-1)}$$

The change in biomass concentration, X_{AUT}, is the result of growth and decay of nitrifiers:

$$\begin{aligned}\frac{dX_{AUT}}{dt} &= \mu_{AUT} \cdot X_{AUT} - b_{AUT} \cdot X_{AUT} \\ &= \left(\mu_{AUT} - b_{AUT}\right) \cdot X_{AUT}\end{aligned} \qquad \text{(E.3-2)}$$

Re-arranging equation E.3-2 and integrating from time zero to time *t*:

$$\frac{dX_{AUT}}{X_{AUT}} = \left(\mu_{AUT} - b_{AUT}\right) \cdot dt$$

$$\int_0^t \frac{dX_{AUT}}{X_{AUT}} = \int_0^t \left(\mu_{AUT} - b_{AUT}\right) \cdot dt \qquad \text{(E.3-3)}$$

$$X_{AUT,t} = X_{AUT,0}\, e^{\left(\mu_{AUT} - b_{AUT}\right) \cdot t}$$

where:

$X_{AUT,t}$ = nitrifier concentration at time t (mg/L)

$X_{AUT,0}$ = nitrifier concentration at time zero (mg/L)

t = time (days)

Substituting equation E.3-3 in equation E.3-1 and integrating from time zero to time *t* again:

$$\frac{dS_{NO}}{dt} = \frac{\mu_{AUT}}{Y_{AUT}} \cdot X_{AUT,0}\, e^{(\mu_{AUT} - b_{AUT})t}$$

$$dS_{NO} = \frac{\mu_{AUT} \cdot X_{AUT,0}}{Y_{AUT}} e^{(\mu_{AUT} - b_{AUT})t} dt$$

$$\int_0^t dS_{NO} = \int_0^t \frac{\mu_{AUT} \cdot X_{AUT,0}}{Y_{AUT}} e^{(\mu_{AUT} - b_{AUT})t} dt \qquad \text{(E.3-4)}$$

$$S_{NO,t} = S_{NO,0} + \frac{\mu_{AUT} \cdot X_{AUT,0}}{Y_{AUT} \cdot (\mu_{AUT} - b_{AUT})} e^{(\mu_{AUT} - b_{AUT})t} - \frac{\mu_{AUT} \cdot X_{AUT,0}}{Y_{AUT} \cdot (\mu_{AUT} - b_{AUT})}$$

$$S_{NO,t} = S_{NO,0} + \frac{\mu_{AUT} \cdot X_{AUT,0}}{Y_{AUT} \cdot (\mu_{AUT} - b_{AUT})} \left\{ e^{(\mu_{AUT} - b_{AUT})t} - 1 \right\}$$

where:

$S_{NO,t}$ = nitrite + nitrate concentration at time t (mg/L)

$S_{NO,0}$ = nitrite + nitrate concentration at time zero (mg/L)

E.4 Simulated Batch Test Response

Simulating the batch test protocol provides useful insights into the high F/M method. This approach was used as the basis for evaluating the factors controlling the response in the test, and for developing the test protocol outlined in Chapter 16.0. Figure E-1 shows the configuration used in simulations. A small volume of mixed liquor from a nitrifying activated sludge plant is withdrawn from the wastage stream, and introduced into an empty batch reactor at time zero. An influent to the batch reactor adds an ammonia-containing stream at the start of the batch test. Simulation allows rapid assessment of a number of features impacting the batch test behavior, including:

- Initial VSS concentration,
- Initial ammonia concentration,
- DO concentration,
- Alkalinity requirements,
- SRT of the system from which sludge is withdrawn,
- Influent TKN/COD to the system from which sludge is withdrawn,
- Maximum specific nitrifier growth rate (μ_{AUT}),
- Nitrifier decay rate (b_{AUT}), and
- Nitrifier half-saturation coefficient (K_{NH}).

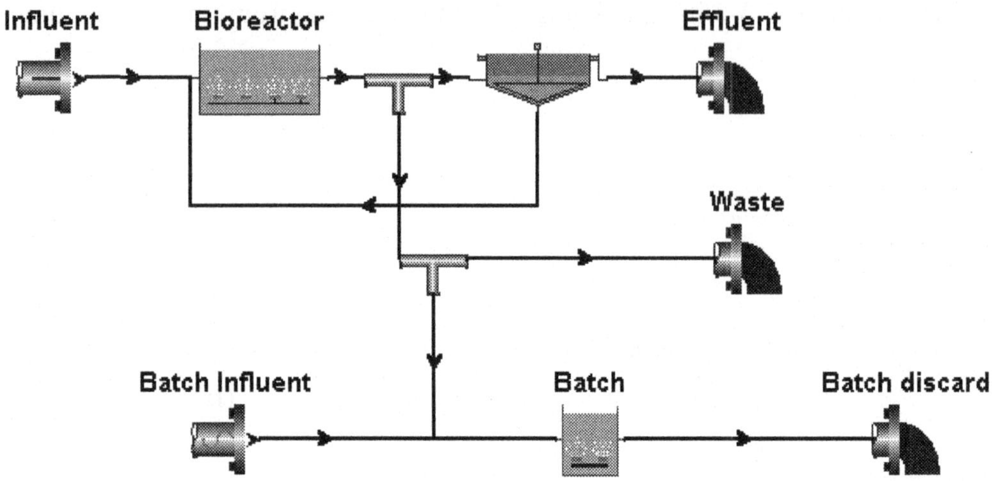

Figure E-1. Configuration for Simulating a High F/M Batch Test

Figure E-2 shows an example of the ammonia and nitrate (nitrification is modeled as a single-step process) concentration response in a simulated batch test. The phase of exponential growth when ammonia is available in excess is very evident. The nitrate concentration plateaus once ammonia is depleted.

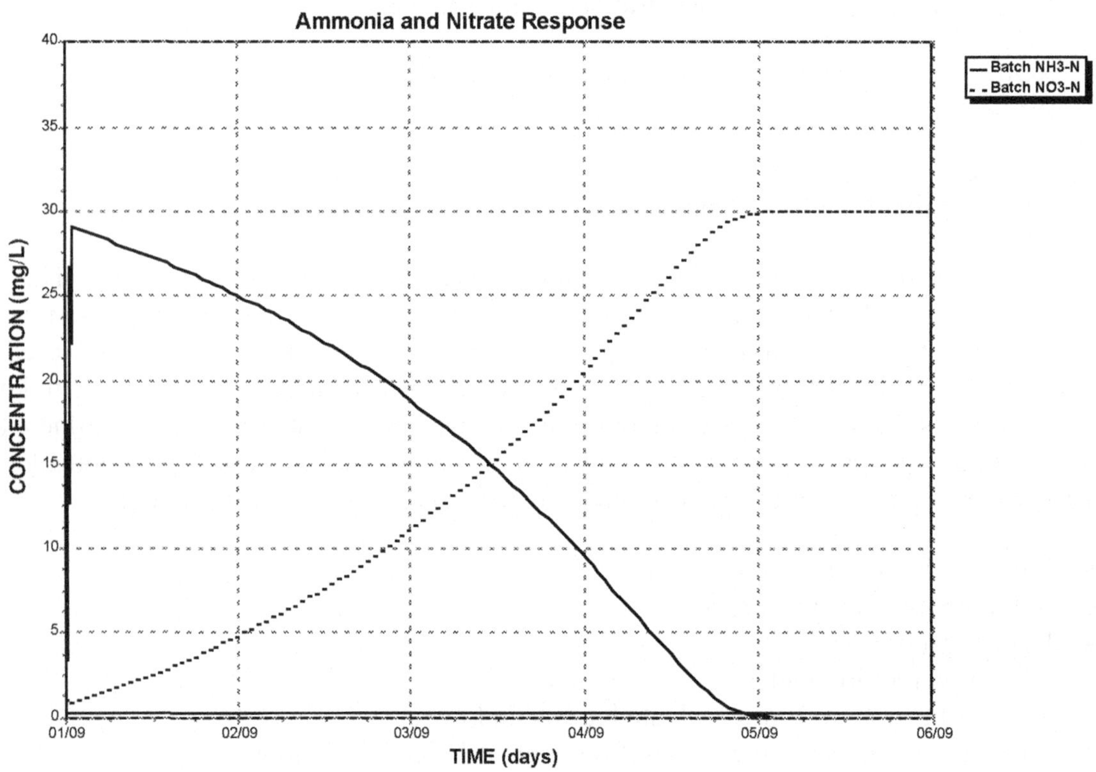

Figure E-2. Example of Ammonia and Nitrate Response in Simulated Batch Test

E.5 Antoniou et al. (1990) Analysis Method

Antoniou et al. (1990) presented equation E.3-4 for the nitrite/nitrate-time response in the following form:

$$S_{NO,t} = \frac{\mu_{AUT} \cdot X_{AUT,0}}{Y_{AUT} \cdot (\mu_{AUT} - b_{AUT})} e^{(\mu_{AUT} - b_{AUT})t} + S_{NO,0} - \frac{\mu_{AUT} \cdot X_{AUT,0}}{Y_{AUT} \cdot (\mu_{AUT} - b_{AUT})} \quad \text{(E.5-1)}$$

Antoniou et al. (1990) stated that "With the exception of an initial start-up time, the sum of the second and third terms in the right hand side of the above equation (equation E.5-1) will be negligible, since they are constants of opposite sign while the first term increases exponentially." Hence:

$$S_{NO,t} \approx \frac{\mu_{AUT} \cdot X_{AUT,0}}{Y_{AUT} \cdot (\mu_{AUT} - b_{AUT})} e^{(\mu_{AUT} - b_{AUT})t} \quad \text{(E.5-2)}$$

Taking logarithms:

$$\ln\left[S_{NO,t}\right] = \ln\left[\frac{\mu_{AUT} \cdot X_{AUT,0}}{Y_{AUT} \cdot (\mu_{AUT} - b_{AUT})}\right] + (\mu_{AUT} - b_{AUT}) \cdot t \quad \text{(E.5-3)}$$

A semi-logarithmic plot of *ln*(nitrite/nitrate) versus time during the exponential growth phase should give a linear plot with slope equal to the effective maximum specific growth rate ($\mu_{AUT} - b_{AUT}$). Antoniou et al. (1990) noted that "a deviation from linearity in the plot is expected for the first few experimental data." That is, data over the initial part of the test should not be included in the data analysis. The degree of non-linearity will depend on a number of factors including the initial nitrite/nitrate and nitrifier concentrations, and the growth rate.

Figure E-3 shows an example of the Antoniou et al. (1990) analysis procedure. This example uses the 20°C data presented by Sözen et al. (1996) (see Table E-1 below). It is evident from the plot that, for the analysis in this case, 4 of the 8 data points must be excluded as the data only "becomes" linear from Day 4. This immediately identifies a shortcoming in the method: namely, that a large portion of the collected data is not utilized as a consequence of the simplifying assumption. For this example, the estimate for ($\mu_{AUT} - b_{AUT}$) is 0.421 /d.

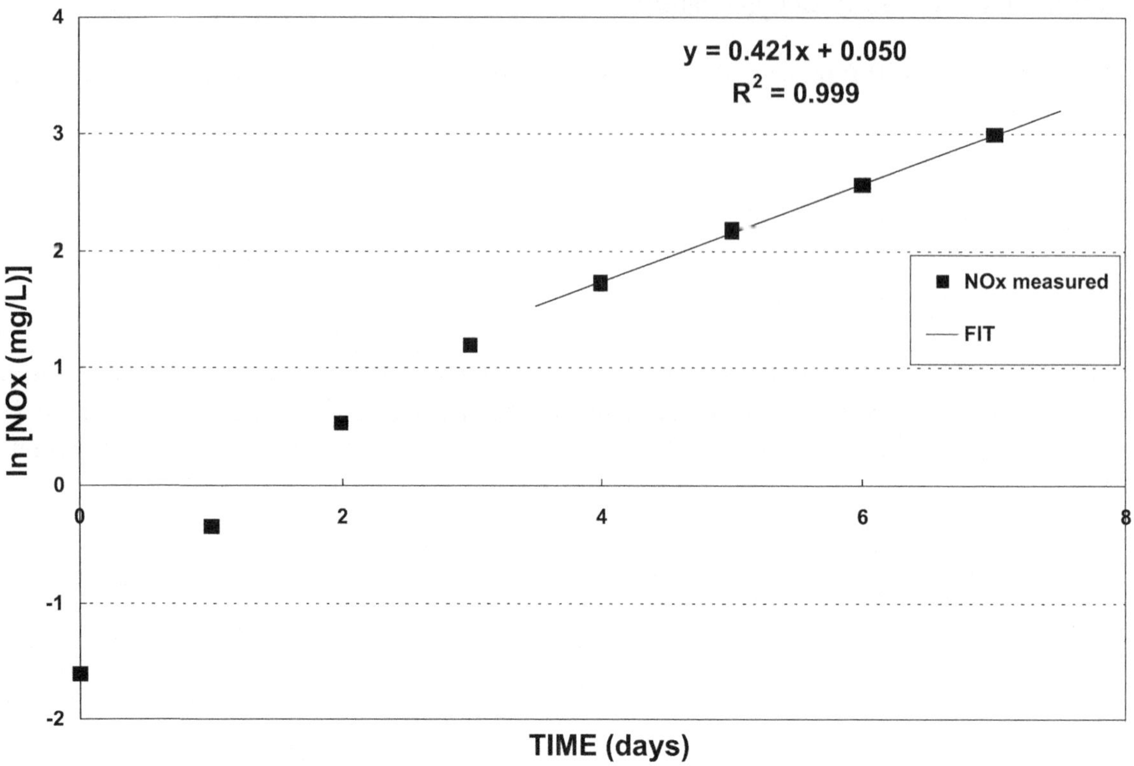

Figure E-3. Example of Antoniou et al. (1990) Analysis

E.6 Sözen et al. (1996) Analysis Method

Sözen et al. (1996) formulated equation E.3-4 as follows:

$$S_{NO,t} = S_{NO,0} + \frac{e^{(\mu_{AUT} - b_{AUT})t} - 1}{k} \qquad \text{(E.6-1)}$$

where:

$$k = \frac{Y_{AUT} \cdot (\mu_{AUT} - b_{AUT})}{\mu_{AUT} \cdot X_{AUT,0}}$$

A non-linear one-dimensional Fibonacci search technique was used to estimate ($\mu_{AUT} - b_{AUT}$) by minimizing the sum of the squares of the errors in the $S_{NO,t}$ estimates; that is, minimizing:

$$\sum_{i=1}^{n}\left(S_{NO,i} - S_{NO,i\,E}\right)^2 \qquad \text{(E.6-2)}$$

where:

$S_{NO,i\,E} =$ *estimate of* $S_{NO,i}$

The method "forces" the solution through the first data point, $S_{NO,0}$. Therefore it can be anticipated that any error in the $S_{NO,0}$ value would bias the solution.

Table E-2 shows an example of the Sözen et al. (1996) analysis procedure for their 20°C data listed in Table E-1. For this example, the estimate for ($\mu_{AUT} - b_{AUT}$) is 0.383 d^{-1}, significantly different to the estimate of 0.421 d^{-1} from the method of Antoniou et al. (1990). A plot of equation E.5-1 is shown together with the experimental data in Fig. E-4.

Table E-1. Batch Test Data from Sözen et al. (1996)

Time (days)	Nitrite/nitrate concentration (mgN/L)	
	20°C	15°C
0	0.2	0.2
1	0.7	0.3
2	1.7	1.2
3	3.3	2.4
4	5.6	3.8
5	8.8	4.8
6	13.0	6.8
7	20.0	9.5

Table E-2. Application of the Sözen et al. (1996) Estimation Method to the 20°C Data Listed in Table E-1

t (days)	NO3 measured	NO3 predicted	Residual	Residual^2
0	0.2	0.20 (forced fit)	0.000	0.000
1	0.7	0.88	-0.179	0.032
2	1.7	1.87	-0.174	0.030
3	3.3	3.33	-0.034	0.001
4	5.6	5.47	0.127	0.016
5	8.8	8.61	0.190	0.036
6	13.0	13.21	-0.209	0.044
7	20.0	19.95	0.049	0.002
			Sum Squares:	0.162
	Parameter estimates:			
	($\mu_{AUT} - b_{AUT}$)	$= 0.383\ d^{-1}$		
	K	$= 0.647$		

Application of the method of Antoniou et al. (1990) effectively only utilized the Day 4 to Day 7 data in this example (see Figure E-3). Table E-3 shows the results of applying the Sözen et al. (1996) analysis procedure to the same data set, but conducting the least squares fit based only on the Day 4 to Day 7 data. In this situation the estimate for ($\mu_{AUT} - b_{AUT}$) is 0.377d^{-1}, still significantly different to the estimate of 0.421 d^{-1} from the method of Antoniou et al. (1990).

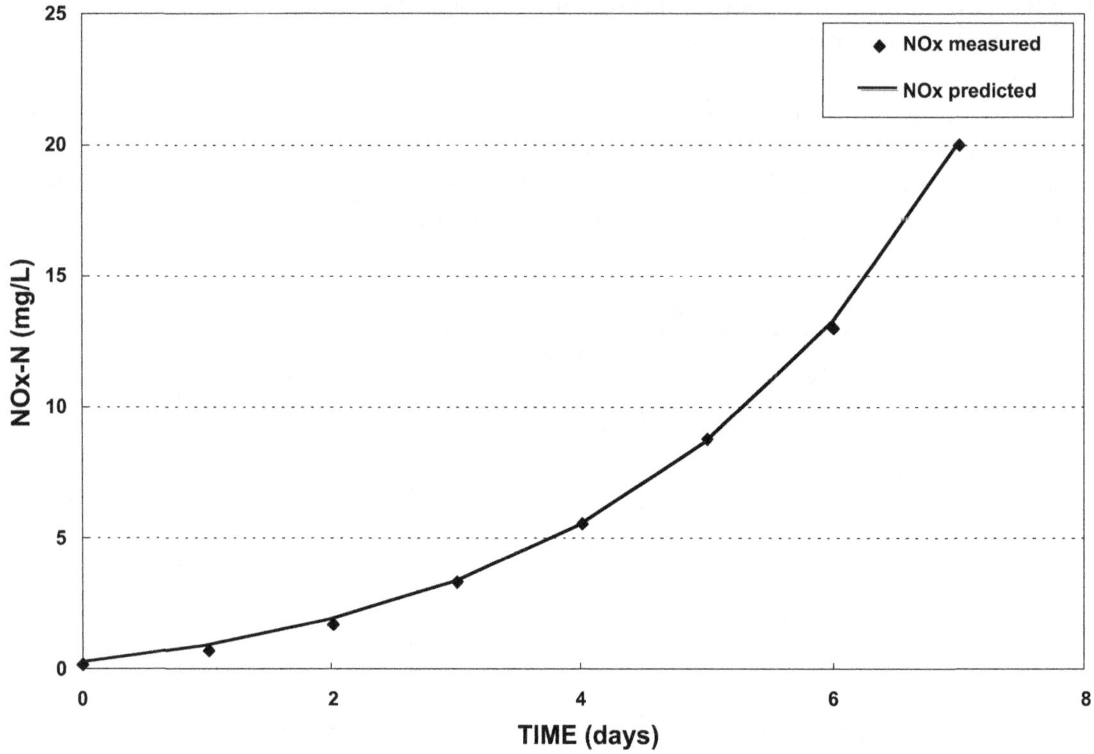

Figure E-4. Plot of Predicted and Measured Nitrite/Nitrate-Time Data from the Sözen et al. (1996) Analysis Method

Table E-3. Application of the Sözen et al. (1996) Estimation Method to the Day 4 to Day 7 Data Listed in Table E-1

t (days)	NO3 measured	NO3 predicted	Residual	Residual^2
0	0.2	0.20 (forced fit)	-	-
1	0.7	0.89	-	-
2	1.7	1.91	-	-
3	3.3	3.38	-	-
4	5.6	5.53	0.067	0.005
5	8.8	8.67	0.130	0.017
6	13.0	13.24	-0.244	0.059
7	20.0	19.91	0.087	0.008
			Sum Squares:	0.088

Parameter estimates:

$(\mu_A - b_N) = 0.377\ d^{-1}$

$K = 0.660$

The solution procedure of Sözen et al. (1996) can be simplified (for the user, at least) by applying the non-linear equation solver functions available in spreadsheet programs such as Microsoft Excel™. The spreadsheet approach is readily implemented as follows (see Appendix H for details):

- The measured nitrite/nitrate-time data are entered in two adjacent columns as shown in Table E-4.
- The predicted nitrite/nitrate concentration is calculated by formula directly from equation E.3-4: That is:

$$S_{NO,t} = S_{NO,0} + \frac{\mu_{AUT} \cdot X_{AUT,0}}{Y_{AUT} \cdot (\mu_{AUT} - b_{AUT})} \left\{ e^{(\mu_{AUT} - b_{AUT})t} - 1 \right\}$$

- Initial estimates are specified for ($\mu_{AUT} - b_{AUT}$), $X_{N,0}$ and $S_{NO,0}$.
- The spreadsheet solver is applied to minimize the sum of the squares of the residuals by adjusting the three parameters: ($\mu_{AUT} - b_{AUT}$), $X_{N,0}$ and $S_{NO,0}$.
- The procedure requires that a value be specified for Y_{AUT}. The magnitude of Y_{AUT} only impacts the estimate of $X_{AUT,0}$; it does not influence the solution for ($\mu_{AUT} - b_{AUT}$) or $S_{NO,0}$.

This approach has the advantage that the solution is not "forced" through the first data point, $S_{NO,0}$, as in procedure of Sözen et al. (1996).

Table E-4 shows an example of this analysis procedure applied to the Sözen et al. (1996) 20°C data listed in Table E-1. For this example, the estimate for ($\mu_{AUT} - b_{AUT}$) is 0.375 d^{-1}, similar to the estimates from the Sözen et al. (1996) method, but still significantly different to the estimate of 0.421 d^{-1} from the method of Antoniou et al. (1990). A plot of equation E.3-4 is shown together with the experimental data in Figure E-5.

Table E-4. Application of the Spreadsheet Least Squares Minimization Method to the 20°C Data Listed in Table E-1

t (days)	NO3 measured	NO3 predicted	Residual	Residual^2
0	0.2	0.09	0.110	0.012
1	0.7	0.79	-0.094	0.009
2	1.7	1.82	-0.119	0.014
3	3.3	3.31	0.010	0.000
4	5.6	5.48	0.119	0.014
5	8.8	8.64	0.160	0.026
6	13.0	13.24	-0.238	0.056
7	20.0	19.93	0.071	0.005
			Sum Squares:	0.136

Specified parameter:

Y_{AUT} = 0.15

Parameter estimates:

($\mu_{AUT} - b_{AUT}$) = 0.375 d^{-1}

$X_{AUT,0}$ = 0.20 mgCOD/L

$S_{NO,0}$ = 0.09 mg/L

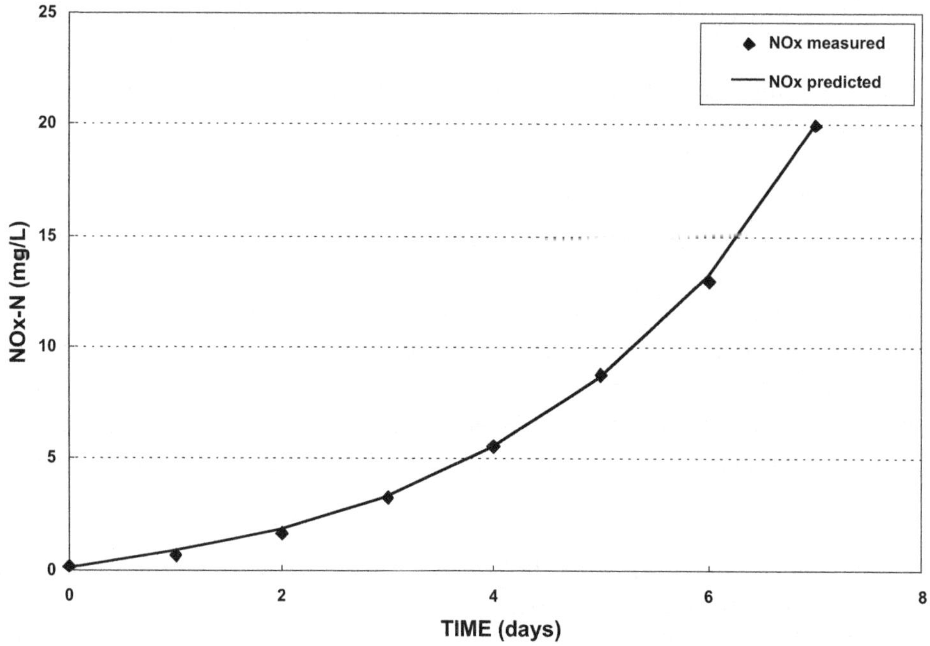

Figure E-5. Plot of Predicted and Measured Nitrite/Nitrate-Time Data from the Spreadsheet Least Squares Minimization Analysis Method

Table E-5 shows the results of applying the spreadsheet least squares minimization analysis procedure to the same data set, but conducting the least squares fit based only on the Day 4 to Day 7 data. In this situation, the estimate for ($\mu_{AUT} - b_{AUT}$) is 0.419 d^{-1}, similar to the estimate of 0.421 d^{-1} from the method of Antoniou et al. (1990). However, the plot of equation E.3-4 in Figure E-6 shows that the model estimates do not fit the data well over the initial part of the test.

Table E-5. Application of the Spreadsheet Least Squares Minimization Method to the Day 4 to Day 7 Data Listed in Table E-1

t (days)	NO3 measured	NO3 predicted	Residual	Residual^2
0	0.2	1.06	-	-
1	0.7	1.61	-	-
2	1.7	2.45	-	-
3	3.3	3.73	-	-
4	5.6	5.67	-0.073	0.005
5	8.8	8.63	0.169	0.029
6	13.0	13.13	-0.128	0.016
7	20.0	19.97	0.032	0.001
			Sum Squares:	0.052
	Specified parameter:			
	Y_{AUT}	= 0.15		
	Parameter estimates:			
	($\mu_{AUT} - b_{AUT}$)	= 0.419 d^{-1}		
	$X_{AUT,0}$	= 0.14 mgCOD/L		
	$S_{NO,0}$	= 1.06 mg/L		

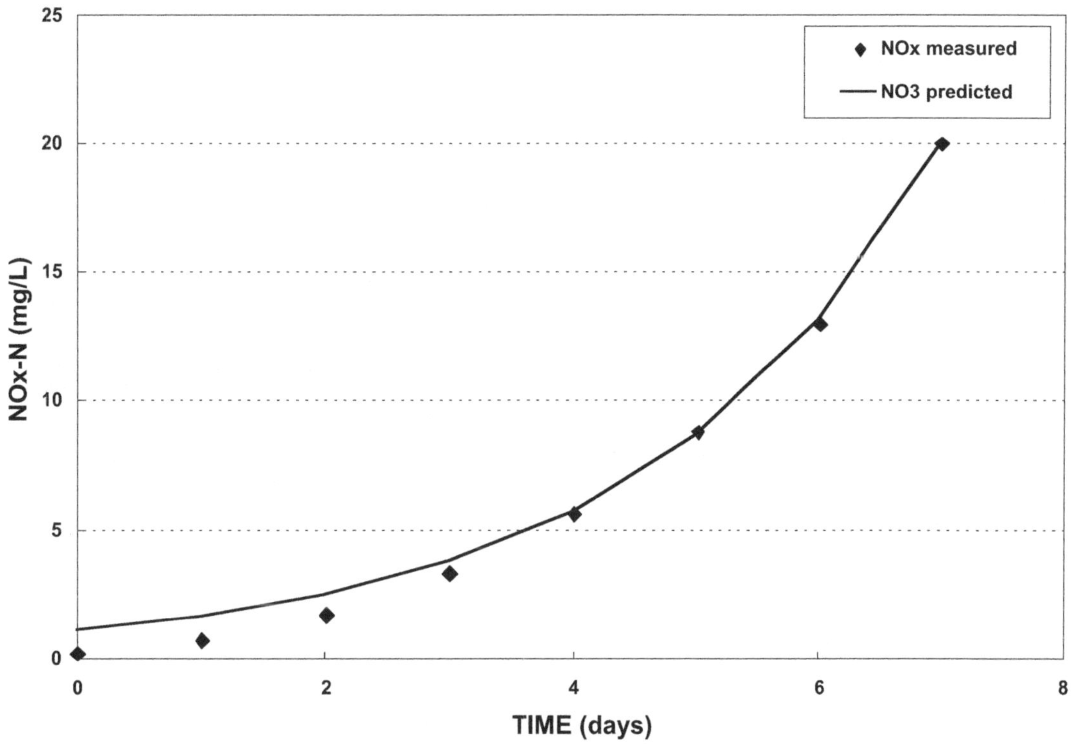

Figure E-6. Plot of Predicted and Measured Nitrite/Nitrate-Time Data from the Spreadsheet Least Squares Minimization Analysis Method When Applied to the Day 4 to Day 7 Data

The experimental and data analysis methods of Antoniou ct al. (1990) and Sözen et al. (1996) were reviewed using several sets of data from high F/M tests. Two further potential concerns were identified:

- Thc possible influence of ammonia concentration in the assumption that growth is not ammonia limited (see equation E.2-1); and
- The potential impact of experimental error in nitrite/nitrate analysis.

These two issues are now considered.

E.7 Influence of Ammonia Concentration

A fundamental requirement in the high F/M bioassay is that the ammonia concentration in the batch test is high enough (relative to the half-saturation coefficient, K_{NH}) to ensure that the nitrification rate is at the maximum. That is, from equation E.2-1:

$$\mu = \mu_{AUT} \frac{S_{NH}}{K_{NH} + S_{NH}} \approx \mu_{AUT} \qquad \text{(E.7-1)}$$

The magnitude of the half-saturation coefficient, K_{NH}, is in the region of 1 mg NH_3-N/L. In the batch test the ammonia concentration is changing with time. However, if the ammonia concentration is, say, 9 mgN/L then the growth rate from equation E.2-1 will be only 90 percent of the maximum. This illustrates that, depending on the exact value of K_{NH}, conducting the test starting at an initial ammonia concentration of less than, say, 20 mg/L likely will impact the estimation results significantly.

Antoniou et al. (1990) and Sözen et al. (1996) used different experimental approaches in an effort to ensure that growth was not ammonia-limited. In each case, the objective was to avoid a too-high ammonia concentration because of concerns that the free ammonia concentration could be inhibitory:

- Antoniou et al. (1990) combined sludge with unchlorinated secondary effluent, and added ammonia to a concentration of 10 mg N/L (and added alkalinity). The ammonia concentration was monitored during tests, and maintained at approximately 10 mg N/L by adding ammonium sulphate as needed.
- Sözen et al. (1996) combined nitrifying mixed liquor with Istanbul domestic wastewater. Examples of test results indicate that nitrification typically was complete when the nitrite/nitrate concentration reached approximately 20 to 25 mg N/L. Therefore it would be reasonable to assume that the ammonia concentration declined from approximately 20 or 25 mg N/L down to zero during the test.

The potential impact of ammonia concentration (relative to the half-saturation coefficient, K_{NH}) was assessed using "ideal" synthetic data generated by simulating two high F/M tests according to the procedure of Sözen et al. (1996). That is, the tests were simulated based on an initial ammonia concentration of 20 mgN/L. Identical operating conditions and model parameters were used in the simulations, except the magnitude of the half-saturation coefficient, K_{NH}, was set at either 1.0 mg/L or 0.05 mg/L. high DO and alkalinity conditions were specified to eliminate growth limitations by these parameters. In each simulation, the maximum specific nitrifier growth rate, μ_{AUT}, was set at 0.32 d^{-1} and the decay rate, b_{AUT}, at 0.04 d^{-1}. That is, ($\mu_{AUT} - b_{AUT}$) was 0.28 d^{-1}. nitrite/nitrate-time data were extracted from the simulation results corresponding to samples being withdrawn for analysis at hourly intervals over an 8-hour period each day of the test. These data sets were used to estimate ($\mu_{AUT} - b_{AUT}$) for each test using the spreadsheet least squares minimization method.

The two sets of simulation-generated "experimental" data are shown in Figure E-7 (for the K_{NH} = 0.05 mg/L case) and Figure E-8 (for the K_{NH} = 1.0 mg/L case). Each plot includes the fit to equation E.3-4 over the exponential growth phase based on the estimated ($\mu_{AUT} - b_{AUT}$). Comparing the figures, it is evident that:

- In each case, the amount of nitrite/nitrate generated is approximately 20 mg N/L, and the ammonia concentration decreases from 20 mg N/L to zero.
- Nitrification is complete after 3 days for the K_{NH} = 0.05 mg/L case, but only after 4 days for the K_{NH} = 1.0 mg/L case. This immediately indicates that the higher K_{NH} value will impact the estimate of ($\mu_{AUT} - b_{AUT}$). That is, the assumption in equation E.2-1 that ammonia is non-limiting for this range of ammonia concentrations is not valid.

- Application of the spreadsheet least squares minimization method yields estimates for ($\mu_{AUT} - b_{AUT}$) of 0.285 d^{-1} for the K_{NH} = 0.05 mg/L case, but only 0.217 d^{-1} for the K_{NH} = 1.0 mg/L case. (c.f. the actual value of 0.280 d^{-1}).

In activated sludge systems treating municipal wastewater, the half-saturation coefficient for the nitrifiers, K_{NH}, likely is in the region of 1.0 mg/L. These results are a clear indication that, in tests where sludge is combined with typical municipal wastewater, ammonia concentrations will not be high enough to avoid the effect of growth rate limitations. Similarly, maintaining the ammonia concentration at say 10 mg/L during the test also will not avoid growth limitation problems.

Note: It is pertinent to note that the ($\mu_{AUT} - b_{AUT}$) values (20°C) reported by Antoniou et al. (1990) and Sözen et al. (1996) typically were in the range of 0.4 to 0.6 d^{-1}. Tests in this study (conducted at higher ammonia concentrations) on mixed liquors from 5 different treatment plants all gave ($\mu_{AUT} - b_{AUT}$) values in the range of 0.85 to 0.95 d^{-1} i.e., significantly higher. This in itself appears to confirm that ammonia limitations were a factor in the Antoniou et al. (1990) and Sözen et al. (1996) tests.

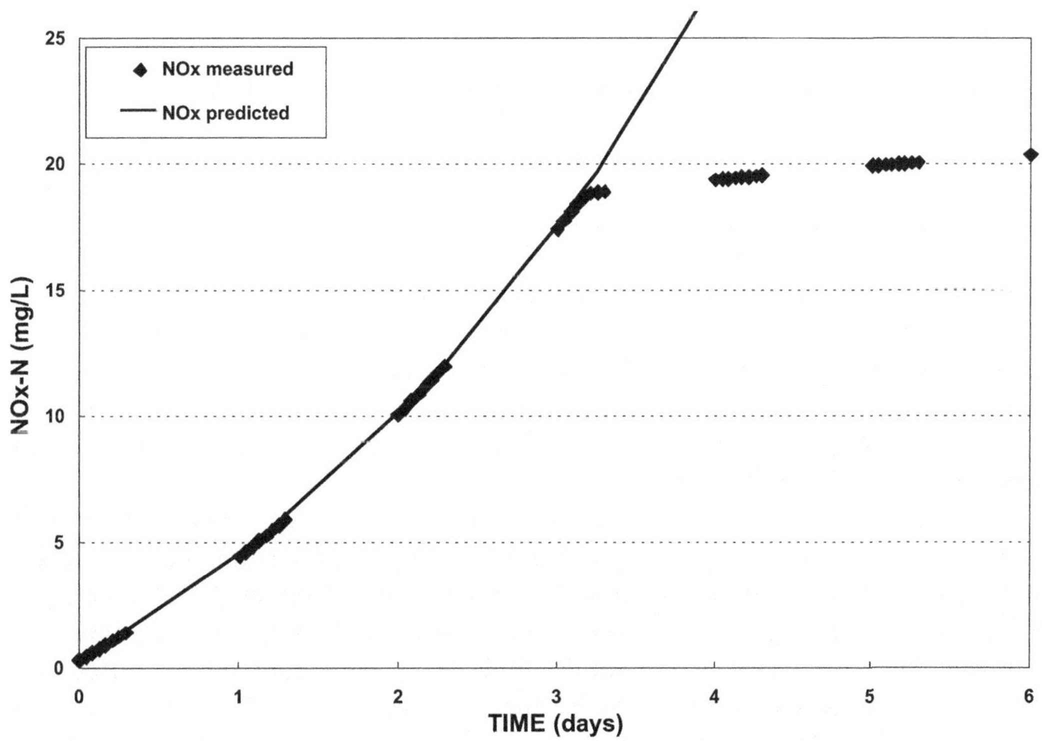

Figure E-7. Plot of Predicted and Measured Nitrite/Nitrate-Time Data from the Simulated High F/M Test (K_{NH} = 0.05 mg/L)

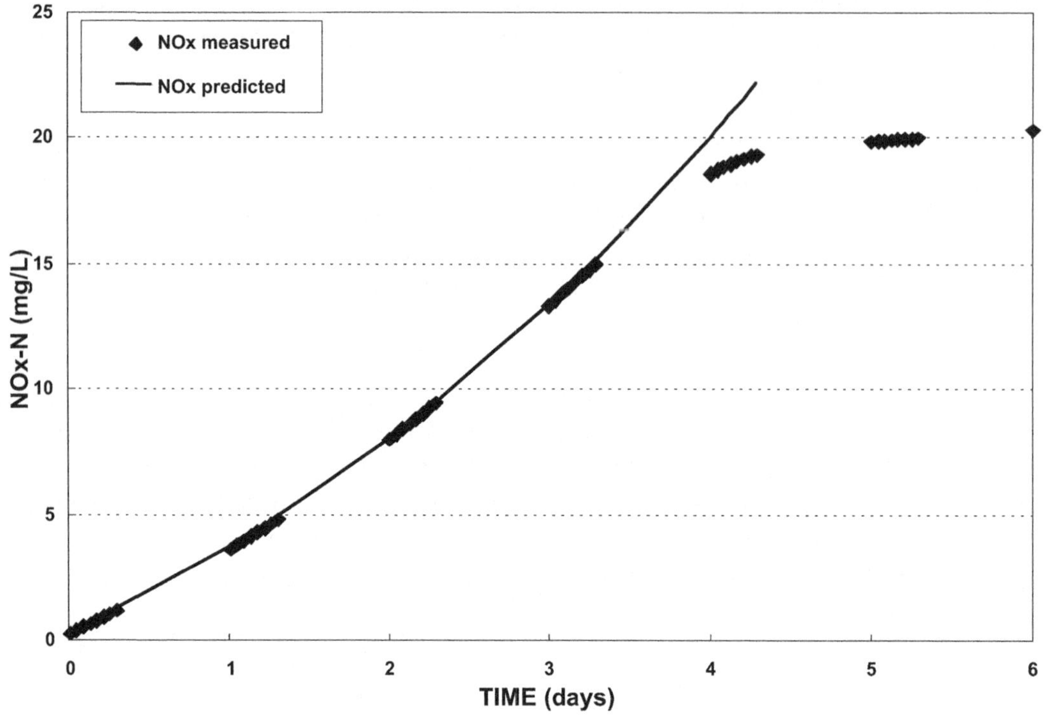

Figure E-8. Plot of Predicted and Measured Nitrite/Nitrate-Time Data from the Simulated High F/M Test (K_{NH} = 1.0 mg/L)

E.8 Influence of Experimental Error in Nitrite/Nitrate Analysis

In the test, samples are withdrawn at intervals, and analyzed for nitrite and nitrate concentration. Obviously this will involve some experimental error. It is of interest to evaluate how these errors will impact the estimation procedure. As an example, the synthetic experimental data shown in Figure E-7 (for the K_{NH} = 0.05 mg/L case that yields the accurate estimate for ($\mu_{AUT} - b_{AUT}$) of 0.285 d^{-1}) were considered. Each nitrite/nitrate value was adjusted by adding a normally-distributed random error with a standard deviation of 1 mg N/L and mean of zero. The resultant experimental data are shown in Figure E-9 together with the fit from equation E.3-4. Application of the spreadsheet least squares minimization method yields an estimate for ($\mu_{AUT} - b_{AUT}$) of 0.464 d^{-1}, almost double the value from analysis of the error-free data. (This difference may appear extremely large from a visual comparison of Figures E-7 and E-9, given that the tests are complete at similar times. The reason for the large difference in the ($\mu_{AUT} - b_{AUT}$) estimate has to do with problems in estimating the initial nitrifier concentration from the poor data; see Appendix H). Further analysis demonstrated that this problem becomes more severe as the number of samples collected during the test decreases.

It can be argued that this analysis is rather simplistic, and that the errors are likely to be smaller in practice (or that the errors likely will increase with increasing nitrite/nitrate concentration). However, this example serves to demonstrate that experimental error in nitrite/nitrate analysis is a very important consideration in the technique.

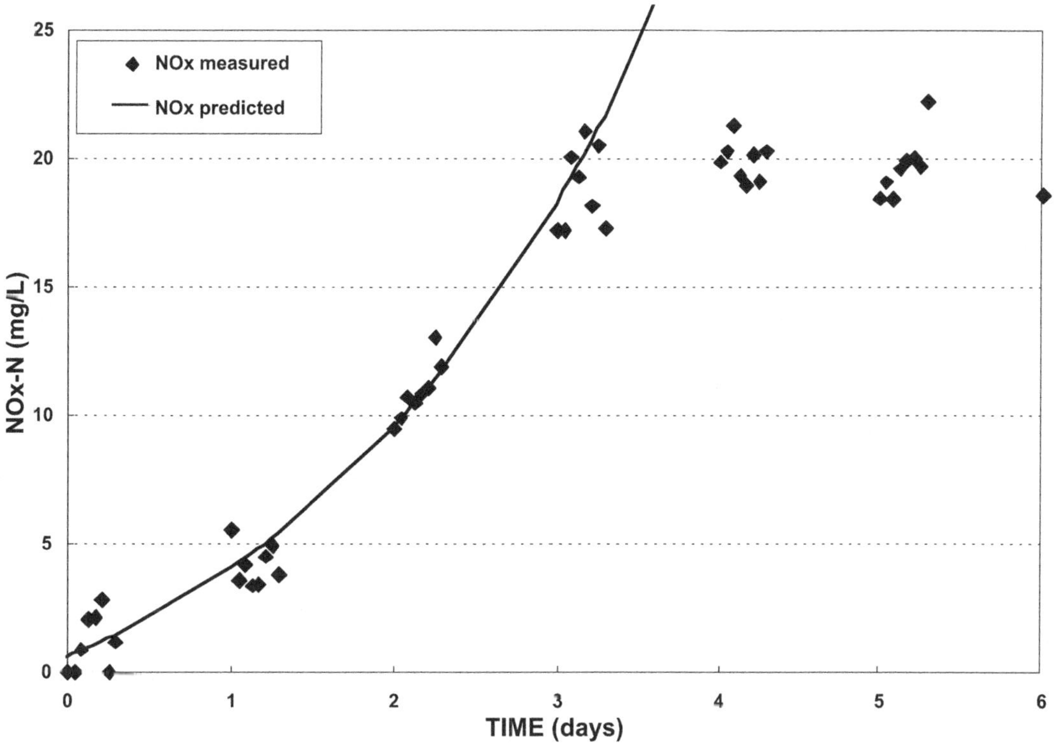

Figure E-9. Plot of Predicted and Measured Nitrite/Nitrate-Time Data from the Simulated High F/M =Test (K_{NH} = 0.05 mg/L) with a Random Error (S.D. = 1 mg/L) Adjustment to the "Ideal" Data from Figure E-7

E.9 Modified High F/M Method

After consideration of the issues discussed above, extensive evaluation of the high F/M method was conducted in simulations and experimentally. This lead to the protocol outlined in Chapter 16.0 and Appendices F and H. The main elements of the proposed protocol are:

- The suggested initial ammonia concentration in the batch test is approximately 120 mgN/L to avoid an ammonia-limited growth rate.
- At a pH of 7, the initial concentration of 120 mg N/L corresponds to an initial free ammonia concentration of approximately 0.5 mg N/L at 20°C. This may raise concerns of potential inhibition problems. However, parallel testing with the once-off initial addition versus intermittent ammonia addition to maintain the concentration at 20 mg N/L resulted in the same estimates of ($\mu_{AUT} - b_{AUT}$).
- A significant benefit of adding ammonia once at the start of the test is that it is simple to check the integrity of the experimental data through a mass balance. The sum of ammonia, nitrite and nitrate concentrations should be near constant throughout the test. This provides a simple means to identify experimental problems (such as ammonia stripping) or analytical problems.
- pH control is crucial in the test. Depending on the means for pH control (see Chapter 16.0 and Appendix F) intermittent addition of alkalinity likely is required.
- The more ammonia/nitrite/nitrate data collected during the test, the greater the confidence in the ($\mu_{AUT} - b_{AUT}$) estimate.

- Fitting equation E.3-4 to the experimental data should only include data from time zero until the ammonia concentration drops to approximately 20 mg N/L. This excludes the problem of possible ammonia limitation towards the end of the test.
- In application of the non-linear least squares estimation, it is useful to check the confidence interval of the estimate and the time-series residual plot (see Appendix H). Experience has shown that problems in a particular test result in a sinusoidal-type residuals plot.

In conclusion, high F/M tests were conducted in parallel with two other test procedures during this study (i.e., Low F/M SBR method and Washout method). The fact that all three procedures yielded the same estimate for ($\mu_{AUT} - b_{AUT}$) perhaps is the strongest confirmation of the high F/M procedure proposed from this study.

E.10 References

Antoniou, P., J. Hamilton, B. Koopman, R. Jain, B. Holloway, G. Lyberatos, and S.A. Svoronos. 1990. Effect of temperature and pH on the effective maximum specific growth rate of nitrifying bacteria. *Water Res.* 24(1):97-101.

Lesouef, A., M. Payraudeau, F. Rogalla, and B. Kleiber. 1992. Optimizing nitrogen removal reactor configurations by on-site calibration of the IAWPRC activated sludge model, *Water Sci. Tech.* 25(6):105-123.

Sözen, S., D. Orhon, and H.A. San. 1996. A new approach for the evaluation of the maximum specific growth rate in nitrification. *Water Res.* 30(7):1661-1669.

APPENDIX F

HIGH F/M METHOD: DETAILS

F.1 Basic Equipment

- 4 liter glass beaker
- Thermometer
- Mixer - propeller or magnetic stirrer
- Air stone and flexible tubing for attachment to aeration source
- Power source with 5-6 outlets
- pH meter
- DO probe and meter (If available, it is useful to attach a data logger or chart recorder to verify that the DO concentration remains greater than 4.0 mg/L during the course of the test.)
- Test reagents: ammonium chloride (NH_4Cl) and sodium bicarbonate ($NaHCO_3$)
- If Hach™ spectrophotometric methods are to be used, equipment and reagents for the following methods are required:

  1. Nitrogen Ammonia HR Test In Tube (0-50 mg/L NH3-N) (Method 10031)
  2. Nitrate HR Test In Tube (0-30 mg/L NO3-N) (Method 10020)
  3. Nitrite Test In Tube (0-0.5 mg/L NO2-N) (Method 8507)

A picture showing the setup for two tests is shown in Figure F-1. Note that some Styrofoam is placed between the beakers and the magnetic stirring plates to minimize heat transfer. The test should be set up in an area where temperature fluctuations are minimal (i.e., ± 2° C).

Figure F-1. Experimental Setup for Two High F/M Tests

F.2 Initial Steps

1. Obtain 1 liter of nitrifying mixed liquor from the treatment plant where the nitrification rate is to be measured. This sample will serve as the seed source in the experiment. Also obtain 5 liters of secondary settling tank effluent to serve as diluent.

2. Measure the temperature of the mixed liquor and diluent samples. Also note the temperature of the location where the test will be conducted. If these are different, they should be equilibrated to the test temperature before commencing the test. Failure to do so introduces additional dynamic behavior in the test since nitrifying bacteria are sensitive to temperature. This temperature related dynamic behavior could confound the test results. To equilibrate the sample and ambient temperatures, gently aerate the samples in the location where the test will be conducted for an appropriate period of time. For example, if the initial sample/ambient temperature difference is significant, several hours may be required.

3. Perform TSS and VSS analyses on the mixed liquor sample. For the sake of accuracy it is recommended that these analyses be performed in triplicate and the results averaged. Note that the remainder of the mixed liquor sample should be aerated gently while waiting for the results of the TSS and VSS analyses.

4. While waiting for the results of the TSS and VSS analyses, perform the following analyses on the diluent:

   i. Nitrate (to ensure that the plant in question is nitrifying)
   ii. Ammonia (used to calculate the amount of ammonia to be added to the test reactor)
   iii. Alkalinity (used to calculate the amount of sodium bicarbonate to add to the test reactor)

   Note that these analyses do not necessarily have to be performed on the day of test commencement. If effluent quality from the plant does not undergo drastic daily changes, these analyses could be performed a day before commencing the test.

5. When the TSS and VSS analyses are complete, use the results to calculate the volume of seed to be added to the test. It has been found that tests work best when they are run at VSS levels of 30 - 35 mg/L. For example:

   Mixed liquor sample TSS : 2,751 mg/L
   Mixed liquor sample VSS : 1,818 mg/L

   For test VSS of 35 mg/L (Total High F/M test volume is 3 L):

$$Seed\ Volume = \frac{Target\ VSS \times Test\ Volume}{Seed\ VSS} = \frac{35\ mg/L \times 3\ L}{1{,}818\ mg/L} = 0.0577\ L \approx 58\ mL$$

   Note that in this example, 55 mL may be an easier volume to measure and would also provide a suitable test VSS level (e.g., 33 mg/L).

6. Once the seed volume has been calculated, the volume of diluent required to achieve a total test volume of 3 L may be calculated. For example, if 55 mL of seed is used 2,945 mL of diluent will be required.

7. Based on the ammonia analysis of the diluent, the amount of ammonium chloride to be added to the test reactor may be calculated as follows:

   Molecular Weight of NH_4Cl : 14 + 4 + 35.5 = 53.5 (note no waters of hydration)
   NH_4 - N content per gram NH_4Cl : 14/53.5 = 0.26 g NH_4 - N / g NH_4Cl

   To calculate the mass in grams of NH_4Cl to add to the High F/M test vessel :

$$Mass\ NH_4Cl = \frac{\dfrac{Target\ NH_4\text{-}N \times Test\ Volume - Diluent\ NH_4\text{-}N \times Diluent\ Volume}{1000\ mg/g}}{0.26\ g\ NH_4\text{-}N / g\ NH_4Cl}$$

   For a test volume of 3 L, target NH_4 - N of 120 mg/L, and diluent NH_4 - N of 0 mg/L :

$$Mass\ NH_4Cl = \frac{\dfrac{120\ mg/L \times 3\ L - 0\ mg/L \times (3\ L - Seed\ Volume)}{1000\ mg/g}}{0.26\ g\ NH_4\text{-}N / g\ NH_4Cl} = 1.38\ g$$

8. Measure the calculated amount of ammonium chloride into a sample dish.
9. At this stage the volumes for the seed sludge and diluent have been calculated. The mass of ammonium chloride also has been calculated. The final steps depend on the method that will be used for pH control. Proceed to either "**pH Control Method A – DO Control with Alkalinity Addition**" or "**pH Control Method B – Continuous Aeration Supplemented with Carbon Dioxide.**"

F.3 pH Control Method A – DO Control with Alkalinity Addition

In addition to the **Basic Equipment** listed above, this pH control method requires a controller to switch the test reactor aeration on or off to maintain the DO between 4 and 6 mg/L. Figure F-1 shows the test using an integrated DO meter/controller.

1. Calculate the potential mass of alkalinity that may be utilized during the course of the test. For example, assuming that the initial NH_4-N concentration in the test reactor will be 120 mg/L:

   Potential mass of Alkalinity to be used (as $CaCO_3$*) :*

   $$Mass\ Alk = Target\ NH_4\text{ - }N \times Test\ Volume \times 7.14\ \frac{mgCaCO_3}{mg\ NH_4 - N}$$

   For a test volume of 3 L, target NH_4 *- N of 120 mg/L :*

   $$Mass\ Alk = 120\ mg/L \times 3\ L \times 7.14 = 2570\ mg\ (as\ CaCO_3)$$

2. Next, calculate the sodium bicarbonate addition that will be required to meet the alkalinity demand over the duration of the test. For example, assuming that the diluent alkalinity was measured to be 120 mg/L as $CaCO_3$ and the potential alkalinity to be consumed is that calculated in the previous step:

$$mmol\ of\ HCO_3^-\ required = \frac{Potential\ Alk.\ Required\ -\ Diluent\ Alk. \times Test\ Volume}{50\ \frac{mg\ CaCO_3}{meq} \times 1 \frac{meq}{mmol\ HCO_3}}$$

$$= \frac{2570\ mgCaCO_3\ -\ 120\ mgCaCO_3/L \times 3\ L}{50\ \frac{mg\ CaCO_3}{meq} \times 1 \frac{meq}{mmol\ HCO_3}} = 44\ mmol\ HCO_3^-$$

Calculate the total theoretical mass of $NaHCO_3$ *:*

$$Mass\ NaHCO_3 = \frac{mmol\ of\ HCO_3^-\ required \times Molecular\ Weight\ of\ NaHCO_3}{1000\ mg/g}$$

$$= \frac{44\ mmol \times 84\ mg/mmol}{1000\ mg/g} = 3.7\ g$$

For this example, a total mass of about 4 g of sodium bicarbonate must be added to the test reactor over the duration of the test.

3. Measure 1 g of sodium bicarbonate into a sample dish.
4. Calibrate the DO probe, and set up the DO controller to maintain the test DO levels between 4 and 6 mg/L.

F.4 pH Control Method B – Continuous Aeration Supplemented with Carbon Dioxide

In addition to the **Basic Equipment** listed above, this pH control method requires equipment for delivery of an air/CO_2 mixture to the test reactor. Ideally this can be done by obtaining a cylinder filled with an air/CO_2 mixture such that the fraction of CO_2 is 1%. Alternately, separate air and CO_2 streams could be mixed in the test reactor aeration line. If the latter method is to be used, it is suggested that the following flow meters be obtained: 0~540 mL/ min flow meter for air, 0~16 mL/min flow meter for CO_2.

1. Calculate the sodium bicarbonate addition required to bring the initial test alkalinity to 250 mg/L (as $CaCO_3$). For example, assuming that the diluent alkalinity was measured to be 120 mg/L as $CaCO_3$:

$$mmol\ of\ HCO_3^-\ required = \frac{(Target\ Alk.\ -\ Diluent\ Alk.) \times Test\ Volume}{50\ \frac{mg\ CaCO_3}{meq} \times 1 \frac{meq}{mmol\ HCO_3}}$$

$$= \frac{(250\ mgCaCO_3/L\ -\ 120\ mgCaCO_3/L) \times 3\ L}{50\ \frac{mg\ CaCO_3}{meq} \times 1 \frac{meq}{mmol\ HCO_3}} = 7.8\ mmol\ HCO_3^-$$

Calculate the mass of $NaHCO_3$ *:*

$$Mass\ NaHCO_3 = \frac{mmol\ of\ HCO_3^-\ required \times Molecular\ Weight\ of\ NaHCO_3}{1000\ mg/g}$$

$$= \frac{7.8\ mmol \times 84\ mg/mmol}{1000\ mg/g} = 0.655\ g$$

For this example, a mass of about 0.655 g of sodium bicarbonate must be added to the test reactor to set the alkalinity to 250 mg/L as $CaCO_3$.

2. Measure 0.655 g of sodium bicarbonate into a sample dish.

F.5 Commencing the Test

1. At this stage, the following 4 test ingredients should be at hand:

i. Seed sludge volume
ii. Diluent volume
iii. Calculated mass of ammonium chloride
iv. Calculated mass of sodium bicarbonate

2. Insert the aeration source into the empty test reactor.
3. Pour the diluent into the test reactor.
4. Pour the seed sludge into the test reactor.
5. Turn on the magnetic stir plate and ensure the mixing bar is functioning properly in the test reactor.
6. Turn on the air source:

- If **Method A** is being used, turn on the DO meter/controller.
- If **Method B** is being used, set the air stream flow rate to about 250 mL/min. Adjust the CO_2 gas flow rate to control pH to about 7.2 (e.g. 3 mL/min).

7. Add the ammonium chloride and sodium bicarbonate to the test reactor. Note that a squeeze bottle filled with distilled/de-ionized water may aid in transferring all of the chemical to the test reactor.
8. Note the time. This is "time zero" for the test.

F.6 Sampling

1. After 15 minutes have elapsed, collect the first sample from the test reactor (the time should be noted each time a sample is taken). Using a graduated pipette, withdraw a 10 mL aliquot and filter this immediately using a 0.45 µm filter. Collect the filtrate in a clean beaker. The temperature in the test reactor should be measured before each sample is taken.
2. Analyze the filtrate for nitrite, nitrate and ammonia.
3. Samples should be extracted from the test reactor and analyzed every four hours over the course of the working day (e.g., early morning, noon and late afternoon).
4. ***pH in the test reactor should be checked each time a sample is taken. The pH should not be allowed to fall outside the range of 7.2 - 8.0. If the pH approaches 7.2, add 0.5 g of sodium bicarbonate. In a typical test, several 0.5 g additions of sodium bicarbonate may be required (keeping in mind the total theoretical amount required over the course of the test).***

Note that during the test, nitrite levels in the test reactor may become high. If Hach spectrophotometric methods are used, take care to eliminate nitrite interference in the nitrate measurement (nitrite interferes with nitrate measurements at concentrations of 12 mg/L and up). Nitrite interference is removed (for nitrite concentrations up to 100 mg/L) by adding approximately 400 mg of urea to 10 mL of sample prior to nitrate analysis. Consult Hach documentation for further details.

A typical test nitrite/nitrate and ammonia versus time response is shown in Figure F-2:

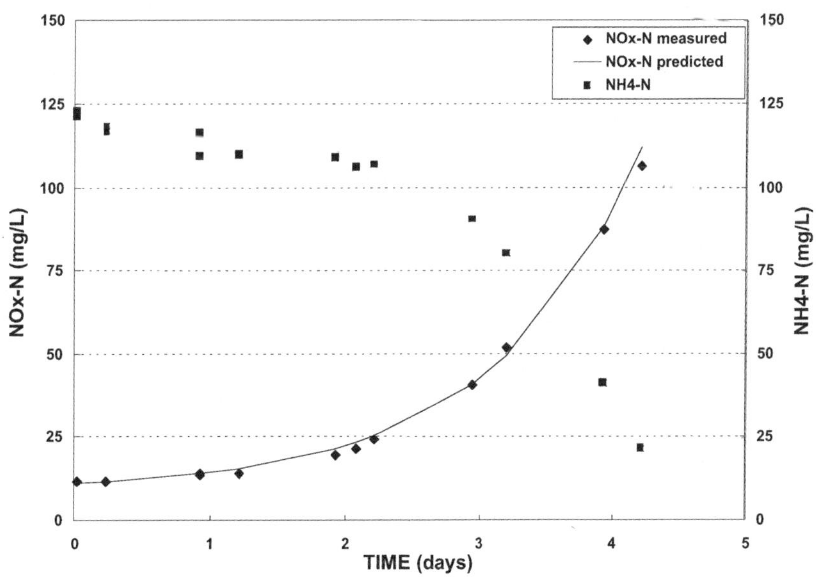

Figure F-2. Typical Test Response

Tests typically run for 5 to 8 days. Data collected after the test reactor ammonia level has dropped to below 25 mg/L may not be useable due to substrate limitation effects on nitrifer growth.

F.7 Estimated Labor Requirements

The following table summarizes the test labor requirements.

Table F-1. Estimated Labor Requirements

Time For Setup and Cleanup	1 day
Time For Data Collection	5 - 8 days
Analytical Load	6 analyses/day for each of nitrite, nitrate, and ammonia
Manpower Required	2 hr/day

The analytical load is based on performing analyses in duplicate. The following example illustrates the calculation of the analytical load for a five day test:

$$nitrite:\ 2\ \frac{analyses}{reactor\ sample} \times 3 \frac{reactor\ sample}{day} \times 5\ day = 30\ analyses$$

$$nitrate:\ 2\ \frac{analyses}{reactor\ sample} \times 3 \frac{reactor\ sample}{day} \times 5\ day = 30\ analyses$$

$$ammonia:\ 2\ \frac{analyses}{reactor\ sample} \times 3 \frac{reactor\ sample}{day} \times 5\ day = 30\ analyses$$

If analyses are to be performed using Hach™ spectrophotometric methods by the person running the test, the manpower requirements likely should be increased to 5 hours per day.

APPENDIX G

WASHOUT METHOD: DETAILS

G.1 Basic Equipment

- 8 liter glass beaker (i.e., the test reactor should be large enough to comfortably maintain an *active volume* of 5 - 6 liters)
- Thermometer
- Mixer - propeller or magnetic stirrer. (The presence of several pieces of equipment inside the reactor (the influent and effluent tubes, aeration supply tube and the DO sensor) might cause a magnetic stirrer bar to be dislodged easily from its rotating condition. A propeller mixer is the preferred mixer in this case)
- Air stone and flexible tubing for attachment to aeration source
- DO probe and meter (If available, it is useful to attach a data logger or chart recorder to verify that the DO concentration remains greater than 4.0 mg/L during the course of the test)
- 8 - 10 L containers for test influent and effluent
- Pump for influent and effluent
- Timer for pump operation (may not be necessary, depending on pump flow rate limits)
- Test reagents: ammonium chloride (NH_4Cl) and sodium bicarbonate ($NaHCO_3$)
- If Hach™ spectrophotometric methods are to be used, equipment and reagents for the following methods are required:

  1. Nitrogen Ammonia HR Test In Tube (0-50 mg/L NH3-N) (Method 10031)
  2. Nitrate HR Test In Tube (0-30 mg/L NO3-N) (Method 10020)
  3. Nitrite Test In Tube (0-0.5 mg/L NO2-N) (Method 8507)

A picture showing the setup for the washout method is shown in Figure G-1. The test should be set up in an area where temperature fluctuations are minimal (i.e., ± 2° C).

Figure G-1. Experimental Setup for the Washout Method

G.2 Initial Steps

1. Estimate the SRT at which the test will be conducted. In order to get a good response, the test should be performed at an SRT that is well below the theoretical minimum (i.e., 0.5 - 0.75 d). Once the test SRT has been selected, this sets the required daily influent flow. For example, for a target SRT of 0.75 days with 6 L of active volume:

   Target SRT of 0.75 d in 6 L active volume

$$Required\ Flow\ Rate = \frac{V}{SRT} = \frac{6\ L}{0.75\ d} = 8\ L/d$$

2. Obtain 10 liters of nitrifying mixed liquor from the treatment plant where the nitrification rate is to be measured. This sample will serve as the source of mixed liquor to be washed out during the experiment. Also obtain enough primary influent to serve as feed for the experiment (based on the daily influent flow rate calculated in the previous step).

3. Measure the temperature of the mixed liquor and primary influent samples. Also note the temperature of the location where the test will be conducted. If these are different, they should be equilibrated to the test temperature before commencing the test. Failure to do so introduces additional dynamic behavior in the test since nitrifying bacteria are sensitive to temperature. This temperature related dynamic behavior could confound the test results. To equilibrate the sample and ambient temperatures, gently aerate the mixed liquor sample in the location where the test will be conducted for an appropriate period of time. The primary influent sample should not be aerated. For example, if the initial sample / ambient temperature difference is significant, several hours may be required. If several days' worth of influent is collected, it should be stored in a cool location (i.e., 4° C). Each day, the appropriate feed volume should be brought to test temperature using a warm water bath.

4. Perform TSS and VSS analyses on the mixed liquor sample. For the sake of accuracy, it is recommended that these analyses be performed in triplicate and the results averaged. Note that the remainder of the mixed liquor sample should be aerated gently while performing the remaining initial steps.

5. Perform the following analyses on the influent:

   i. Ammonia (used to calculate the amount of additional ammonia to add to the test influent)
   ii. Alkalinity (used to calculate the amount of sodium bicarbonate to add to the test influent)

   Perform the following analyses on the mixed liquor:

   i. Ammonia (used to calculate the amount of additional ammonia to add to the reactor at the beginning of the test)
   ii. Alkalinity (used to calculate the amount of sodium bicarbonate to add to the reactor at the beginning of the test)

   Note that these analyses do not necessarily have to be performed on the first day of the test. If plant influent and mixed liquor characteristics do not undergo drastic daily changes, these analyses could be performed a day before commencing the test.

6. Set up the influent pump. This may be a simple task if a pump is available that can be set to a low flow rate (such as that calculated in Step 1). However, pumps with higher flow rates may need to be set up to operate intermittently. The following example illustrates pump calibration and setup for this scenario, assuming that 8 L/d must be pumped through the reactor:

Measure pump flow rate by monitoring time required to pump a known volume:

93.7 sec required to pump 365 mL, therefore

$$Pump\ Q = \frac{0.365\ L}{96.71\ s} \times \frac{60\ s}{1\ min} \times \frac{60\ min}{1\ hr} \times \frac{24\ hr}{1\ d} = 326\ L/d$$

Next calculate the total time the pump must run each day to deliver 8 L:

$$Total\ Daily\ Pump\ Time = \frac{8\ L}{326\ L/d} \times \frac{24\ hr}{1\ d} \times \frac{60\ min}{1\ hr} = 35.33\ min$$

If the pump is to be run once per hour for 24 hours:

$$Pump\ Time\ Per\ Hour = \frac{Total\ Daily\ Pump\ Time}{24\ hr} = \frac{35.33\ min}{24\ hr} \times \frac{60\ s}{1\ min} = 88\ s/hr$$

So turning the pump on for 88 seconds each hour will pump a total daily volume of 8 L

Note that a pump also must be set up to draw effluent from the reactor at the same rate that influent is pumped in. An alternative is to use a dual-head peristaltic pump such as the one shown in Figure G-1. The pump(s) should be connected to a timer to operate for the calculated length of time each hour.

7. Set up the pump controller to cycle the pump on and off each hour for the calculated time interval.

8. Add the calculated daily feed volume (i.e., from Step 1) to the influent feed container.

9. The reactor influent ammonia level should be approximately 50 mg N/L. If the influent ammonia analysis indicates lower levels, extra ammonia (e.g., in the form of ammonium chloride) should be added to each feed batch. The amount may be calculated as follows:

$$Molecular\ Weight\ of\ NH_4Cl:\ 14 + 4 + 35.5 = 53.5\ (note\ no\ waters\ of\ hydration)$$

$$NH_4\text{ - }N\ content\ per\ gram\ NH_4Cl:\ 14/53.5 = 0.26\ g\ NH_4\text{ - }N / g\ NH_4Cl$$

To calculate the mass in grams of NH_4Cl to add to each feed batch :

$$Mass\ NH_4Cl = \frac{\dfrac{(Target\ Influent\ NH_4\text{ - }N\text{ - }Measured\ Plant\ Influent\ NH_4\text{ - }N) \times Feed\ Volume}{1000\ mg/g}}{0.26\ g\ NH_4\text{ - }N / g\ NH_4Cl}$$

For a feed volume of 8 L, target NH_4 - N of 50 mg/L, and measured influent NH_4 - N of 36.8 mg/L :

$$Mass\ NH_4Cl = \frac{\dfrac{(50\ mg/L - 36.8\ mg/L) \times 8\ L}{1000\ mg/g}}{0.26\ g\ NH_4\text{ - }N / g\ NH_4Cl} = 0.406\ g$$

10. Add the calculated amount of ammonium chloride to the influent feed container.

11. The initial reactor ammonia level also should be approximately 50 mg N/L. If the reactor ammonia analysis indicates lower levels, extra ammonia (e.g. in the form of ammonium chloride) should be added to the reactor. The amount may be calculated as follows:

$$Molecular\ Weight\ of\ NH_4Cl:\ 14 + 4 + 35.5 = 53.5\ (note\ no\ waters\ of\ hydration)$$

$$NH_4\text{ - }N\ content\ per\ gram\ NH_4Cl:\ 14/53.5 = 0.26\ g\ NH_4\text{ - }N / g\ NH_4Cl$$

To calculate the mass in grams of NH_4Cl to add to the reactor initally :

$$Mass\ NH_4Cl = \frac{\dfrac{(Target\ Reactor\ NH_4\text{ - }N\text{ - }Measured\ Mixed\ Liquor\ NH_4\text{ - }N) \times Reactor\ Volume}{1000\ mg/g}}{0.26\ g\ NH_4\text{ - }N / g\ NH_4Cl}$$

For a reactor volume of 6 L, target NH_4 - N of 50 mg/L, and mixed liquor NH_4 - N of 5.0 mg/L :

$$Mass\ NH_4Cl = \frac{\dfrac{(50\ mg/L - 5.0\ mg/L) \times 6\ L}{1000\ mg/g}}{0.26\ g\ NH_4\text{ - }N / g\ NH_4Cl} = 1.039\ g$$

12. Add the calculated amount of ammonium chloride to the test reactor. *Note that ammonium chloride only is added to the reactor once, at the commencement of the test.* Once the test has started, suitable reactor ammonia levels will be achieved as a result of the high influent ammonia level.

13. The reactor influent alkalinity level should be approximately 350 mg/L as $CaCO_3$. If the influent alkalinity analysis indicates lower levels, extra alkalinity (e.g, in the form of sodium bicarbonate) should be added to the reactor. The amount may be calculated as follows (assuming a measured plant influent alkalinity of 250 mg/L as $CaCO_3$):

$$mmol\ of\ HCO_3^-\ required = \frac{(Target\ Influent\ Alk.\ -\ Measured\ Influent\ Alk.) \times Feed\ Volume}{50\ \frac{mg\ CaCO_3}{meq} \times 1 \frac{meq}{mmol\ HCO_3}}$$

$$= \frac{(350\ mgCaCO_3/L\ -\ 250\ mgCaCO_3/L) \times 8\ L}{50\ \frac{mg\ CaCO_3}{meq} \times 1 \frac{meq}{mmol\ HCO_3}} = 16.0\ mmol\ HCO_3^-$$

Calculate the mass of $NaHCO_3$:

$$Mass\ NaHCO_3 = \frac{mmol\ of\ HCO_3^-\ required \times Molecular\ Weight\ of\ NaHCO_3}{1000\ mg/g}$$

$$= \frac{16.0\ mmol \times 84\ mg/mmol}{1000\ mg/g} = 1.344\ g$$

14. Add the calculated amount of sodium bicarbonate to the influent feed container.

15. The initial reactor alkalinity level should be approximately 250 mg/L as $CaCO_3$. If the reactor alkalinity analysis indicates lower levels, extra alkalinity (e.g., in the form of sodium bicarbonate) should be added to the reactor. The amount may be calculated as follows (assuming a measured mixed liquor alkalinity of 120 mg/L as $CaCO_3$):

$$mmol\ of\ HCO_3^-\ required = \frac{(Target\ Alk.\ -\ Mixed\ Liquor\ Alk.) \times Reactor\ Volume}{50\ \frac{mg\ CaCO_3}{meq} \times 1 \frac{meq}{mmol\ HCO_3}}$$

$$= \frac{(250\ mgCaCO_3/L\ -\ 120\ mgCaCO_3/L) \times 6\ L}{50\ \frac{mg\ CaCO_3}{meq} \times 1 \frac{meq}{mmol\ HCO_3}} = 15.6\ mmol\ HCO_3^-$$

Calculate the mass of $NaHCO_3$:

$$Mass\ NaHCO_3 = \frac{mmol\ of\ HCO_3^-\ required \times Molecular\ Weight\ of\ NaHCO_3}{1000\ mg/g}$$

$$= \frac{15.6\ mmol \times 84\ mg/mmol}{1000\ mg/g} = 1.310\ g$$

16. Add the calculated amount of sodium bicarbonate to the test reactor. *Note that sodium bicarbonate only is added to the reactor once, at the commencement of the test.* Once the test has started, suitable reactor alkalinity levels will be maintained via influent alkalinity addition.

17. At this stage, the feed and reactor volumes have been set. Masses of ammonium chloride and sodium bicarbonate for addition to the test reactor and influent feed batch have been calculated.

G.3 Commencing the Test

1. Insert the DO probe into the empty reactor. The probe should be clamped to the side of the reactor. Insert the aeration source into the empty reactor.

2. Using clamps, insert the various pump hoses into their respective vessels (i.e., influent pump intake into the influent feed container, influent pump output into the test reactor, effluent pump intake into the test reactor, effluent pump output into the effluent container).

3. Pour 6 L of mixed liquor into the test reactor.

4. Turn on the reactor mixer.

5. Turn on the air source and DO meter/controller.

6. Allow 2 - 3 minutes to elapse to ensure that the reactor contents are mixed well.

7. Turn on the influent and effluent pump(s). If a timer is being used, turn it on also.

"Time zero" for the test occurs when flow through the reactor commences. Steps 3 to 7 only should be performed just prior to "time zero." Therefore, if a timer is being used to cycle the pump(s) on and off, Steps 3 to 7 should be performed just prior to commencement of the first timer cycle.

G.4 Sampling

1. An influent sample should be taken from each daily feed batch and filtered and analyzed for ammonia.

2. Reactor samples should be collected from the effluent pump output. This is best accomplished by temporarily removing the effluent pump output from the effluent container and placing it in a graduated cylinder. Note the time and the volume of sample taken. Filter 10 mL of the effluent sample immediately using a 0.45 μm filter. Collect the filtrate in a clean beaker.

3. Analyze the filtrate for nitrite, nitrate and ammonia.

4. Perform TSS and VSS analyses on the remaining reactor sample.

5. Reactor samples should be collected and analyzed four times over the course of the working day. pH and temperature in the test vessel should be recorded each time a sample is taken.

Note that during the test, nitrite levels in the test vessel may become high. If Hach spectrophotometric methods are used, take care to eliminate nitrite interference in the nitrate measurement (nitrite interferes with nitrate measurements at concentrations of 12 mg/L and up). Nitrite interference is removed (for nitrite concentrations up to 100 mg/L) by adding approximately 400 mg of urea to 10 mL of sample prior to nitrate analysis. Consult Hach documentation for further details.

At the end of the first 24-hour period, prepare a new batch of feed and add it to the influent container (remember to supplement with ammonium chloride and sodium bicarbonate). Measure and record the total volume of effluent, and empty the effluent container.

A typical test NOx versus time response is shown in Figure G-2. Tests typically run for 5 to 8 days.

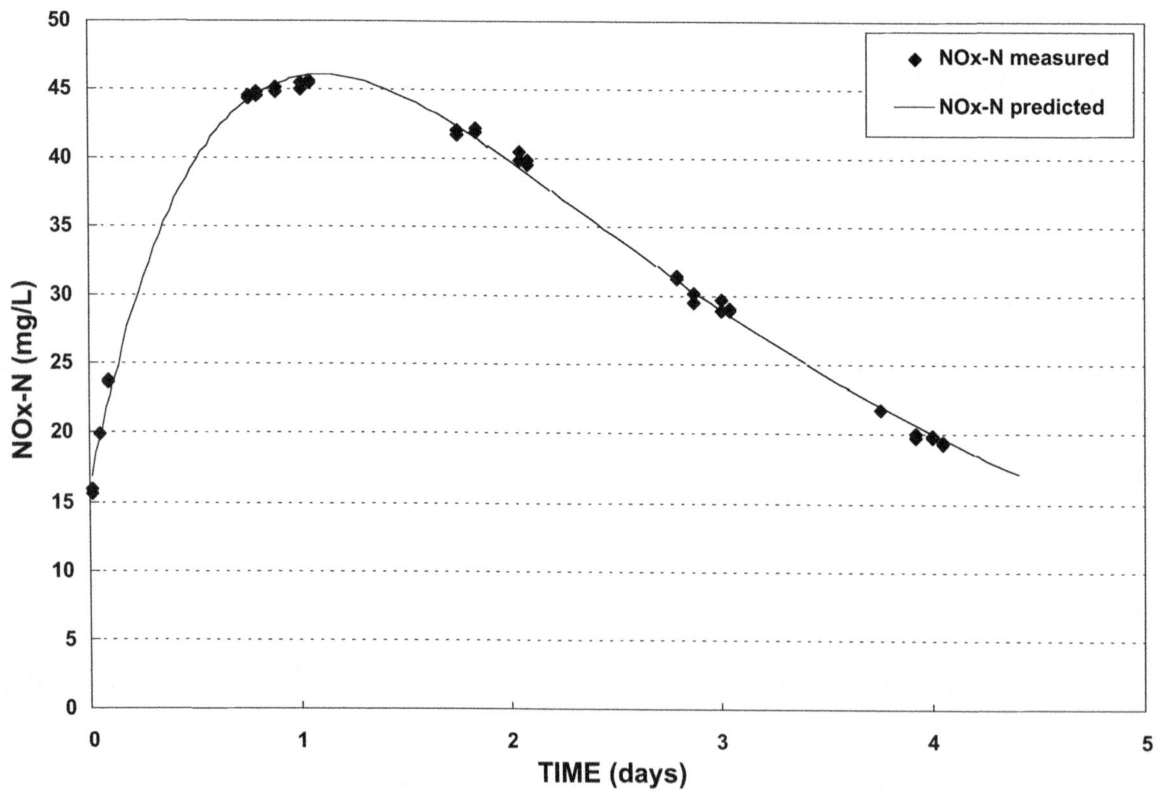

Figure G-2. Typical Test Response

G.5 Estimated Labor Requirements

The following table summarizes the test labor requirements.

Table G-1. Estimated Labor Requirements

Time For Setup and Cleanup	1 day
Time For Data Collection	5 to 8 days
Analytical Load	8 analyses/day for each of nitrite and nitrate, 10 analyses/ day for ammonia
Manpower Required	2 hr/day

The analytical load is based on performing analyses in duplicate. The following example illustrates the calculation of the analytical load for a five day test:

$$nitrite:\ 2\ \frac{analyses}{reactor\ sample} \times 4 \frac{reactor\ sample}{day} \times 5\ day = 40\ analyses$$

$$nitrate:\ 2\ \frac{analyses}{reactor\ sample} \times 4 \frac{reactor\ sample}{day} \times 5\ day = 40\ analyses$$

$$ammonia:\ \left(2\ \frac{analyses}{reactor\ sample} \times 4 \frac{reactor\ sample}{day} + 2\ \frac{analyses}{influent\ batch} \times 1 \frac{influent\ batch}{day}\right) \times 5\ day$$

$$= 50\ analyses$$

If analyses are to be performed using Hach™ spectrophotometric methods by the person running the test, the manpower requirements likely should be increased to 7 hours per day.

APPENDIX H

PARAMETER ESTIMATION METHOD FOR DETERMINING NITRIFICATION RATES

H.1 Background

This appendix provides background on the methods used for parameter estimation and statistical analysis in determination of the nitrifier maximum specific growth rate (μ_{AUT}). An example is given which demonstrates the application of the methods to high F/M batch test data. These methods also are applicable to washout test data. Equations that should be used in the analysis of washout test data are given at the end of this appendix.

H.2 Non-Linear Regression Analysis

The following is a general model for the observed response of a time dependant variable $\mathbf{y_i}$;

$$y_i = f(t_i, \beta) + \varepsilon_i; \quad i = 1,2,3,...,n \tag{H.2-1}$$

where:

$t_i = the\ independent\ variable\ (time)$
$\beta = vector\ of\ parameters$
$\varepsilon_i = errors$
$n = the\ number\ of\ observations$

The errors (ε_i) are considered to be measurement errors. In the method of least squares, a set of parameters $\underline{\beta}$, is chosen to minimize the sum of squares function, as follows:

$$S = \sum_{i=1}^{n} \left[y_i - f(t_i, \underline{\beta}) \right]^2; \quad i = 1,2,3,...,n \tag{H.2-2}$$

If the model in equation H.2-1 is linear with respect to the parameters (see the discussion on sensitivity coefficients below), then the parameters that minimize S can be found by simultaneously solving the set of equations found when:

$$\frac{dS}{d\beta_j} = 0; \quad j = 1,2,...,p \qquad \text{(H.2-3)}$$

where:

$p = the\, number\, of\, parameters$

Most spreadsheet programs will calculate parameter estimates for the familiar model of the straight-line relationship y = mx+b, where explicit equations for the slope, m, and the intercept, b, are used. However, in the case where equation H.2-1 is non-linear with respect to the parameters, an iterative search technique must be used to minimize S. These techniques are described in a number of references on parameter estimation, including Beck and Arnold (1977) and Draper and Smith (1966).

An important concept that appears in relation to a number of facets of parameter estimation is that of sensitivity coefficients. If the mathematical model $f(t_i, \beta)$ is considered, the first derivative of f with respect to β_j is called the sensitivity coefficient for β_j. If the sensitivity coefficient is denoted as X_j, then

$$X_j = \frac{\partial f}{\partial \beta_j} \qquad \text{(H.2-4)}$$

Note that one means of checking whether a model is linear with respect to its parameters is by way of the sensitivity coefficients. For the model f above, f is linear with respect to the parameters if all of the sensitivity coefficients are *not* functions of any of the model parameters, β_j.

H.2.1 Example

The time-dependent concentration of nitrite/nitrate ($S_{NO,t}$ = nitrite + nitrate) in a high F/M batch test can be expressed as follows:

$$S_{NO,t} = S_{NO,0} + \frac{\mu_{AUT} \cdot X_{AUT,0}}{Y_{AUT} \cdot \left(\mu_{AUT} - b_{AUT}\right)} \cdot \left[e^{\left(\mu_{AUT} - b_{AUT}\right)t} - 1\right] \qquad \text{(H.2.1-1)}$$

where:

$s_{NO,t} = oxidized\, nitrogen\, concentration\, at\, time\, t$
$s_{NO,0} = oxidized\, nitrogen\, concentration\, at\, time\, 0$
$Y_{AUT} = nitrifier\, yield\, coefficient$
$b_{AUT} = nitrifier\, decay\, rate$
$\mu_{AUT} = \max imum\, specific\, nitrifier\, growth\, rate$
$X_{AUT,0} = initial\, nitrifier\, concentration$

If the values of Y_{AUT} and b_{AUT} are assumed known (such as from independent experiments), then there are three parameters that require estimation from a batch test. These are, $S_{NO,0}$, μ_{AUT}, and $X_{AUT,0}$ (Note: the value assumed for the Y_{AUT} parameter impacts the estimate of $X_{AUT,0}$ but does not alter the estimate of μ_{AUT}.) The sensitivity coefficients for these parameters are:

$$\frac{\partial S_{NO,t}}{\partial \mu_{AUT}} = \frac{X_{AUT,0} \cdot e^{(\mu_{AUT}-b_{AUT})t}}{Y_{AUT}\left(\mu_{AUT}-b_{AUT}\right)} \cdot \left(\mu_{AUT}t - \frac{b_{AUT}}{\mu_{AUT}-b_{AUT}}\right) + \frac{b_{AUT} \cdot X_{AUT,0}}{Y_{AUT}\left(\mu_{AUT}-b_{AUT}\right)^2} \quad \text{(H.2.1-2)}$$

$$\frac{\partial s_{NO,t}}{\partial X_{AUT,0}} = \frac{\mu_{AUT} \cdot e^{(\mu_{AUT}-b_{AUT})t}}{Y_{AUT}\left(\mu_{AUT}-b_{AUT}\right)} \cdot \left(e^{(\mu_{AUT}-b_{AUT})t} - 1\right) \quad \text{(H.2.1-3)}$$

$$\frac{\partial s_{NO,t}}{\partial s_{NO,0}} = 1 \quad \text{(H.2.1-4)}$$

The expressions for the sensitivity coefficients contain the parameters μ_{AUT} and $X_{AUT,0}$, indicating that the model is non-linear in terms of these parameters. Therefore, an iterative search technique is required to estimate the model parameters. A practical means of achieving this is through a spreadsheet package with an optimization tool. Equation H.2.1-1 is used to calculate a value for $S_{NO,t}$ for each time point that a measurement is available. The residuals are calculated at each time, and then the sum of the squares of the residuals (S) is calculated. The spreadsheet optimization tool is configured to adjust the values of the three estimated parameters to minimize S. Table H-1 illustrates the layout of a spreadsheet for this purpose.

Table H-1. Suggested Spreadsheet Layout for Non-Linear Regression

Specified Parameters				
Y_{AUT}				
b_{AUT}				
Estimated Parameters	**Initial Guess**			
μ_{AUT}				
$X_{AUT,0}$				
$S_{NO,0}$				
Time	**$S_{NO,t}$ Measured**	**$S_{NO,t}$ Estimated**	**Residual**	**(Residual)^2**
t_1				
t_2				
…,				
t_n				
				Σ(Residual)^2

H.3 Confidence Intervals

To calculate confidence intervals for the parameters, β, an estimate of the variance of the parameter estimates is required. The first step is to obtain an estimate for the variance of the measurement errors, $V(\varepsilon_i)$,

$$V(\varepsilon_i) = \sigma^2 \quad \text{(H.3-1)}$$

An estimate of σ^2 can be obtained from the residual sum of squares (S) as follows:

$$s^2 = \frac{S}{n-p} \qquad \text{(H.3-2)}$$

where:

$E(s^2) = \sigma^2$

(i.e., s^2 is an unbiased estimator of σ^2 and can be used in place of σ^2).

Note that the above holds true as long as the following assumptions about ε are reasonable:

- $E(\varepsilon) = 0$ (i.e. the mean of the measurement errors is 0)
- $V(Y_i|\beta) = \sigma^2$ (i.e. the errors have a constant and common variance)
- The errors are uncorrelated

Also required is the covariance matrix of the parameters which is defined as follows:

$$cov(\beta) \approx (X^T X)^{-1} \cdot s^2 = P \qquad \text{(H.3-3)}$$

where:

X = the (p x n) sensitivity matrix

X^T = the (n x p) transpose of the sensitivity matrix

P = the (p x p) covariance matrix

The above expression is considered to be an approximation for cov(β) because the estimate s^2 is used in place of σ^2, and because the sensitivity coefficients are functions of β.

The sensitivity matrix is expressed as follows:

$$X = \begin{bmatrix} X_{11} & \cdot & \cdot & \cdot & X_{1p} \\ \cdot & \cdot & & & \cdot \\ \cdot & & \cdot & & \cdot \\ \cdot & & & \cdot & \cdot \\ X_{n1} & \cdot & \cdot & \cdot & X_{np} \end{bmatrix} = \begin{bmatrix} \frac{\partial f_1}{\partial \beta_1} & \cdot & \cdot & \cdot & \frac{\partial f_1}{\partial \beta_p} \\ \cdot & \cdot & & & \cdot \\ \cdot & & \cdot & & \cdot \\ \cdot & & & \cdot & \cdot \\ \frac{\partial f_n}{\partial \beta_1} & \cdot & \cdot & \cdot & \frac{\partial f_n}{\partial \beta_p} \end{bmatrix} \qquad \text{(H.3-4)}$$

The approximate 100(1-α) percent confidence interval for each estimated parameter, $\underline{\beta}_k$, is then calculated as follows:

$$\beta_k \pm P_{kk}^{1/2} \cdot t_{\alpha/2}(v); \quad k = 1,2,\ldots,p \tag{H.3-5}$$

where:

P_{kk} = the kth diagonal element of P, which is representative of the variance of $\underline{\beta}_k$

v = the degrees of freedom

= n − p

$t_{\alpha/2}(v)$ = the t-statistic for an $\alpha/2$ tail area with v degrees of freedom

H.3.1 Example

Figure H-1 shows the results from a simulated high F/M batch test where noise has been added to the measurements. The data has been entered into a spreadsheet for estimation of the parameters as shown in Table H-2, with the $S_{NO,t}$ estimates calculated using equation H.2.1-1.

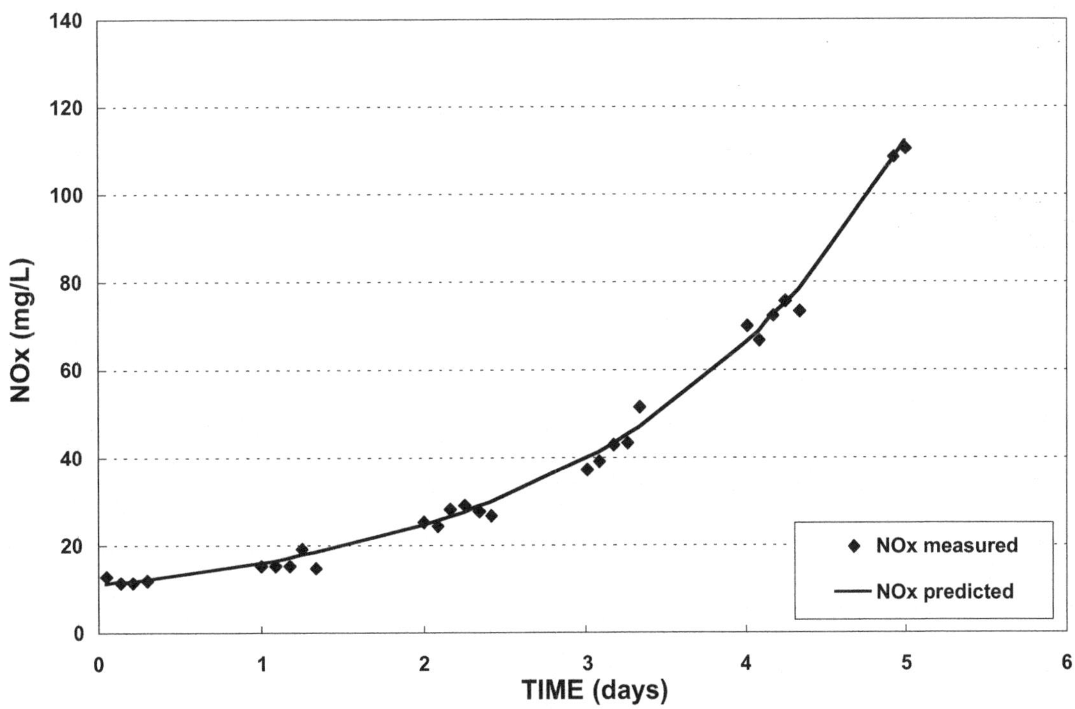

Figure H-1. Results of a Simulated High F/M Batch Test for Nitrifier Growth Rate Estimation

Table H-2. Spreadsheet for Parameter Estimation in High F/M Batch Test Example

Known Parameters				
Y_{AUT} (mg/mg)	0.15	Number of Parameters (p)	3	
b_{AUT} (d-1)	0.15	Number of observations (n)	27	
Estimated Parameters				
μ_{AUT}	0.71			
$X_{AUT,0}$	0.78			
$S_{NO,0}$	10.9			
Time (days)	**$S_{NO,t}$ Measured**	**$S_{NO,t}$ Estimated**	**Residual**	**(Residual)^2**
0.042	13.11	11.01	2.106	4.437
0.125	11.53	11.33	0.204	0.042
0.208	11.47	11.66	-0.192	0.037
0.292	12.23	12.01	0.214	0.046
1.000	15.52	15.77	-0.256	0.065
1.083	15.61	16.32	-0.713	0.508
1.167	15.31	16.90	-1.588	2.523
1.250	19.52	17.50	2.021	4.084
1.333	14.84	18.13	-3.285	10.792
2.000	25.51	24.38	1.129	1.276
2.083	24.77	25.34	-0.574	0.329
2.167	28.22	26.34	1.875	3.515
2.250	29.25	27.40	1.854	3.436
2.333	28.10	28.50	-0.397	0.157
2.417	26.76	29.65	-2.892	8.366
3.000	37.56	39.44	-1.877	3.523
3.083	39.21	41.11	-1.905	3.630
3.167	42.88	42.87	0.013	0.000
3.250	43.50	44.71	-1.207	1.457
3.333	51.75	46.64	5.113	26.146
4.000	70.17	65.77	4.397	19.331
4.083	67.04	68.70	-1.665	2.772
4.167	72.82	71.78	1.045	1.092
4.250	75.92	74.99	0.921	0.849
4.333	73.35	78.37	-5.012	25.116
4.917	108.74	106.93	1.812	3.284
5.000	110.68	111.83	-1.142	1.305
			ε(Residual)^2=	128.118

From equation H.3-2, the error variance is estimated as follows:

$$s^2 = \frac{S}{n-p}$$
$$= \frac{128.118}{27-3}$$
$$= 5.338$$

The sensitivity matrix for this example is:

$$X = \begin{bmatrix} \frac{\partial s_{NO,t_0}}{\partial \mu_{AUT}} & \frac{\partial s_{NO,t_0}}{\partial X_{AUT,0}} & \frac{\partial s_{NO,t_0}}{\partial s_{NO,0}} \\ \cdot & \cdot & \cdot \\ \cdot & \cdot & \cdot \\ \cdot & \cdot & \cdot \\ \frac{\partial s_{NO,t_n}}{\partial \mu_{AUT}} & \frac{\partial s_{NO,t_n}}{\partial X_{AUT,0}} & \frac{\partial s_{NO,t_n}}{\partial s_{NO,0}} \end{bmatrix}$$

and the transpose is

$$X^T = \begin{bmatrix} \frac{\partial s_{NO,t_0}}{\partial \mu_{AUT}} & \cdot\ \cdot\ \cdot & \frac{\partial s_{NO,t_n}}{\partial \mu_{AUT}} \\ \frac{\partial s_{NO,t_0}}{\partial X_{AUT,0}} & \cdot\ \cdot\ \cdot & \frac{\partial s_{NO,t_n}}{\partial X_{AUT,0}} \\ \frac{\partial S_{NO,t_0}}{\partial S_{NO,0}} & \cdot\ \cdot\ \cdot & \frac{\partial S_{NO,t_n}}{\partial S_{NO,0}} \end{bmatrix}$$

These matrices utilize equations H.2.1-2, H.2.1-3, and H.3.1-4 to calculate the partial derivatives. For this example, the computed sensitivity matrix is:

$$X = \begin{bmatrix} 0.22 & 0.20 & 1 \\ 0.70 & 0.61 & 1 \\ 1.23 & 1.04 & 1 \\ 1.82 & 1.50 & 1 \\ 9.63 & 6.33 & 1 \\ 10.98 & 7.04 & 1 \\ 12.43 & 7.78 & 1 \\ 14.01 & 8.55 & 1 \\ 15.71 & 9.36 & 1 \\ 35.08 & 17.41 & 1 \\ 38.39 & 18.64 & 1 \\ 41.94 & 19.94 & 1 \\ 45.75 & 21.29 & 1 \\ 49.83 & 22.71 & 1 \\ 54.21 & 24.19 & 1 \\ 94.66 & 36.78 & 1 \\ 102.12 & 38.94 & 1 \\ 110.09 & 41.2 & 1 \\ 118.59 & 43.57 & 1 \\ 127.66 & 46.05 & 1 \\ 225.18 & 70.67 & 1 \\ 241.17 & 74.44 & 1 \\ 258.18 & 78.39 & 1 \\ 276.26 & 82.53 & 1 \\ 295.49 & 86.87 & 1 \\ 468.35 & 123.63 & 1 \\ 499.52 & 129.92 & 1 \end{bmatrix}$$

The covariance matrix (P), is calculated from equation H.3-3. With the current example, this matrix is as follows:

$$P = \begin{bmatrix} 6.99e\text{-}04 & -2.62e\text{-}03 & 1.73e\text{-}02 \\ -2.62e\text{-}03 & 9.95e\text{-}03 & -7.04e\text{-}02 \\ 1.73e\text{-}02 & -7.04e\text{-}02 & 8.37e\text{-}01 \end{bmatrix}$$

The confidence intervals are computed using the diagonal values of the P -matrix. For a 95 percent confidence interval with 24 degrees of freedom (i.e., n - p),

$$100(1-\alpha)\% = 95\%$$
$$\alpha = 0.05$$
$$t_{\alpha/2}(v) = t_{0.05/2}(24)$$
$$= 2.064$$

Then the approximate 95 percent confidence interval for μ_{AUT} is

$$\mu_A \pm P_{11}^{1/2} \cdot 2.064 = 0.71 \pm (6.99e-04)^{1/2} \cdot 2.064$$
$$= 0.71 \pm 0.0546$$
$$or$$
$$0.655 \langle \mu_A \langle 0.765$$

A plot that graphically illustrates the point estimate and confidence interval for this example (and another synthetic data set) is shown in Figure H-2.

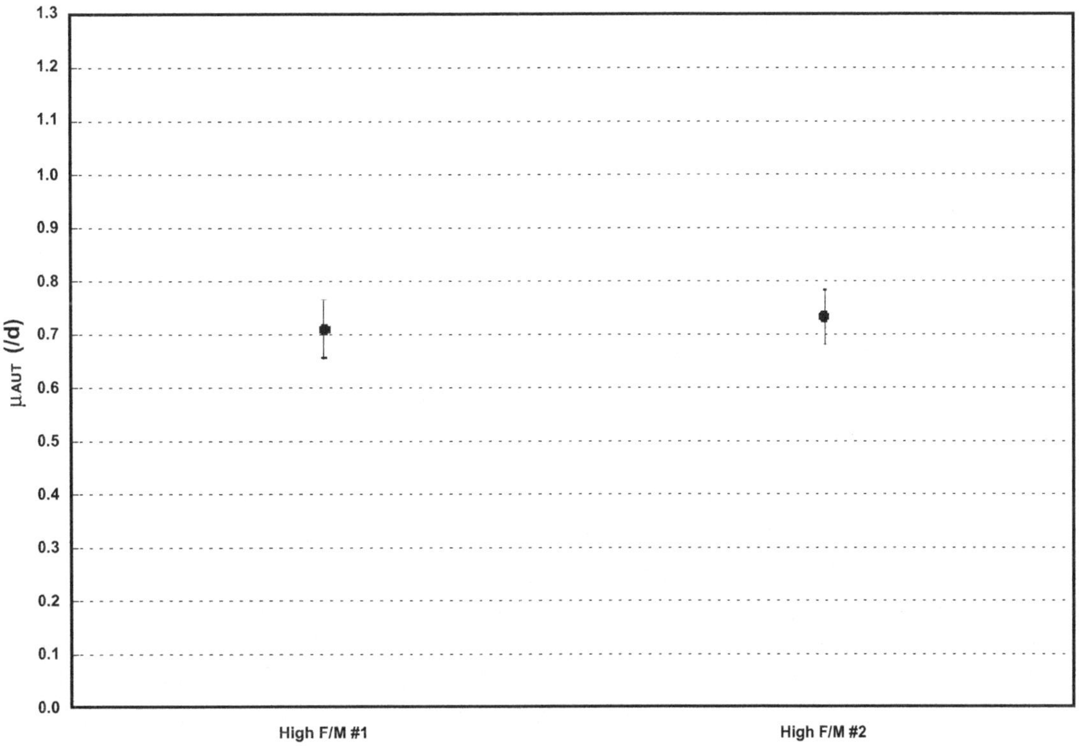

Figure H-2. Confidence Interval Plot for a Pair of Synthetic High F/M Tests

H.4 Diagnostics

This section discusses two simple diagnostic checks on the model fit.

H.4.1 Analysis of Residuals

When modeling a time-dependent response such as in the nitrification rate testing discussed here, a plot of the residuals versus time can uncover potential model inadequacies, and violations of the standard assumptions used in the regression analysis. When the model adequately represents the response, there should be a random pattern of the residuals around the time axis. Any systematic patterns, suggest that the model is not representing all of the phenomena impacting the response. This might be the case, for example, if the impact of temperature changes during an experiment is not included in the model. Also, drifts in a chemical standard or instrument calibration during an experiment may be apparent in a residual plot.

One of the important standard assumptions in the application of ordinary least squares is that the error variance is constant through the entire range of the response. However, it is common for the error to increase as the response increases. This is the case, for example, when an instrument has a constant percentage error rather than a constant absolute error. In the case where the value of a measured response increases with time, the residuals versus time plot may show an increasing spread in the residuals around the time axis. In such a case, the higher values of the response should be weighted less than lower values when fitting the model.

H.4.2 Parameter Correlations

The P-matrix can be used to calculate the parameter correlation matrix, r. The ij^{th} element of the matrix is computed as follows:

$$r_{ij} = \frac{P_{ij}}{(P_{ii}P_{jj})^{1/2}} \qquad \text{(H.4.2-1)}$$

where P is an element of the P -matrix as given by equation H.3-3.

Note that for r, all of the diagonal elements equal 1, and all of the off-diagonal elements are between -1 and 1. The matrix is also symmetric. If any of the off-diagonal elements are greater than 0.9 (or less than -0.9), the two parameters corresponding to the element are considered to be highly correlated. This may be a result of the structure of the model, or a poor experimental design. The danger of highly correlated parameters is that it may be difficult to obtain accurate individual estimates of the parameters. Instead, there may be a number of combinations of the parameters that minimize the sum of squares. Also, a model with highly correlated parameters is more sensitive to noise in the measured response. If high correlations between parameters are found, it is advisable to attempt the estimation with different starting values to ensure that the same optimum solution is found. Increasing the number of observations, and experimental replicates are also advisable to reduce the sensitivity of the estimates to measurement noise.

H.4.3 Example

A plot of residuals versus time is shown for the high F/M batch test example in Figure H-3. The residuals appear to be randomly distributed. Therefore, from the behavior of the residuals there is no reason to suspect the adequacy of the model, or the standard assumptions.

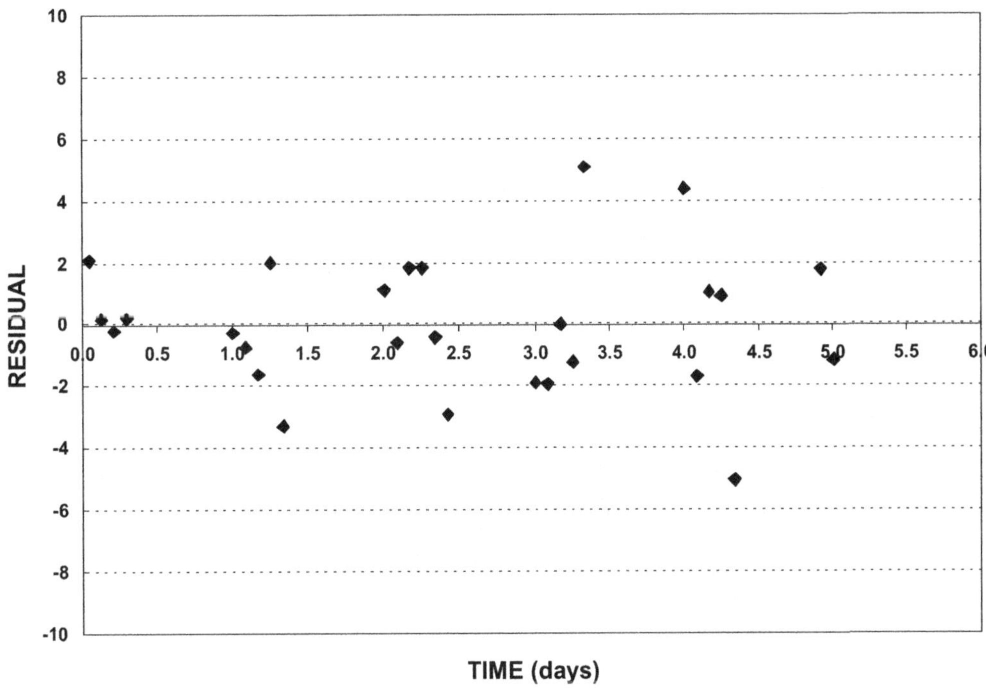

Figure H-3. Residuals Plotted versus Time for theHigh F/M Nitrifier Growth Rate Estimation Example

The correlation matrix elements were computed using equation H.4.2-1, and using the P-matrix determined previously. For example, the r_{13} element (which is the correlation between μ_{AUT} and $S_{NO,0}$), is

$$r_{13} = \frac{P_{13}}{(P_{11}P_{33})^{1/2}}$$

$$= \frac{.0173}{(0.000699 * 0.837)^{1/2}}$$

$$= 0.715$$

The complete r -matrix is,

$$r = \begin{bmatrix} 1.0 & -0.993 & 0.715 \\ -0.993 & 1.0 & -0.772 \\ 0.715 & -0.772 & 1.0 \end{bmatrix}$$

The correlation between $X_{AUT,0}$ and μ_{AUT} is quite high. This likely is due to the form of the model, in which a product of the two parameters appears in one of the terms. Therefore, high quality data and experimental replicates are very important for obtaining accurate parameter estimates.

H.5 Application of Methods to Washout Test Data

The approach detailed in this appendix is applicable to washout test data. Certain of the equations are different from those used in the analysis of high F/M data because the equation that predicts the time-dependent concentration of nitrite/nitrate ($S_{NO,t}$ = nitrite + nitrate) in a washout test incorporates a flow term. That is, for a washout test equation H.2.1-1 above changes to:

$$S_{NO,t} = \frac{\mu_{AUT} \cdot X_{AUT,0}}{Y_{AUT} \cdot \left(\mu_{AUT} - b_{AUT}\right)} \cdot \left[e^{\left(\mu_{AUT} - b_{AUT} - \frac{Q}{V}\right)t} - e^{-\left(\frac{Q}{V}\right)t} \right] + S_{NO,0} \cdot e^{-\left(\frac{Q}{V}\right)t} \qquad \text{(H.5-1)}$$

where:

$S_{NO,t}$ = *oxidized nitrogen concentration at time t*
$S_{NO,0}$ = *oxidized nitrogen concentration at time* 0
Y_{AUT} = *nitrifier yield coefficient*
b_{AUT} = *nitrifier decay rate*
μ_{AUT} = *maximum specific nitrifier growth rate*
$X_{AUT,0}$ = *initial nitrifier concentration*
Q = *flow through washout test reactor*
V = *washout test reactor volume*

As in the high F/M test, if the values of Y_{AUT} and b_{AUT} are assumed to be known, there are three parameters that require estimation from the washout test. These are, $S_{NO,0}$, μ_{AUT} and $X_{AUT,0}$. The sensitivity coefficients for these parameters take the following form:

$$\frac{\partial S_{NO,t}}{\partial \mu_{AUT}} = \frac{X_{AUT,0} \cdot e^{(\mu_{AUT} - b_{AUT} - \frac{Q}{V})t}}{Y_{AUT}\left(\mu_{AUT} - b_{AUT}\right)} \cdot \left(\mu_{AUT} t - \frac{b_{AUT}}{\mu_{AUT} - b_{AUT}} \right) + \frac{b_{AUT} \cdot X_{AUT,0} \cdot e^{-(\frac{Q}{V})t}}{Y_{AUT}\left(\mu_{AUT} - b_{AUT}\right)^2} \qquad \text{(H.5-2)}$$

$$\frac{\partial s_{NO,t}}{\partial X_{AUT,0}} = \frac{\mu_{AUT}}{Y_{AUT}\left(\mu_{AUT} - b_{AUT}\right)} \cdot \left(e^{(\mu_{AUT} - b_{AUT} - \frac{Q}{V})t} - e^{-(\frac{Q}{V})t} \right) \qquad \text{(H.5-3)}$$

$$\frac{\partial S_{NO,t}}{\partial S_{NO,0}} = e^{-(\frac{Q}{V})t} \qquad \text{(H.5-4)}$$

The analysis of washout test data is identical to that of high F/M data, with equation H.5-1 substituted for equation H.2.1-1 and equations H.5-2, H.5-3, and H.5-4 substituted for equations H.2.1-2, H.2.1-3, and H.2.1-4, respectively.

H.6 References

Beck, J.V., and K.J. Arnold. 1970. *Parameter Eestimation in engineering and science.* New York, NY: John Wiley & Sons, Inc.

Draper, N.R., and H. Smith. 1966. *Applied regression analysis.* New York, NY: John Wiley & Sons, Inc.

APPENDIX I

ESTIMATION OF SRT AT FULL-SCALE PLANTS

I.1 Background

An important factor in simulating a wastewater treatment plant is to have an accurate account of the sludge mass inventory in the process and an accurate estimate of the sludge age or solids residence time (SRT). In practice, problems are often encountered in estimating and controlling SRT. This appendix will provide a review of the concepts and calculations involved with estimating sludge mass inventory and calculating and controlling SRT.

I.2 Definitions

The total sludge mass inventory in an activated sludge process is comprised of the sludge residing in the bioreactor as well as the sludge in the clarifier:

$$\begin{aligned} Total\ Sludge\ Mass = & \ Bioreactor\ Sludge\ Mass \\ & + Clarifier\ Sludge\ Mass \end{aligned}$$

For a BNR process, the bioreactor sludge mass will reside in the various zones which may or may not have a uniform solids concentration, depending on the configuration:

$$\begin{aligned} Total\ Sludge\ Mass = & \ Anaerobic\ Mass \\ & + Anoxic\ Mass \\ & + Aerobic\ Mass \\ & + Clarifier\ Sludge\ Mass \end{aligned}$$

Sludge wasting is comprised of the sludge that is intentionally wasted for SRT control, and the solids that are unintentionally wasted in the effluent:

$$\begin{aligned} Sludge\ Wasting = & \ Intentional\ Wasting \\ & + Unintentional\ Wasting \end{aligned}$$

The SRT is calculated as follows:

$$SRT = \frac{Total\ Sludge\ Mass}{(Sludge\ Wasting)/d}$$

A target SRT is used to calculate the daily waste mass as follows:

$$Intentional\ Waste\ Mass/day = \frac{Total\ Sludge\ Mass}{Target\ SRT} - Unintentional\ Waste\ Mass$$

The volume to waste is:

$$Volume\ to\ Waste = \frac{Waste\ Mass}{Waste\ Concentration}$$

I.3 Example – Calculation of Waste Sludge for SRT Control

This example will show sludge mass inventory and SRT control calculations for the simple BNR process shown in Figure I-1. The process is configured to maximize excess biological phosphorus removal and includes an anoxic zone for denitrification of the return activated sludge (RAS) before it enters the anaerobic zone. The following is a step-by-step calculation of the sludge mass inventory and the waste volume required to maintain the target SRT. The data to be used in the calculations are shown in Table I-1. The TSS concentrations can be considered for this example to be data collected on a given day (more discussion on monitoring for SRT control will be provided following the example).

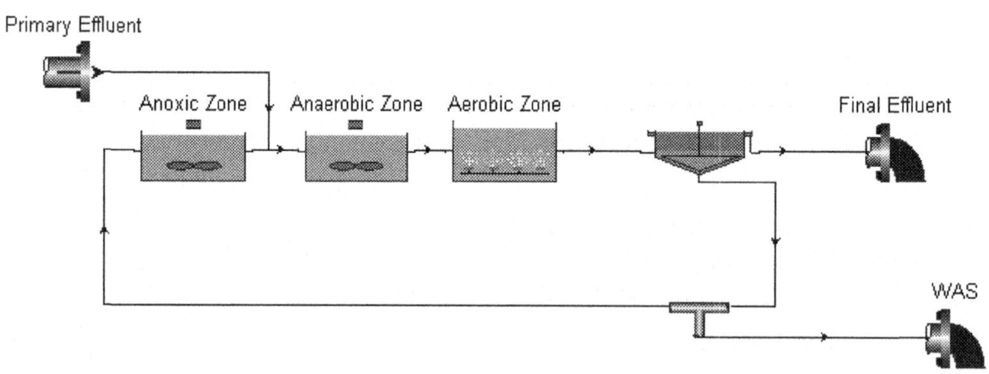

Figure I-1. BNR Process Used in SRT Control Example

Table I-1. Plant Data Used in SRT Control Example

Parameter	Value	Units
Primary Effluent Flow	40,000	m^3/d
Anoxic MLSS	6,000	mg/L
Anaerobic MLSS	2,000	mg/L
Aerobic MLSS	2,000	mg/L
RAS/WAS Concentration	6,000	mg/L
Clarifier TSS Concentration	1,500	mg/L
Effluent TSS Concentration	12	mg/L
Anoxic Volume	1,750	m^3
Anaerobic Volume	3,500	m^3
Aerobic Volume	8,750	m^3
Clarifier Volume	14,000	m^3
Target SRT	10	days

1. Calculate Sludge Mass

$$Sludge\ Mass = Volume \times MLSS \times unit\ conversion\ factors$$

$$AnoxicMass = 1750\ m^3 \times 6000 \frac{mg}{L} \times \frac{1}{1000} \frac{kg/mg}{m^3/L} = 10,500\ kg$$

$$Anaerobic\ Mass = 3500\ m^3 \times 2000 \frac{mg}{L} \times \frac{1}{1000} \frac{kg/mg}{m^3/L} = 7000\ kg$$

$$Aerobic\ Mass = 8750\ m^3 \times 2000 \frac{mg}{L} \times \frac{1}{1000} \frac{kg/mg}{m^3/L} = 17,500\ kg$$

$$Clarifier\ Mass = 14,000\ m^3 \times 1500 \frac{mg}{L} \times \frac{1}{1000} \frac{kg/mg}{m^3/L} = 21,000\ kg$$

$$Total\ Sludge\ Mass = 10,500 + 7000 + 17,500 + 21,000 = 56,000\ kg$$

2. Calculate Unintentional Wasting

$$Unintentional\ Waste\ Mass = Effluent\ Flow \times Effluent\ TSS \times unit\ conversion\ factors$$

$$= 40,000 \frac{m^3}{d} \times 12 \frac{mg}{L} \times \frac{1}{1000} \frac{mg/kg}{m^3/L}$$

$$= 480 \frac{kg}{d}$$

3. Calculate Intentional Waste Mass

$$Intentional\ Waste\ Mass / d = \frac{56,000\ kg}{10\ d} - 480\frac{kg}{d}$$
$$= 5120\frac{kg}{d}$$

4. Calculate Volume to Waste

(Note: Wastage is from the RAS)

$$Waste\ Volume = \frac{Waste\ Mass}{Waste\ Concentration} \times unit\ conversion\ factor$$
$$= \frac{5120\ kg / d}{6000\ mg / L} \times 1000 \frac{mg / kg}{L / m^3}$$
$$= 853\frac{m^3}{d}$$

I.4 Additional Considerations for SRT Estimation

SRT estimation is an important aspect both in simulation of an activated sludge process and in a basic understanding of the behavior of the system. This section presents some important additional considerations when analysing plant data to account for the sludge mass inventory and to estimate the operating SRT.

I.4.1 Sludge Mass in the Clarifier

The preceding calculations considered the sludge mass in the secondary clarifier as part of the total sludge mass to be included in SRT calculations. In many plants this is not the practice because it is assumed that it is only the bioreactor solids that are involved in the biochemical reactions and process performance. However, there are a number of reasons why the clarifier solids should be monitored as part of the overall monitoring of plant sludge inventory. First, from a process performance point of view, the impact of significant sludge mass in the clarifiers can include the following:

- A higher overall SRT will be required for nitrification;
- There is the potential for denitrification in the clarifiers; and,
- There is the potential for phosphorus release in the clarifiers of a BNR process if anaerobic conditions occur in the sludge blanket.

When simulating a plant with a large clarifier sludge mass, the use of a reactive settler model should be considered to account for the above interactions.

From the point of view of obtaining an accurate estimate of the plant SRT:

- If MLSS and RAS TSS measurements are conducted once per day for SRT control, a significant portion of the mass can potentially shift between the clarifiers and bioreactors over a diurnal cycle and the grab sample measurements of TSS concentrations may provide a poor representation of the mass;
- A large sludge inventory in the clarifiers may result in intermittent excursions in effluent solids concentrations, which go unmeasured.

Measurement of the clarifier blanket TSS concentration requires core samples of the clarifier solids to be removed with a sludge judge. The core sample is then emptied into a bucket and mixed before removing a sample for analysis. Depending on the sludge blanket profile in the clarifier, it may be necessary to take several core samples across the clarifier as shown, for example, in the reactangular clarifier in Figure I-2.

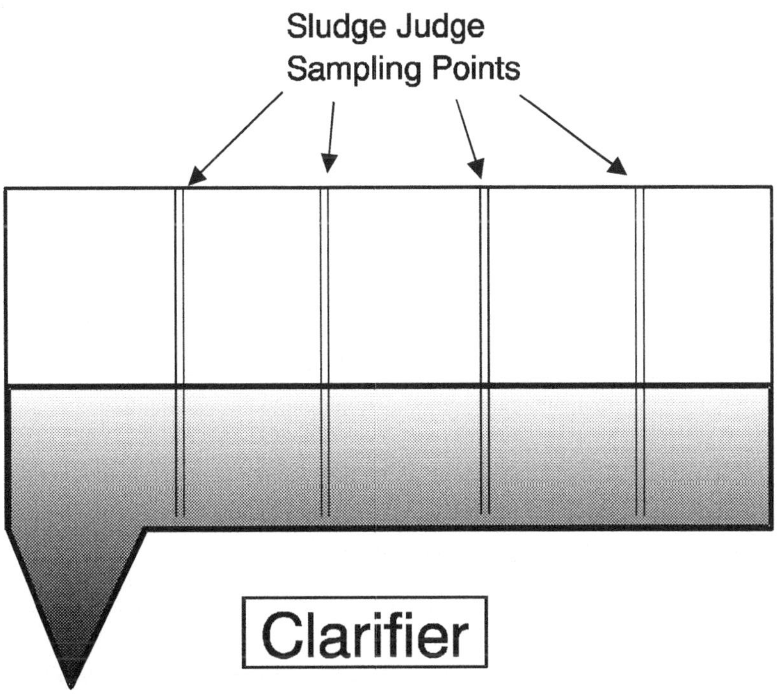

Figure I-2. Illustration of Sampling in a Rectangular Clarifier for Measurement of the Clarifier Solids Concentration

I.4.2 Additional Sampling Problems

There are a number of considerations to ensure that the measurements for SRT estimation and for simulator calibration reflect the true conditions in the treatment process. These are presented in Chapter 21.0.

2003 WERF SUBSCRIBERS

Wastewater Utility

Alabama
Montgomery Water Works & Sanitary Sewer Board

Alaska
Anchorage Water & Wastewater Utility

Arizona
Gila Resources
Glendale, City of, Utilities Department
Mesa, City of
Phoenix Water Services Department
Pima County Wastewater Management

Arkansas
Little Rock Wastewater Utility

California
Calaveras County Water District
Central Contra Costa Sanitary District
Contra Costa Water District
Crestline Sanitation District
Delta Diablo Sanitation District
Dublin San Ramon Services District
East Bay Municipal Utility District
El Dorado Irrigation District
Escondido, City of
Fairfield-Suisun Sewer District
Irvine Ranch Water District
Las Virgenes Municipal Water District
Lodi, City of
Los Angeles, City of
Los Angeles County, Sanitation Districts of
Napa Sanitation District
Orange County Sanitation District
Palo Alto, City of
Riverside, City of
Sacramento Regional County Sanitation District
San Diego Metropolitan Wastewater Department, City of
San Francisco, City & County of
San Jose, City of
Santa Barbara, City of
Santa Rosa, City of
South Bayside System Authority
South Orange County Wastewater Authority
Union Sanitary District
West Valley Sanitation District

Colorado
Boulder, City of
Colorado Springs, City of
Littleton/Englewood Water Pollution Control Plant
Metro Wastewater Reclamation District, Denver

Connecticut
New Haven, City of, WPCA

District of Columbia
District of Columbia Water & Sewer Authority

Florida
Broward, County of
Fort Lauderdale, City of
Gainesville Regional Utilities
JEA
Kissimmee, City of, Department of Water Resources
Miami-Dade Water & Sewer Authority
Orange County Utilities Department
Orlando, City of
Reedy Creek Improvement District
Seminole County Environmental Services
St. Petersburg, City of
Stuart Public Utilities
Tallahassee, City of
Tampa, City of
West Palm Beach, City of

Georgia
Atlanta Wastewater Services
Augusta, City of
Clayton County Water Authority
Cobb County Water System
Columbus Water Works
Fulton County
Gwinnett County Department of Public Utilities
Macon Water Authority
Savannah, City of

Hawaii
Honolulu, City and County of

Idaho
Pocatello, City of

Illinois
American Bottoms Wastewater Treatment Plant
Dupage, County of
Greater Peoria Sanitary District
Kankakee River Metropolitan Agency
Metropolitan Water Reclamation District of Greater Chicago
Wheaton Sanitary District

Indiana
Fort Wayne, City of
Indianapolis, City of

Iowa
Ames, City of
Des Moines Metro Wastewater Reclamation Authority

Kansas
Johnson County Unified Wastewater Districts
Unified Government of Wyandotte County/Kansas City, City of

Kentucky
Louisville & Jefferson County Metropolitan Sewer District

Louisiana
Sewerage & Water Board of New Orleans

Maine
Bangor, City of
Portland Water District

Maryland
Anne Arundel County Bureau of Utility Operations
Howard County Department of Public Works
Washington Suburban Sanitary Commission

Massachusetts
Boston Water & Sewer Commission
Upper Blackstone Water Pollution Abatement District

Michigan
Ann Arbor, City of
Detroit, City of
Grand Rapids, City of
Holland Board of Public Works
Lansing, City of
Owosso Mid-Shiawassee County WWTP
Saginaw, City of
Wayne County Department of Environment
Wyoming, City of

Missouri
Independence, City of
Kansas City Missouri Water Services Department
Little Blue Valley Sewer District
Metropolitan St. Louis Sewer District

Nebraska
Lincoln Wastewater System

Nevada
Henderson, City of

New Jersey
Passaic Valley Sewerage Commissioners

New York
New York City Department of Environmental Protection

North Carolina
Cary, Town of
Charlotte/Mecklenburg Utilities
Durham, City of
Metropolitan Sewerage District of Buncombe County
Orange Water & Sewer Authority

Ohio
Akron, City of
Butler County Department of Environmental Services
Columbus, City of
Metropolitan Sewer District of Greater Cincinnati
Northeast Ohio Regional Sewer District

Oklahoma
Tulsa, City of

Oregon
Clean Water Services
Eugene/Springfield Water Pollution Control
Water Environment Services

Pennsylvania
Philadelphia, City of
University Area Joint Authority, State College

South Carolina
Charleston Commissioners of Public Works
Mount Pleasant Waterworks & Sewer Commission
Spartanburg Sanitary Sewer District

Tennessee
Cleveland, City of
Knoxville Utilities Board
Murfreesboro Water & Sewer Department
Nashville Metro Water Services

Texas
Amarillo, City of
Austin, City of
Dallas Water Utilities
Denton, City of
El Paso Water Utilities
Fort Worth, City of
Gulf Coast Waste Disposal Authority
Houston, City of
San Antonio Water System
Trinity River Authority

Utah
Salt Lake City Corporation

Virginia
Alexandria Sanitation Authority
Arlington, County of
Fairfax County Virginia
Hampton Roads Sanitation District
Henrico, County of
Hopewell Regional Wastewater Treatment Facility
Loudoun County Sanitation Authority
Lynchburg Regional WWTP
Prince William County Service Authority
Richmond, City of

Washington
Edmonds, City of
Everett, City of
King County Department of Natural Resources
Seattle Public Utilities

Wisconsin
Green Bay Metro Sewerage District
Madison Metropolitan Sewerage District
Milwaukee Metropolitan Sewerage District
Racine, City of
Sheboygan Regional Wastewater Treatment
Wausau Water Works

Australia
Sydney Water Corporation
Water Corporation of Western Australia

Canada
Toronto, City of, Ontario

Mexico
Servicios de Agua y Drenaje de Monterrey, I.P.D.

United Kingdom
Yorkshire Water Services Limited

Stormwater Utility

California
Monterey, City of†
Santa Rosa, City of†

Colorado
Boulder, City of†

Georgia
Griffin, City of†

Kansas
Overland Park, City of†

Kentucky
Louisville & Jefferson County Metropolitan Sewer District†

Maine
Portland Water District†

North Carolina
Cary, Town of†
Charlotte, City of/Mecklenburg County†

Pennsylvania
Philadelphia, City of†

State

Fresno Metropolitan Flood Control District, Calif.†
Urban Drainage & Flood Control District, Colo.†

Corporate

ADS Environmental Services
The ADVENT Group Inc.
Alan Plummer & Associates
Alden Research Laboratory
Alpine Technology Inc.
Aquateam–Norwegian Water Technology Centre A/S
BaySaver Inc.†
BioVir Inc.
Black & Veatch
Boyle Engineering Corporation
BPR CSO
Brown & Caldwell
Burns & McDonnell
CABE Associates Inc.
The Cadmus Group
Camp Dresser & McKee Inc.
Carollo Engineers Inc.
Carpenter Environmental Associates Inc.
CDS Technologies Inc.
Chemtrac Systems Inc.
CH2M HILL
Clancy Environmental Consultants Inc.
Damon S. Williams Associates, LLC
Earth Tech Inc.
Ecolab Water Care Services
EMA Inc.
The Eshelman Company Inc.
Finkbeiner, Pettis, & Strout Inc.
Frontier Geosciences Inc.
ftn Associates Inc.
Gannett Fleming Inc.
GE Betz
Golder Associates Inc.
Greeley and Hansen LLC
The HACH Company
Hazen & Sawyer, P.C.
HDR Engineering Inc.
HNTB Corporation
HydroQual Inc.
Insituform Technologies Inc.
Institute for Environmental Technology & Industry, Korea
Jacobson Helgoth Consultants Inc.
Jason Consultants Inc.
Jordan, Jones, & Goulding Inc.
KCI Technologies Inc.
Kelly & Weaver, P.C.
Kennedy/Jenks Consultants
Komline Sanderson Engineering Corporation
Lawler, Matusky & Skelly Engineers, LLP
Limno-Tech Inc.
Lombardo Associates Inc.
Malcolm Pirnie Inc.
McKim & Creed
MWH
New England Organics
Odor & Corrosion Technology Consultants Inc. (OCTC)
ONDEO Degremont Inc.
PA Government Services Inc.
Parametrix Inc.
Parsons
Post, Buckley, Schuh & Jernigan
The RETEC Group
R.M. Towill Corporation
Ross & Associates Ltd.
Royce Technologies
SAIC Maritime Technical Group
Sear-Brown
Sierra Environmental Services
SYNAGRO
Tetra Tech Inc.
Trojan Technologies Inc.
URS Corporation
USFilter NATC
Vortechnics Inc.†
Wade-Trim Inc.
Weston Solutions Inc.
Woodard & Curran
WRc/D&B, LLC
WWETCO, LLC†

Industry

American Electric Power
BP
ChevronTexaco Energy Research & Technology Company
The Coca-Cola Company
Dow Chemical Company
DuPont Company
Eastman Chemical Company
Eastman Kodak Company
Merck & Company Inc.
ONDEO Services
Procter & Gamble Company
Severn Trent Services Inc.
Shell Global Solutions
Thames Water Plc
United Water Services LLC

† Charter Stormwater Subscribers

Note: List as of 1/03

WERF Product Order Form

As a benefit of joining the Water Environment Research Foundation, subscribers are entitled to receive one complimentary copy of all final reports and other products. Additional copies are available at cost (usually $10). To order your complimentary copy of a report, please write "free" in the unit price column. WERF keeps track of all orders. If the charge differs from what is shown here, we will call to confirm the total before processing.

Name _______________ Title _______________

Organization _______________

Address _______________

City _______________ State _______ Zip Code _______ Country _______

Phone _______________ Fax _______________ Email _______________

Stock #	Product	Quantity	Unit Price	Total

Postage & Handling	
VA Residents Add 4.5% Sales Tax	
Canadian Residents Add 7% GST	
TOTAL	

Method of Payment:

(All orders must be prepaid.)

❑ Check or Money Order Enclosed

❑ Visa ❑ Mastercard ❑ American Express

Account No. _______________ Exp. Date _______

Signature _______________

Shipping & Handling:

Amount of Order	United States	Canada & Mexico	All Others
Up to but not more than:	Add:	Add:	Add:
$20.00	$5.00*	$8.00	50% of amount
30.00	5.50	8.00	40% of amount
40.00	6.00	8.00	
50.00	6.50	14.00	
60.00	7.00	14.00	
80.00	8.00	14.00	
100.00	10.00	21.00	
150.00	12.50	28.00	
200.00	15.00	35.00	
More than $200.00	Add 20% of order	Add 20% of order	

*minimum amount for all orders

Note: Please make checks payable to the Water Environment Research Foundation.

To Order (Subscribers Only):

Log on to www.werf.org and click on the "Product Catalog."

Phone: (703) 684-2470
Fax: (703) 299-0742.

WERF
Attn: Subscriber Services
601 Wythe Street
Alexandria, VA 22314-1994

To Order (Non-Subscribers):

Non-subscribers may be able to order WERF publications either through WEF (www.wef.org) or IWAP (www.iwapublishing.com). Visit WERF's website at www.werf.org for details.

IWA Publishing's authorised EU representative for General Product Safety Regulations is Diane D'Arras, 15 rue Duret, 75116 Paris, France, e-mail: safety@iwap.co.uk.

Printed and bound by CPI Group (UK) Ltd, Croydon, CR0 4YY
17/09/2026
02218793-0012